LATE CAINOZOIC PALAEOCLIMATES OF THE SOUTHERN HEMISPHERE

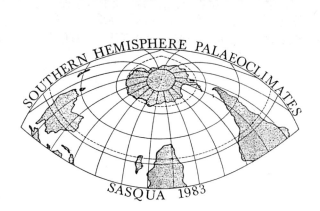

PROCEEDINGS OF AN INTERNATIONAL SYMPOSIUM HELD BY THE SOUTH AFRICAN SOCIETY FOR QUATERNARY RESEARCH / SWAZILAND / 29 AUGUST – 2 SEPTEMBER 1983

Late Cainozoic Palaeoclimates of the Southern Hemisphere

Edited by
J.C.VOGEL
CSIR, Pretoria, South Africa

With the assistance of:
NICOLINE BASSON / URSULA VOGEL / ANNEMARIE FULS

A.A.BALKEMA / ROTTERDAM / BOSTON / 1984

SPONSORS OF THE SYMPOSIUM AND WORKSHOPS

Swaziland National Trust Commission
Gold Fields of South Africa Limited
Anglo American and De Beers Chairman's Fund
Rand Mines (Mining & Services) Limited
Ubombo Ranches Limited
Gencor Development Fund
Noordwes Koöperasie Beperk
Cooperative Scientific Programmes, CSIR
University of the Witwatersrand

ORGANIZING COMMITTEE

Dr J.C.Vogel: President of SASQUA
Prof. T.C.Partridge: Convenor
Dr D.Price-Williams: Local organizer
Mr P.B.Beaumont
Prof. J.A.Coetzee
Dr N.Lancaster
Dr J.M.Maguire
Dr R.R.Maud
Dr M.K.Seely

Four of the papers in this volume were set by Dr Janette Deacon, Stellenbosch.
The rest were set by typists of the CSIR, Pretoria, under the supervision of the editor.

ISBN 90 6191 554 6

© 1984 A.A.Balkema, P.O.Box 1675, 3000 BR Rotterdam, Netherlands

Distributed in USA & Canada by: A.A.Balkema Publishers, P.O.Box 230, Accord, MA 02018

Printed in the Netherlands

Contents

Preface	IX
Theme song of the Symposium	XI
Opening address *Hugues Faure*	1

Palaeoclimatology

Climatic evolution in the Southern Hemisphere and the equatorial region during the Late Cenozoic *H.Flohn*	5
A climatic model of the Last Glacial/Interglacial transition based on palaeotemperature and palaeohydrological evidence *S.P.Harrison, S.E.Metcalfe, F.A.Street-Perrott, A.B.Pittock, C.N.Roberts & M.J.Salinger*	21
An analytical climate model: Application to the Southern Hemisphere Quaternary period *Bruce Denness*	35

South America

Late Cainozoic glacial variations in South America south of the equator *J.H.Mercer*	45
Late Quaternary climates of Chile *Calvin J.Heusser*	59
Testing the Late Quaternary climatic record of southern Chile with evidence from fossil Coleoptera *A.C.Ashworth & J.W.Hoganson*	85
Late Glacial glaciation and the development of climate in southern South America *W.Lauer & P.Frankenberg*	103
The palaeohydrology of the Late Pleistocene Lake Tauca on the Bolivian Altiplano and recent climatic fluctuations *Albrecht Kessler*	115
Late Quaternary climatic changes in the basin of Lake Valencia, Venezuela, and their significance for regional paleoclimates *L.Peeters*	123

Australasia

New Zealand climate: The last 5 million years *M.J.Salinger*	131
The changing face of the evidence: An examination of proxy data for climate change during the Late Pleistocene in New Zealand *J.M.Soons*	151

Palaeobotanical evidence for changes in Miocene and Pliocene climates in New Zealand   159
D.C.Mildenhall & D.T.Pocknall

Holocene climates of the Vestfold Hills, Antarctica, and Macquarie Island   173
John Pickard, P.M.Selkirk & D.R.Selkirk

Glacial age environments of inland Australia   183
J.M.Bowler & R.J.Wasson

Evolution of Australian landscapes and the physical environment of aboriginal man   209
J.B.Firman

Late Quaternary climatic change — Evidence from a Tasmanian speleothem   221
A.Goede & M.A.Hitchman

Southern Africa

Late Quaternary environments in South Africa   235
K.W.Butzer

The evidence from northern Botswana of Late Quaternary climatic change   265
H.J.Cooke

Geomorphic evidence of Quaternary environmental changes in Etosha, South West Africa/Namibia   279
Uwe Rust

The occurrence of ferricrete at Witsand in the south-eastern Kalahari   287
T.H.van Rooyen & E.Verster

Cainozoic palaeosols in the Naboomspruit area, Transvaal   295
E.Verster & T.H.van Rooyen

Lake level fluctuations during the last 2000 years in Malawi   305
R.Crossley, S.Davison-Hirschmann, R.B.Owen & P.Shaw

Late Quaternary palaeoenvironments in the Transvaal on the basis of palynological evidence   317
L.Scott

Environmental changes since 32 000 BP at Kathu Pan, northern Cape   329
Peter B.Beaumont, E.M.van Zinderen Bakker, Sr. & John C.Vogel

Correlation of palaeoenvironmental data from the Late Pleistocene and Holocene deposits at Boomplaas cave, southern Cape   339
H.J.Deacon, J.Deacon, A.Scholtz, J.F.Thackeray, J.S.Brink & John C.Vogel

Investigations on archaeological charcoals from Swaziland, using SEM techniques   353
Juliet Prior

Micromammalian population dynamics and environmental change: The last 18 000 years in the southern Cape   361
D.M.Avery

Climatic change and mammalian fauna from Holocene deposits in Wonderwerk cave, northern Cape   371
J.F.Thackeray

Late Pleistocene environmental changes and implications for the archaeological record in southern Africa   375
H.J.Deacon & J.F.Thackeray

Evidence for Late Quaternary climatic change in southern Africa: Summary of the proceedings of the SASQUA Workshop held in Johannesburg, September 1983   391
Janette Deacon, N.Lancaster & L.Scott

The southern deserts

Ancient ergs of the Southern Hemisphere   407
D.S.G.Thomas & A.S.Goudie

Late Quaternary palaeoenvironments in the desert dunefields of Australia   419
R.J.Wasson

Aridity in southern Africa: Age, origins and expression in landforms and sediments   433
N.Lancaster

The development of the Namib dune field according to sedimentological and geomorphological evidence   445
Helga Besler

A reappraisal of the Cenozoic stratigraphy in the Kuiseb valley of the central Namib desert   455
J.D.Ward

Radiocarbon dating of speleothems from the Rössing cave, Namib desert, and palaeoclimatic implications   465
Klaus Heine & Mebus A.Geyh

African faunal record

Horses, elephants and pigs as clues in the African later Cainozoic   473
H.B.S.Cooke

Paleoclimatic framework for African hominid evolution   483
N.T.Boaz & L.H.Burckle

The Terminal Miocene Event: A critical environmental and evolutionary episode?   491
C.K.Brain

Biogeography of Miocene-Recent larger carnivores in Africa   499
A.Turner

Preliminary radiometric ages for the Taung tufas   507
John C.Vogel & T.C.Partridge

Climatic change and evolution — Evidence from the African faunal and hominid sites
Summary of the proceedings of the SASQUA Workshop held in Johannesburg, September 1983   515
Phillip V.Tobias

Preface

In 1981 the Council of the South African Society for Quaternary Research decided that the time had come for a meeting to be devoted specifically to the palaeoclimates of the Southern Hemisphere. For several years now climatologists have been aware of the major role that Antarctica and the surrounding oceans play in regulating atmospheric circulation patterns - even across the equator. In addition, evidence from deep sea cores has recently been taken to indicate that climatic change in the south actually preceded that in northern regions by some 3000 years. This would further stress the possible predominance of the Southern Hemisphere in determining global climate. The suspected phase shift was highlighted at the CLIMANZ conference of Australian and New Zealand quaternarists in February 1981, but the local data did not seem to substantiate the supposition.

The significance that such a time-lag would have for our understanding of global changes in climate, however, warrants thorough investigation and for this reason alone it was considered well worth while to bring together and explain the data from the different regions of the Southern Hemisphere. This issue would seem to be especially relevant now that Post-glacial warming has been correlated with an increase in the carbon dioxide content of the atmosphere, and the implications which this finding may have for evaluating the future effects of industrial CO_2 in the environment. The world-wide effects on climate that the marked el Niño event of 1982/83 appears to have had, furthermore focussed attention on the interconnection of climatic phenomena throughout the Southern Hemisphere.

The outcome of SASQUA's decision was that some 130 participants from 29 different countries assembled at the Royal Swazi Spa on 29 August 1983 to discuss 'Late Cainozoic Palaeoclimates of the Southern Hemisphere'. The Swaziland National Trust Commission acted as host for the Symposium. Although international meetings in the past had, on occasion, paid attention to palaeoclimatic studies in the Southern Hemisphere, this conference was the first to be devoted solely to the topic.

The importance that the Executive of the International Union for Quaternary Research, INQUA, assigns not only to the Southern Hemisphere as such, but specifically to the advancement of Quaternary research in Africa, was reflected by the presence of the Committee members at both the Symposium and the Workshops.

For obvious geographic reasons the southern African subcontinent was better represented than were other regions, but this was perhaps an advantage since the environmental changes that have taken place in Africa are as yet less well understood than those in South America, Australia and New Zealand. The specific venue of the conference in Africa also prompted the inclusion of a session on the important faunal and early hominid collections from Africa so that the relationship of these data with those of Late Cainozoic climatic change could be reviewed. The theme was further pursued at a workshop on 'Evolution and Climatic Change: evidence from the African faunal and hominid sites' that was held at the University of the Witwatersrand in Johannesburg after the symposium, on 3 September 1983. The exciting prospect of the linkage between environmental climatic change and evolutionary events that was discussed at the meeting has already given rise to the planning of further workshops where palaeoclimatologists and palaeontologists will explore the theme more thoroughly.

A second post-conference workshop on the 'Evidence for Late Quaternary climatic change in southern Africa' was also held at the University of the Witwatersrand, on 3 and 5 September 1983. At this meeting the presently available evidence for the subcontinent was presented and discussed extensively. The documentation and deliberations clearly revealed the achievements and shortcomings of Quaternary studies in the region, and provided a basis on which to plan future research. Summaries of the proceedings and deliberations of both workshops are included in this volume..

The papers presented at the symposium were refereed and corrected, and finally edited at the National Physical Research Laboratory of the CSIR in Pretoria with the competent assistance of Nicoline Basson, Ursula Vogel and Annemarie Fuls. The camera-ready copy was produced mainly by Santie Mörsner, Irma Uytenbogaardt and Rita Howcroft to whom the Editor wishes to express his sincere appreciation. We trust that this volume will provide a useful summary of the present state of our knowledge on the topic of past climatic change for some years to come.

Pretoria, May 1984   J C Vogel.

Theme song of the Symposium

CLIMATE-TIME

(to the tune of "Clementine")

H Basil S Cooke

Global circulation patterns
Were surveyed by Dr Flohn,
  But El Nino messed the scene-oh
Leaving sands and theories blown.

CHORUS: Changing climates, changing climates,
      In the southern hemisphere,
        Chilly! Baking!
        Humid! Arid!
      Drive us all to drinking beer.

In the white Antarctic wasteland
Denton dotes on volume change,
  Finds the ice has over-ridden
All the Transantarctic range.

CHORUS:

Wasson wanders from Canberra
In Australia's desert sands,
  And is buried by the pellets
Blowing from the dried-out pans.

CHORUS:

Tim Partridge and John Vogel
Tackle travertines from Taung,
  Show that Australopithecus
Died before he could be born.

CHORUS:

Alan Ashworth and Cal Heusser
Cored the soil in Andes range,
  But the beetles ate the pollen
And just cancelled out the change.

CHORUS:

Dave Price-Williams and his helpers
Earn the thanks of one and all,
  For it's due to all their efforts
That we really had a ball!

CHORUS: Changing climates, changing climates
       In the southern hemisphere,
         Chilly! Baking!
         Humid!   Arid!
       Drive us all to drinking beer.

Opening address

HUGUES FAURE
President of INQUA

Monsieur le Deputé Premier Ministre, Monsieur le Président, Mesdames et Messieurs, chers collègues et amis:

On behalf of the International Union for Quaternary Research, INQUA, I would like to thank the Swaziland National Trust Commission and especially SASQUA and its president, Dr Vogel, for all the time and help he and his organizing committee have given to make it possible for the officers of INQUA to attend this meeting. I know that there were many problems to overcome, especially those of communication, visas, etc. We have already experienced a well-organized field trip through coastal Natal, Zululand and Swaziland, for which we thank Drs Maud, Price-Williams and Goudie very much. The considerable interest of the sites and the lively discussions that ensued, hold much promise that the Symposium, and the following Workshops and Excursions will prove to be equally rewarding. When all is over, we shall, all of us, not only have enriched our knowledge of the Quaternary, but also have made many new friends.

Many INQUA meetings are continuously taking place all around the world - hardly a month passes without at least one such event being staged somewhere - and I would like to explain why the INQUA Executive Committee specifically chose the occasion of the Swaziland Symposium to have one of its infrequent meetings. There are, of course, several reasons for this of which I shall mention but two:

Firstly INQUA would like to emphasize the importance it attaches to palaeoclimatic studies in the Southern Hemisphere. Fifty years ago, when INQUA was formed, members were primarily occupied with the Quaternary of glaciated Europe. As time passed, interest gradually expanded to incorporate the other regions of the world. Then, six years ago, the Union demonstrated its interest in the Southern Hemisphere when our past president Jane Soons was elected. The current attention being paid to the Southern Hemisphere is a very normal evolution: climatic change can only be understood on a global scale, and we need to know what is happening on both the hemispheres if we are to reconstruct past changes in atmospheric circulation patterns. If teleconnections are real and if climatic changes in the Southern Hemisphere preceded those in northern countries by some 3000 years, as has been suggested, one may even reason that Quaternarists should concentrate on the south - especially with respect to predicting future changes in climate!

The second reason for our presence here is to demonstrate our special interest in the African continent - this occasion will indeed be the first time the INQUA Executive is meeting on this continent. Africa is the only continent that extends far across the Equator into both hemispheres and we feel confident that this unique geographic situation has the potential of providing new insights into past climates. In future it may prove to be as significant as the oceans and the evidence derived from deep-sea drilling have been. One example of important information that this continent is producing, is the field studies and drilling projects in African lakes such as the CEGAL Project. Detailed palaeoenvironmental data based on an accurate time-scale are being derived from this work.

To further Quaternary studies in the developing countries of Africa, INQUA has established a special group under the chairmanship of Dr Salif Diop from Senegal. We are fortunate that he is also present here, and he will be glad to answer any questions you may have on the topic.

In past millennia development has taken place as a result of the vigorous interaction between man and his environment, and it is our conviction that future social improvement would profitably be based on the thorough understanding of our environment that is obtained from Quaternary research.

At present UNESCO, which is represented here by the Director of the Earth Sciences Division, Dr Vladimir Sibrava, is designing a new project entitled "Quaternary Geology for Survival". We anticipate that this project will lead to significant advances in the years to come and will help all African countries to progress on the way to achieving peaceful development.

Finally I would like to mention a personal reason why I am glad to be here today. When I last visited this part of the world I was a guest at the V SASQUA biennial conference. On this occasion I was introduced to the excellent research that is being done locally on the Quaternary. I am looking forward to hearing of the more recent advances in this field and to renewing my acquaintance with our colleagues of southern Africa.

Ladies and Gentlemen, I would like to express our appreciation to the organizers of this "International Symposium on the Late Cainozoic Palaeoclimates of the Southern Hemisphere" for their efforts to stage this meeting, and also to the speakers who are about to make their contributions to what promises to be a highly significant conference. MERCI à tous!

Palaeoclimatology

Climate evolution in the Southern Hemisphere and the equatorial region during the Late Cenozoic

H. FLOHN
Meteorologisches Institut, Universität Bonn, Germany

ABSTRACT. Since the atmosphere above the Antarctic is much colder than that above the Arctic (due to the different heat budgets), the Southern Hemispherc (SH) tropospheric circulation is stronger and expands across the equator. In contrast with the weaker standing eddies in SH mid-latitudes, large longitudinal deviations exist in the equatorial region (Southern Oscillation). A large-scale monsoon system develops in the tropical latitudes of both hemispheres during their respective summers, but it is much weaker in the SH. A physical interpretation of the quasi-stationary upper troughs above the Pacific and the Atlantic, in the SH, is still lacking. The role of coastal and equatorial upwelling (concentrated between 0-10 °S) in the control of the CO_2 and H_2O balance is outlined. This air-sea exchange of greenhouse gases may be responsible for abrupt climatic changes on the 100 year time scale.

After a short review of major climatic events since the Eocene/Oligocene transition, the unipolar glaciation of the Antarctic during the Late Miocene and Early Pliocene is described together with the strongly asymmetric climatic pattern prevailing during this period. Prolonged upwelling in the Pacific and Atlantic south of the equator should have enhanced aridity in the 0-20 °S belt. During Pleistocene glaciations an increase in the equator-pole temperature gradient caused more intense circulation, stronger upwelling and thus a drop in the atmospheric CO_2 and H_2O contents. During the last interglacial and the Holocene moist period, on the other hand, atmospheric circulation and equatorial upwelling were weakened, resulting in a rise in CO_2 and H_2O content.

PRESENT CLIMATIC PATTERNS

For a better understanding of climatic history (and possibly also the climatic future), it is useful to consider some essential features of the present climatic pattern and to understand their geophysical background. Since atmosphere and ocean (at least the upper mixed layer, with its thin ice cover in polar regions) are interacting parts of the climate system, both components and their interaction should be considered together.

1. Hemispheric asymmetry
The (incorrect) assumption of hemispherically symmetric climatic pat-

terns has frequently been used by model designers. It is, however, inconsistent with the available data, even for the central Pacific. The different heat budgets and temperatures of both polar caps cause an asymmetry, which is demonstrated by the latitudinal pattern of isotherms and circulation belts, especially near the equator. Aerological data demonstrate that the tropospheric air above Antarctica, as an annual average, is about 11 °C colder than above the central Arctic Ocean (Flohn 1967, 1978). This difference increases to 25 °C during July and nearly disappears during December-February. The result is a shift of the meteorological equator (defined as the belt of maximum frequency of rainfall and thunderstorms and of the reversal of the meridional surface wind component) towards 6 °N as an annual average (Fig. 1).

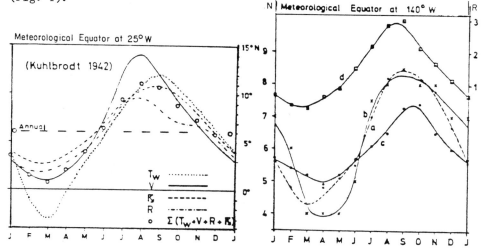

Figure 1. Seasonal variations of the meteorological equator over the Atlantic and Pacific (Flohn 1967; Ramage et al. 1981).

During July its average position is near 12 °N (8 °N over the central Pacific, but 28 °N in NW India). The most striking difference is the strong baroclinic, eastward directed "Equatorial Counter-Current" between 4 ° and 8 °N in the Pacific and Atlantic, with only occasionally a weak counterpart in the Southern Hemisphere (SH). While the subtropical and polar front jet streams are separated during some seasons in the SH (Van Loon et al. 1972), they frequently merge in the Northern Hemisphere (NH).

During the 1979 First Global Geophysical Experiment (FGGE) data coverage was much better than ever before. The total kinetic energy of the atmosphere (Lyne et al. 1982) is nearly equal over both hemispheres during January, but much stronger over the SH during July; this holds particularly true for its zonal component. The contribution of standing eddies is much smaller over the SH (Lyne et al. 1982). The same holds true for the wave numbers 1-4 of the travelling waves (Bengtsson et al. 1982). This revises earlier results (Van Loon et al. 1972) and confirms essentially the results of daily analyses over six years at Melbourne (Trenberth 1980, 1981).

The contrast between Arctic and Antarctic drift ice is especially important (Polar Group 1980); while 70-80 % of the Arctic drift ice is permanent, this value drops to about 15 % (mostly in the Weddell Sea) in the SH, where relatively thin seasonal sea ice with many more leads and polynyas than in the Arctic prevails (Budd 1982; Weller 1982). Quasi-permanent large-scale polynyas have been observed in the SH, e.g. NE of the Weddell Sea (Carsey 1980; Parkinson 1983), while in the Arctic only relatively small, orographically fixed polynyas (North Water and others at the outlets of Baffin Bay) exist (Dey 1979, 1980, 1981). They act as powerful local sources of heat and water vapour during winter.

2. Longitudinal asymmetries

While the zonal temperature differences in the NH troposphere amount to about 50 % of the latitudinal differences (due to the irregular pattern of land and sea and the mountains), their role in the SH is distinctly smaller, but not negligible. The cold source of the Weddell Sea with its high ice producing capacity causes a remarkable northward shift of atmospheric and ocean isotherms in the Atlantic section (40°W - 60°E, including the western Indian Ocean) in contrast with the warmer Pacific, reaching or surpassing 10° in latitude (Van Loon et al. 1972; Trenberth 1980, 1981) and reflected in the seasonal ice cover. Recent model studies suggest that its cause lies in the configuration of the Antarctic continent (Mechoso 1981).

Even more important is a large-scale longitudinal anomaly in the SH tropics. This is caused mainly by the unusually warm sea surface temperatures (SST) of 29-30 °C in the "maritime continent" of Indonesia and the western Pacific, in contrast with values of 20-25 °C along the western coast of South America and (during the northern summer) along the eastern coast of Africa (Ramage 1968). This causes an elongated zonal "Walker circulation" (WC) along 0-5 °S (Bjerknes 1969) (Fig. 2), with marked variations in time (Southern Oscillation, El Niño),

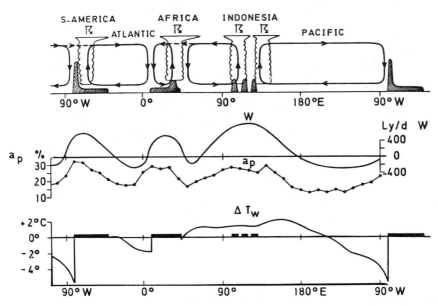

Figure 2. Zonal Walker circulation in the equatorial belt (Flohn 1971, 1975). W = heat budget of an atmospheric column, a_p = planetary albedo, ΔT_W = longitudinal anomalies of sea surface temperatures.

which in turn is correlated (probably via strong spatial and time fluctuations of rainfall and latent heat release) with many far distant anomalies. Recent empirical studies (Rasmussen et al. 1982) and model computations (Philander 1981; McCreary 1983) have contributed to a better understanding of these phenomena but have also left some gaps. This holds especially true after the recent strong and quite unusual El Niño, since July 1982, which has now (August 1983) come to an end. It is now clear that its effect extends over the whole globe and reaches up into the stratosphere (Arkin 1982; Quiroz 1983). One of the most important features is the variability of oceanic upwelling in equatorial and coastal regions (discussed further on), which is also more frequent and intense south of the Equator. Occurrences of oceanic upwelling have been rather infrequent and ineffectual since 1976.

3. Seasonal monsoons and their geophysical causes

The largest anomaly of the general circulation in the NH is the summer monsoon system extending from the Atlantic off W Africa, 25°W, across Africa and Southern Asia, up to 150°E, i.e. beyond the Philippines. This is a consequence of the heat source above the elevated Tibetan Plateau (Hahn et al. 1975; Flohn 1966, 1968; Krishnamurti 1971) which, at 150-200 mbs (12-14 km), causes temperatures of up to 10-11°C higher than above the Equator.

The system consists of a long belt of low-level "equatorial westerlies" and, between 400 and 100 mbs, of a Tropical Easterly Jet, both extending over about the same area. Its cross-circulation in the exit region above the Sahara is responsible for the prevailing subsidence (and thus aridity) of the Saharan belt 15-23°N in contrast with other areas in this latitude (Fig. 3) (Flohn 1966, 1968).

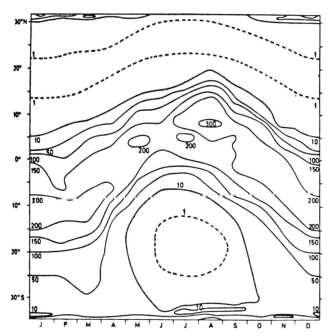

Figure 3. Seasonal variation of rainfall (mm) along 22°E (Strüning et al. 1969).

During southern summer/northern winter the area between Australia/New Guinea and East Africa is characterized by a similar but weaker and less regular monsoon system. Its driving force appears to be the release of latent heat above southern Indonesia and New Guinea, here enforced by the South Pacific Convergence. Recent studies during the FGGE Winter MONEX Period of 1978/79 (Sumi et al. 1981; Davidson et al. 1983), with good data coverage, have shown the prevalence of easterly/westerly winds at 200 mb/850 mb. The source of heat was then centered near $5°C$, $165°E$ above the West Pacific. Two branches of the Intertropical Convergence Zone (ITCZ) along $10-15°S$ and $5°N$ are easily recognizable at the surface and 850 mb. Remarkably enough, this seasonal wind system - as shown by the upper wind observations at Gan (Maledives), Mahé (Seychelles) and Diego Garcia at $0-8°S$ - reaches right across the Indian Ocean and the African continent (Strüning et al. 1969; St Hastenrath 1974) towards the Atlantic. This indicates a shift of the thermal equator towards $10°S$ over more than $100°$ longitude; no local heat source can be recognized over the Indian Ocean. A much smaller counterpart exists, also during the southern summer, above South America and this is caused by the elevated Altiplano in southern Peru and Bolivia, near $15°S$, $65°W$ (Kreuels et al. 1975) (Fig. 4).

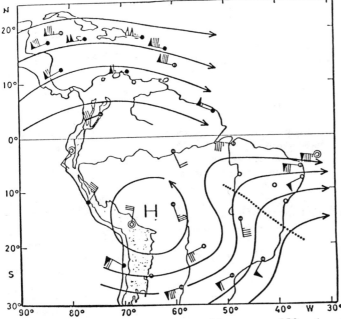

Figure 4. Winds and streamlines at 150 mb above South America, S-Summer (Kreuels et al. 1975).

Figure 5 shows the seasonal shift of the 200 mb easterlies along $32°E$ between Cairo and Durban (Strüning et al. 1969).

In such considerations one should not forget the 7 % increase of (extraterrestrial) solar heating of the earth due to the present position of the perihelion during January, which is also responsible for the annual variation of tropopause temperatures in the tropics (Reid et al. 1981) and the frequent penetration of Cb clouds with their water vapour into the stratosphere (Newell et al. 1981).

Recent investigations (Tanaka 1981) have indicated that this "winter"

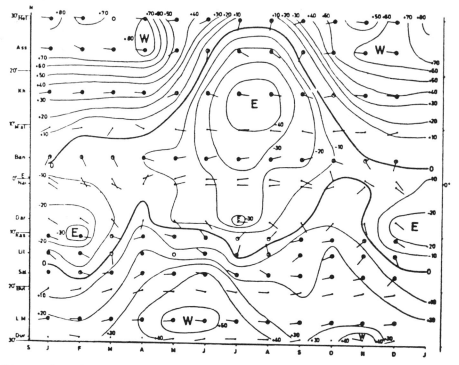

Figure 5. Seasonal shift of the monsoonal easterlies at 150 mb above Africa (32°E) (Strüning 1969).

monsoon system is also time-dependent: a correlation of + 0.94 (18 pairs) exists between the intensity of the 150 mb easterlies at Singapore and the Walker Circulation Index, thus connecting these two circulation types. Table 1 gives a comparison of estimated parameters of these circulations.

Table 1. Tropical circulation systems.

	Abbrev.	Width[1]	ΔT (°C)	Temp.[2] gradient	Mass-[3] transport	Direction of circulation
Hadley Circulation	HC	30	4-6	1.3-?	50	meridional
Walker Circulation	WC	80-120	3-5	~0.3	10-20	zonal(along 0-5°S)
Summer Monsoon (NH)	NMC	30-35	10-12	2-4	100-180	meridional
Summer Monsoon (SH)	SMC	20-30	3-5	0.6-1.5	~50	meridional

[1] horizontal extension in °lat. (or °long. at Eq.)
[2] intensity increase by latent heat release, ΔT/1000 km
[3] in kg m^{-1} s^{-1}

4. Quasi-stationary high-tropospheric troughs and cloud-bands

The operational use of satellite data in daily weather maps has demonstrated that, similar to the NH, and perhaps even more regularly, quasi-stationary high-tropospheric (500-200 mb) troughs regularly modify the zonal wind belts in the SH, especially above the South Pacific (Van Loon et al. 1972; Streten 1970, 1972) and the South Atlantic. The South Pacific Convergence Zone (SPCZ) regularly runs from about 35°S, 135°W diagonally above Samoa towards New Guinea and New Britain, west of the date line, merging with the southern branch of the ITCZ and strengthening its convective activity. In some seasons it may shift westwards to the line Fiji-Solomon Islands or New Hebrides, but here daytime local circulation may affect the large-scale features, which are confirmed by analyses of the in-flight observations from airplanes (Sadler 1975) (near 250 mb). A similar band above the South Atlantic is also known (Van Loon et al. 1972; Trenberth 1980, 1981; Flohn 1971, 1975); it runs from 35°S, 20°W via 25°S, 35°W towards the Brazil coast and is prolonged during the southern summer even towards the Amazon river (Kreuels et al. 1975). A similar feature in the southwestern Indian Ocean across Madagascar (Van Loon et al. 1972; Flohn 1971, 1975) is apparently much less frequent. A statistical verification of its frequency and of the special conditions of its occurrence and seasonal variation is needed. Observations from geostationary satellites suggest that cirrus clouds may blow out from the frequent convective cloud clusters in the ITCZ region along the troughs in the high-level westerlies (Thépennier et al. 1983).

In spite of many recent general circulation model studies, the physical background of the quasi-stationary nature of these troughs is not yet completely understood. Thirty years ago, the South Atlantic trough was interpreted as orographically fixed by the Andes (Boffi 1949), similar to the trough above eastern North America. Data from 500 mb (Trenberth 1980) confirm this idea for the SH summer only. However, this concept certainly also applies to the NH, where the mechanical effect of mountains is enlarged by the concentration of the release of latent heat in some oceanic areas. But the more frequent SPCZ lacks such a relation and no convincing physical interpretation is known as yet. An essential feature of all these "oblique" troughs is their marked tilt equatorwards of the subtropical jet and their extension to the equator itself (in the NH to 10°N). A remarkable feature is the intensity of the high-tropospheric SH westerlies concentrated in the eastern hemisphere, especially over the western Indian Ocean. An evaluation of the three first planetary waves (Trenberth 1980) - with wave number 1 predominating - indicates a complete reversal of the quasi-stationary anomalies between 35°S and 55°S, which is related to this velocity core over the Indian Ocean.

MAJOR CLIMATIC EVENTS DURING THE LATE CENOZOIC

If we look at the last part of earth's history (some 1 m.y.) and assume the solar radiation to be constant, large-scale climatic patterns are controlled by the following parameters:

1. position and extension of continents, mountains and oceans, as caused by slow horizontal and vertical plate motions;

2. spatial and seasonal variations of surface albedo, especially due to snow and ice, but also to vegetation;

3. atmospheric composition, especially the content of H_2O, CO_2, O_3 and other infrared-absorbing trace gases.

The history of the present pattern is taken as starting with a strong global event at the Eocene/Oligocene transition around 38 m.y. BP (Frakes 1979, 1982; Kennett 1982). At this time Antarctic glaciation gradually began to develop, while warm-temperate moist climates prevailed in high Arctic latitudes. After the opening of Drake Street and the isolation of Antarctica by a northward drift of Australia and the formation of a strong circum-Antarctic ocean current, the glaciation of East Antarctica intensified and was completed during the Mid-Miocene (~14 m.y. BP) (Tingey 1982; Truswell 1982; Kennett 1982). During this long period a circumpolar Antarctic Current developed, with gradual cooling at its southern fringe, and the contemporaneous linkage of the Arctic Ocean to the Atlantic caused an additional flux of relatively warm intermediate water towards Antarctica (Schnitker 1980; Blanc et al. 1980) which may have contributed to this effect. During the Middle and probably Late Miocene, no definite signs of a large-scale NH glaciation (except for Alaskan mountain glaciers at about 9 m.y. BP) have been found as yet, while the Antarctic ice-sheet grew to its peak at the end of the Miocene, when West Antarctica was also ice-covered (Kerr 1981).

During the Early and Mid-Pliocene (5 to 3.5 m.y. BP) evidence for the Arctic is apparently controversial: while in northernmost Siberia (including the islands on the large Siberian shelf), in northern Alaska and in the Canadian archipelago convincing evidence for a mixed boreal forest exists (Frenzel 1968; Hopkins 1971), deep-sea cores from the Central Arctic Ocean were interpreted as showing increasing glacial evidence (Clark 1980, 1982). A quantitative validation of these data, however, seems to be conclusive only after the Mid-Pliocene. (This controversy will be discussed further on.) All other available data indicate no large-scale NH continental glaciation before 3.5 m.y. BP (Shackleton et al. 1977; Zubakov et al. 1983; Kennett 1982). This gradual evolution has been linked to the simultaneous final closure of the isthmus of Panama (Keigwin 1978; Webb 1976) which forced the enormous warm water masses of the westward flowing Equatorial Current to deviate northward and to intensify the Gulf Stream. Increased evaporation thus produced sufficient atmospheric moisture to build the first large ice domes in the Greenland-Iceland-Baffinland area.

In addition to this evolution, it should be mentioned that the lifting of the Tibetan highlands had not started before the end of the Pliocene (about 2.0 to 1.8 m.y. BP), shortly before the (paleomagnetic) Olduvai event (Li et al. 1979; Opdyke et al. 1979), after which the giant NH summer monsoon system evolved gradually to its present extension and intensity.

The consequences of these geological/geophysical processes for climatic patterns will be discussed in the final section of this paper.

OCEANIC UPWELLING AND ATMOSPHERIC CONSTITUENTS

The phenomenon of upwelling cool water from the thermocline region and the intermediate water below it, with temperatures between 25 and 14°C along some tropical coasts and along the Pacific and Atlantic equator is well known, as are its climatic consequences: stabilization of air, suppression of convective clouds and rainfall, and downward flux of sensible heat. However, some other consequences have been found only recently (Fig. 6): suppression of evaporation (Henning et al. 1980;

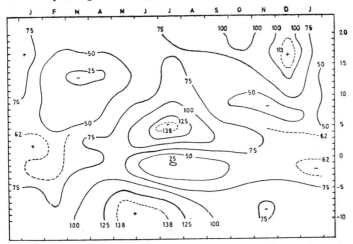

Figure 6. Ocean evaporation (mm/month) along shipping route Europe-South Africa (Henning et al. 1980).

Weber 1981) (in spite of higher relative humidity, except in the case of very strong winds such as the Somali jet) and high consumption of atmospheric CO_2 by the photosynthesis of algae in the nutrientrich upwelling water, in spite of its supersaturation with CO_2-gas and inorganic carbon (Newell et al. 1978; Bacastow 1981, 1980; Angell 1981).

The frequent change between upwelling cool water and downwelling warm water (= El Niño) in the equatorial oceans is caused by a complex interaction between atmospheric and oceanic dynamics (Rasmussen et al. 1982; Horel 1982; Philander 1981, 1983; McCreary 1983). During warm water episodes the H_2O and CO_2 content of the air increases, while upwelling has the opposite effect. Since the frequency and intensity of upwelling increases with the strength of the trade winds (indeed with the square of wind speed), a large-scale feedback mechanism has been proposed (Flohn 1982, 1983) (Table 2), which controls the global atmospheric content of both water vapour and carbon dioxide.

Table 2. Natural hemispheric climatic feedback system.

Initial change in polar regions	cooling	warming
Meridional temperature gradient	stronger	weaker
Latitude of subtropical anticyclones	equatorward	poleward
Intensity of Hadley cell (trades)	strong	weak
Prevailing mode (tropical oceans)	upwelling	downwelling
Equatorial sea surface temperature	cold	warm
Atmospheric content of CO_2 and H_2O	decreasing	increasing
producing further	cooling	warming

Since both gases are transparent in the visible part of the spectrum, but absorb terrestrial infrared radiation, their combined action contributes significantly to the greenhouse effect. Investigations on the CO_2 content of the air bubbles enclosed in continental ice cores have shown that during the last glaciation the CO_2 content dropped to 180-200 ppm, as compared with 265-295 ppm during the 19th century and as much as 340 ppm in 1982 (Berner et al. 1980; Delmas et al. 1980; Neftel et al. 1982). Parallel changes of water vapour content are suggested by the prevalence of arid climates during glacial periods and of humid climates during the Holocene (about 12 000 - 6000 BP). The actual rate of this process involves 1-2 Gt (10^{15} g) of carbon per year, or about 20-40 % of the present annual input of carbon from fossil fuels into the climatic system. This large contribution to the carbon cycle suggests a major role of this air-sea exchange in the abrupt global-scale climatic change on the 100 year time scale (Flohn 1982, 1983; Oeschger et al. 1983).

PALEOCLIMATIC STAGES AND OCEANIC UPWELLING

1. Late Miocene and Pliocene

A world-wide cooling event during the Mid-Miocene (c.15-14 m.y. BP) coincided with an episode of high volcanism, especially around the Pacific (Kennett et al. 1977). During that period bottom water temperatures decreased rather rapidly by about 5 °C. Evidence for repeated coolings has been found in an equatorial core (Woodruff et al. 1981), suggesting orbitally caused fluctuations of a growing Antarctic ice sheet. During the Late Miocene, NE Siberia, Alaska and presumably all of northern Canada were covered by a rich, diversified coniferous forest which can today only grow at summer temperatures near 12 °C (Wolfe 1978, 1980). This excludes the existence of a permanent ice cover over the Arctic Ocean. Local glacial evidence in Alaska at 9 m.y. BP is generally assumed to be restricted to mountains. It also seems to coincide with a sharp cooling event in E Iceland (Denton et al. 1969; Mudie et al. 1983). However, marine data from Iceland evaluated by F Strauch (1972) suggest annual SST values between 8 and 13 °C until about 4 m.y. BP. This also precludes a large ice cap before the Mid-Pliocene. In Central Europe and the North Sea abundant evidence is available indicating a warm-temperate climate (Buchardt 1978; Schwarzbach 1968). Evidence from Africa (Maley et al. 1980) indicates a marked asymmetry: while the southern Sahara was covered with tropical humid/semi-humid forests, the Zaire Basin and South Africa were dry, sometimes even fully desertic.

During the Messinian (6.5-5 m.y. BP), repeated Mediterranean desiccation, cyclic sedimentation in Sicily and in the equatorial Pacific, together with a eustatic drop of global sea level, suggests growth (Tingey 1982; Truswell 1982) and orbitally induced fluctuations of the Antarctic ice (Adams 1977; McKenzie et al. 1979). During this time, the Arctic climate had hardly changed: a rich mixed forest (with hazel) was observed on the Alaskan coast (66 °N), with a thermophilous insect fauna, such as found now in British Columbia (48-54 °N) (Hopkins et al. 1971), probably simultaneous with the highest volume of the Antarctic ice (Adams et al. 1977; McKenzie et al. 1979). Russian investigations (Frenzel 1968; Zubakov et al. 1983) indicate that NE Siberia and the islands on the Siberian shelf were covered by rich coniferous forest until the Late Pliocene, about 2.5 m.y. BP. This seems to be inconsistent with evidence of ice-rifted material on the bottom of the Arctic Ocean (Clark 1980, 1982) from probably as early as the latest Miocene (5.4 m.y. BP). However, the occurrence of icebergs carrying some glacial debris from local continental glaciations into the sea certainly does not necessitate the occurrence of a permanent ice cover; the present situation around Newfoundland may serve as an example. Along the Nile valley the Late Pliocene was relatively cool and wet (McCauley et al. 1982). Here and in the Chad basin aridity did not set in before the Pliocene/Pleistocene boundary, coinciding with the marked uplift of the Tibetan highlands, which then caused the above-mentioned tropical easterly jet-stream, with its efficient cross-circulation resulting in subsidence and aridity (Flohn 1966, 1968, 1971, 1975). Recent data from southern Ethiopia (7 °N) (Bonnefille 1983) suggest a cool, but drier climate in a period of global cooling during the Pretiglian (2.4-2.5 m.y. BP).

Considered together, these results indicate a period of about 10 m.y. between Mid-Miocene and Early Pliocene, in which a unipolar glaciation of Antarctica existed, while the Arctic was most probably essentially ice-free. Since during this period the Panama isthmus was not yet closed (Keigwin 1978; Webb 1976), a large equatorial ocean current ran from West Africa westwards to Indonesia, i.e. over about 230 ° longitude. The temperature contrast between the Arctic and Antarctic must have been much larger than now, at least during the Messinian, indicating a strong hemispheric asymmetry of climatic belts and a shift of the meteorological equator (i.e. the annually averaged position of the ITCZ) towards about 10-12 °N (now 6 °N). Over the ocean, this should have led to a nearly permanent transgression of the SH circulation across the equator. This resembles the present situation during the northern summer and suggests all-year upwelling south of the equator, producing a shift or an extension of the SH subtropical arid belt towards the equator, while the ITCZ rain belt would have been centered at 5-15 °N (cf. Table 3). If we extrapolate from recent observations of the role of oceanic upwelling in the exchange of CO_2 and H_2O through the sea-air interface, the south-tropical belt (0-20 °S) should have had an arid or semi-arid, relatively cool climate.

The role of continental drift is small but not negligible. The available maps (Smith et al. 1977; Habicht 1979) for the Late Tertiary are not quite coherent, but both the African and South American plates seem to have been displaced slightly towards north (between 0 and 6 °

Table 3. Approximate latitude of climatic belts during unipolar glaciation.

Climatic belt	Latitudes	
	NH	SH
Snow and ice climates	–	65-90 °
Subpolar cyclonic belt	65-90 °	50-65 °
Temperate belt: surface westerlies	45-65 °	25-50 °
Subtropical winter rains	40-45 °	
Subtropical arid belt	25-40 °	5-25 °
Tropical summer rains	15-25 °	5 °S-5 °N
Tropical rain belt	5-15 °	–

latitudes), since 10-15 m.y. BP, with a velocity of about a few cm/year, and perhaps with a weak rotation. Within the limits of error this effect could have contributed to the observed shift of the ITCZ (Sarnthein et al. 1982) in addition to the increased asymmetry of climatic zonation on a unipolar glaciated earth. After the Mid-Miocene (15 m.y. BP), alternating dry and wetter episodes have been inferred from the ocean cores (Sarnthein et al. 1982), while coastal upwelling along NW Africa did not exist, indicating a weak and sluggish Hadley cell in the NH. Due to the stronger SH circulation, the upwelling region and the Benguela Current along the Angola-Namibia coast may have developed earlier, even before the Late Miocene (Siesser et al. 1978). Two further important new publications have not been taken into account: a review of the evolution of marine paleoceanography during the Miocene (Keller et al. 1983) and a discussion of the Pliocene climatic evolution, mainly on the basis of NH data, written jointly by a geologist and a climatologist (Zubakov and Borzenkova 1983, in Russian).

2. Late Pleistocene and Holocene

One of the most important discoveries in paleoclimatology has been the occurrence of variations in the CO_2 content (as mentioned earlier) of air bubbles in Greenland and Antarctic ice cores (Berner et al. 1980; Delmas et al. 1980; Neftel et al. 1982). The confirmed results of 180-220 ppm during the period 20 000 - 13 000 BP, described by Ruddiman et al. (1981) present, for the first time, an adequate interpretation of the parallel climatic trends in both polar zones (Oeschger et al. 1983) which are indeed inconsistent with the orbital variations. Perhaps the most intriguing result are the rapid changes of 80-100 ppm within about 100 years that occurred three times during the Late Pleistocene/Holocene transition (beginning of Bölling, beginning and end of Younger Dryas), obviously synchronous with rapid variations of the O18/O16 ratio which depends on temperature. If these fluctuations are considered in the light of the present knowledge of the relationship between CO_2 growth rate, Pacific upwelling (Newell et al. 1978; Bacastow et al. 1981, 1980; Angell 1981) and intensity of the Hadley cell, the working hypothesis of a rapid global-scale feedback process between atmospheric/oceanic circulation and the CO_2 content (and probably also that of H_2O) in the air (Flohn 1982, 1983, and in press) seems to be promising and merits further investigation.

In the case of the Pleistocene glaciations, the extension of seasonal or permanent drift ice towards the equator, which probably reached latitudes 44°N and 46°S in the Atlantic 18 000 years ago (Ruddiman et al. 1981; Hays 1978), causes an intensification of extratropical westerlies, a shift of the subtropical belts towards the equator and thus an intensification of both Hadley cells. Since the hemispheric asymmetry of the climate (as discussed earlier) was consequently much weaker and perhaps even absent in the Atlantic, the vorticity of the wind stress vector and its effects on the vertical component of the oceanic drift (Flohn, in press) should have caused intensified upwelling varying seasonally on both sides of the equator. Prevailing westerly winds of both monsoon systems straddle the equator over the Indian Ocean and cause convergence of the Ekman flow with downwelling. This situation should have been similar to the present, in spite of the fact that the NH monsoon system weakened 18 000 years ago (Duplessy

1982; Prell et al. 1980; Van Campo et al. 1982).

During the glacial maxima – here mainly during maximum extension of Arctic and Antarctic drift ice during the second half of the glaciation (Ruddiman et al. 1981) – upwelling had indeed intensified along the coast of Angola and Namibia (Embley et al. 1980) and may have extended seasonally even beyond the equator. This also involves suppression of oceanic evaporation and prevalence of more arid climates in the equatorial zone, where the rain forests had been reduced to a few mountain islands even in the Amazon and Congo/Zaire basin and perhaps also in Indonesia.

During a warm interglacial period – between about 12 000 and 6000 BP – the situation tended to be the reverse, since the orbital elements (precession) produced an insolation maximum during the northern summer, when the monsoon-triggering role of the northern continents is largest (Kutzbach et al. 1981, 1982; Rossignol-Strick 1983). Due to the eccentricity of the earths's orbit, a position of the perihelion in the northern summer provides up to 7 % more solar energy in the NH than at present (perihelion Jan. 4). This leads to an increased thermal contrast between continents and oceans and thus to a strengthening of the NH summer monsoon (Kutzbach et al. 1981, 1982). A perihelion date during the southern winter should only slightly change the meridional temperature gradient in the SH and consequently the intensity of circulation.

Since some evidence exists for an early retreat of the seasonal subantarctic drift-ice (Hays 1978), the period 12 000 – 6000 BP should have been characterized by a decrease of the equator-pole temperature difference and hence by a weaker zonal circulation in both the SH and the NH, but by a stronger NH monsoon (Duplessy 1982; Prell et al. 1980; Van Campo et al. 1982; Kutzbach et al. 1981, 1982). Thus the frequency and intensity of warm water episodes (El Niño) (Rasmussen et al. 1982; Horel 1982; Philander 1981, 1983; McCreary 1983) should have increased and the trade winds weakened. This would have led to a rather humid period with relatively high CO_2 concentrations, which also accelerated the ice retreat in the NH after the Younger Dryas interlude with its abrupt beginning and end (Duplessy et al. 1981). Regarding the SH circulation, the orbital effect of an increased meridional temperature gradient may have been balanced (or outweighed) by the retreat of drift ice and the shrinking of the cold air above it. In South Africa available evidence from this period points to a more humid climate in the south and west (Heine 1982; Rust et al. 1981), but to a drier period with open grassland in the Transvaal (Scott 1982).

It should be mentioned that by evaluating a purely orbital Monsoon Index, Rossignol-Strick (1983) assumes that in some cases intense tropical monsoons may coincide with glacial (or cold interstadials) in the NH. The last interglacial, however, should have similar conditions to those prevailing during the Holocene, with one noteworthy exception: right in the centre of the warmest phase (5e, about 125 000 BP) part of the West Antarctic ice sheet collapsed and produced a 5-7 m rise in global sea level together with a regional cooling of the southern oceans (Aharon et al. 1980; Duplessy 1978). Due to the time-distance between this event and the sudden transition 5e/5d, an immediate consequence in the NH, as proposed by Flohn (1974) and Hollin (1980), seems to be rather unlikely, and the cooling should not have lasted more than a few centuries (as in the case of Younger Dryas in the northern Atlantic).

REFERENCES

Adams CG et al. 1977. Nature 269:383-386.
Aharon P et al. 1980. Nature 283:649-653.
Angell JK 1981. Mon. Weather Rev. 109:230-243.
Arkin PA 1982. Mon. Weather Rev. 110:1393-1404.
Bacastow RB and Keeling CD 1981. In: World Climate Programme, Conference on Analysis and Interpretation of Atmospheric CO_2 data, Bern. p. 109-112.
Bacastow RB and Keeling CD 1980. Science 210:66-68.
Bengtsson L et al. 1982. Bull. Am. meteor. Soc. 63:277-303.
Berner W, Oeschger H and Stouffer B 1980. Radiocarbon 22:227-235.
Bjerknes J 1969. Mon. Weather Rev. 97:163-172.
Blanc PL et al. 1980. Nature 283:553-555.
Boffi JA 1949. Bull. Am. meteor. Soc. 30:242-247.
Bonnefille M 1983. Nature 303:487-491.
Buchardt B 1978. Nature 275:121-123.
Budd WF 1982. Aust. meteor. Mag. 30:265-272.
Carsey FD 1980. Mon. Weather Rev. 108:2032-2044.
Clark DL 1980. Spec. Pap. geol. Soc. Am. 181.57 pp.
Clark DL 1982. Nature 300:321-325.
Davidson NE et al. 1983. Mon. Weather Rev. 111:496-516.
Delmas RJ et al. 1980. Nature 284:155-157.
Denton G and Armstrong RL 1969. Am. J. Sci. 267:1121-1142.
Dey B 1979. Arctic and Alpine Research 11:229-242.
Dey B 1980. J. Glaciol. 25:425-438.
Dey B 1981. J. geophys. Res. 86:3223-3235.
Duplessy JC 1978. In: J Gribbin (ed.), Climatic Change. Cambridge Univ. Press, Cambridge. p. 46-67.
Duplessy JC et al. 1981. Palaeogeogr. Palaeoclimatol. Palaeoecol. 35:121-144.
Duplessy JC 1982. Nature 295:494-498.
Embley RW and Morley JJ 1980. Mar. Geol. 36:183-204.
Flohn H 1966. Z. Meteor. 17:316-320.
Flohn H 1967. Ann. meteor. N.F. 3:76-80.
Flohn H 1968. Colorado State Univ. Atmos. Sci. Paper 120.
Flohn H 1971. Bonner met. Abh. 15:55 pp.
Flohn H 1974. Quat. Res. 4:385-404.
Flohn H 1975. Bonner met. Abh. 21:82 pp.
Flohn H 1978. In: EM van Zinderen Bakker (ed.), Antarctic Glacial History and World Palaeoenvironments. Balkema, Rotterdam. p. 3-13.
Flohn H 1982. J. meteor. Soc. Japan 60:268-273.
Flohn H 1983. In: A Street-Perrott et al. (eds.), Variations in the Global Water Budget. Reidel, Dordrecht. p. 403-418.
Flohn H, in press.
Frakes LA 1979. Climates throughout Geologic Time. Amsterdam. 310 pp.
Frakes LA 1982. Aust. meteor. Mag. 30:175-179.
Frenzel B 1968. Science 161:637-649.
Habicht J 1979. Am. Assoc. Petr. Geol., Studies in Geology 9.
Hahn DG and Manabe S 1975. J. atmos. Sci. 32:1515-1541.
Hastenrath St. 1974. Bonner met. Abh. 16:353-360.
Hays JD 1978. In: EM van Zinderen Bakker (ed.), Antarctic Glacial History and World Palaeoenvironments. Balkema, Rotterdam. p. 57-71.
Heine K 1982. Palaeoecol. Africa 15:53-76.
Henning D and Flohn H 1980. Contrib. Atmos. Phys. 53:430-441.
Hollin JT 1980. Nature 281:629-633.

Hopkins DM et al. 1971. Palaeogeogr. Palaeoclimatol. Palaeoecol. 9:211-231.
Horel JD 1982. Mon. Weather Rev. 110:1863-1878.
Keigwin LD 1978. Geology 6:630-634.
Keller G and Barron JA 1983. Bull. geol. Soc. America 94:590-613.
Kennett JP et al. 1977. J. volcan. geotherm. Res. 2:145-164.
Kennett J 1982. Marine Geology Prentice Hall Inc. xv + 813 pp.
Kerr RA 1981. Science 213:427-428.
Kreuels R, Fraedrich K and Ruprecht E 1975. Met. Rdsch. 28:17-24.
Krishnamurti TN 1971. J. atmos. Sci. 28:1342-1347.
Krishnamurti TN 1971. J. appl. Meteor. 10:1066-1096.
Kutzbach JE et al. 1981. Science 214:59-61.
Kutzbach JE et al. 1982. J. atmos. Sci. 39:1177-1188.
Lajoie FA 1972. Aust. meteor. Mag. 20:207-216.
Li JJ et al. 1979. Scientia Sinica 22:1314-1328.
Lyne WH et al. 1982. Q. J. R. meteor. Soc. 108:575-594.
Maley J, Servant M, Butzer KW et al. 1980. In: MAJ Williams and H Faure (eds.), The Sahara and the Nile. Balkema, Rotterdam. xvi + 607 pp.
McCauley JF et al. 1982. Science 218:1004-1020.
Mechoso CR 1981. Mon. Weather Rev. 109:2131-2139.
McCreary JP 1983. Mon. Weather Rev. 111:370-387.
McKenzie JA et al. 1979. Palaeogeogr. Palaeoclimatol. Palaeoecol. 29:125-141.
Mudie PJ and Helgason J 1983. Nature 303:689-692.
Neftel R et al. 1982. Nature 295:220-223.
Newell RE et al. 1978. Pure and Appl. Geophys. 116:351-371.
Newell RE and Gould-Stewart S 1981. J. atmos. Sci. 38:2789-2796.
Oeschger H and Dansgaard W et al. 1983. In: A Ghazi (ed.), Palaeoclimatic Research and Models. Reidel, Dordrecht. p. 95-107.
Opdyke ND et al. 1979. Palaeogeogr. Palaeoclimatol. Palaeoecol. 27:1-34.
Parkinson CL 1983. J. phys. Oceanogr. 13:501-511.
Philander SGH 1981. J. phys. Oceanogr. 11:176-189.
Philander SGH 1983. Nature 302:295-301.
The Polar Group 1980. Reviews of Geophysics and Space Physics 18:525-543.
Prell L et al. 1980. Quat. Res. 14:309-336.
Quiroz RS 1983. Mon. Weather Rev. 111:143-154.
Ramage CS 1968. Mon. Weather Rev. 96:365-370.
Ramage CS et al. 1981. J. geophys. Res. 86:6580-6598.
Rasmussen EM and Carpenter TH 1982. Mon. Weather Rev. 11:354-384.
Reid GC and Gage KS 1981. J. atmos. Sci. 38:1928-1936.
Rossignol-Strick M 1983. Nature 304:46-49.
Ruddiman WF and McIntyre A 1981. Palaeogeogr. Palaeoclimatol. Palaeoecol. 35:145-214.
Ruddiman WF and McIntyre A 1981. Science 212:617-627.
Ruddiman WF and McIntyre A 1981. Quat. Res. 16:125-134.
Rust U and Schmidt H 1981. Mitt. geogr. Ges. München 66:141-174.
Sadler JC 1975. The Upper Tropospheric Circulation over the Global Tropics. Dept. of Meteorol., Univ. of Hawaii, UHMET-75-05.
Sarnthein M et al. 1982. In: M von Rad et al. (eds.), Geology of the Northwest African Continental Margin. Springer, Berlin-Heidelberg. p. 545-604.
Schnitker D 1980. Earth-Science Reviews 18:1-20.
Schwarzbach M 1968. Z. Dt. geol. Ges. 118:33-68.
Scott L 1982. Quat. Res. 17:339-370.
Shackleton NJ and Opdyke ND 1977. Nature 270:216-219.

Siesser WG and Coetzee JA 1978. In: EM van Zinderen Bakker (ed.), Antarctic Glacial History and World Predicaments. Balkema, Rotterdam. p. 105-113 and 115-127.
Smith AJ and Briden JC 1977. Mesozoic and Cenozoic Paleoenvironmental Maps, Cambridge Univ. Press.
Strauch F 1972. Z. Dt. geol. Ges. 123:163-177.
Streten NA 1970. Aust. meteor. Mag. 18:31-38.
Strüning JO and Flohn H 1969. Bonner met. Abh. 10:56 pp.
Sumi A and Murakami T 1981. J. meteor. Soc. Japan 59:625-645.
Tanaka M 1981. J. meteor. Soc. Japan 59:825-831.
Thépennier RM and Cruette D 1983. Abstract from IAMAP-WMO Symposium, Paris (Aug-Sept). p. 329-333.
Tingey RJ 1982. Aust. meteor. Mag. 30:181-189.
Trenberth KE 1980. Mon. Weather Rev. 108:1378-1389.
Trenberth KE 1981. J. atmos. Sci. 38:2585-2605.
Truswell EM 1982. Aust. meteor. Mag. 30:169-173.
Van Campo E et al. 1982. Nature 296:56-59.
Van Loon H et al. 1972. Meteorology of the Southern Hemisphere. Met. Monogr. Am. meteor. Soc. 35.
Webb SD 1976. Paleobiology 2:220-234.
Weber KH 1981. Diploma thesis, Univ. Bonn.
Weller G 1982. Aust. meteor. Mag. 30:163-168.
Wolfe JA 1978. Am. Scientist 66:694-703.
Wolfe JA 1980. Palaeogeogr. Palaeoclimatol. Palaeoecol. 30:313-323.
Woodruff F et al. 1981. Science 212:665-668.
Zubakov WA and Borzenkova II 1983. Paleoklimaty Pozdnego Kainozoya. Leningrad. 216 pp.

A climatic model of the Last Glacial/Interglacial transition based on palaeotemperature and palaeohydrological evidence

S.P.HARRISON, S.E.METCALFE & F.A.STREET-PERROTT
School of Geography, Oxford, UK

A.B.PITTOCK
CSIRO, Mordialloc, Australia

C.N.ROBERTS
University of Technology, Loughborough, UK

M.J.SALINGER
University of East Anglia, Norwich, UK

ABSTRACT. At present the atmospheric circulation in the two hemispheres is markedly asymmetric, due to the much stronger meridional temperature gradient in the Southern Hemisphere. In this paper we argue that adjustments of the two hemispheric circulations to variations in boundary conditions, notably the extent of ice and snow in high latitudes, resulted in fluctuations in the intensity and annual-average location of major circulation features such as the subtropical anticyclones and the equatorial trough. Changes in the meridional temperature gradient in each hemisphere are reconstructed from zonally-averaged palaeotemperature data. The circulation trends predicted by the model are in good agreement with palaeohydrological evidence from Africa, western Eurasia and Australasia for the periods 18 000, 9000 and 6000 BP.

INTRODUCTION

Ever since the development of the glacial theory in the nineteenth century, there has been speculation about the effects of variations in the ice and snow cover of the Northern and Southern Hemispheres on the general atmospheric circulation. Two extreme points of view soon emerged. One (the 'concertina' model) considered that the displacements of the circulation belts in the two hemispheres were essentially synchronous (Willett 1953) whereas the other (the 'shunting' model) predicted that glacial events in the Northern and Southern Hemispheres occurred in antiphase (Imbrie and Imbrie 1979: ch.6). Recent work suggests that the thermal history of the two hemispheres and hence the interaction of their atmospheric circulations has been much more complex than was at first anticipated (Nicholson and Flohn 1980).

In this exploratory paper we develop a conceptual, annual-average model of climatic changes during the last 25 000 years, based on existing palaeotemperature estimates. The predictions of the model are tested against independent palaeohydrological data. We shall concentrate on shifts in the latitude and intensity of the major circulation features, touching only briefly on the important topic of longitudinal asymmetry.

PRESENT-DAY ATMOSPHERIC CIRCULATION PATTERNS

A schematic representation of the present meridional (north-south) structure of the general circulation provides an appropriate starting point for our model. An idealized annual-average hemispheric circulation is composed of the following components:

a. An equatorial trough (ET) which usually contains an inter-tropical convergence zone (ITCZ) located on or near the equator, with variable winds and frequent rains.

b. A trade-wind belt between the ET and approximately 30° latitude, with predominant surface easterly winds, especially over the oceans.

c. A subtropical anticyclonic belt (STA), consisting of ridges of high pressure between about 30° and 40°, with an upper baroclinic zone on its poleward flank.

d. A westerly wind belt between 40° and 60° with travelling mid-latitude depressions.

e. A subpolar low-pressure belt with travelling cyclonic centres at about 60° latitude.

f. A polar anticyclone with easterly outblowing winds.

The intensity and mean latitude of the ET vary seasonally at present as a function of the absolute and relative vigour of the two hemispheric circulations, respectively (Flohn 1967; Newell 1973). The absolute vigour of each hemispheric circulation is largely determined by its tropospheric temperature gradient, whereas the relative vigour depends on the thermal contrast (the difference in meridional temperature gradient) between the two hemispheres. This can be expressed as

$$\phi_{ET} = f(\Delta T_{nh} - \Delta T_{sh})$$

where ϕ is the latitude of the ET and ΔT is the mean equator-pole temperature gradient.

At present the annual-average ΔT_{sh} in the 700-300 mb layer is about 39°C compared with 27°C for ΔT_{nh}. Surface values are less easily compared due to the great altitude of the Antarctic ice dome (Flohn 1967). The greater magnitude of ΔT_{sh} ensures that the southern circulation is much stronger than the northern circulation. This displaces the ET north of the geographical equator so that the various circulation belts occur closer to the equator in the Southern Hemisphere. ϕ_{ET} (the meteorological equator) varies seasonally between 0° and almost 15°N, its annual average being about 6°N (Flohn 1980).

$$\tan \phi_{STA} = \frac{H\,(\partial\theta/\partial z)}{R\,(\partial\theta/\partial y)}$$

where H is a scale height, R is the radius of the earth, $\partial\theta/\partial z$ is the mean vertical lapse rate of equivalent potential temperature, and $\partial\theta/\partial y$ is the mean meridional gradient of equivalent potential temperature. The equivalent potential temperature is the air temperature corrected to a standard pressure and water-vapour content.

This relationship has been substantiated for the 700-300 mb layer by Korff and Flohn (1969) and for the 850-300 mb layer by Pittock (1973, 1974), and is illustrated by the present-day mean annual ϕ_{STA} of about 31°S in the Southern Hemisphere and 37°N in the Northern Hemisphere (Flohn 1967). The meridional temperature gradient also determines the intensity of the STA and the westerlies; a strong hemispheric temperature gradient resulting in a narrower but more intense anticyclonic belt, and stronger and more zonal westerlies (see below), than a weaker hemispheric temperature gradient.

The generalized pattern of circulation belts described above gives rise to marked latitudinal variations in the distribution of mean annual precipitation, P, and runoff, P-E, where E is mean annual evaporation. In both hemispheres, maxima of P-E are associated with zones a and d-e, and minima with zone b-c (Peixóto and Oort 1983). The subtropical P-E minimum intensifies and moves towards the equator in winter. The equatorial P-E maximum migrates seasonally between $0°-10°S$ and $10°N$, with a mean latitude of about $5°N$ (Peixóto and Oort 1983). Its intensity is greatest in June-August due to the strong heating of the Northern Hemisphere continents.

A second major aspect of the general circulation is the east-west pattern of stationary long waves, or persistent ridges and troughs (Flohn 1971; Pittock 1978). These are of two types: features that appear on long-term mean maps, such as the Rossby waves in the westerlies and the 'Walker Circulation'; and seasonally reversing circulations, notably the tropical monsoons (Holton 1979).

The wavelength, λ, of a stationary Rossby wave is given by

$$\lambda = 2\pi \ (u/\beta)^{\frac{1}{2}}$$

where u is the mean zonal wind speed and $\beta = df/dy$ where f is the Coriolis parameter $2\Omega \sin \phi$ (Lamb 1972). Since u varies with $\partial\theta/\partial y$, a larger meridional temperature gradient will lead to an increased persistence of lower wave-number configurations, e.g. 3 or 4 waves around the hemisphere rather than 4 or 5. In mid-latitudes, different longitudinal sectors may, therefore, experience opposing climatic trends as the meridional temperature gradient changes (van Loon and Williams 1976).

The monsoonal circulations are direct thermal circulations between land areas and surrounding oceans (Webster 1981). In summer, the land heats more rapidly than the oceans, leading to surface convergence with an influx of moist unstable air and precipitation, while in winter the land is cooler than the oceans, giving atmospheric subsidence and surface outflow over the land. These circulations may be modified by variations in land albedo, ocean temperatures, atmospheric humidity and cloudiness.

The Walker Circulation is an east-west, thermally driven circulation between the eastern tropical Pacific, which is normally cool, as a result of ocean currents and upwelling, and the warmer north Australian-Indonesian region (Troup 1965; Bjerknes 1969). Year-to-year variations in the strength of this circulation are related to the 'Southern Oscillation' (Troup 1965; Julian and Chervin 1978) and to fluctuations in the trade winds and oceanic upwelling off the coast of Peru (Wyrtki 1975). Both the tropical monsoons and the Walker Circulation interact dynamically with the westerlies.

FACTORS CAUSING LONG-TERM CHANGES IN ATMOSPHERIC CIRCULATION

The above description of the atmospheric circulation demonstrates its sensitivity to changes in atmospheric boundary conditions, since the 'controlling' temperature gradients are largely determined by boundary conditions such as ice limits, sea-surface temperature patterns, albedo, etc.

During the Last Glaciation, the strong cooling in the higher latitudes of both hemispheres led to an increase in ΔT_{nh} and ΔT_{sh}, and to a steepening of the temperature gradients across mid-latitudes (CLIMAP 1981). As argued above, the result should have been an equatorward compression of the circulation belts in both hemispheres, together with a stronger and more zonal westerly circulation (Flohn 1953; Lamb 1961; Williams 1978). The reverse has been postulated for times of maximal warmth in high latitudes (Lamb 1961; Manabe and Wetherald 1980).

However, the nature and extent of snow and ice cover at the Last Glacial Maximum were markedly different in the two hemispheres (CLIMAP 1981; Kukla 1981). In the Northern Hemisphere, the equatorward shift of the southern limit of snow and ice was about 20° in summer, reflecting the presence of the large ice sheets, and 10°-15° in winter when increased snow cover and, to a lesser extent, pack ice were responsible. In contrast, the equatorward advance of the ice limit in the Southern Hemisphere was controlled by the increase in sea ice and was smaller: about 13° in summer and 10° in winter. This disparity should have resulted in a displacement of the ET towards, or even south of, the equator (Tricart 1956; Newell 1973).

A further contrast in the behaviour of the two hemispheres is apparent in the timing of temperature maxima and minima. A variety of continental and oceanic evidence suggests that temperature trends in the Southern Hemisphere preceded changes in the Northern Hemisphere ice sheets and, hence, Northern Hemisphere temperatures by up to several thousand years (Hays et al. 1976; Chappell and Grindrod 1983). Systematic phase shifts, therefore, are to be expected in the palaeohydrological record.

Precise, quantitative application of the Smagorinsky criterion, to evaluate the effects of past variations in temperature on the two hemispheric circulations, required that allowance be made for changes in vertical lapse rate $\partial\theta/\partial z$ due to a changed moisture regime, and that $\partial\theta/\partial y$ be determined over an appropriate height and latitude range. As pointed out by Flohn (1980), clouds also introduce discontinuities into the lapse rate below the 300 mb level and so it is difficult to apply the criterion in a rigorous or quantitative manner. It has been argued that climatic fluctuations do not significantly alter the vertical lapse rate in tropical and subtropical latitudes, which is normally close to the moist adiabatic lapse rate (Webster and Streten 1978). In the case of $\partial\theta/\partial y$, changes in atmospheric moisture will amplify the effects of changes in surface boundary conditions. As a first approximation, therefore, we ignore possible variations in $\partial\theta/\partial z$ and use temperature, T, instead of θ when considering climatic changes. Flohn (1967) suggested that the Smagorinsky criterion should be applied in the latitude range spanning the subtropical baroclinic zone, where $\partial T/\partial y$ reaches its maximum value. Since $\partial T/\partial y$ is comparatively small between 0° and 20°N and S (Newell et al. 1972), we assume that variations in ϕ_{ET} have had a negligible effect on the temperature gradients in mid-latutudes.

In the following sections, we use palaeotemperature data to derive estimates of zonal-mean temperature departures in the latitude belts 30°-60°N, 15°N-15°S and 30°-60°S, T_{nh}, T_{eq} and T_{sh}, since 25 000 BP. From these we calculate the meridional temperature gradients $\Delta T_{nh} = T_{eq} - T_{nh}$ and $\Delta T_{sh} = T_{eq} - T_{sh}$, and the thermal contrast between the two hemispheres, $\Delta T_{nh} - \Delta T_{sh}$. Surface temperatures are employed in the absence of information about the palaeotemperature structure of the middle and upper troposphere. The results allow us to make qualitative inferences about long-term changes in ϕ_{STA}, ϕ_{ET} and the intensity of the westerly circulation in each hemisphere.

Changes in surface temperature should also have influenced the intensity of the hydrological cycle. Modelling studies (Manabe and Wetherald 1980) suggest that warmer global surface temperatures increase overall global evaporation and precipitation. Conversely, numerical models using ice-age boundary conditions have tended to show lower evaporation rates and reduced precipitation, especially over land (Williams 1978). It must be remembered, however, that these models assume fixed glacial-age ocean temperatures, thus eliminating possibly important feedback processes.

The temperature anomalies T_{eq}, T_{nh} and T_{sh} are used below as a general guide to the relative amounts of precipitation generated by the equatorial trough, the onshore trade-wind regimes and the westerlies. The importance of convective precipitation over land, however, also depends on the temperature contrast between land and sea (Manabe and Hahn 1977), which will be discussed more fully in a future paper, and on the degree of seasonality (Kutzbach and Otto-Bliesner 1982), which is not considered in our model.

The predictions of our model are tested by comparison with palaeohydrological data for three key time horizons: 18 000, 9000 and 6000 BP. In this analysis a combination of lake-level and other palaeohydrological data are employed. As Street-Perrott and Roberts (1983) have shown, latitudinal variations in the relative extent of lakes closely reflect the meridional distributions of P-E. We have assumed that the same holds true for the other types of palaeohydrological evidence discussed.

METHODS

1. Palaeotemperature data

The palaeotemperature curves shown in Fig. 1 were compiled from published and unpublished sources. As far as possible, continuous curves (derived from coleopteran faunas, O18/O16 ratios in speleothems, or statistical calibration of pollen or marine microfossil assemblages) were used, supplemented where necessary by other estimates based on pollen or macrofossils, snowline data, D/H ratios or O18/O16 ratios in temperate lake sediments. Estimates based on O18/O16 ratios in oceanic or ice cores were excluded. Indicators of seasonal (e.g. summer) temperatures were only included where seasonal variations in temperature depression appear to have been slight (e.g. between 15°N and S).

Separate curves were obtained for the land areas and for the oceans, and then area-weighted to produce a zonal-mean curve for each latitude belt. In order to reduce the bias introduced into the continental data set by the uneven distribution of information, the data points were aggregated into 10 regions. The regional means were then averaged to produce zonal-mean curves for the land areas. In the case of sea-surface temperature (SST), there were large differences in the amplitude of the glacial/interglacial temperature change between different oceanic sectors. In an attempt to reduce the resulting bias, the zonal-average SST departures at 18 000 BP (CLIMAP 1976) were used to calculate a weighting factor which was then applied to the data from each

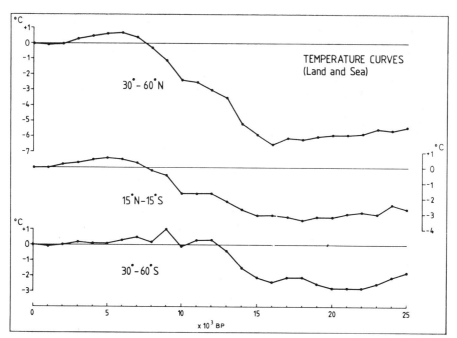

Figure 1. Curves showing the variations in T_{nh}, T_{eq} and T_{sh} since 25 000 BP, expressed as departures from present-day values in °C. The modern area-weighted annual-average surface temperatures in these latitude belts are 10.7°C, 25.7°C and 9.8°C, respectively (from data in Newell et al. 1972).

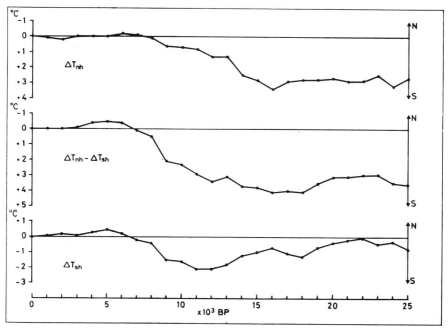

Figure 2. Curves showing the variations in ΔT_{nh}, $\Delta T_{nh} - \Delta T_{sh}$ and T_{sh} since 25 000 BP, expressed as departures from present conditions in °C. The corresponding modern values are 15.0°C, -0.9°C and 15.9°C, respectively (from data in Newell et al. 1972).

oceanic region in order to derive a zonal-mean SST curve for each latitude belt.

2. Palaeohydrological data

Two study areas were selected to test the palaeoclimatic inferences derived from the model. These are Africa-western Eurasia (20°W-60°E, 35°S-40°N), and Australia - New Zealand.

Our main source of palaeohydrological data was the Oxford lake-level data bank (Street-Perrott and Roberts 1983). Only water-level fluctuations dated by geochronometric methods are included. The original data have been standardized to yield estimates of lake status, an index of relative depth, at 1000 year intervals. Lake status is coded in three classes: high, intermediate and low. In this study, the data are presented in the form of differences in lake status from today, as follows:
- Much wetter than present (a difference of 2 status classes, i.e. between high and low).
- Wetter than present (a difference of 1 status class, e.g. between high and intermediate)
- No difference in status
- Drier than present (a difference of 1 status class)
- Much drier than present (a difference of 2 status classes).

For Africa-western Eurasia, the data bank includes 111 lake basins. For Australasia, only 26 basins are available. Additional data on moisture conditions at 18 000, 9000 and 6000 BP in Australia and New Zealand have therefore been obtained from published work on pollen, ostracodes and diatoms.

RESULTS

1. Palaeotemperature curves

Figure 1 shows that zonal-mean temperature departures from present values in the belts 30°-60°N, 15°N-15°S and 30°-60°S. T_{nh} was already low at 25 000 BP, decreasing slowly to a minimum of -6.5°C at 16 000 BP. Note, however, that the averaging process has concealed large oscillations in temperature in some regions, particularly NW Europe and the NE Atlantic, between 14 000 and 10 000 BP. Conditions were warmer than today between 7000 and 3000 BP, since when they have remained similar to the present.

T_{eq} shows a prolonged minimum (-2.9 to -3.4°C) from 23 000 to 15 000 BP. It then rises irregularly, approaching modern values by 9000 BP and peaking around +0.6°C at 5000 BP. A word of caution is appropriate here. The continental data from this belt are dominated by high-altitude sites. If our assumption about the constancy of lapse rates in the tropics is incorrect, then the amplitude of the variations in T_{eq} (Fig. 1) is exaggerated, possibly by as much as 0.5-1.0°C at 18 000 BP.

The T_{sh} curve is significantly different from the other two. From a value of -1.8°C at 25 000 BP it falls to a minimum of -2.8°C at 22 000-20 000 BP and then rises in undulating fashion, reaching the zero line around 12 500 BP. Temperatures oscillated above present values until 3000 BP. The peak of warmth (+1.0°C) occurred at 9000 BP, 3000 years before the maximum in T_{nh}.

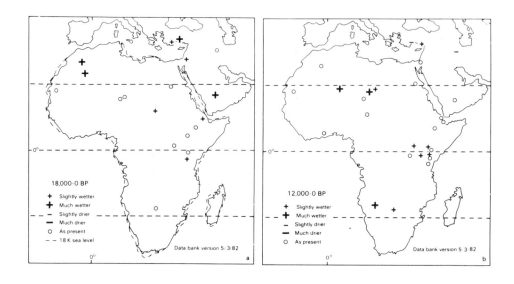
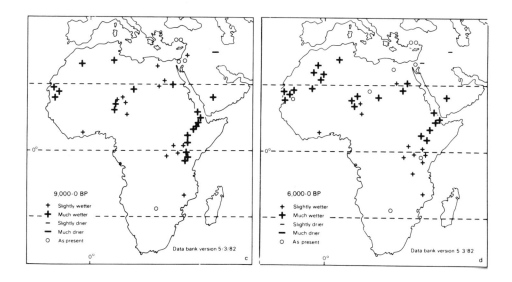

Figure 3. Moisture conditions in Africa and western Eurasia, expressed as departures from those of the present day: (a) 18 000 BP; (b) 12 000 BP; (c) 9000 BP; (d) 6000 BP.

Figure 2 shows the variations in ΔT_{nh}, ΔT_{sh} and $\Delta T_{nh} - \Delta T_{sh}$ since 25 000 BP. In each case, northward displacements of the climatic belts can be inferred from points lying above the zero line, and southward displacements from points below it. It is important, however, not to view these shifts as linear functions of ΔT.

ΔT_{nh} is large and positive throughout the period 25 000-16 000 BP, implying a large equatorward displacement of ϕ_{STA} and the westerlies. Stronger and more zonal westerly flow is indicated. A pronounced cooling in ocean temperatures between 30° and 60°N, combined with an even larger cooling of the continents, suggests reduced precipitation over land. After 16 000 BP the westerlies should have weakened and retreated irregularly polewards, reaching their most northerly position around 6000 BP.

In contrast, between 25 000 and 19 000 BP, ΔT_{sh} exhibits values close to those of the present, indicating strong, zonal westerly flow over the Southern Ocean. If the glacial lowering of T_{eq} has indeed been exaggerated, then the westerlies may actually have been more vigorous and situated further north than today. The temperature lowering over the southern continents was significantly greater than over the ocean, implying a reduction in precipitation, especially over land.

After a slight oscillation around 18 000-16 000 BP, ΔT_{sh} decreases to a minimum value of $-2.1°C$ at 12 000-11 000 BP. We infer that the interval 15 000-9000 BP was characterized by a poleward shift in ϕ_{STA}, with a weaker and more meridional circulation in mid-latitudes. After 9000 BP, ΔT_{sh} increases again, suggesting an equatorward shift and reintensification of the southern westerlies.

The thermal contrast between the two hemispheres ($\Delta T_{nh} - \Delta T_{sh}$) is large and positive (implying a net southward shift of ϕ_{ET} relative to that of today) throughout the period 25 000-12 000 BP. This curve is entirely independent of T_{eq}. After 12 000 BP, $\Delta T_{nh} - \Delta T_{sh}$ decreases to a minimum at 5000 BP, when the ET should have reached its most northerly position. Since T_{eq}, and the temperature contrast between land and sea, follow the same general trend, the tropical monsoonal circulations should have been weakest during the glacial and strongest during the period 9000 BP to the present.

2. Palaeohydrological maps

Africa and western Eurasia

Figure 3 shows the distribution of moisture anomalies in the region at 18 000, 9000 and 6000 BP. A map for 12 000 BP (Fig. 3b) is also included to illustrate the contrast with 9000 BP.

The map for 18 000 BP (Fig. 3a) indicates conditions moister than those prevailing today along the southern and eastern shores of the Mediterranean and adjacent to the Red Sea, whereas the northern intertropical belt was relatively dry. This pattern is in accord with an equatorward displacement of the westerlies and ϕ_{STA} (Nicholson and Flohn 1980). Data from the Southern Hemisphere are too scarce to permit any inferences about the ET or the southern westerlies.

By 12 000 BP conditions in the tropics were becoming generally wetter, i.e. more monsoonal. The distribution of enlarged lakes (Fig. 3b) was more or less symmetrical about the equator, indicating a more southerly location of the ET than at present. A dry zone existed at $23°-37°N$, which suggests that the northern westerlies had begun to retreat polewards.

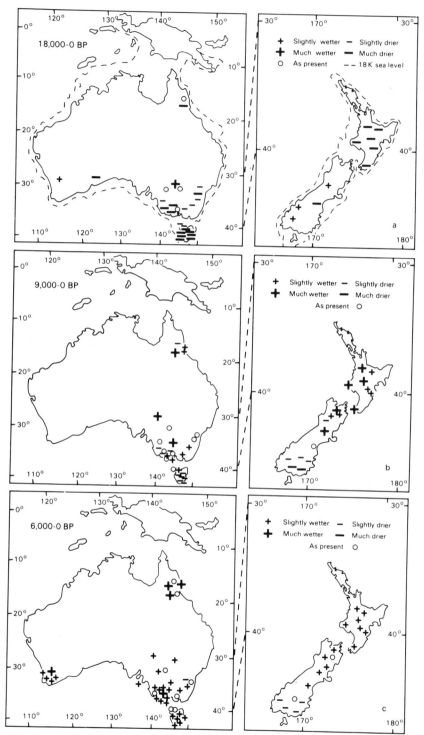

Figure 4. Moisture conditions in Australia and New Zealand, expressed as departures from those of the present day: (a) 18 000 BP; (b) 9000 BP; (c) 6000 BP.

By 9000 BP (Fig. 3c), the climate in northern intertropical Africa was much moister than it is today. This implies a stronger summer-monsoon circulation. The scarcity of data points south of the Equator makes it difficult to determine ϕ_{ET}, although a mean location close to, or slightly north of, the present one seems probable (Fig. 3c). The map for 6000 BP (Fig. 3d) is generally similar to Fig. 3c, but provides clearer evidence for a northward displacement of the ET (Nicholson and Flohn 1980). At this time the annual-average ϕ_{ET} seems to have coincided roughly with its present value in June-August. The fall in lake levels in East Africa between 9000 and 6000 BP is consistent with a slight equatorward shift in the southern STA.

Australasia

The palaeoenvironmental data from New Zealand for 18 000, 9000 and 6000 BP are described by Salinger (this volume). At 18 000 BP (Fig. 4a), a strengthened westerly circulation is indicated, together with a weaker Walker Circulation over the Pacific. An equatorward shift in ϕ_{STA} (Pittock 1973) is borne out by the widespread occurrence of dry conditions in the present summer rainfall areas of Queensland and SE Australia. The suggestion of increased wetness in Western Australia is also consistent with a stronger and more zonal westerly circulation. Although this inference conflicts with the occurrence of dry conditions on the windward coasts of Victoria and Tasmania, several lines of argument tend to reduce the apparent discrepancy. Firstly, the lowering of sea level would have enhanced the continentality of these areas (Fig. 4a). Secondly, stronger westerly flow was apparently associated with an increased frequency of winds blowing parallel to the coast rather than onshore (Derbyshire 1971; Bowler and Wasson, this volume). Thirdly the greater cooling of the land compared with the adjacent Southern Ocean would lead to reduced convective precipitation from westerly disturbances.

At 9000 BP (Fig. 4b), the evidence from New Zealand indicates a much more meridional circulation than is prevalent today, enhanced precipitation in North Island and the northern part of South Island contrasting with drier conditions than those of the present in the south (Salinger, this volume). This pattern of moisture anomalies is consistent with weaker westerlies and a stronger Walker Circulation. In Australia the combination of wetter conditions in Queensland, New South Wales and eastern Victoria, with drier conditions on the western fringes of Victoria and Tasmania, also suggests a weaker westerly circulation and a poleward migration of the STA (Pittock 1973). All the evidence from Australasia indicates an equatorward readvance of the circulation belts between 9000 and 6000 BP (Fig. 4b,c). Although Queensland was still very moist, a drying trend in New South Wales and eastern Victoria suggests a reduction in summer precipitation, linked to a northward migration of ϕ_{STA} (Pittock 1973). Increased wetness in western Australia, western Victoria and western Tasmania implies a stronger westerly circulation, as does the dissipation of the anomaly pattern seen at 9000 BP in New Zealand. This picture is consistent with a slight weakening of the Walker Circulation over the Pacific.

CONCLUSIONS

In this paper we have presented an empirical model to explain the major features of the palaeohydrological record since 25 000 BP. This model does not consider the ultimate causes of climatic change, but takes as its starting point the zonal-mean temperature departures in the equatorial belt and the mid-latitudes, as reconstructed from palaeotemperature data. The temporal trends exhibited by lake levels in different latitude belts (Street-Perrott and Harrison, in press) are in excellent general agreement with the reconstructed variations in ΔT_{nh} and ΔT_{sh}. Where differences exist between the two, they may in part be attributable to variations in seasonality (Kutzbach and Otto-Bliesner 1982), which is not included in our annual-average model.

The maps of moisture anomalies in Africa, western Eurasia and Australasia at 18 000, 9000 and 6000 BP presented in this paper confirm that significant shifts in ϕ_{STA} and ϕ_{ET} have occurred through time, in good accordance with the trends in ΔT_{nh} and ΔT_{sh}, although the magnitude of the departures from present conditions is not always predicted correctly. We conclude that the two hemispheric circulations have interacted in a complex way. During the Last Glaciation, the circulation belts in both hemispheres were compressed equatorwards, but by differing amounts. In contrast, all the climatic zones were displaced southwards at 12 000 BP, and northwards at 6000 BP, due to the time lag between the thermal maxima in the two hemispheres.

ACKNOWLEDGEMENTS

We are indebted to H Flohn and SE Nicholson, whose papers on hemispheric asymmetry stimulated our interest in this topic.

REFERENCES

Bjerknes J 1969. Atmospheric teleconnections from the Equatorial Pacific. Mon. Weath. Rev. 97:163-172.
Bowler JM and Wasson RJ, this volume. The last glacial cycle in Australia.
Chappell JMA and Grindrod A (eds.), 1983. Proceedings of the First CLIMANZ Conf. Dept. Biogeogr. Geomorph. ANU, Canberra, 2 vols.
CLIMAP Project Members 1976. The surface of the ice-age earth. Science 191:1131-1137.
CLIMAP Project Members 1981. Seasonal reconstructions of the earth's surface at the last glacial maximum. Geol. Soc. Am. Map and Chart MC-36.
Flohn H 1953. Studien über die atmosphärische Zirkulation in der letzten Eiszeit. Erdkunde 7:266-275.
Flohn H 1967. Bemerkungen zur Asymmetrie der atmosphärischen Zirkulation. Ann. Met. N.F. 3:76-80.
Flohn H 1971. The Tropical Circulation Pattern. Bonn. met. Abh. 15. 55pp.
Flohn H 1980. Possible climatic consequences of a man-made global warming. Int. Inst. Appl. Syst. Anal. Rept. (RP-80-30). Laxenburg, Austria. 80pp.
Hays JD, Imbrie J and Shackleton NJ 1976. Variations of the Earth's orbit: pacemaker of the ice ages. Science 194:1121-1132.
Holton JR 1979. An Introduction to Dynamic Meteorology. Academic Press, New York.
Imbrie J and Imbrie P 1979. Ice Ages: Solving the Mystery. Macmillan Press, London. 224pp.

Julian PR and Chervin RM 1978. A study of the Southern Oscillation and Walker Circulation phenomenon. Mon. Weath. Rev. 106:1433-1451.

Korff HC and Flohn H 1969. Zusammenhang zwischen dem Temperaturgefälle Äquator-Pol und den planetarischen Luftdruckgürteln. Ann. Met. N.F.4: 163-164.

Kukla G 1981. Climatic role of snow covers. In: Allison (ed.), Sea Level, Ice and Climatic Change. Int. Ass. Hydrol. Sci. Publ. 131. p.79-107.

Kutzbach J and Otto-Bliesner BL 1982. The sensitivity of the African-Asian monsoonal climate to orbital parameter changes for 9000 years BP in a low-resolution general circulation model. J. atm. Sci. 38: 1177-1188.

Lamb HH 1961. Fundamentals of climate. In: AEM Nairn (ed.), Descriptive Palaeoclimatology. Wiley-Interscience, New York. p. 3-44.

Lamb HH 1972. Climate: Present, Past and Future. Vol. 1: Fundamentals and Climate Now. Methuen, London. 613pp.

Manabe S and Hahn DG 1977. Simulation of the tropical climate of an Ice Age. J. geophys. Res. 82:3889-3911.

Manabe S and Wetherald RJ 1980. On the distribution of climate change resulting from an increase in CO_2 content of the atmosphere. J. atm. Sci. 37:99-118.

Newell RE 1973. Climate and the Galapagos Is. Nature 245:91-92.

Newell RE, Kidson JW, Vincent DG and Boer GJ 1972. The General Circulation of the Tropical Atmosphere. Vol. 1. 258pp.

Nicholson SE and Flohn H 1980. African environmental and climatic changes and the general atmospheric circulation in Late Pleistocene and Holocene. Climatic Change 2: 313-348.

Peixóto JP and Oort AH 1983. The atmospheric branch of the hydrological cycle. In: A Street-Perrott, M Beran and R Ratcliffe (eds.), Variations in the Global Water Budget. D Reidel, Dordrecht. p. 5-65.

Pittock AB 1973. Global meridional interactions in stratosphere and troposphere. Q.J.R. met Soc. 99:424-437.

Pittock AB 1974. Global interactions in stratosphere and troposphere. In: Proc. Int. Conf. on Structure, Composition and General Circulation of the Upper and Lower Atmosphere and Possible Anthropogenic Perturbations. p. 716-720.

Pittock AB 1978. Patterns of variability in relation to the general circulation. In: AB Pittock et al. (eds.), Climatic Change and Variability: A Southern Perspective. Cambridge University Press, Cambridge. p. 167-179.

Salinger MJ, this volume. New Zealand climate: the last 5 million years.

Smagorinsky J 1963. General circulation experiments with the primitive equations. Appendix B: A non-dimensional form of the baroclinic instability criterion. Mon. Weath. Rev. 91:159-162.

Street-Perrott FA and Harrison SP, in press. Temporal variations in lake levels since 30 000 BP - an index of the global hydrological cycle. Am. Geophys. Un. Maurice Ewing Ser. 4.

Street-Perrott FA and Roberts N 1983. Fluctuations in closed lakes as an indicator of past atmospheric circulation patterns. In: A Street-Perrott, M Beran and R Ratcliffe (eds.), Variations in the Global Water Budget. D Reidel, Dordrecht. p. 331-345.

Tricart J 1956. Tentative de correlation des periodes pluviales africaines et des periodes glaciaires. C.r. somm. Soc. Géol. Fr. 9-19: 164-167.

Troup AJ 1965. The Southern Oscillation. Q.J.R. met. Soc. 91: 490-506.

Van Loon H and Williams J 1976. The connection between trends of mean temperature and circulation at the surface: Part I. Winter. Mon. Weath. Rev. 104:365-380.

Webster PJ 1981. Monsoons. Scient. Am. 245:108-118.

Webster PJ and Streten NA 1978. Late Quaternary climates of tropical Australasia; interpretations and reconstructions. Quat. Res. 10: 279-309.

Willett WC 1953. Atmospheric and oceanic circulation as factors in glacial-interglacial changes of climate. In: H Shapley (ed.), Climatic Change, Evidence, Causes and Effects. Harvard University Press, Cambridge. p.51-71.

Williams J 1973. A brief comparison of model simulations of glacial period maximum atmospheric circulation. Palaeogeogr. Palaeoclimatol. Palaeoecol. 25:191-198.

Wyrtki K 1975. El Niño - the dynamic response of the equatorial Pacific Ocean to atmospheric forcing. J. phys. Oceanogr. 5:572-584.

An analytical climate model: Application to the Southern Hemisphere Quaternary period

BRUCE DENNESS
University of Newcastle upon Tyne, UK

ABSTRACT. By considering various time-series data over a range of time scales from 3000 million to 1 million years a quantitative analytical climatic model is evolved in the form of a sine series of long period components. This is then extrapolated to higher frequencies and compared successfully with observed climatic time series for the Quaternary of the Southern Hemisphere.

INTRODUCTION

Over recent decades the realization has been growing that climate changes both regionally and globally on a time scale from as little as a year to over the tens of thousands of years of a glacial epoch and the millions of years of the geological time scale. Examples of concern are to be seen in the many national (Lamb 1972; Bryson 1977) and international (Bach et al. 1980; WMO 1975) research programmes that attempt to elucidate the problem, for a problem it is. For example, Faure and Gac (1981) report that in the short term hundreds of thousands of people die in semi-regularly recurring droughts in the Sahel; the production of corn in the American Midwest is said by Silcock (1981) to be sensitive to a global change of $1\,^\circ C$ to the tune of about 11%, and that in Kazakhstan to about 20% (Bach 1978); while the Interdepartmental Group of Climatology (1980) notes that the UK Department of Energy recognizes an annual budgetary implication of about £200 million for the same change.

It is the object of this paper to illustrate that the long-term climatic changes are equally well represented by the analytical model for the Southern and the Northern Hemisphere.

The literature abounds with time series describing the variation of a multitude of climatic indices on every conceivable time scale. Most writers conclude that climatic change has taken place and many of them such as Hayes et al. (1976) and Gray (1975) attempt to deduce by means of computers the periodicities of the changes through spectral analysis. Until recently no one, not even Milankovitch (1938), had succeeded in establishing the elusive relationship which would enable climatic change on all time scales to be embraced by one single model. If this were possible, so that the model could match time-series data on a scale of billions of years as well as millions, thousands, and a few years, it would represent not only an interesting step towards the culmination of so much effort in unravelling the past but also a soundly based predictive method for the future. Recently Denness (in press) introduced such a model in relation to the Quaternary of south-east Asia, using primarily Northern Hemisphere data for a range of time scales covering climatic variation over periods from about one million years down to a few decades.

A CLIMATIC MODEL: GENERATING THE MODEL

Several hundred examples of climate related time-series data were examined in an attempt to approach such a model. Though the tangled path through this information to the construction of the model described below was in itself an exciting detective story (Denness 1981), it is sufficient here to take the complete model and show its ability to match, in hindcast, measured time-series at appropriate time scales. This is done graphically by superimposing selected time-series data onto computer drawn, moving-average plots from the "fundamental" equation:

$$G(t) = \sum_{n=N(T)}^{\alpha} A(T) a^n \cdot \sin b^{1-n\pi} \left(\frac{t}{T}\right), \text{ zero registered at time } T_0$$

Where $G(t)$ = a time-based climate index,
$A(T)$ = the amplitude of a reference periodicity T,
$N(T)$ = the reference integer for periodicity T,
a,b = absolute constants, here taken as 0.84 and 0.50 respectively,
n = an integer, i.e. the reference number of a particular sine component,
and t = time in years.

We are thus seeing the variation of the sum of a series of sine curves, each successively smaller component being 0.84 times the amplitude of its more fundamental neighbour and of half the period. Over the range of time scales required to establish the climate model during the Phanerozoic, more than 12 components are needed to secure variation with a period as gross as a few billion years and as sensitive as a million years. This analytical model is here seen to represent global temperature variation with consequent regional implications for other climate-related data.

TESTING THE MODEL OVER THE GEOLOGICAL TIME SCALE

In the subsequent figures a moving average technique has been applied to $G(t)$ before plotting the derived $G_{ma}(t)$ to remove the higher frequencies from each plot. This was done in an attempt to secure compatibility of sensitivity between the climate model and the measured data. To achieve this, the time interval for each diagram was divided into 600 portions for each of which $G(t)$ was calculated and then subjected to a moving average analysis over successive groups of 100 points, resulting in the following plots which consequently use 500 data points each. In order that each moving-average plot should commence at the present time (zero BP) it was necessary to use the predictive quality of the equation to project 50 data points into the future for each time scale. Consequently the degree of success in matching the latest (nearest zero BP) 10% of the plots to the observed data could be seen as a preliminary assessment of these predictive capabilities.

The technique generally used to overlay the observed time-series data, all of which have been culled from the literature, was to photograph the figure and place the negative in an enlarger so that its image could be superimposed at the appropriate time scale onto the $G_{ma}(t)$ graph which had been previously prepared by x-y plotter. Apart from errors involved in tracing the image by hand, this preserves the accuracy of the measured data and renders it readily recognizable from the original source. The advantage of preserved integrity is offset to some degree by the consequent inconsistencies of ordinate scale. However, no serious attempt is made here to describe absolute temperature (or other) amplitudes but merely to demonstrate the consistent compilation of a regular sine series with ordered periodicities. In addition some data required processing to be compatible with the common presentation; this is noted where appropriate.

The reader is referred to the original publications for additonal information concerning the reasons for supposing that the variables are climatically related. Equally, the original publications are the best sources of reference on dating accuracy which the writer here notes to be somewhat irregular between the various records. The first two figures use essentially global data while the others are more local, as indicated.

3000 m.y. BP to the present

Figure 1 describes the progress of three variables over the period 3000 m.y. BP to the present. These are estimates of global temperature and precipitation by Frakes (1979) and oxygen isotope ratios measured on chert by Knauth and Epstein (1976). These data have been treated brutally by comparing maxima and minima at every point possible for the isotope data and every point possible up to about 600 m.y. BP (80% of the record) for the temperature and precipitation data. For the more recent period the sensitivity of the data exceeds that of the intended interpretation on this time-scale and have also been reduced photographically from the original pseudo-logarithmic time scale used by Frakes so that not even relative amplitudes could be preserved later than that time.

There appears to be a near-coincidence of maxima and minima from each of the measured time series with those (arrowed) in the $G_{ma}(t)$ plot. This applies equally to the long period variation between 3000 m.y. to 600 m.y. BP and to the shorter period variation since 600 m.y. BP for the isotope data. The curves show that it is merely the record of the data that has become more frequent in the more recent period; this does not necessarily imply more rapid variation of the variables. Nevertheless, accepting all the isotope data and all the low frequency temperature and precipitation data (and, indeed, higher frequency temperature data to some extent), the near coincidence of four maxima and four minima is evident between the observed time series and the model. Both the long period of variation of about 1.125×10^9 years and the shorter of about 5.623×10^8 years are available for comparison.

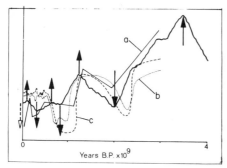

Figure 1. Timescale 0-3000 m.y. BP.
a) Oxygen isotope ratio for cherts
b) Estimated global temperature
c) Estimated global precipitation

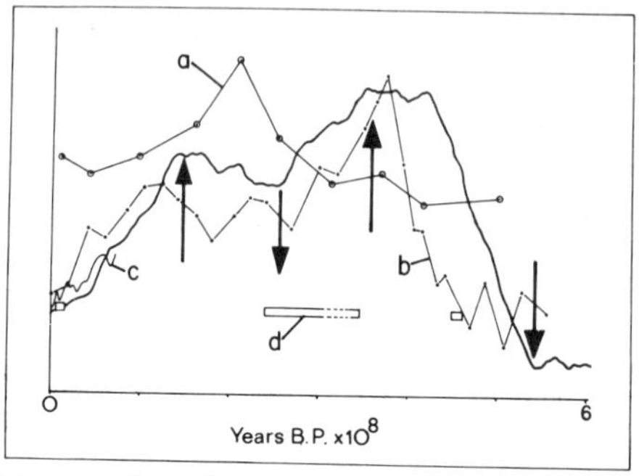

Figure 2. Timescale 0-600 m.y. BP.
a) Area of continental platforms covered by evaporites
b) Solar radiation absorption at Earth's surface
c) Arctic surface water temp and
d) Glacial epochs

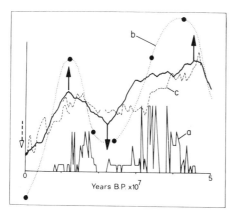

Figure 1-5. Comparison of climatic indices and model on different timescales.
The thick line on each figure is the model prediction.
Note the coincidence of maxima and minima of the observed time series with those of the model (arrowed).

Figure 3. Timescale 0-5 m.y. BP.
a) Fish debris in sand fraction off South Africa
b) Carbonate compensation depth in South Atlantic
c) Oxygen isotope ratio for foramenifera in southern oceans

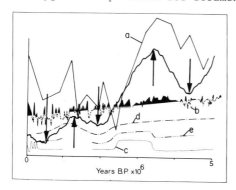 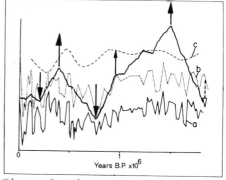

Figure 4. Timescale 0-5 m.y. BP.
a) Foramenifera (N.Pachyderma) abundance for sub. Antarctic
b) Oxygen isotope ratio for N.Atlantic
c) Sediment temp. for Weddell Sea
d) Faunal index for New Zealand
e) Estimated mean for c & d

Figure 5. Timescale 0-1.8 m.y. BP.
a) Oxygen isotope ratio for Tropical Atlantic
b) Volume of glacial ice
c) Oxygen isotope ratio for the Challenger Plateau

600 m.y. BP to the present

This leads to the consideration of Fig. 2 which again illustrates three types of primary, climate-related variables but now over the time scale from 600 m.y. BP to the present, essentially the Phanerozoic. The figure shows the secular change of global evaporite volume according to Gordon (1975), the variation of solar radiation absorption deduced by Burnett (1982), and the occasion of major glacial epochs noted by Whyte (1977) and comparable, relatively recent, high latitude marine water temperatures described by Savin et al. (1975). These exhibit three minima and two maxima which are reasonably consistent with those of the major period seen on the $G_{ma}(t)$ plot (large arrows). That the depiction of glacial conditions and water temperature trend is broadly in agreement with the minima of the $G_{ma}(t)$ curve, needs no further comment. The approximate correspondence of increasing occurrence of evaporite with maxima on the model is also to be expected. The solar radiation absorption takes account of differing albedo anticipated from the drifting of crustal plates known from palaeomagnetic studies. It is, therefore, an indirect measure of heat absorption from which global temperature is inferred and seen to be in approximate agreement with the $G_{ma}(t)$ plot.

50 m.y. BP to the present

With definite observation of a period of variation of about 2.812×10^8 yrs and the tentative observation of shorter period variations of the order of 1.406×10^8 yrs and 7.029×10^7 yrs, it is appropriate to move to Fig. 3 which shows the variation of three more climate-related variables over the period 5×10^7 yrs to the present. In addition to the continuing higher frequency variation of $G_{ma}(t)$, this diagram also shows the observed secular variation of carbonate compensation depth (CCD) in the South Atlantic described by Le Pichon et al. (1978), the oxygen isotope ratio in planktonic foraminifera in the southern oceans presented by Sclater (1978) and the proportion of fish debris in the sand fraction off South Africa determined by Melguen and Thiede (1974). The general association of the next shorter period (arrowed) in the $G_{ma}(t)$ series at 3.515×10^7 yrs with the variation of all the observed variables is seen clearly.

5 m.y. BP to the present

Time-series describing the observed variation of climatic indices over the anticipated periods 1.757×10^7 and 8.787×10^6 years could not be traced in the available literature. However, Fig. 4 restores the correlation of the $G_{ma}(t)$ series with foraminiferal abundance from the sub-Antarctic, determined by Kennett and Vella (1975), a further oxygen isotope record from North Atlantic sediment prepared by Shackleton and Cita (1979), and the variation of Weddell Sea Sediment temperature described by Anderson (1972) and of New Zealand fauna noted by Hornibrook (1971) from 5×10^6 yrs to the present. Collectively these records draw attention to minima (arrowed) at about 4.3×10^6, 1.9×10^6 and 0.5×10^6 yrs with interposed maxima (arrowed), all superimposed on a generally falling trend with the passage of time. This is consistent with the behaviour of $G_{ma}(t)$ and draws attention to a periodicity of 2.197×10^6 yrs (especially in the general trend available from simultaneous consideration of the sediment temperature and fauna variation curves), while hinting at the longer period of 4.393×10^6 yrs through the isotope records (maxima emphasized in solid form above a falling mean) and foraminiferal abundance.

APPLYING THE MODEL TO THE QUATERNARY OF THE EQUATORIAL REGION AND THE SOUTHERN HEMISPHERE

The model has already been shown to comprise a series of components with periods down to 2.197×10^6 yrs. A continuation of the series should lead, therefore, to the inclusion of the next two shorter period components with variations of period 1.098×10^6 and 5.492×10^5 yrs respectively. If the model is general both over different time scales and in different regions, these should be evident in observed Quaternary climatic time series for any part of the Southern Hemisphere.

Figure 5 depicts the progress of two separate records of oxygen isotope ratio, one from the Challenger Plateau off New Zealand, as reported by Shackleton and Kennett (1975) and the other from the tropical Atlantic recorded by Van Donk (1976). In addition the volume of glacial ice, which would have a Southern Hemisphere component, a fact also noted by Van Donk, is also presented superimposed with the others on the $G_{ma}(t)$ plot from 1.8×10^6 yrs to the present. The more sensitive isotope ratio plot and that for ice volume clearly indicate two primary maxima (arrowed) corresponding approximately to those of the $G_{ma}(t)$ plot at about 0.4×10^6 and 1.5×10^6 yrs with intermediate and neighbouring minima (arrowed) offset in the saw-toothed pattern typical of the $G_{ma}(t)$ plot at all time scales. This could suggest support for a variation with period of about 1.098×10^6 yrs as does the less sensitive isotope ratio curve, which also indicates a further maximum (arrowed) at about 0.95×10^6 yrs and neighbouring minima representing the observation of a higher frequency variation with a period 5.492×10^5 yrs. This degree of sensitivity is sufficient to demonstrate that the extrapolation of the model to higher frequencies is able to compare with gross climatic variation of the Southern Hemisphere and the equatorial region during the Quaternary.

DISCUSSION

Seventeen observational time-series have thus been used in the figures to illustrate the compatibility of the sine series model with measured climate-related indices. These are less than a third of the time-series used to test the validity of the model in hindcast over this range of geological time scales. It should also be noted that three of the sine series (the 3000 million year general temperature record, the 50 million year CCD record and the 5 million year North Atlantic oxygen isotope ratio record) were among those used to derive the sine series; the rest were not, and are thus genuine tests of the model.

Through the figures periodicities of climatic indices have been demonstrated from a "fundamental" of 1.125×10^9 yrs with the superposition of higher frequency variations at periods ever halving, and amplitudes reducing by about 0.84 for each successive component. Consideration of both period and amplitude is necessary to retain compatibility across time scales, i.e. between each figure. The inclusion of longer period variation of $G(t)$ was necessary in the preparation of Fig. 1 in order to provide the correct trend of $G_{ma}(t)$ to match the observed time-series. However, the observed data are not sufficiently extensive to permit a matching of these periods. Their mention here is to invite consideration of the degree of fundamentality of the 1.125×10^9 year period.

With the illustration of observed data which are consistent with 10 of 12 essential components of the model used in the preparation of the figures it is to be hoped that the palaeoclimatic literature for the geological time scale will soon provide evidence in support of those two periods not substantiated here. However, the use of Fourier analysis to approach these periods is questionable since, if this model is indeed appropriate, there is no linear datum axis about which the series is based nor can the whole of the influential components be contained in the sampling interval.

CONCLUSION

Observational evidence supports the establishment of an analytical climatic model that applies equally to gross climate variations in both the Northern and Southern Hemispheres. Climate variations over billions of years and also less than a million years can be described by the model.

REFERENCES

Anderson JB 1972. The marine geology of the Weddell Sea. Sediment. Res. Lab. Rep. 36, Fla. State Univ. 222 pp.
Bach W 1978. Carbon Dioxide, Climate and Society. In: J Williams (ed.). Pergamon, Frankfurt.
Bach W, Pankrath J and Williams J 1980. Interactions of Energy and Climate. Reidel Publ. Co., London. 569 pp.
Bryson RA 1977. Climates of Hunger. Univ. Wis. Press, Madison. p. 31-44.
Burrett CF 1982. Phanerozoic land-sea and albedo variation as climate controls. Nature 296:54-56.
Denness B 1981. How to build an ocean. Proc. IEEE Conf. Oceans '81, Boston: 341-344.
Faure H and Gac Y-Y 1981. Nature 291:475.
Frakes LA 1979. Climates throughout geologic time. Elsevier Scient. Publ. Co., Amsterdam. p. 260-263.
Gordon WA 1975. Distribution by latitude of Phanerozoic evaporite deposits. J. Geol. 83:671-684.
Gray BM 1975. Weather 30:359-368.
Hays JD, Imbrie J and Shackleton NJ 1976. Variations in the earth's orbit: pacemaker of the ice ages. Science 194:1121-1132.
Hornibrook N de B 1971. New Zealand Tertiary climate. N.Z. Geol. Surv. Rep. 47. 19 pp.
Inter-Departmental Group on Climatology 1980. Climate Change. H.M.S.O., Cardiff. 19 pp.
Kennett JP and Vella P 1975. Initial Reports Deep Sea Drilling Project 29:769-799.
Knauth LP and Epstein S 1976. Hydrogen and oxygen isotope ratios in nodular and bedded cherts. Geochim. cosmochim. Acta 40:1095-1108.
Lamb HH 1972. British Isles weather types. Geophys. Mem. 16 (116). H.M.S.O., London. 85 pp.

Le Pichon X, Melguen M and Sibuet JC 1978. A schematic model of the evolution of the South Atlantic. In: H Charnock and G Deacon (eds.), Advances in Oceanography. Plenum Press, New York. p. 1-48.

Melguen M and Thiede J 1974. Facies distribution and dissolution depths of surface sediment components from the Vema Channel and the Rio Grande Rise (Southwest Atlantic Ocean). Mar. Geol. 17:341-353.

Milankovitch M 1938. Astronomische Mittel zur Erforschung der erdgeschichtlichen Klimate. In: B Gutenberg (ed.), Handbuch der Geophysik 9. Berlin. p. 593-698.

Savin SM, Douglas RG and Stehli FG 1975. Bull. geol. Soc. Am. 86:1499-1510.

Sclater JG 1978. The marine geosciences. In: J Charnock and G Deacon (eds.), Advances in Oceanography. Plenum Press, New York. p. 307-338.

Shackleton NJ and Cita MB 1979. Oxygen and carbon isotope stratigraphy of benthic foraminifera at Site 397: detailed history of climate change during the late Neogene. Initial Reports Deep Sea Drilling Project 47:433-445.

Shackleton NJ and Kennett JP 1975. Late Cenozoic oxygen and carbon isotope changes at DSDP Site 284. Initial Reports Deep Sea Drilling Project 29:801-807.

Silcock B 4th Jan. 1981. Focus, Sunday Times, London, p. 13.

Van Donk J 1976. ^{18}O record of the Atlantic Ocean for the entire Pleistocene Epoch. In: RM Cline and JD Hays (eds.), Investigation of Late Quaternary Palaeo-oceanography. Geol. Soc. Am. Mem. 145:147-163.

Whyte M 1977. Turning points in Phanerozoic history. Nature 267:679-682.

World Meteorological Organization 1975. Proc. WMO/IAMAP Symp. on long-term climatic fluctuations, Norwich (WMO-No. 421). World Meteorological Organization, Geneva. 503 pp.

South America

Late Cainozoic glacial variations in South America south of the equator

J.H.MERCER
Ohio State University, Columbus, USA

ABSTRACT. Glacial variations in South America south of the equator have been radiometrically dated by both the (whole-rock) K/Ar and C14 methods. The K/Ar-dated record, derived mainly from southern Argentina, shows that glaciations on a scale comparable to those in the Pleistocene occurred near the Miocene-Pliocene transition, during the mid-Pliocene c. 3.6 m.y. BP, and repeatedly after 2 m.y. BP. Other glaciations may have occurred that have not been recognized in the geological record. During the Last Glacial Age, glaciers in southern South America, and probably also in tropical South America, were largest at a time beyond the range of C14 dating - perhaps about 73 000 years ago during marine isotopic stage 4. In Chile, two successively less extensive glaciations that peaked about 19 500 and 15 000 - 14 500 BP were separated by an interval of deglaciation. These two advances and the intervening recession occurred at about the same time as similar events in the Northern Hemisphere. After the final full-glacial advance 15 000 - 14 500 BP, glaciers receded rapidly, and did not readvance regionally until Neoglacial time. No equivalent of the Younger Dryas stade 11 000 - 10 000 BP in Europe has been detected. In Peru a glacial advance that reached its limit about 14 000 BP was, in the one valley where it has been dated, the greatest of at least the last 40 000 years. The following deglaciation was temporarily reversed by a readvance culminating about 11 000 BP; unlike the preceding maximum c. 14 000 BP, this had no counterpart in southern South America, and apparently preceded the European Dryas stade by several centuries.

INTRODUCTION

The Andean Cordillera extends along the western side of the entire South American continent. Glaciers occur throughout, the largest being the two Patagonian icefields south of 46°S. The snowline is highest - nearly 6000 m - in the arid zone between 23° and 30°S, falling northward to about 4500 m on the equator and southward to about 1000 m at 55°S in Tierra del Fuego. South of 46°S, icefield outlet glaciers reach sea level on the west side, and calve into lakes at low elevations on the east. During glacial ages, glaciers at low latitudes reached further down the mountainsides or, on upland plateaus, coalesced to form extensive icefields or ice caps. In southern South America glaciers spread onto the lowlands west of the mountains between

39° and 43° S, reaching the outer coast further to the south. On the east side of the cordillera south of 39° S, glaciers extended beyond the mountain front onto the lowlands, and reached the sea south of 51° S during the severest glaciations.

This paper is concerned chiefly with radiometrically-dated advances and recessions. Glacial variations have been dated by the radiocarbon and the (whole-rock) potassium/argon methods, giving records of events before c. 1 m.y. BP and after c. 25 000 BP ago. Neither method gives satisfactory results for the intervening interval. Except for one site in Bolivia, all the K/Ar age determinations are from the Patagonian plains of Argentina east of the cordillera between 47° and 52° S (Fig. 1). All dates given here have been revised according to the time scale of LaBrecque et al. (1977) with corrected K/Ar constants calculated by Mankinen and Dalrymple (1979). They therefore differ slightly from those given by Fleck et al. (1972), Mercer et al. (1975), Mercer (1976), and Clapperton (1979). Radiocarbon dating of full-glacial events has so far been accomplished only between 40° 30'S and 42° 30'S in Chile, and near 13° 45'S in Peru. Events during deglaciation (after the final full-glacial readvance) have been dated in both Argentina and Chile south of 40° S, and in Peru between 13° S and 14° S.

Figure 1. Location map, Chile and Argentina south of latitude 45°S.

THE POTASSIUM/ARGON-DATED RECORD, LATE MIOCENE TO PLEISTOCENE

South of 40° S and east of the cordillera, basaltic lava flows of Late Cainozoic age cover extensive areas of the Patagonian plains in Argentina. In places till or outwash gravels are interbedded with or covered by lava flows, from which whole-rock K/Ar ages have been obtained. In Bolivia till is covered by an ignimbrite flow from which biotite crystals have been separated and dated. In this way glaciations have been dated at between 4.6 and 7 m.y. BP, c. 3.6 m.y. BP, and repeatedly after c. 2 m.y. BP in Argentina, and at before c. 3.36 m.y. BP in Bolivia.

Late Miocene or earliest Pliocene glaciation in Argentina

The oldest known glacial deposits are in Argentina south of Lago Buenos Aires, where at 46° 47 'S, 71° 40'W, elevation 1550 m, as much as 40 m of till, covered by basaltic lava flows 4.5 - 5 m.y. old and covering flows 6.75 - 7 m.y. old, crop out in escarpment faces (Fig. 2). No weathering horizons are visible but they may be present as the till is poorly exposed through much of its thickness (Mercer and Sutter 1982). Today the nearest large glaciers are in the central cordillera 100 km to the west.

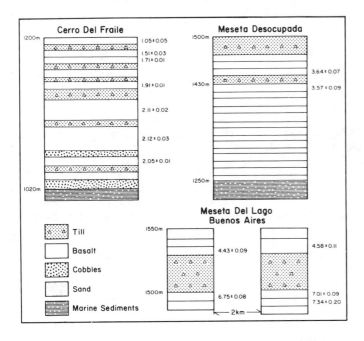

Figure 2. Diagrammatic sections of interbedded till and basaltic lava flows, southern Argentina. At a third site in the Meseta del Lago Buenos Aires, basalt 5.05 + 0.07 m.y. old covers till (reproduced from Annual Reviews of Earth and Planetary Sciences, vol. 11, p. 95, 1983).

Mid-Pliocene and early Pleistocene glaciations in Argentina

In southern Argentina, glacial deposits of mid-Pliocene age are exposed in escarpment faces north of Lago Viedma where, at 49° 28'S, 72° 25'W, elevation 1500 m, a single bed of till lies between basaltic flows with indistinguishable ages of c. 3.6 m.y. BP (Fig. 2). Till, as much as 70 m thick in places, and in which several weathering horizons are visible, covers the plateau surface (Mercer et al. 1975; Mercer 1976: 130).

South of Lago Argentino, a sequence of six tills of latest Pliocene and early Pleistocene age is exposed (50° 35'S, 72° 40'W, elevation 1 200 m); the oldest till is covered by a flow 2.12 m.y. old and the youngest lies between flows 1.05 and 1.51 m.y. old (Fleck et al. 1972; Mercer 1976: 138) (Fig. 2). This important site, the only one known in Patagonia with several tills of different ages, was first described by Feruglio (1944). Twenty-five km to the east (50° 27'S, 72° 17'W, elevation 1240 m), till lies between two lava flows with indistinguishable ages: 2.02 ± 0.09 m.y. BP above, and 2.01 ± 0.10 BP below (Mercer 1976: 138).

Pre-late Pliocene glaciation in Bolivia

In Bolivia near La Paz (16° S), according to Clapperton (1979), an ignimbrite dated at c. 3.36 m.y. BP covers till. Clapperton (1979) believes that the till may be the same age as that near Lago Viedma in Argentina. This is the only glacial deposit in South America between the equator and 46° S that has dating control by K/Ar.

Discussion

Glaciations, on a much more extensive scale than that prevailing today, affected southern South America some time between c. 4.6 and 7 m.y. BP, c. 3.6 m.y. BP, and repeatedly after c. 2 m.y. BP. Evidence for cool and cold climates at these times is widespread outside South America. Many workers have inferred glacial buildup (from lowering of sea level or change in the oxygen isotopic composition of microfaunal casts in ocean cores) and cooling at the end of the Miocene; Mercer and Sutter (1982) discuss the evidence. Ciesielski (1975: 649), Kennett and Vella (1975: 779), and Keany (1978: 46) note a cold episode that affected the Southern Ocean about 3.6 m.y. BP. Ice-rafted detritus in marine sediments of the North Atlantic (Berggren 1972) and change in the isotopic composition of ocean water (Shackleton and Opdyke 1977) show that the northern ice sheets repeatedly built up after c. 3 m.y. BP.

The late Miocene or earliest Pliocene glaciation in Argentina is less closely bracketed by radiometric ages than are the glaciations during the mid-Pliocene and early Pleistocene. The possible age range is more than 2 m.y. - between 4.6 and 7 m.y. ago. If major climatic variations are assumed to have been simultaneous at similar latitudes on either side of the South Pacific, the glaciation(s) may have been coeval with a cooling of Pleistocene severity in New Zealand during the Kapitean Age (Kennett and Watkins 1974: 1393), dated by Loutit and Kennett (1979: 1198) at 6.2 - 5.3 m.y. BP.

Although the absence in southern Argentina of known glacial deposits during the c. 1 m.y. interval before 3.6 m.y. BP is not positive evidence that no glaciations occurred, warmer climates are known to have prevailed outside Argentina during at least part of this interval. For example, in New Zealand Kennett and Watkins (1974: 1393) conclude that much warmer conditions in the early Pliocene followed the severe cooling at the end of the Miocene. Ciesielski (1975: 649), Kennett and Vella (1975: 779) and Keany (1978: 46) note pronounced and rapid warming of the Southern Ocean between late Miocene and mid-Pliocene cooling.

Opinions differ about climatic trends in the Southern Hemisphere during the c. 1 m.y. interval after 3.6 m.y. BP. Ciesielski and Wise (1977: 205) conclude that cold conditions continued in high southern latitudes, whereas Bandy et al. (1971: 18) and Kennett and Vella (1975: 780) believe that most of the Gauss chron (3.4 - 2.5 m.y. BP) was warm. Recent Antarctic geological studies support the concept of southern hemispheric warmth some time during the last half of the Pliocene. Till at high elevations in the Transantarctic Mountains contains abundant marine microfossils which are not found earlier than about 3 - 3.5 m.y. BP in cores from the Southern Ocean (Webb et al., in press). This fossiliferous till is believed to have been carried by ice from former marine embayments inland of the Transantarctic Mountains. These embayments, which are now occupied by thick ice grounded below sea level (e.g., in the Wilkes Basin) would have been deglaciated when temperatures rose too high for such marine-based ice to persist (Mercer 1983: 115). At the end of the warm interval falling temperatures allowed marine-based ice to build up once more in the East Antarctic basins. This ice was incorporated into the ice sheet, which then expanded to cover the Transantarctic Mountains, carrying the sediments to their summits. Although previously the East Antarctic Ice Sheet was generally thought to have changed little since the middle Miocene, Shackleton and Cita (1979: 443) suggested from totally independent evidence that the Antarctic Ice Sheet may have halved in volume some time between 3 and 4 m.y. BP. No major glaciation could have occurred in southern South America during these late Pliocene intervals of marine sedimentation in East Antarctica.

LATE PLEISTOCENE GLACIATIONS BEYOND THE RANGE OF C14 DATING

Southern Argentina and Chile

Caldenius (1932: 154), in his classic study of the glacial deposits of Patagonia, noted four end moraine belts. He disputed the matching of these by some earlier workers with the four Alpine glaciations (Günz-Mindel-Riss-Würm) of Penck and Brückner, and he concluded from their state of preservation that the three inner moraine belts correspond to the Daniglacial, Gotiglacial and Finiglacial stages in Europe, which are now known to have been deposited within the last 20 000 years. However, radiometric dating shows that most of the moraines are much older than Caldenius thought. East of Lago Buenos Aires (at 46°38'S, 70°45'W), the moraine that he equated with the Daniglacial moraines of Europe (that is, with the Brandenburg moraine, c. 20 000 BP [Cepek 1965]) is covered by lava 177 000 ± 57 000 years old (Mercer 1982a: 36). It can, therefore, be no younger than the penultimate glaciation, and may be older.

The outermost moraine belt, which Caldenius (1932) thought to be older than the last glaciation, is evidently much older than 177 000 BP. Mercer (1976: 138) showed that the most extensive glaciation occurred after 1.2 m.y. BP, because its sediments covered lava of that age, and he also described inconclusive field evidence that would have dated the maximum glaciation rather closely to between 1 and 1.2 m.y. BP. However, further fieldwork has shown that this evidence for a minimal age of 1 m.y. is permissive only.

In the Chilean lake region west of Lago Llanquihue (41 °S), Porter (1981) describes three till sheets that extend to the west of the outermost Llanquihe moraines; they are distinguished by weathering characteristics from each other and from till of the last glaciation.

Peru

In the Cordillera Vilcanota, eastern cordillera of Peru, about 10 m of peat underlies a horizon dated at c. 25 000 BP, at a site at 4400 m elevation just outside a lateral moraine of the Last Glacial Maximum and about halfway between the present ice front and the outermost moraines (Mercer 1982b) (see below). The maximum glaciation is, therefore, inferred to have taken place at a time beyond the range of conventional C14 dating. However, the well-preserved state of the moraines and the slight weathering of exposed clasts suggest that the glaciers may have reached their outermost and lowest positions early in the last glaciation, that is, probably at the peak of global ice volume during marine stage 4 c. 73 000 BP (Shackleton and Opdyke 1973: 45).

THE LAST GLACIATION

Chile

Radiocarbon dating of full-glacial advances of the last glaciation (Llanquihue Glaciation) in Chile has been obtained between 40 ° and 43 ° S, where the ice fronts reached the humid vegetated lowlands but did not extend to the coast. Moraines of the last glaciation - that is, those formed during the interval covered by marine stages 2, 3 and 4 - are clearly distinguishable from older moraines by their virtual absence of weathering. Three advances took place: the Llanquihue I, II and III. The first and second were separated, probably, by about 50 000 years, the second and third by about 5000 years. The outermost unweathered Llanquihue I moraines are more than 56 000 years old by C14 dating (Stuiver et al. 1975); Mercer (1983: 122) suggests that they may be c. 73 000 years old, formed during the peak of global ice volume of marine stage 4 (Shackleton and Opdyke 1973: 45).

The later, rather less extensive Llanquihue II advance lies within the reach of reliable C14 dating. West of Lago Rupanco at 40 ° 53'S, 72 ° 36'W (Fig. 3), an end moraine covers peat 19 450 ± 350 years old (I-5679) (Mercer 1972: 118, 1976: 152). This is still the most unequivocal age determination for the Last Glacial Maximum during marine stage 2 yet obtained from the Southern Hemisphere. It is supported by a dated sample from Lago Llanquihue; west of the lake (41 ° 07'S, 73 ° 02'W) the Llanquihue II end moraine covers outwash gravel containing a peat fragment 20 100 ± 500 years old (RL-116: Mercer 1972, 1976: 152), giving a minimum age for the moraine. These dates show that the

Figure 3. Southern part of Chilean lake region and northern part of Isla Chiloé between latitudes 40°S and 42°30'S, showing sites of C14-dated samples mentioned in text.

Llanquihue II advance was broadly equivalent in time to the Late Wisconsin advances in eastern North America.

Shortly after the Llanquihue II maximum, the glacier in the Lago Llanquihue basin was receding. Near-basal peat from a spillway of an ice-marginal lake was dated at 17 370 ± 670 BP (RL-120) (Mercer 1972, Mercer 1976: 153). Later, basal peat in contact with blue clay at this site was dated at 18 170 ± 650 BP (GX-5274). Basal peat from a nearby spillway was dated at 18 900 ± 370 BP (UW-418; Porter 1981: 277). The ice continued to recede and a single age determination suggests that by 16 270 ± 360 BP (RL-113) the Lago Llanquihue glacier had receded from the lake, allowing it to drain from its eastern end during the Varas Interstade, which was approximately coeval with the Erie Interstade in eastern North America (Mercer 1972; Mörner and Dreimanis 1973: 120). Mercer (1972: 118, 1976: 153) concluded that at the culmination of this recession the Lago Llanquihue glacier had shrunk to less than half its maximum length. Readvancing Llanquihue III ice later closed this eastern outlet, causing the lake level to rise; but the c. 2500-year spread in the ages of dated samples (13 200 - 15 700 BP) from the surface beneath the lake sediments makes reconstruction of events problematic. Both glacial variations and volcanic activity may have affected the lake level; the eastern outlet is today blocked by volcanic debris. Porter (1981: 279) concluded that two pulses of glacial readvance, at c. 15 000 - 14 500 and after 13 200 BP, had dammed the lake at slightly different levels, thereby accounting for the spread in ages. Mercer (1976: 155) suggested that the readvance had culminated c. 13 000 BP, and that the older dated samples had resulted from truncation of the peat by wave erosion. However, he emphasized that the date of the readvance would remain in doubt until an end moraine formed during the readvance had been dated.

More reliable dating for the final full-glacial readvance (Llanquihue III), based on less equivocal field evidence, has now been obtained from the east side of Isla Chiloé northeast of Castro (42°21'S,73°39'W) (Fig. 3). At this site till covers outwash gravel that rests on peat; the peat covers an older till. The site is close to a subdued ridge, with abundant large boulders, that is thought to mark the limit of the readvance. Wood samples from the peat-gravel interface are 14 355 ± 700 (GX-8686), 14 970 ± 210 (I-12 996) and 15 600 ± 560 (GX-9978) years old, pointing to a date of c. 14 500–15 000 BP for the glaciation of the site. In view of the evidence for major deglaciation in southern South America before 12 800 BP (see below), such a date for the preceding major readvance seems more reasonable than does a date of c. 13 000 BP. It is also similar to dates for the final major readvance in other parts of the world: the Erie Lobe of the Laurentide Ice Sheet in Ohio, after 14 780 ± 192 BP and before 14 300 ± 450 BP (Mayewski et al. 1981: 95), the Des Moines Lobe in Iowa, some time between 14 150 and 13 775 BP (Ruhe 1969), the Cordilleran Ice Sheet in Washington State between 14 500 and 14 000 BP (Mullineaux et al. 1965), 14 500 – 14 000 BP in New Zealand (Suggate and Moar 1970), and c. 14 000 BP in Peru (see below).

Peru

The only age determinations for the Last Glacial Maximum in Peru come from one glaciated valley in the eastern cordillera: the Upismayo valley on the north side of the Cordillera Vilcanota (Figs. 4 and 5). A group of sharp-crested, closely nested end moraines terminate between 4300 and, probably, c. 4100 m; the outermost moraines survive only as lateral moraines some distance back from the former terminal positions. At one site (13°44'S, 71°17'W), the glacier at its maximum extent was encroaching onto a peat bog. A stream channel through the moraine has exposed overridden and visibly disturbed peat.

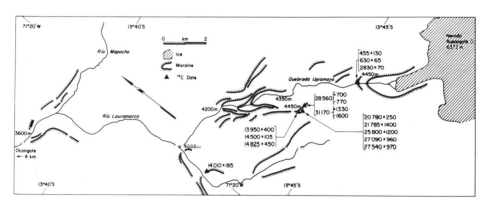

Figure 4. Cordillera Vilcanota, eastern cordillera of Peru; principal end moraines and sites of C14-dated samples in the valley descending northwest from Nevado Ausangate towards Ocongate village.

Two samples were dated at 28 560 ± 700 and 31 170 +1330 -1600 BP (DIC-681 and DIC-677) (Mercer and Palacios 1977: 602). At c. 1 m depth in the peat bog beyond the moraine a 50 cm-thick layer of fine rock fragments was interpreted as sediment redeposited from the moraine at and shortly after its formation. About 10 m of peat underlies the inorganic layer. Peat immediately beneath the sharp contact was inferred to slightly predate the formation of the moraine. Samples were dated at 21 785 ± 1400 (Beta-1556), 25 800 ± 1200 (GX-4917), 27 090 ± 960 (Beta-1555), and 27 540 ± 970 BP (GX-8080) (Mercer 1979: 116, 1982b: 139). A thin lens of peat interbedded with the rock fragments was dated at 20 780 ± 250 BP (Beta-1554). This was thought to give the best approximation of the age of the moraine. However, an excavation made later into the front of the moraine exposed the original surface of the peat, arched up by the pressure of the advancing ice at the time it reached its maximum. Three age determinations, from two laboratories, show that the moraine covered the bog c. 14 000 BP: 13 950 ± 400 and 14 500 ± 105 BP (GX-8081 and Beta-1725) for peat in contact with till, and 14 825 ± 450 BP (GX-8189) for peat 10 cm below the contact (Mercer 1982b: 139). In an adjacent valley, a thin lens of peat is interbedded with till in an end moraine; the peat is 14 010 ± 185 years old (I-9623) (Mercer and Palacios 1977: 602).

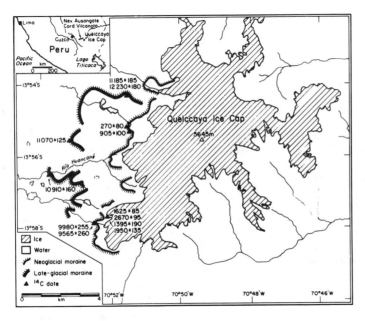

Figure 5. Quelccaya Ice Cap, eastern cordillera of Peru; principal end moraines formed during the last 11 000 years, and sites of C14-dated samples.

Thus the age of the outermost lateral moraine in the Upismayo valley is apparently c. 14 000 BP, not 20-25 000 BP as previously thought. The origin of the band of fine rock fragments is unknown. The great thickness of peat - c. 10 m - underlying the horizon dated at c. 25 000 BP shows that the glacier has not extended past its 14 000 BP position for, probably, tens of thousands of years.

The glacier in the Upismayo valley formed several massive, closely nested end moraines not far inside its 14 000 BP position. The dates of the glacial oscillations that formed these moraines are not known.

DEGLACIATION

Southern Chile and Argentina

Very rapid deglaciation followed the final late-glacial readvance in southern South America. Lago Ranco in Chile, which Mercer (1976: 155) showed was deglaciated by c. 12 000 BP, is now known to have been ice-free by 13 000 BP. A cut bank of the Río Caunahue (40°08'S, 72°15'W) (Fig. 3) exposes sediments deposited in the ice-free lake. Organic silt grades down to laminated inorganic clay; wood from successively lower levels in the silt dates from 12 200 ± 400 BP (GX-2935)(Mercer 1976: 155), 12 810 ± 190 BP (I-12 295)(Mercer, in press) and 13 900 ± 560 (GX-9979). The last sample was obtained from close to the gradational contact between organic silt and underlying clay. Ashworth and Hoganson (this volume) obtained their palaeoclimatic reconstructions, based on fossil beetles, from this site. Similar dates have been obtained for the deglaciation of the Patagonian Andes. South of 46°S several piedmont lakes on the east side of the mountains now drain westward to the Pacific Ocean through trans-cordilleran valleys. In full and late-glacial times these valleys were ice-filled, and the lakes or the glaciers occupying them drained eastward to the Atlantic Ocean. The peat-filled former spillway at the eastern end of glacial lake Pueyrredón (47°19'S, 70°58'W) (Fig. 1) had been abandoned by 12 800 ± 445 BP (GX-8682) (Mercer, in press). North of the Strait of Magellan the former spillway at the eastern end of glacial lake Otway (52°48'S, 71°05'W) had been abandoned by 12 460 ± 190 BP (I-3512), pointing to extensive deglaciation of the cordillera by that time (Mercer 1970: 19).

Glaciers continued to recede after 12 800 BP. Two C14 dates show that by 11 000 BP the Tempano Glacier (48°45'S, 74°00'W), an outlet on the west side of the southern Patagonian icefield, had withdrawn behind its AD 1968 borders, where it remained until Neoglacial time. It did not readvance between 11 000 and 10 000 BP, the interval corresponding to the European Younger Dryas stade (Mercer 1970: 14). This behaviour conforms to the conclusions, based on fossil beetle stratigraphy, of Ashworth and Hoganson (this volume); they infer a climate similar to today's throughout Younger Dryas time. However, it conflicts with climatic reconstruction from pollen stratigraphy in southern Chile, which shows a pronounced reversion to colder and wetter climate at that time (Heusser and Streeter 1980; Heusser, this volume).

Peru

By 12 200 BP, peat had started to accumulate only 2 km from the present western edge of the Quelccaya Ice Cap, which then cannot have been much larger than it is today (Fig. 5). However, the ice cap then re-expanded, the associated outwash sediment covering peat 11 460 ± 165 years old (I-8210). Three age determinations suggest that this readvance culminated c. 11 000 BP (Mercer and Palacios 1977: 603), that is, several centuries earlier than the Younger Dryas readvance in Europe (Mercer 1979: 119). Whether or not the ice cap remained expanded during Younger Dryas time is not known. If it did, the following shrinkage was rapid. About 500 m beyond the present ice margin, and directly beneath the outermost Neoglacial end moraine, age determination of two samples of basal peat in a 15 m-thick exposure of interbedded peat and outwash sand are 9980 ± 255 and 9565 ± 260 BP (DIC-685 and GX-4933).

NEOGLACIATION

Southern Chile and Argentina

Mercer (1982a) summarizes the chronology of Neoglaciation in southern South America. On both sides of the cordillera glaciers readvanced regionally during three intervals reaching their maxima c. 4600 - 4200 BP, probably 2700 - 2000 BP, and during recent centuries. During each interval some glaciers reached their outermost Neoglacial positions, but on the whole the first readvance was the greatest (Mercer 1976: 157, 1982a: 37).

The first two readvances are best dated on the east side of the cordillera in Argentina. At its Neoglacial maximum the glacier on the east side of Cerro San Lorenzo (47°39'S, 72°15'W) reached about 2.5 km beyond its AD 1967 position, forming a small moraine-dammed lake. A rooted stump beneath lacustrine clay at the edge of this lake was drowned 4590 ± 115 BP (I-2208) (Mercer 1968: 104, 1982a: 38). Further south the Upsala Glacier, which in AD 1963 calved into Lago Argentino at 50°02'S, 73°18'W, was about 10 km further forward at its Neoglacial maximum. About 5 km in front of its AD 1963 terminus, a log in till was dated at 2310 ± 120 BP (I-988) (Mercer 1965: 404, 1982a: 38).

Peru

Neoglacial readvances in Peru are poorly dated. In the Cordillera Vilcanota, the glacier in the Upismayo valley, whose full-glacial variations were described above, readvanced in late Neoglacial time, bulldozing into a peat bog at 4550 m elevation. Samples from the top of c. 1.5 m of peat, covering coarse sand and covered by sand and gravel, exposed arched up in the distal face of this well-vegetated moraine, have been dated at 630 ± 65 and 455 ± 130 BP (DIC-678 and GX-4925), and a sample from the base of the peat has been dated at 2830 ± 70 BP (DIC-682). Closer to the present ice front, massive, closely nested, nearly bare moraines are evidently considerably younger, and probably date from recent centuries.

On the western side of the Quelccaya Ice Cap, no early Neoglacial moraines have been identified. The outermost Neoglacial moraines generally contain abundant peat in large lumps, but the original surface of the bog has not been recognized with certainty in these fragments. The ages of two dated samples - 270 ± 80 and 950 ± 100 BP (I-9 624) and I-8 441) - suggest that the moraines were formed during recent centuries (Mercer and Palacios 1977: 603).

Between the c. 11 000 BP and recent advances, the ice cap receded behind its present borders. Locally, receding ice at the southern end of the ice cap in 1976 and 1977 was uncovering peat c. 50 cm thick. Age determinations of the top and bottom of the peat are 1625 ± 85 and 2670 ± 95 BP (I-9625 and DIC-680), and 1395 ± 190 and 1950 ± 135 BP (GX-4930 and GX-4932).

CONCLUSIONS

Southern South America contains the best radiometrically-dated record of pre-late Pleistocene glaciations in the Southern Hemisphere. Its only counterpart in the Northern Hemisphere is the sequence of interbedded glacial sediments and basaltic lava flows of Pliocene to Pleistocene age in Iceland. In southern South America the late Miocene to mid-Pleistocene glacial record is incomplete, but as it stands it is compatible with variations of temperature of surface waters of the Southern Ocean as inferred from microfossil assemblages in ocean cores, and with geological evidence for changes in the extent of the Antarctic Ice Sheet.

During the last glacial age in southern South America, and perhaps in low latitudes also, glaciers were largest at a time beyond the range of C14 dating, probably during marine stage 4 c. 73 000 BP. This suggests that the Southern Hemisphere was then colder than it was c. 20 000 years ago, when global ice volume peaked. During the interval accessible to conventional C14 dating, the major glacial advances at c. 19 500 and 15 000 - 14 500 BP in middle latitudes of South America were contemporaneous with major glacial advances elsewhere in northern and southern middle latitudes. If, as is generally believed, the major climatic changes during the Pleistocene were caused primarily by variations in the Earth's orbital parameters, the simultaneous timing of glacial variations in both hemispheres is unexpected and remains to be explained satisfactorily. Furthermore, orbital variations do not seem to account for the globally-recognized final full-glacial readvance between 15 000 and 14 000 BP, recognized not only in middle latitudes of both hemispheres, but also in the tropics. This readvance occurred when the level of summer solar radiation in northern middle and high latitudes had been rising for c. 10 000 years, was still rising, and was higher than it is today (Mercer, in press).

The rapid and uninterrupted deglaciation that followed the readvance 15 000 - 14 500 BP in southern South America, and which reduced the ice cover to interglacial extent by 11 000 BP, contrasts strikingly with the course of events round the northern North Atlantic Ocean, where readvances and standstills repeatedly interrupted the general recession until the end of the Younger Dryas stade c. 10 000 BP. This probably reflects the simpler mode of shrinkage of the Antarctic Ice Sheet compared to the Laurentide and Scandinavian ice sheets, and also the much smaller difference in the Southern Hemisphere between full-glacial and full-interglacial ice volume.

For the millennium 11 000 - 10 000 BP, during which conditions of almost full-glacial severity returned to northwest Europe for the last time during the Younger Dryas stade, the continuing conflict between interpretations of the glacial and fossil beetle evidence (which show no cooling in southern South America at that time) on the one hand, and of the pollen stratigraphy (from which much cooler and wetter conditions have been inferred) on the other, has not yet been resolved. A resolution of the problem is needed in order to understand the nature of cooling at the time - was it a local or global event? The answer will have important implications for hypotheses of the mechanisms of climatic change.

ACKNOWLEDGEMENTS

Field investigations by the author were supported by The National Science Foundation, Divisions of Earth Sciences, Atmospheric Sciences, and Polar Programs. Contribution no. 488 of the Institute of Polar Studies, The Ohio State University.

REFERENCES

Ashworth AC and Hoganson JW, this volume. Was there a climatic change in southern Chile about the same time as the Younger Dryas in Europe?

Bandy OL, Casey RE and Wright RC 1971. Late Neogene planktonic zonation, magnetic reversals, and radiometric dates, Antarctic to the tropics. Am. geophys. Union Antarct. Res. Ser. 15:1-26.

Berggren WA 1972. Late Pliocene-Pleistocene glaciation. In: AS Laughton, WA Berggren et al. (eds.), Initial Reports of the Deep Sea Drilling Project, 12. GPO, Washington. p. 953-63.

Caldenius CC 1932. Las glaciaciones cuaternarias en la Patagonia y Tierra del Fuego. Geogr. Ann. 14:1-164.

Cepek AG 1965. Geologische Ergebnisse der ersten Radiokarbondatierungen von Interstadialen im Lausitzer Urstromtal. Geologie 14:625-57.

Ciesielski PF 1975. Biostratigraphy and paleoecology of Neogene and Oligocene silicoflagellates from cores recovered during Antarctic Leg 28, Deep Sea Drilling Project. In: DE Hayes, LA Frakes et al. (eds.), Initial Reports of the Deep Sea Drilling Project, 28. GPO, Washington. p. 625-64.

Ciesielski PF and Wise SW 1977. Geologic history of the Maurice Ewing Bank of the Falkland Plateau (Southwest Atlantic sector of the Southern ocean) based on piston and drill cores. Mar. Geol. 25:175-207.

Clapperton CM 1979. Glaciation in Bolivia before 3.27 Myr. Nature 277:375-77.

Feruglio E 1944. Estudios geológicos y glaciológicos en la región del Lago Argentino (Patagonia). Boln Acad. Nac. Ciencias (Córdoba) 37:1-208.

Fleck RJ, Mercer JH, Nairn AEM and Peterson DN 1972. Chronology of Late Pliocene and Early Pleistocene glacial and magnetic events in southern Argentina. Earth Planet. Sc. Letters 16:15-22.

Heusser CJ, this volume. Late Quaternary climates of Chile.

Heusser CJ and Streeter SS 1980. A temperature and precipitation record of the past 16 000 years in southern Chile. Science 210:1345-7.

Keany J 1978. Paleoclimatic trends in Early and Middle Pliocene deep-sea sediments of the Antarctic. Mar. Micropaleontol. 3:35-49.

Kennett JP and Vella P 1975. Late Cenozoic planktonic foraminifera and paleoceanography at DSDP Site 284 in the cool subtropical South Pacific. In: JP Kennett, RE Houtz et al. (eds.), Initial Reports of the Deep Sea Drilling Project, 29. GPO Washington. p. 769-82.

Kennett JP and Watkins ND 1974. Late Miocene-Early Pliocene paleomagnetic stratigraphy, paleoclimatology and biostratigraphy in New Zealand. Bull. geol. Soc. Am. 85:1385-98.

La Brecque JL, Kent DV and Cande SC 1977. Revised magnetic polarity time scale for Late Cretaceous and Cenozoic time. Geology 5:330-35.

Loutit TS and Kennett JP 1979. Application of carbon isotope stratigraphy to Late Miocene shallow marine sediments, New Zealand. Science 204:1196-99.

Mankinen EA and Dalrymple GB 1979. Revised geomagnetic polarity time scale for the interval 0-5 m.y. BP. J. geophys. Res. 84:615-26.

Mayewski PA, Denton GH and Hughes TJ 1981. Late Wisconsin ice sheets of North America. In: GH Denton, TJ Hughes (eds.), The Last Great Ice Sheets. John Wiley and Sons, New York. p. 67-178.

Mercer JH 1965. Glacier variations in southern Patagonia. Geogr. Rev. 55:390-413.
Mercer JH 1968. Variations of some Patagonian glaciers since the Late-Glacial. Am. J. Sc. 266:91-109.
Mercer JH 1970. Variations of some Patagonian glaciers since the Late-Glacial: II. Am. J. Sc. 269:1-25.
Mercer JH 1972. Chilean glacial chronology 20 000 - 11 000 carbon-14 years ago: some global comparisons. Science 176:1118-20.
Mercer JH 1976. Glacial history of southernmost South America. Quat. Res. 6:125-66.
Mercer JH 1979. Chronology of the last glaciation in Peru. Boln Soc. Geológica Perú 61:113-20.
Mercer JH 1982a. Holocene glacier variations in southern South America. Striae 18:35-40.
Mercer JH 1982b. The last glacial-deglacial hemicycle in Peru. Am. Quat. Ass. Conf. 7th Abstr. p. 139.
Mercer JH 1983. Cenozoic glaciation in the Southern Hemisphere. Ann. Rev. Earth Planet. Sc. 11:99-132.
Mercer JH, in press. Simultaneous climatic change in both hemispheres and similar Arctic and Antarctic interglacial warming: implications for Milankovitch theory and CO_2-induced warming. Proc. Ewing Symp., Lamont-Doherty Geol. Obs., Nov. 1982.
Mercer JH, Fleck RJ, Mankinen EA and Sander W 1975. Southern Patagonia: glacial events between 4 MY and 1 MY ago. In: RP Suggate, MM Cresswell (eds.), Quaternary studies. R. Soc. New Zealand, Bull. 13. Wellington, N.Z. p. 223-30.
Mercer, JH and Palacios O 1977. Radiocarbon dating of the last glaciation in Peru. Geology 5:600-4.
Mercer JH and Sutter JF 1982. Late Miocene-earliest Pliocene glaciation in southern Argentina: implications for global ice sheet history. Palaeogeogr. Palaeoclimatol. Palaeoecol. 38:185-206.
Mörner NA and Dreimanis A 1973. The Erie Interstade. In: RF Black, RP Goldthwait and HB Willman (eds.), The Wisconsinan Stage. Geol. Soc. Am. Mem. 136. p. 107-134.
Mullineaux DR, Waldron HH and Rubin Meyer 1965. Stratigraphy and chronology of Late Interglacial and early Vashon Glacial time in the Seattle area, Washington. US geol. Surv. Bull.:1194-O.
Porter, SC 1981. Pleistocene glaciation in the southern Lake District of Chile. Quat. Res. 16:263-92.
Ruhe RF 1969. Quaternary landscapes in Iowa. Iowa State University Press, Ames, Iowa.
Shackleton NJ and Cita MB 1979. Oxygen and carbon isotope stratigraphy of benthic foraminifers at Site 397: detailed history of climatic change during the late Neogene. In: U. von Rad, WBF Ryan et al. (eds.), Initial Reports of the Deep Sea Drilling Project, 46(1). GPO, Washington. p. 433-45.
Shackleton NJ and Opdyke ND 1973. Oxygen isotope and paleomagnetic stratigraphy of Equatorial Pacific core V28-238. Quat. Res. 3:39-55.
Shackleton NJ and Opdyke ND 1977. Oxygen isotope and palaeomagnetic evidence of early Northern Hemisphere glaciation. Nature 270:216-19.
Stuiver M, Mercer JH and Moreno HR 1975. Erroneous date for Chilean glacial advance. Science 188:73-4.
Suggate RP and Moar NT 1970. Revision of the chronology of the Late Otira Glacial. New Zealand J. Geol. Geophys. 13:742-46.
Webb P-N, Harwood DM, McKelvey BC, Mercer JH and Stott LD, in press. Cenozoic marine sedimentation and ice volume variation on the East Antarctic craton. Geology.

Late Quaternary climates of Chile

CALVIN J.HEUSSER
New York University, New York, USA

ABSTRACT. Six fossil pollen records selected from multiple stratigraphic sections represent Late Quaternary climatic sequences over 20° of latitude extending to southernmost Chile. The records were factor analyzed with temperature and precipitation quantified for the past 43 000 years from sections at Alerce (41°25'S) and Taiquemó (42°10'S) by applying regression equations, which relate modern pollen in surface samples and present-day temperature and precipitation, to the fossil pollen data. Climate, on the whole, was found to be colder and drier from about 37 000 to 14 000 BP than it was between 43 000 and 37 000 BP. At about 13 000 years ago, warming with some increase inhumidity took place and lasted until around 11 000 BP. Thereafter, cooler and wetter episodes are evident between approximately 11 000 and 10 000 BP, 5000 and before 3200 BP, after 3200 and before 900 BP, and between about 350 BP and the present; the interval from around 9400 to 7400 BP was warmer and less humid.

Times of cooler and wetter climate after 10 000 BP generally coincide with episodes of Holocene glacial advance in the Andes Mountains. San Rafael Glacier (46°40'S), for example, advanced on at least three occasions between about 5000 and 4000 BP, between 3740 and 500 BP, and during the 19th century, after having been in a state of recession at 6850 BP. During the Late Glacial, the last major glacier advance took place around 13 000 BP or before, and afterwards glaciers pulled back into the Andes until the time of Neoglaciation which began after 6850 BP. Fossil beetle assemblage data, which indicate uniform conditions from 14 635 to 4525 BP, support the interpretation of a protracted interval of ameliorated climate made from glacier observations. A pollen record recently described from Rucañancu (39°33'S), however, conflicts with these climatic implications during the interval of 11 000 to 10 000 BP and reconfirms previous evidence for cooler and wetter conditions in a dated sequence from between 10 440 and 10 000 BP. Thus, although glacier readvance has not been documented, Late Glacial climate characteristic of high montane elevations today on the west slope of the Andes was in effect evidently until 10 000 BP.

The climate of Chile is controlled by the Pacific anticyclone in the north and at mid-latitudes by cyclones in the belt of westerly winds. Whereas in the south, the westerlies bring heavy precipitation all year, central Chile is summer-dry and affected by cyclonic storms mostly in winter. The pollen record from summer-dry Laguna de Tagua Tagua (34°30'S) which dates from over 45 000 BP indicates that climate at the time of the Last Glaciation was more humid and colder than at present. By comparison, climate at Puerto Williams (54° 56'S), on subantarctic Isla Navarino near the southern tip of South America, and at Puerto del Hambre (53° 36'S) was cold and dry during the Late Glacial. These data imply greater northerly influence of the westerlies at this time, accompanied by a weakening of the Pacific anticyclone.

Evidence suggests that major Late Quaternary climatic changes related to first-order changes in Chile occurred with much the same timing at comparable latitudes elsewhere in the Southern Hemisphere, thus pointing towards general uniformity in the fluctuations of atmospheric circulation. Polar fronts at the time of the Last Glaciation, when the West Antarctic Ice Sheet reached its maximum between 21 000 and 17 000 BP, were located at lower latitudes than at present. In Chile, this probably amounted to a shift of $5° - 7°$ of latitude, whereas the oceanic fronts shifted $2°$ and $10°$. Sea surface temperatures in summer off the Chilean coast were at least $4°C$ lower than now, when on land, temperatures were depressed $6° - 8°$. Climatic differences evident in time-stratigraphic pollen records may be caused by latitudinal variations of atmospheric and oceanic frontal systems associated with the westerly wind belt.

INTRODUCTION

The wealth of dated stratigraphic records for the reconstruction of climate during the Late Quaternary makes Chile one of the more important land masses in the Southern Hemisphere. The broad and continuous latitudinal extent of Chile embraces a wide range of climates, while its poleward penetration of the Southern Ocean is beyond that of any other major land mass outside Antarctica. The climate of this part of South America is strongly influenced by the southern westerlies and by the topographic barrier imposed by the Andes Mountains.

Deposits left by glaciers originating in the Andes (Caldenius 1932; Laugenie 1971; Mercer 1976, 1982, 1983; Heusser and Flint 1977; Porter 1981) form a primary source from which to assess Late Quaternary climate. Evidence placed times of glacier advance, reflecting an apparent response to colder climate, before 20 000 BP and later between 20 000-19 000 and 15 000-13 000 BP, followed by wastage until the late Holocene when Neoglacial advances culminated at 4500-4000 and 2700-2000 BP and during recent centuries.

Another important source is the fossil pollen records which document the response of vegetation to major climatic variations and closely follow the more obvious trends of climate indicated from glacial evidence (Auer 1933, 1950, Heusser 1966, 1974, 1981, 1983a, in press; Villagran 1980; Markgraf and Bradbury 1982). By their latitudinal spread, the records include unglaciated regions, while their dated sequences cover a milder interstadial before about 37 000 BP, the colder Late Pleistocene afterwards up until around 10 000 BP, and the more temperate Holocene. Examination of these records in this paper is with a view toward tracing past Chilean climate and placing the data in the context of comparable Southern Hemisphere records.

MODERN CLIMATES AND VEGETATION OF CHILE

The meridional configuration of Chile, extending between approximately 18° and 56°S, accounts for a succession of climates and vegetation zones which range from tropical to polar (Fig. 1). Climatic gradients are affected also by the topography, particularly by the formidable Andean cordillera, and by the proximity of the Pacific Ocean and the cold water of the West Wind Drift and Humboldt Current. The annual range of precipitation is pronounced, from less than 50 mm in the north to over 7000 mm in the south, while the temperature range is mostly

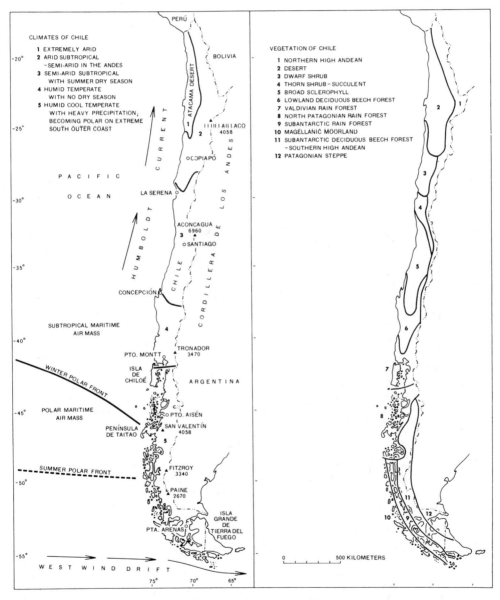

Figure 1. Map of Chile showing physical features, climatic provinces (Miller 1976), and vegetation zones (modified from Schmithüsen 1960).

less than 10 °C (Table 1). The arid north of Chile (about 18° - 31° S) is under the influence of stable air of the Pacific anticyclone, whereas poleward, cyclonic storms of the westerlies become increasingly important, especially beyond 42 °S (Lamb 1959; Taljaard 1972; Miller 1976; Zamora and Santana 1979).

Coastal northernmost Chile, north of Copiapó in the sector of the Atacama Desert, is extremely arid. On approaching the Andes at this latitude, semi-arid conditions develop from summer showers and

Table 1. Climatological data from selected stations in Chile[a]

Station	Av temperature		Av precipitation				
	Jan	July	Autumn	Winter	Spring	Summer	Total
	(°C)			(percentage)			(mm)
Copiapó	20.9	11.9	18	71	7	-	28
La Serena	18.3	11.7	22	68	8	-	110
Valparaíso[b]	17.6	11.5	25	62	12	-	671
Santiago	20.6	8.0	24	58	15	-	360
Chillán	21.9	9.1	27	50	16	6	1033
Concepción	17.8	9.1	28	51	16	5	1338
Los Angeles	20.6	8.3	28	48	17	6	1285
Temuco	17.0	7.8	30	42	18	10	1345
Valdivia	17.1	7.8	29	44	18	9	2510
Pro. Montt	15.3	7.6	28	35	21	16	1960
Melinka	13.3	7.6	25	37	22	16	4277
Pto. Aisén	14.0	4.6	29	30	21	20	2868
Cabo Raper	11.6	6.6	26	26	22	23	1979
Pto. Eden	11.6	2.8	24	28	28	20	3586
Guarelo	-	-	27	25	22	26	7330
Evangelista	8.7	4.4	27	24	24	25	2454
Bahía Felix	-	-	27	23	24	26	4428
San Miguel	-	-	24	22	19	27	1570
San Isidro	9.0	2.9	29	21	22	28	877
Pta. Arenas	10.4	1.0	32	27	20	20	439
Pto. Williams	8.6	1.5	33	26	15	26	553

[a] Data from Almeyda and Sáez (1958), and Zamora and Santana (1979).
[b] Station is nearby Placilla.

thunderstorms, a result of moisture brought in from the interior by northeast winds. The region is a wasteland (Reiche 1907; Schmithüsen 1956) with vegetation scattered about oases (Prosopis tamarugo, Geoffroea decorticans), where there is coastal drizzle (Euphorbia lactiflua), and in the Andes (Stipa frigida, Fabiana bryoides, Adesmia hystrix). Southward, beginning in the vicinity of La Serena, the effect of winter rain is apparent by the increase of thorn shrubs and trees (Acacia caven) and succulents (Eulychnia acida). The northernmost cloud forests (Aextoxicon punctatum, Myrceugenia correaefolia, Drimys winteri var. chilensis), which receive additional moisture from fog drip, are locally established in the coastal cordillera (Muñoz and Pisano 1947). Coastal fog increases between 30° and 40°S, as the eastward moving air crosses the cold Humboldt Current, with colder conditions prevailing on the coast than inland during the summer.

Climate of central Chile (31° - 42°S) is subtropical to temperate and typically winter-wet because of the increased frequency of frontal systems moving northward from the belt of westerlies (Fig. 2; Table 1). Precipitation is greater in the Andes and on the Pacific side of

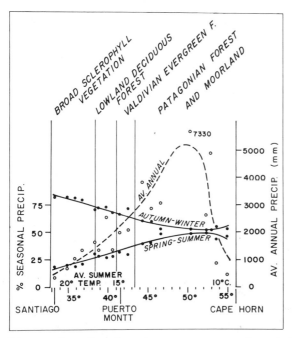

Figure 2. Trends of average annual and % seasonal precipitation and summer (January) temperature in Chile (33° - 55°S) from data prepared by Almeyda and Sáez (1958), and Zamora and Santana (1979).

the coastal cordillera than in the central valley; on uppermost slopes of the Andes, it ranges from 400 mm in the north to around 5000 mm in the south, accompanied by a marked increase in cloud cover (Almeyda and Sáez 1958). At about Concepción (near 37°S), climate is no longer summer-dry, and temperatures inland are less contrasted with the immediate coast. Broad-sclerophyll vegetation (Cryptocarya alba, Peumus boldus, Maytenus boaria) at this latitude changes to lowland deciduous forest, which is represented by southern beech (Nothofagus obliqua) and other trees (Laurelia sempervirens, Persea lingue). Vegetation zones rise altitudinally northward in the cordillera where additional species of beech (N. glauca, N. alpina) and several gymnosperms (Austrocedrus chilensis, Podocarpus andinus, Araucaria araucana) grow.

Southern Chile (42° to about 56°S) is positioned athwart the main stream of the westerlies, so that storm passages are incessant throughout the year. Powerful and persistent wind associated with the storms gives rise to the descriptive appellations of "roaring forties" and "screaming fifties" for the region. The polar front is located between 45° and 50°S, and rainfall is extremely heavy, as exemplified by weather data at Guarelo (Fig. 3) where the maximum recorded is almost

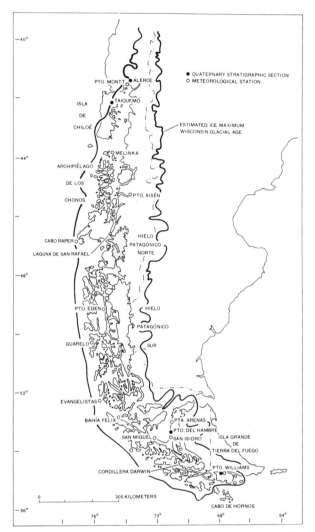

Figure 3. Locations in southern Chile (40° - 56°S) of meteorological stations and Quaternary section sites discussed in the text. Extent of Wisconsinan glaciation is from Hollin and Schilling (1981).

8500 mm. Above an elevation of about 1000 m, precipitation is usually in the form of snow; at Laguna de San Rafael (46° 40'S), for example, snow in summer after a frontal passage often blankets the mountain front down to 1200 m (Heusser 1960). Freezing conditions aloft at so low an altitude cause excessive snowfall and the formation of icefields and glaciers in the Andes. The Hielo Patagónico Norte (46° 30' - 47° 30'S) and Hielo Patagónico Sur (48° 15' - 51° 20'S) are features of the cordillera, feeding glaciers that descend into both Chile and Argentina (Lliboutry 1956; Mercer 1967). The San Rafael Glacier, its terminus at Laguna de San Rafael, is nearest to the equator of any tidewater glacier in the world. Progressive decrease in moisture follows storm movement to the east and southeast across Argentina.

Valdivian evergreen rain forest (Nothofagus dombeyi, N. nitida,

Eucryphia cordifolia), extending from the neighborhood of Puerto Montt (41° 25'S) south on the mainland and on Isla de Chiloé to 43° 20'S, but also at higher altitudes in both the Andes and coastal cordillera where Fitzroya cupressoides grows, is rich with lianas, epiphytes, mosses, and liverworts. North Patagonian rain forest (Weinmannia trichosperma, Laurelia philippiana, Saxegothaea conspicua, Podocarpus nubigenus) follows to 47° 30'S and is succeeded, in turn, by subantarctic rain forest (Nothofagus betuloides, Pilgerodendron uvifera), which forms mostly discontinuous stands in protected places in the magellanic moorland or tundra as far as Cape Horn (Skottsberg 1916; Godley 1960; Pisano 1977). Moorland (Donatia fascicularis, Astelia pumila, Schoenus antarcticus) dominates the outer coastal region of southernmost Chile in a windy and cold climate with summer temperatures averaging around 8°- 9°C. Rain forest to the east and northeast is bounded by subantarctic deciduous forest (Nothofagus pumilio, N. antarctica) which grades successively into steppe (Festuca gracillima). Compared with the humid coast, the climate of the steppe, with a greater temperature range and with precipitation averaging only 300 mm annually, is more continental.

POLLEN RECORDS

Six pollen records of Late Quaternary stratigraphic sections were selected from 32 C14-dated records to form the basis for climatic reconstruction. Sites chosen from between 54° 56'S and 34° 30'S (Figs. 3-4) are 12 730 to over 45 000 years old. Multivariate pollen data are expressed as factors, highlighting leading covarying taxa, the

Figure 4. Locations in central Chile (32° - 43° S) of the meteorological stations and Quaternary section sites discussed in the text. Extent of Wisconsinan glaciation is from Hollin and Schilling (1981).

result of Q-mode factor analysis using CABFAC (Imbrie and Kipp 1971). The number of pollen taxa as variables in the basic data sets used in the analyses, between 6 and 20, differs from site to site as a result of changing floristic representation in the records. The factors shown account for 93-97 % of the variance. Factors are advantageous over relative frequency because they stress ecological relationships and simplify presentation of the pollen data.

Puerto Williams (54° 56'S, 67° 38'W; elev. 21 m). As shown by the high loading of the Empetrum-Ericaceae-Gramineae factor (zone PW-3), treeless vegetation apparently prevailed for over 2500 years of the Late Glacial between about 12 730 and 10 080 BP (Fig. 5). Before this, increased loading of the Nothofagus factor, represented by any or all of its species, N. betuloides, N. pumilio, and N. antarctica, suggests an interval (zone PW-4) when tree cover had expanded. This increase in the cover of beech need not have been local, as surface studies show that beech pollen can be wind transported close to 200 km in Patagonia so as to appear in large numbers when pollen production by local species is comparatively low. After about 10 000 BP, Nothofagus became prominent and continued so throughout the remainder of the record. Loading by the Empetrum-Ericaceae-Gramineae factor is low until about 5000 BP (zone PW-2) but later increased, reaching its highest values during the past 2500 years (zone PW-1).

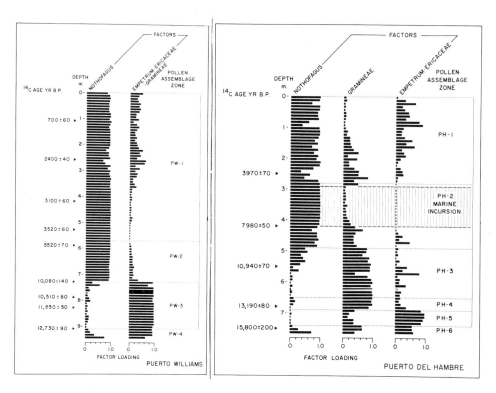

Figure 5. Loading of principal pollen factors in section at Puerto Williams.
Figure 6. Loading of principal pollen factors in section at Puerto del Hambre.

Climate, with the spread of tundra in the region, was colder and drier (zone PW-3), following what appears to have been a somewhat warmer and more humid interval (zone PW-4). Conditions became warmer and wetter after 10 000 BP, so that the extent of closed forest was greater 5000 years ago (zone PW-2). Since that time, as Empetrum returned (zone PW-1), the record is seen to reflect an apparently colder and drier climate. Empetrum rubrum, a species with a broad ecological amplitude, forms significant cover at the section site. This is possibly because the acidity of the bog sediments simulates physiological dryness corresponding to the more xeric localities on the upland where it grows. With the Gramineae, Empetrum occupies tracts beyond the treeline, both in alpine areas and in the steppe.

Puerto del Hambre (53° 36'S, 70° 55'W; elev. 4 m). The record of over 15 800 years is of greater antiquity than Puerto Williams (Fig. 6). Loading trends, distinguished from three factors, follow the previous pattern (Fig. 5). The Gramineae and Empetrum-Ericaceae combine to again imply treeless landscape during much of the Late Glacial; Nothofagus loading, which is higher late in zone PH-3 and in zones PH-4 and PH-6, is associated with some increase in the areal extent of beech. The slight rise of Nothofagus around 13 190 BP (zone PH-4) appears correlative with the basal increase of this factor before 12 730 BP at Puerto Williams (zone PW-4; Fig. 5). After about 10 000 BP, when Gramineae loading dropped sharply above 4.9 m, the Nothofagus factor is highest (zone PH-2) and continues to be high with few fluctuations until after the interval of marine incursion. Increased loading by the Gramineae at first and afterwards by Empetrum-Ericaceae characterizes the record of the last approximately 5000 years (zone PH-1). Moore (1978) in a study of seeds and other macrofossils in deposits at Cueva del Milodón (51° 35'S) finds Gramineae with Empetrum at about 12 400 BP, later dominance by Nothofagus betuloides reaching a maximum about 7000 BP, and lastly, an increase of N. pumilio after 5600 BP.

Climatic interpretation of the Puerto del Hambre and Puerto Williams records differs little, if at all, where they overlap chronologically. Earlier than about 13 000 BP at Puerto del Hambre, climate was tundra-like and comparatively cold and dry, except for the milder episode around 15 800 BP. The marine incursion at about 5000 BP in the section is representative of regional coastal submergence whereby sea level rose to a level at least 3.5 m higher than today (Porter et al., in press).

Taiquemó (42° 10'S, 73° 36'W; elev. 170 m) and Alerce (41° 25'S, 72° 54'W; elev. 107 m). These records (Figs. 7-8) are presented in tandem because together they bring out the detail of the Late Quaternary. Taiquemó emphasizes the Pleistocene between about 43 000 and 11 000 BP, and Alerce treats the Late Glacial and Holocene of the past approximately 16 000 years. The sites are located in Valdivian rain forest, 105 km apart, and in comparable climatic settings (Heusser 1966; Heusser and Flint 1977).

Nothofagus, the first factor at Taiquemó, is associated with Gramineae, Myrtaceae-Podocarpus, and Fitzroya factors. The species of Podocarpus is P. nubigenus of the rain forest. Moderate loading of the Gramineae in the early part of the record between about 43 000 and 37 000 BP (zone T-8), suggests forests of limited extent apparently dominated by beech (N. dombeyi type: Heusser 1971). From 37 000 to 25 000 BP (zone T-7), higher loading of Gramineae, coupled with lower values of Myrtaceae-Podocarpus and Fitzroya, mark the beginning of a cooling trend that continued in effect and intensified after about

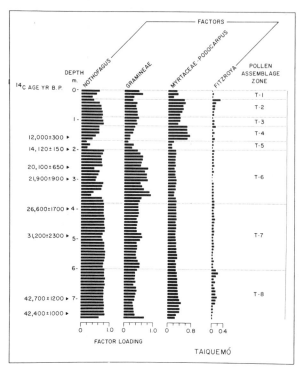

Figure 7. Loading of principal pollen factors in section at Taiquemó.

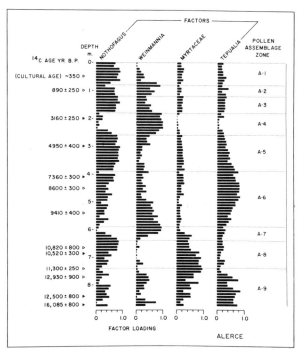

Figure 8. Loading of principal pollen factors in section at Alerce; C14-ages shown by open triangles are from nearby sites that are stratigraphically correlated.

25 000 until 14 000 BP when Gramineae loading was highest (zone T-6). Beech became restricted at this time, as the altitude of the tree line lowered and tundra developed near sea level on Isla de Chiloé. Summer temperature is shown to have been about 6 °C lower than present from the evidence of Lycopodium fuegianum, now distributed near 54 °S in subantarctic Chile (Heusser 1972). Another subantarctic species, Drapetes muscosus, growing today south of 47° 25'S (Skottsberg 1916; Moore 1974), does not occur in the record after about 12 000 BP (zone T-5). Late in the Pleistocene, climate improved, as is indicated by the high loading of Myrtaceae (zone T-4). During the Holocene (zone T-3 through zone T-1), Valdivian rain forest species entered the record or became better represented while species identified with the Pleistocene diminished or disappeared.

Alerce factors include Valdivian rain forest taxa Weinmannia, Myrtaceae, and Tepualia with the primary factor, Nothofagus (N. dombeyi type). The record displays throughout widespread fluctuation of values. Peak loading of Nothofagus (zones A-8, A-5, A-3, and A-1) alternates, by and large, with Weinmannia (zones A-7, A-4 and A-2) and these together alternate with Tepualia (zones A-9 and A-6). The rise in loading of the Myrtaceae factor after 12 500 to a peak at about 11 300 BP (zone A-9) compares with the behaviour of Myrtaceae-Podocarpus at Taiquemó with which the factor is correlated.

Mean January temperature (°C) and mean annual precipitation (mm) were quantified (Fig. 9) using regression equations for the Alerce record and for the record from Taiquemó until around 11 000 BP (Heusser and Streeter 1980; Heusser et al. 1981). The equations, which relate modern pollen in surface samples and present-day temperature and precipitation, were applied to the fossil pollen data. Modern sample sites, distributed between 41° and 55°S, consisted of 59 localities for estimating temperature and 57 for precipitation. Meteorological data were supplied by weather stations located in the sample range (Almeyda and Sáez 1958). Seven terms used in the temperature equation explain 82 % of the variance (the standard error of estimate = 0.96 °C); nine terms for the precipitation equation account for 86 % of the variance (standard error of estimate = 415 mm). Limitations of the equations apply to peak temperature and precipitation values, which appear to be excessive, and also to the lower temperatures, but trends are believed to be correct.

Results show climate to have been relatively mild and wet between 43 000 and 31 000 BP, compared with the cold and drier interval from 31 000 to 14 000 BP (Fig. 9). At about 13 000 BP, climate became warmer, accompanied by an increase in humidity, which continued to approximately 11 000 BP. Following this, episodes that were cooler and wetter date from between 11 000 and 10 000 BP, 5000 and before 3200 BP, after 3200 and before 900 BP, and between 350 BP and the present. The warmest and least humid interval of the Holocene lasted from around 9400 to 7400 BP.

Episodes of cooler and wetter climate over the past 10 000 years implied by these results correspond to intervals of glacier growth in the Andes (Mercer 1976, 1982). The San Rafael Glacier (46° 40'S), at Laguna de San Rafael (Fig. 3), is one of the glaciers that advanced on at least three occasions during the Holocene: between 5000 and 4000 BP, between 3740 and 500 BP, and during the 19th century (Heusser 1960, 1964).

No evidence of glacial readvance has been found in the Andes during Late Glacial time after about 13 000 BP, although the cooler and wetter

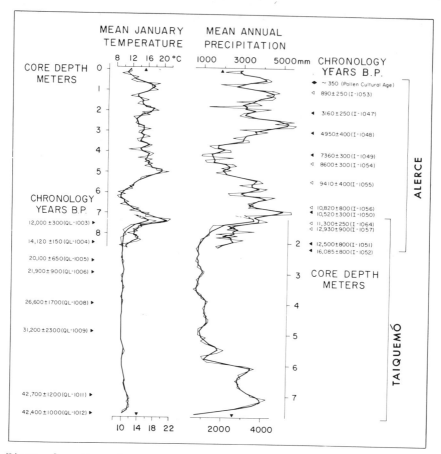

Figure 9. Mean January temperature and mean annual precipitation at Alerce and Taiquemó for the last approximately 43 000 years from Heusser, Streeter and Stuiver (1981). Three-point moving averages of the data are shown by bold lines. Triangles on the temperature and precipitation scales are modern means; C14 ages are also indicated by triangles (open triangles are from nearby sites that are stratigraphically correlated).

trend established between 11 000 and 10 000 BP (Fig. 9) is suggestive of glacier growth. Mercer (1982) concluded that Pleistocene glaciers readvanced for the last time around 13 000 BP or before and did not become active again until Neoglaciation began after 6850 BP. Hoganson and Ashworth (1981, 1982), lending support to Mercer's conclusion, found climate to be relatively uniform from 14 635 to 4525 BP in a study of fossil beetle assemblages located at 40°S. But their data, implying a uniform climate similar to that at the present time, are questionable because the time span of their study includes Pleistocene glaciation and maximum Holocene warmth followed by Neoglacial cooling, when climatic changes were considerable and are documented from various sources in the southern Andes (Markgraf and Bradbury 1982). Despite the apparent absence of evidence for glacial readvance 11 000 to 10 000 years ago in Chile, glaciers were active in New Zealand in this time range (Burrows and Gellatly 1982). Glaciers there were smaller then,

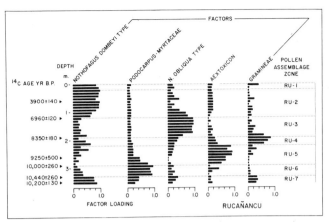

Figure 10. Loading of principal pollen factors in section from Rucañancu.

compared with their size earlier in the Pleistocene, but were larger than at any time subsequently.

Rucañancu (39° 33'S, 72° 18'S; elev. 290 m). Vegetation of the site, located at the contact of Valdivian rain forest on the lower slopes of the Andes and the lowland deciduous beech forest to the west, is reflected by five leading factors (Fig. 10). Profiles at the beginning suggest enclaves of forest species, among which Nothofagus dombeyi type prevailed, growing in terrain dominated by Gramineae (zone RU-7). This was followed by conspicuous loading of the Podocarpus-Myrtaceae factor up until 10 000 BP (zone RU-6). Podocarpus is represented by P. andinus, a species that grows today at montane elevations around and somewhat below 1200 m in the Andes (Schmithüsen 1956) where summer temperature is figured to average 5° - 8°C lower and precipitation annually to be some 2000 mm heavier than at Rucañancu (Heusser, in press). Because P. andinus today seems to be adapted to greater summer drought than N. dombeyi, these data imply an incipient decrease in summer precipitation while temperature was low. They also corroborate evidence for cooler and wetter climate shown by the Taiquemó-Alerce records at the close of the Late Glacial (Fig. 9).

The succession of high loadings by Aextoxicon (zone RU-5), Gramineae (zone RU-4), and Nothofagus obliqua type expresses warmer climate that was at first humid, as it was during the Late Glacial, followed by dryness at about 8350 BP and later by a return of greater humidity. The climate was cooler and more humid after about 7000 BP, as loading by N. dombeyi type become higher (zone RU-2). It seems clear that this shift occurred in response to the expansion of Valdivian rain forest and its replacement of lowland deciduous beech forest about Rucañancu. Surface increase of Gramineae (zone RU-1) follows from human settlement. Although this record does not reveal the detail of Holocene climatic change shown at Alerce (Fig. 9), its trends agree with the major variations at that site.

Laguna de Tagua Tagua (34° 30'S, 71° 10'W; elev. 200 m). This site located in subtropical broad sclerophyll vegetation, is the oldest continuous depositional record, dating over 45 000 BP, of the six sites chosen for analysis. Lugana de Tagua Tagua, occupying a basin formed by Andean laharic deposits, apparently originated during the middle

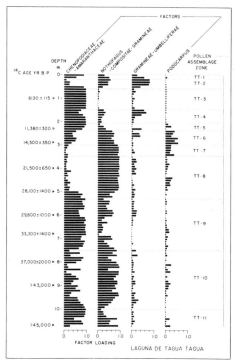

Figure 11. Loading of principal pollen factors in section from Laguna de Tagua Tagua; C14 ages shown by open triangles are from nearby sites that are stratigraphically correlated.

Wisconsinan Glaciation and was in existence until the mid-19th century when it was artificially drained (Heusser 1983b). The factor diagram (Fig. 11) contrasts the profile of Chenopodiaceae-Amaranthaceae, the first factor, and the profile of the second factor, Nothofagus-Compositae-Gramineae. Remaining factors, Gramineae-Umbelliferae and Podocarpus, show a loading pattern that at times closely resembles the second factor. Loading of Chenopodiaceae-Amaranthaceae is apparently in response to more xeric conditions, whereas the loading shown by the other factors is related to wetter climate. Water levels at the laguna no doubt fluctuated in accordance with precipitation/evaporation trends.

Climate during the Pleistocene was evidently cool and drier before 43 000 BP (zone TT-11), between about 36 000 and 28 500 BP (zone TT-9), and after about 14 500 BP (zones TT-6 and TT-5). Increase of beech (Nothofagus dombeyi type), included in the Nothofagus-Compositae-Gramineae factor, is reason to consider the intervals dating from over 43 000 to about 36 000 BP (zone TT-10) and from 28 500 to 14 500 BP (zone TT-7) to have been cool and comparatively wet. Loading by the Podocarpus factor, depicting the high montane P. andinus, at about 18 000 BP (zone TT-7) suggests that climate, although cool, was tending to become summer-dry. This condition was increasingly accentuated later, as implied by the progressive loading of Chenopodiaceae-Amaranthaceae from about 14 500 to 11 380 BP while loading of Podocarpus was high (zone TT-6). Species of N. dombeyi type beech (N. dombeyi, N. pumilio, and N. antarctica) and P. andinus today range in the Andes of central Chile but not north to the latitude or at the elevation of Laguna de Tagua Tagua (Donoso 1978). When loading of these species occurs in the record, mean summer temperature probably fell by as much as 5°C and was accompanied by a rise in precipitation of at least 1000 mm. The climate at the close of the Pleistocene (zone TT-5) is regarded as cool and dry.

The Holocene was warm and probably driest for an extended interval centered at about 6000 BP (zone TT-3) and has been comparatively warm and dry recently (zone TT-1). More mesic conditions prevailed following the Pleistocene (zone TT-4) and late in the Holocene (zone TT-2). Additional chronological control is needed for the Holocene but the succession of climatic fluctuations resembles, in the main, the sequence interpreted at Rucañancu (Fig. 10).

LATE QUATERNARY CLIMATES

Interpretations of the climates of Chile, based on factor analysis of the pollen records distributed over some 20° of central and southern latitudes (Fig. 12), point to certain relationships and differences for the region. Climate throughout was warmer during the Early Holocene and colder during the Late Holocene. Conditions were humid during the warmer millennia, except in the north where an interval of dryness developed and became protracted in the northernmost record. Beginning 5000-6000 BP, and in one instance as early as 7000 BP, succeeding cooler climate was also more humid with the exception of high-latitude Chile which appears to have been drier. In the Alerce record, the Late Holocene includes two warmer and drier episodes. The episodes do not show up in other records, but this may be because sampling intervals were not close enough in the sections with thinner Holocene deposits.

Pleistocene climate (before about 10 000 BP, Fig. 12) was mostly cold and drier in the south of Chile. At lower latitudes after around 14 000 years ago, it was milder and more humid before ultimately turning colder. At Taiquemó, relative cold and dryness prevailed before 14 000 BP to about 37 000 BP, when it was milder and wetter during an interstade dating back to at least 43 000 BP. This interstade at Laguna de Tagua Tagua in central Chile, dated from about 36 000 to before 43 000 BP, is posed between intervals of cold, drier climate, one of which is older than 45 000 years and the other follows to about 28 500 BP. At the laguna after 28 500 BP wetness during most of the remainder of the Pleistocene contrasts with dryness shown by records from the south of Chile.

The data from central Chile corroborate other evidence at lower latitudes for cooler, wetter climate during the Late Pleistocene (Paskoff 1977). Biota that had migrated in response to pluvial conditions became extinct in the Holocene or were reduced to disjunctive populations (Hoffsetter and Paskoff 1966; Herm 1969; Paskoff 1971; Ochsenius and Ochsenius 1977; Villagrán and Armesto 1980; Troncoso et al. 1980). Intensification of the westerlies during the Pleistocene, involving movement northward of the polar front by 5° - 7° or more of latitude, accompanied by greater cloudiness, is considered a most likely mechanism to account for the biotic and other observed changes (Paskoff 1970, 1977; Hastenrath 1971; Caviedes 1972; Caviedes and Paskoff 1975). Temperature depression amounting to as much as 6°-8°C in summer is implied from pollen data at Taiquemó, Rucañancu, and

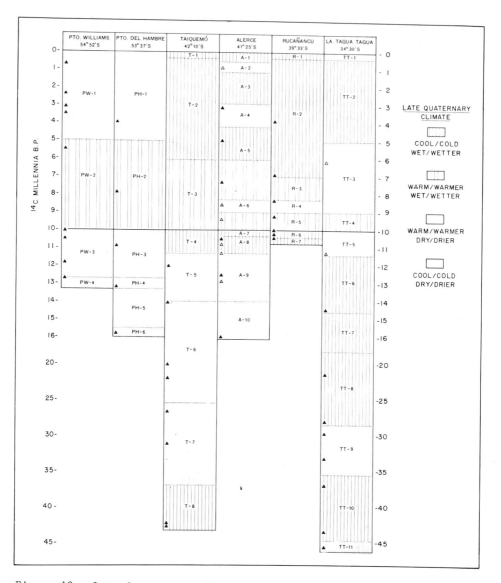

Figure 12. Late Quaternary climate of central and southern Chile based on pollen evidence. Dotted lines conform with pollen assemblage zone boundaries (Figs. 5-8, 10-11); bold line approximates the Holocene-Pleistocene boundary; and triangles represent C14-dated horizons (open triangles from nearby stratigraphically correlated sections). Note time scale change in the Pleistocene.

Lugana de Tagua Tagua. Data from the marine realm show the Humboldt Current was colder, mostly as a result of upwelling, by at least 4 °C at the time of the Last Glacial Maximum (CLIMAP Project Members 1981).

For the lower temperature of central Chile to be accompanied by greater effective precipitation while southern Chile became drier, it is postulated that the polar maritime air mass (Fig. 1) spread northward, reducing the influence of subtropical maritime air associated with the Pacific anticyclone. Expansion of circumpolar circulation at this time probably caused the continental antarctic air mass to influence southern Chile, especially as the apron of sea ice about Antarctica enlarged, and accounts for the drier, colder climate when tundra thrived at Puerto Williams and at Puerto del Hambre in the Late Pleistocene. But in the south of Chile sufficient moisture was apparently available at times to cause glaciers to advance across Tierra del Fuego and the adjacent mainland (Fig. 3).

LATE QUATERNARY CLIMATIC RECORDS FROM OTHER PARTS OF THE SOUTHERN HEMISPHERE

Records of land masses located mostly at middle latitudes (Fig. 13) bear certain relationships to one another and to the climatic sequences from Chile. Coordinated with data from marine cores, they furnish ground for speculation regarding the development of Late Quaternary climatic events. In drawing comparisons, it is important to observe

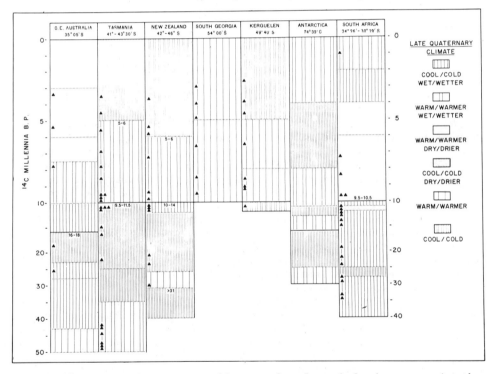

Figure 13. Late Quaternary climate of selected land masses in the Southern Hemisphere from pollen records discussed in the text. Pollen zone boundaries are dotted lines with C14 ages shown by triangles; bold lines within the sections approximate the Holocene-Pleistocene boundary. Note time scale change in the Pleistocene.

that some reservation is necessary because of the unevenness of chronological control, the influence of local topographic conditions, and the tentative nature and reliability of certain source data.

The early Holocene was warmer everywhere after about 10 000 BP. Where the Pleistocene boundary is interpreted to occur at 16 000 or

18 000 BP as at Lake George (35°05'S) in southeast Australia, cool conditions with temperatures 2°- 4°C lower than today's continued until around 10 000 years ago (Bowler et al. 1976; Singh et al. 1981a, 1981b; Singh, in press); in east Antarctica (74°39'S), an oxygen isotope climatic record from an ice core (Lorius et al. 1979) implies warming after 16 000 BP that peaked between 11 000 and 8000 BP, following a cold dry interval between 13 000 and 11 000 BP. Cooler trends begin as early as 6000 BP and as late as 4000 BP with the exception of Lake George, which has remained warm and become drier after about 7500 BP, and east Antarctica, which was apparently colder from 8000 to 4000 years ago. Age determination on shells from the Antarctic Peninsula, in accordance with the data from east Antarctica, indicates less ice around 8000 BP (Sugden and Clapperton 1980). In South Africa (24°26'S), a temperate climate has been warm up to the present day after having been cooler between 4000 and 2000 BP (Scott 1982).

The Holocene of Tasmania (41°- 43°30'S) was warmer and wetter following a rise in temperature that began 11 500-9500 BP (Macphail 1979; Colhoun et al. 1982). Later, after a more equable climate between 8000 and 5000 BP, an increase in the incidence of frost and drought is evident. In the Holocene of New Zealand (42°- 46°S), dated at 10 000 BP and also at 14 000 BP (Lintott and Burrows 1973; Suggate and Moar 1974), a shift from cool and moist to warm, moist conditions is followed after 6000 BP by cooler intervals indicated by pollen evidence at 5000 and 2000 BP and by glacier advances after 5400 BP (Cranwell and von Post 1936; Moar 1961, 1971, 1973a, 1982; McIntyre and McKellar 1970; McGlone and Moar 1977; Burrows 1979; Burrows and Gellatly 1982). At least seven glacial episodes are known, three of which show good correlation with intervals when advances culminated in Chile (Mercer 1982). The pollen record from subantarctic Kerguelen (Young and Schofield 1973), situated farther poleward than either Tasmania or New Zealand, also traces a change from warmer to colder climate around 5000 years ago. This change, however, is not apparent in pollen data from subantarctic South Georgia (Barrow 1978).

Late Pleistocene climate in all sectors was relatively cold. In southeast Australia, temperature depression is estimated to have been 8°- 10°C between about 23 000 and 16 000 years ago at the time of the Last Glacial Maximum (Singh, in press). Colder and drier climate was preceded by a cooler, wetter interval that began around 43 000 BP and, before this, by more equable interstadial climate. Tasmania and New Zealand were also drier when the Last Glacial Maximum took place. Earlier than 35 000 BP in Tasmania, temperatures moderated and precipitation was heavier (Colhoun and Goede 1979; Colhoun et al. 1982; van de Geer et al., in press), while in New Zealand, an interstadial is recognized later from before 31 000 to about 26 000 BP (Moar 1973b, 1980; Moar and Suggate 1975). This interstadial may be expressed by warmer conditions inferred in Antarctica before 25 000 BP (Lorius et al. 1979); the last temperature drop there occurred around 16 000 years ago, but dating is imprecise and instead may have occurred earlier so as to coincide with the time of maximum of the West Antarctic Ice Sheet between about 21 000 and 17 000 BP (Stuiver et al. 1981). In South Africa (24°26'-30°19'S), cooler and wetter climate, with mean temperature 5°- 6°C lower than the present between about 25 000 and 12 500 BP, is followed by relatively warm and dry episodes and, lastly, by cooler and more humid climate after 11 000 BP at the close of the Pleistocene; before 25 000 until some time after

34 400 years ago, data suggest greater dryness preceded by a more humid interval that lasted in parts of the region until 28 000 BP (van Zinderen Bakker 1957, 1976, in press; Coetzee 1967; Heine 1982; Butzer et al. 1978; Vogel 1982; Scott 1982).

Radiolaria and planktonic foraminifera contained in marine cores from the South Atlantic and Indian Oceans identify Late Quaternary positions of the oceanic Antarctic Polar Front and Subtropical Convergence, which represent boundaries of cool, low-salinity subpolar water and warm, subtropical water of high salinity (Hays et al. 1976; Bé and Duplessy 1976; Williams 1976; Prell et al. 1979, 1980; Morley and Hays 1979). According to the data, the boundaries were several degrees of latitude farther north during the Last Glacial Maximum but not equally displaced in all sectors of the ocean. The Antarctic Polar Front moved $5° - 10°$ and the Subtropical Convergence $2° - 5°$ of latitude. Because the boundaries roughly identify the northern edge of the westerly wind belt, these data with the evidence from the land point to a mid-latitude shift of polar frontal systems throughout the Southern Hemisphere. Moreover, the irregularity of the geographic extent of influence of the fronts during the Late Quaternary probably accounts for at least some of the dissimilarity in the climatic records. It is shown, for example, that glaciation of the subantarctic islands at the time of the last glaciation was regulated by the limit of the Antarctic Polar Front (Hays et al. 1976).

SUMMARY AND CONCLUSIONS

1. Cold and wet climate of southern Chile, characterized by cloudiness and powerful wind, is influenced by the westerly wind belt and by cyclonic storms originating in the trough of low pressure associated with the subpolar maritime air mass. Relatively stable air of the subtropical Pacific anticyclone maintains summer-dry climate in central Chile and aridity in the far north. Cold subantarctic water of the West Wind Drift bathes the southern coast while the Humboldt Current travelling northward cools subtropical and tropical coastal Chile.

2. During the Late Pleistocene at the time of the Last Glacial Maximum, the polar front at the boundary of the subpolar and subtropical air masses was displaced northward $5° - 7°$ or more of latitude. This resulted in greater precipitation, lower temperature, and increased cloudiness in central and northern Chile. Estimates indicate that middle latitudes cooled $5° - 8°C$ and precipitation rose by 1000-2000 mm. Conditions in southern Chile were apparently drier and under the influence of the continental antarctic air mass.

3. Late Pleistocene fluctuations in the location of the polar front are implied by fossil pollen records dating back over 45 000 years. Climate was at times comparatively cold and wet, cold and dry, and warm and wet. Somewhat milder climate is evident in some records after 14 000 BP, followed by a cooler interval until about 10 000 BP. Summer-dry conditions apparently developed in central Chile after about 14 500 BP.

4. Holocene climate at the beginning was warm and humid and continued so in the south until 5000-6000 BP. Climate northward became dry for less than a millennium about 9000 BP, except in central Chile where dryness continued until 5000 BP. Cooler and more humid conditions, first apparent close to 7000 years ago, generally prevailed during later millennia, with the exception of southernmost Chile which appears to have been cooler and drier. Climatic variations evident in one

detailed record occur, by and large, in concert with fluctuations in the glacial record. It is believed that wetter Holocene intervals at lower latitudes were times when storms originating in the belt of southern westerlies travelled farther north and with greater frequency than at present.

5. A review of records from other parts of the Southern Hemisphere shows Late Pleistocene climate to have been generally cold and dry at higher latitudes when lower latitudes were cold and wet. In Tasmania, a milder and wetter interstadial before 35 000 BP corresponds to an interstadial in Chile.

6. Holocene climatic changes show local variations. Records reveal early warming which at lower latitudes became comparatively protracted. The Late Holocene in Tasmania and New Zealand after 5000-6000 BP was, at least at times, cooler than it was earlier; episodes when glaciers advanced in Chile show good agreement with episodes in New Zealand. The record from Antarctica, the least reliable chronologically, implies a mid-Holocene cooling trend followed by milder climate.

7. Late Pleistocene northward movement of the polar front, evident from the land records, also obtains from deep-sea data. Oceanic boundaries in the Southern Ocean, represented by the Antarctic Polar Front and Subtropical Convergence, shifted $2° - 10°$ of latitude toward the equator at the time of the Last Glacial Maximum. This reflects the change of geographic influence of the westerly wind belt. Deep-sea data in conjunction with the land records indicate that shifting of frontal systems varied considerably from place to place and may explain certain inconsistencies in the climatic records from Chile as well as from other land masses.

ACKNOWLEDGMENTS

The work was supported by National Science Foundation grants to New York University with field travel provided by the Chilean Empres Nacional del Petroleo and Servicio Nacional de Geología y Minería. Thanks are expressed to LE Heusser (Lamont-Doherty Geological Observatory of Columbia University) for assistance with data reduction and to EM van Zinderen Bakker (University of the Orange Free State, Bloemfontein), N T Moar (New Zealand Department of Scientific and Industrial Research, Christchurch), EA Colhoun (University of Tasmania, Hobart), and G Singh (Research School of Pacific Studies, Canberra) for current views on Late Quaternary climates and preprints of unpublished papers.

REFERENCES

Almeyda A, E and F Sáez F. 1958. Recopilación de datos climáticos de Chile y mapas sinópticos respectivos. Ministerio de Agricultura, Santiago. 195 pp.

Auer V 1933. Verschiebungen der Wald- und Steppengebiete Feuerlands in postglazialer Zeit. Acta Geographica 5:1-313.

Auer V 1958. The Pleistocene of Fuego-Patagonia. Part II. The history of the flora and vegetation. Annls Academiae Scientiarum Fennicae III. Geologica Geographica 50:1-239.

Barrow CJ 1978. Postglacial pollen diagrams from South Georgia (sub-Antarctic) and West Falkland Island (South Atlantic). J. Biogeogr. 5:251-274.

Bé AWH and Duplessy JC 1976. Subtropical Convergence fluctuations and Quaternary climates in the middle latitudes of the Indian Ocean. Science 194:419-422.

Bowler JM, Hope GS, Jennings JN, Singh G and Walker D 1976. Late Quaternary climates of Australia and New Guinea. Quat. Res. 6: 359-394.

Burrows CJ 1979. A chronology for cool-climate episodes in the Southern Hemisphere 12,000-1000 yr B.P. Palaeogeogr. Palaeoclimatol. Palaeoecol. 27:287-347.

Burrows CJ and Gellatly AF 1982. Holocene glacier history in New Zealand, In: W. Karlén (ed.), Holocene Glaciers. Striae 18. p. 41-47.

Butzer K, Stuckenrath R, Bruzewicz AJ and Halgren DM 1978. Late Cenozoic paleoclimates of the Gaap Escarpment, Kalahari Margin, South Africa. Quat. Res. 10:310-339.

Caldenius CC 1932. Las glaciaciones cuaternarias en la Patagonia y Tierra del Fuego. Geografiska Annaler 14:1-164.

Caviedes C 1972. On the paleoclimatology of the Chilean littoral. Iowa Geographer 29:8-14.

Caviedes C and Paskoff R 1975. Quaternary glaciations in the Andes of north-central Chile. J. Glaciol. 70:155-170.

CLIMAP Project Members 1981. Seasonal reconstructions of the earth's surface at the Last Glacial Maximum. Geol. Soc. Am. Map and Chart Series MC-36.

Coetzee JA 1967. Pollen analytical studies in east and southern Africa. Palaeoecol. Africa 3:1-146.

Colhoun EA, van de Geer G and Mook WG 1982. Stratigraphy, pollen analysis, and paleoclimatic interpretation of Pulbena Swamp, northwestern Tasmania. Quat. Res. 18:108-126.

Colhoun EA and Goede A 1979. The Late Quaternary deposits of Blakes Opening and the middle Huon Valley, Tasmania. R. Soc. London Phil. Trans. B 286:371-395.

Cranwell LM and von Post L 1936. Post-Pleistocene pollen diagrams from the Southern Hemisphere. Geografiska Annaler 18:308-347.

Donoso Z,C 1978. Dendrologia arboles y arbustos chilenos. Universidad Chile Facultad Ciencias Forestales Departamento Silvicultura Manual 2. 142 pp.

Godley EJ 1960. The botany of southern Chile in relation to New Zealand and the Subantarctic. Proc. R. Soc. B 152:457-475.

Hastenrath SL 1971. On the Pleistocene snow-line depression in arid regions of the South American Andes. J. Glaciol. 10:255-267.

Hays JD, Lozano JA, Shackleton N and Irving G 1976. Reconstruction of the Atlantic and Indian Ocean sectors of the 18 000 BP. Antarctic Ocean, pp. 337-372. In: RM Cline and JD Hays (eds.), Geol. Soc. Am. Mem. 145. 464 pp.

Heine K 1982. The main stages of the Late Quaternary evolution of the Kalahari region, southern Africa. Palaeoecol. Africa 15:53-76.

Herm, D. 1969. Marines Pliozän und Pleistozän in Nord- und Mittel-Chile unter besonderer Berücksichtigung der Entwicklung der Mollusken-Faunen. Zitteliana 2:1-158.

Heusser CJ 1960. Late Pleistocene environments of the Laguna de San Rafael area, Chile. Geogr. Rev. 50:555-577.

Heusser CJ 1964. Some pollen profiles from the Laguna de San Rafael area, Chile. In: LM Cranwell (ed.), Ancient Pacific Floras. The Pollen Story. Univ. Hawaii Press, Honolulu. 114 pp. p. 95-114.

Heusser CJ 1966. Late Pleistocene pollen diagrams from the Province of Llanquihue, southern Chile. Am. Phil. Soc. Proc. 110:269-305.

Heusser CJ 1971. Pollen and Spores of Chile. Univ. Arizona Press, Tucson. 167 pp.

Heusser CJ 1972. On the occurrence of Lycopodium fuegianum during late-Pleistocene interstades in the Province of Osorno, Chile. Bull. Torrey Botanical Club 99:178-194.

Heusser CJ 1974. Vegetation and climate of the southern Chilean lake district during and since the last interglaciation. Quat. Res. 4:290-315.

Heusser CJ 1981. Palynology of the last interglacial-glacial cycle in mid-latitudes of southern Chile. Quat. Res. 16:293-321.

Heusser CJ 1983a. Quaternary palynology of Chile. Quaternary of South America and Antarctic Peninsula 1: 5-22.

Heusser CJ 1983b. Quaternary pollen record from Laguna de Tagua Tagua, Chile. Science 219: 1429-1432.

Heusser CJ, in press. Late-glacial-Holocene pollen sequence in the northern lake district of Chile. Quat. Res.

Heusser CJ and Flint RF 1977. Quaternary glaciations and environment of northern Isla Chiloé, Chile. Geology 5:305-308.

Heusser CJ and Streeter SS 1980. A temperature and precipitation record of the past 16,000 years in southern Chile. Science 210: 1345-1347.

Heusser CJ, Streeter SS and Stuiver M 1981. Temperature and precipitation record in southern Chile extended to 43,000 yr ago. Nature 294: 65-67.

Hoffstetter R and Paskoff R 1966. Présence des genres Macrauchenia et Hippidion dans la faune pléistocène du Chili. Bull. Mus. Histoire Naturelle Paris 4:476-490.

Hoganson JW and Ashworth AC 1981. Late glacial climatic history of southern Chile interpreted from Coleopteran (beetle) assemblages. Geol. Soc. Am. Abstr. 13:474.

Hoganson JW and Ashworth AC 1982. The late-glacial climate of the Chilean lake region implied by fossil beetles. Proc. Third N. Am. Paleontol. Conv. 1:251-256.

Hollin JT and Schilling DH 1981. Late Wisconsin-Weichselian mountain glaciers and small ice caps. In: GH Denton and TJ Hughes (eds.), The Last Great Ice Sheets. Wiley, New York. 484 pp. p. 179-206.

Imbrie J and Kipp NG 1871. A new micropaleontological method for quantitative paleoclimatology: application to a late Pleistocene Caribbean core. In: KK Turekian (ed.), Late-Cenozoic Ice Ages. Yale Univ. press, New Haven, 606 pp. p. 71-181.

Lamb HH 1959. The southern westerlies: a preliminary survey; main characteristics and apparent associations. Q. J. R. met. Soc. 85: 1-23.

Laugenie C 1971. Elementos de la cronología glaciar en los Andes chilenos meridionales. Cuadernos Geográficos del Sur 1:7-20.

Lintott WH and Burrows CJ 1973. A pollen diagram and macrofossils from Kettlehole Bog, Cass, South Island, New Zealand. New Zealand J. Botany 11:269-282.

Lliboutry L 1956. Nieves y glaciares de Chile. Ediciones de la Universidad de Chile, Santiago, 471 pp.

Lorius C, Merlivat L, Jouzel J and Pourchet M 1979. A 30,000-yr isotope climatic record from Antarctic ice. Nature 280:644-648.

Macphail MK 1979. Vegetation and climates in southern Tasmania since the last glaciation. Quat. Res. 11:306-341.

Markgraf V and Bradbury JP 1982. Holocene climatic history of South America. In: J Mangerud, HJB Birks and KD Jaeger, (eds.), Chronostratigraphic Subdivision of the Holocene. Striae 16. p. 40-45

McGlone MS and Moar NT 1977. The Ascarina decline and post-glacial change in New Zealand. New Zealand J. Botany 15:485-490.

McIntyre JD and McKellar IC 1970. A radiocarbon dated post glacial pollen profile from Swampy Hill, Dunedin, New Zealand. New Zealand J. Geology and Geophysics 13:346-349.

Mercer JH 1967. Southern Hemisphere Glacier Atlas. US Army Natick Lab. tech. Rep. 67-76-ES, 325 pp.

Mercer JH 1976. Glacial history of southernmost South America. Quat. Res. 6:125-166.

Mercer JH 1982. Holocene glacier variations in southern South America, In: W. Karlén (ed.), Holocene Glaciers. Striae 18. p. 35-40.

Mercer JH 1983. Cenozoic glaciation in the Southern Hemisphere. A. Rev. Earth Planet. Sciences 11:99-132.

Miller A 1976. The climate of Chile. In: W Schwerdtfeger (ed.), Climates of Central and South America. World Survey of Climatology 12. Elsevier, Amsterdam. 532 pp. p. 113-145.

Moar, NT 1961. Contributions to the Quaternary history of the New Zealand flora 4. Pollen diagrams from the Western Ruahine Ranges. New Zealand J. Science 4:350-359.

Moar NT 1971. Contributions to the Quaternary history of the New Zealand flora 6. Aranuian pollen diagrams from Canterbury, Nelson and North Westland, South Island. New Zealand J. Botany 9:80-145.

Moar NT 1973a. Contributions to the Quaternary history of the New Zealand flora 7. Two Aranuian pollen diagrams from central South Island. New Zealand J. Botany 11:291-304.

Moar NT 1973b. Late Pleistocene vegetation and environment in southern New Zealand. Palaeoecol. Africa 11:179-198.

Moar NT 1980. Late Otiran and early Aranuian grassland in central South Island. New Zealand J. Ecology 3:4-12.

Moar NT 1982. Chronostratigraphy and the New Zealand post-glacial, In: J Mangerud, HJB Birks and KD Jaeger (eds.), Chronostratigraphic Subdivision of the Holocene. Striae 16. p. 10-16.

Moar NT and Suggate RP 1978. Contributions to the Quaternary history of the New Zealand flora 8. Interglacial and glacial vegetation in the Westport District, New Zealand. New Zealand J. Botany 17: 361-387.

Moore DM 1974. Catalogo de las plantas vasculares nativas de Tierra del Fuego. An. Inst. Patagonia 5:105-121.

Moore DM 1978. Post-glacial vegetation in the south Patagonian territory of the giant ground sloth, Mulodon. Botan. J. Linnean Soc. 77: 177-202.

Morley JJ and Hays JD 1979. Comparison of glacial and interglacial oceanographic conditions in the South Atlantic from variations of calcium carbonate and radiolarian distributions. Quat. Res. 12: 396-408.

Muñoz P, C and E Pisano V 1947. Estudio de la vegetacion y flora de los parques nacionales de Fray Jorge y Tallinay. Agricultura Técnica 7:71-190.

Ochsenius C and Ochsenius F 1979. Biogeographical context of the pluvial lakes of the Atacama Desert during the late Pleistocene. Tropic of Capricorn. Proc. First int. Congress Pacific Neogene Stratigr. COSIUGS. Tokyo. p. 169-171.

Paskoff R 1970. Recherches géomorphologiques dans le Chili semi-aride. Biscaye Frères, Bordeaux. 420 pp.

Paskoff R 1971. Edad radiometrica del mastodonte de Los Vilos. Noticiario Mensual Museo Nacional Historia Natural (Santiago) 177:11.

Paskoff R 1977. Quaternary of Chile: the state of research. Quat. Res. 8:2-31.

Pisano V, E 1977. Fitogeografía de Fuego-Patagonia chilena. I. Comunidades vegetales entre las latitudes 52 y 56 S. An. Inst. Patagonia 8:121-250.

Porter SC 1981. Pleistocene glaciation in the southern lake district of Chile. Quat. Res. 16:263-292.

Porter SC, Stuiver M and Heusser CJ, in press. Holocene sea-level changes along the Strait of Magellan and Beagle Channel, southernmost South America. Quat. Res.

Prell WL, Hutson WH and Williams DF 1979. The Subtropical Convergence and Late Quaternary circulation in the southern Indian Ocean. Marine Micropaleontol. 4:225-234.

Prell WL, Hutson WH, Williams DF, Bé AWH, Geitzenauer K and Molfino B 1980. Surface circulation of the Indian Ocean during the Last Glacial Maximum, approximately 18,000 yr B.P. Quat. Res. 14:309-336.

Reiche K 1907. Grundzüge der Pflanzenverbreitung in Chile. Die Vegetation der Erde 8:1-374.

Schmithüsen J 1956. Die räumliche Ordnung der chilenischen Vegetation. Bonner Geographische Abh. 17:1-86.

Scott L 1982. A late Quaternary pollen record from the Transvaal bushveld, South Africa. Quat. Res. 17:339-370.

Singh G, in press. Late Quaternary vegetation and lake level record from Lake George, New South Wales. CLIMANZ. Research School of Pacific Studies, ANU Publn BG.

Singh G, Opdyke ND and Bowler JM 1981a. Late Cainozoic stratigraphy, paleomagnetic chronology and vegetational history from Lake George, NSW. J. Geol. Soc. Australia 28:435-452.

Singh G, Kershaw AP and Clark R 1981b. Quaternary vegetation and fire history in Australia. In: AM Gill, RA Groves and IR Noble (eds.), Fire and Australian Biota. Australian Acad. Science, Canberra. p. 25-54.

Skottsberg C 1910. Die Vegetationsverhältnisse längs der Cordillera de los Andes S. von 41 S. Br. Ein Beitrag zur Kenntnis der Vegetation in Chiloé, Westpatagonian, dem andinen Patagonien und Feuerland. Kungl. Svensk. Vet. Akad. Handl. 56:1-366.

Stuiver M, Denton GH, Hughes TJ and Fastook JL 1981. History of the marine ice sheet in West Antarctica during the last glaciation: a working hypothesis. In: GH Denton and TJ Hughes (eds.), The Last Great Ice Sheets. Wiley, New York. 484 pp. p. 319-436.

Sugden DE and Clapperton CM 1980. West Antarctic ice sheet fluctuations in the Antarctic Peninsula area. Nature 286:378-381.

Suggate RP and Moar NT 1970. Revision of the chronology of the late Otira glacial. New Zealand J. Geol. Geophys. 13:742-746.

Taljaard JJ 1972. Synoptic meteorology of the Southern Hemisphere, In: CW Newton (ed.), Meteorology of the Southern Hemisphere. Met. Monogr. 13. Am. met. Soc. Boston, 263 pp. p. 139-213.

Troncoso A, Villagrán C and Muñoz M 1980. Una nueva hipótesis acerca del origen y edad del bosque de Fray Jorge (Coquimbo, Chile). Boln Museo Nacional Historia Natural (Santiago) 37: 117-152.

van de Geer G, Colhoun EA and Mook WG, in press. Mowbray and Broadmeadows Swamps, northwestern Tasmania: stratigraphy, pollen analysis and paleoclimatic interpretation. Quat. Res.

van Zinderen Bakker EM 1957. A pollen analytical investigation of the

Florisbad Deposits (South Africa). Proc. III Pan-Afr. Congress Prehistory, Livingstone, pp. 56-57.

van Zinderen Bakker EM 1976. The evolution of late-Quaternary palaeoclimates of southern Africa. Palaeoecol. Africa 9:160-202.

van Zinderen Bakker EM, in press. The Late Quaternary history of climate and vegetation in east and southern Africa. Bothalia.

Villagrán C 1980. Vegetationsgeschichtliche und pflanzensoziologische Untersuchungen im Vicente Pérez Rosales Nationalpark (Chile). Dissertationes Botanicae 54: 1-165.

Villagrán C and Armesto JL 1980. Relaciones floristicas entre las comunidades relictuales del Norte Chico y la zona central con el bosque del sur de Chile. Boln Museo Nacional Historia Natural (Santiago) 37: 87-101.

Vogel JC 1982. The age of the Kuiseb River silt terrace at Homeb. In: JC Vogel, EA Voigt and TC Partridge (eds.), S. Afr. Soc. Quat. Res. Proc. VI Biennial Conf. (Pretoria). Palaeoecol. Africa 15. Balkema, Rotterdam. p. 201-209.

Williams DF 1976. Late Quaternary fluctuations of the polar front and Subtropical Convergence in the southeast Indian Ocean. Marine Micropaleontol. 1:363-375.

Young SB and Schofield EK 1973. Pollen evidence for late Quaternary climate changes on Kerguelen Islands. Nature 245:311-312.

Zamora M, E and A Santana A 1979. Caracteristicas climaticas de la costa occidental de la Patagonia entre las latitudes $46°\ 40'$ y $56°\ 30'S$. An. Inst. Patagonia 10:109-154.

Testing the Late Quaternary climatic record of southern Chile with evidence from fossil Coleoptera

A.C. ASHWORTH
North Dakota State University, Fargo, USA

J.W. HOGANSON
North Dakota Geological Survey, Grand Forks, USA

ABSTRACT. From quantitative paleoclimatological data derived from mathematical transfer functions applied to a pollen profile from Alerce, southern Chile, Heusser infers that Post-glacial warming was interrupted by a cold and very wet episode between 11 000 - 9500 BP. To test the pollen-inferred paleoclimatic interpretation, a method was devised for inferring climatological data from fossil Coleoptera assemblages. Fossil Coleoptera were analysed from a cutbank section on the Rio Caunahue, Lago Ranco, southern Chile, that spans the interval 12 810 to 10 000 BP. The site is at a similar elevation to that of the Alerce site and is sufficiently close to have had a similar climatic history. The Caunahue Coleoptera imply that the climate was relatively constant during the entire early Post-glacial. No evidence was found of a major climatic deterioration. This result is supported by pollen evidence from the Caunahue section, by pollen evidence from adjacent regions of Argentina, and indirectly by the absence of moraines that surely would have formed if the climatic deterioration had been as cold and wet as has been proposed.

INTRODUCTION

The Quaternary stratigraphic record of temperate South America is one of the few in the Southern Hemisphere that lends itself to climatic interpretation. The proximity of Antarctica, whose climatic history it must reflect, places the region in a key geographic location for paleoclimatic studies. Also, the type of detailed climatic record being compiled can be compared with those from both Europe and North America to counter the heavy bias of the Northern Hemisphere data. In recent years climatic reconstructions for the Late Quaternary of the Lake Region of southern Chile have been attempted from glacial geomorphological evidence (Heusser and Flint 1977; Mercer 1972, 1976, 1982, 1984; Laugenie and Mercer 1973; Porter 1981), pollen analysis (Heusser 1966, 1974, 1981; Heusser and Streeter 1980; Heusser, Streeter and Stuiver 1981; Villegran 1980) and fossil Coleoptera studies (Hoganson and Ashworth 1982). The Holocene climatic record for the region in the context of the continent has recently been reviewed by Markgraf and Bradbury (1982).

The Last Glacial Maximum in the Lake Region of Chile culminated about 19 500 BP (Mercer 1976), approximately synchronous with the Glacial

Maxima of the Northern Hemisphere. The deglaciation that followed was punctuated by one (Mercer 1976) or perhaps two (Porter 1981) glacial readvances. The timing of the last advance has been the subject of considerable speculation. A date of 13 000 BP (Mercer 1976; Porter 1981) had been proposed, but Mercer (1984) has recently revised this date to 14 500 BP. In an earlier paper we implied that deglaciation had been completed before 14 635 BP (Hoganson and Ashworth 1982). However, a recently obtained date of 13 900 ± 560 BP (Mercer, this volume) from a lower stratigraphic horizon at the same site suggests that deglaciation commenced no earlier than 14 000 BP. Mercer (1976) believes that in the following millennia glaciers rapidly retreated to their present positions which they reached by 11 000 BP.

The paleotemperature curve inferred by Heusser (1974) from pollen analysis generally conforms with the glacial geological evidence. The notable exception is for the interval 11 000 - 10 000 BP when low temperatures are inferred from pollen at a time when glaciers are thought to have been smaller than they are today. Heusser (1974) made no inferences about precipitation, which left the possibility open that the shrunken state of the glaciers resulted from their starvation.

Heusser and Streeter (1980) and Heusser, Streeter and Stuiver (1981) recently compiled temperature and precipitation curves (based on pollen-climate transfer functions) for the past 43 000 years in the Lake Region. The inferred temperature trends are similar to those that Heusser (1974) described previously. The period of cooling after 11 000 BP is characterized by an even greater lowering of temperature, at least 6 °C. Differences are the absence of a warm interval, the Varas Interstade, at 16 000 BP and high temperatures at about 11 300 BP immediately preceding the cold phase. Evidence of high temperatures is in agreement with the glacial geological evidence that glaciers shrank very rapidly into the mountains and were behind their present margins at this time. The most interesting feature of Heusser and Streeter's (1980) precipitation curve is the very wet episode that accompanies the cold period at 11 000 - 9500 BP. If their reconstruction is valid, glaciers must have undergone a major expansion, an inference that cannot be reconciled with the glacial evidence.

The pollen-inferred cold and wet phase between 11 000 and 9500 BP is approximately coeval with the Younger Dryas Stade of Europe, an emphatic climatic reversal in the warming trend after the Glacial Maximum. The Younger Dryas is such a marked event that it is assumed by many Quaternary researchers to have had a global effect. For example, a global event was an implicit assumption in the choice of 10 000 BP as the Pleistocene-Holocene boundary. A cold episode of equivalent age from regions as far apart as Europe and Chile would certainly lend support to this assumption.

Clearly, the importance of determining unequivocally whether the interval 11 000 - 9500 BP in southern Chile was warm or cold extends far beyond South America. In order to test the conflicting hypotheses we have developed a basis for inferring climatic information from fossil Coleoptera in the Chilean Lake Region. The extent to which this works and some of the preliminary results are the subject of this paper.

COLEOPTERA AS CLIMATIC INDICATORS

Quaternary Coleoptera (beetles) and their use as paleoclimatic indicators have been the subject of a number of excellent reviews by Coope (1967, 1970, 1975, 1977, 1978, 1979). Organisms used for paleoclimatic analyses must meet a number of general criteria, the more important of which are a good fossil record and a recognizable response to climatic change. The following commentary describes how Coleoptera meet these requirements and how methods had to be modified for South America.

Fossil abundance is a function of species diversity, individual frequency, and preservation. When JBS Haldane was reportedly asked what he could infer about God's creations from his biological studies, he replied that "God must have had an inordinate fondness for beetles". His witty if irreverent reply aptly describes the diversity of this group of organisms that is richer in species than any other, and that occurs in every continental habitat with the exception of those in the polar regions. The great diversity of the Coleoptera makes them particularly valuable as paleoenvironmental indicators. Beetles are also frequently abundant as individuals and density is often particularly high in the wet habitats where beetles are best preserved. For example, one sweep of an aquatic net in the weedy margins of a lake may yield hundreds of individuals.

One of the principal ways that beetles differ from other insects is that they possess robust, completely encasing skeletons of sclerotized chitin. Chitin is reasonably resistant to abrasion and to both chemical and biological breakdown, given a non-oxidizing environment. The combination of diversity, individual frequency, and preservability of chitin makes fossil beetle fragments common in shallow water and bog sediments.

The relationship between beetles and climate is very complex and poorly understood. The Coleoptera are poikilothermic animals and consequently their existence is at the mercy of external factors. A fundamental assumption of coleopterists is that climate controls distribution and that the most important factors are temperature and moisture. Non-climatic factors are also known to influence distribution, but as Coope (1977) points out, factors such as availability of food, competition for space, and disease, are density-dependent and tend to be important where the climatic conditions for existence are optimal. The role played by these factors is expected to decrease at the periphery of range, and the limits of distribution are more probably controlled by climate. Notable exceptions are species that feed on particular plants, or the dung of particular mammals. Even for these species, however, climatic controls are likely to be the most important secondary factors. Coope (1967, 1977) describes specific ways in which climate may influence the existence of individuals, but empirical data on the controlling temperatures and moisture content are mostly unavailable. In the absence of this type of data the range of temperature and other climatic factors within which a species exists is inferred from its geographic or elevational range. This information, usually reported as a climatic average, has the unfortunate tendency of being treated as a control of distribution. Lindroth (1956) was very emphatic about the need to be cautious in the use of secondarily-derived climatic information, a point also stressed by Coope (1977).

The derivation of this type of data involves numerous assumptions, the most important of which is that the distribution reflects the full climatic range that the species can occupy. This type of assumption is clearly an overgeneralization. Another type of problem that might be envisaged (for example, the nonsensical grouping of species from different climatic regimes) has not yet materialized in practice.

The demonstration that Pleistocene Coleoptera species are the same as those now living (Coope 1978) enables climatic inferences to be transferred from the present to the past in a straightforward uniformitarian manner. The process assumes that physiological constancy accompanies morphological constancy. This cannot be put to the test but is supported by considerable circumstantial evidence. The strongest evidence is from groups of species of unrelated ecologies that repeatedly occur together at different times. These occurrences cannot possibly be accounted for by continuous adaptation to changing environments.

Coope (1977) proposes that the constancy of species in both morphology and physiology may result from environmental instability, an ironic departure from earlier interpretations. He postulates that the expansion and contraction of range occurring with each environmental change led to short episodes of isolation followed by periods of convergence. This continual movement is viewed as a stirring mechanism that kept gene pools homogeneous and ensured constancy. Species constancy, however, may be the norm of the fossil record. Eldridge and Gould (1972) certainly make a strong argument for it, describing long periods of stasis punctuated by occasional evolutionary events, simply the consequence of allopatric speciation. Clearly, this does not rule out Coope's hypothesis but suggests that more fundamental mechanisms may be involved in maintaining the integrity of species through very long periods.

The rapidity with which beetles respond to climatic events makes them potentially valuable in the early detection of climatic change. Coope and Brophy (1972) presented evidence that beetles responded considerably more rapidly to a period of Late Glacial warming in the British Isles than did trees. Their conclusion is logical, and considering the very different dispersal rates involved, entirely according to expectation. Unfortunately, the results of their study tend to be oversimplified so that beetles are perceived as always responding faster than plants. This generalization is simply untenable. The mobility of beetles is often stressed, but the difference in arrival time between different species of beetles can be as great as it is between beetles and any other group of organisms, including trees. Arrival time will depend on ecology and the ability to disperse: habitat generalists that fly well will arrive before habitat specialists that are flightless.

The difference in arrival time will in every instance be magnified by the distance that the species have to travel. The time-transgressive nature of response makes it difficult to infer the timing of climatic events, and for this reason Coope (1981) argued that local extinctions resulting from climatic change would be better marker horizons. This makes theoretical sense but is impractical. Firstly, it is impossible to prove extinction and, secondly, extinction on the scale that might be inferred is not a common phenomenon, at least if judged by the existing Quaternary fossil beetle record.

The interpretation of climate from fossil beetle assemblages is clearly a complicated process but has been demonstrated to work in both Europe (Coope 1977) and North America (Morgan et al. 1983). One of the reasons why successful interpretations have been possible in these regions is that the systematics and distribution of the existing fauna are reasonably well known. However, this is not the situation in most other parts of the world. The systematics of the fauna of Chile are reasonably well known compared to other areas of South America, but even so, geographic distributional data are inadequate for paleoclimatic analyses. Fortunately, Chile is a country of considerable relief and climate plays a major role in the elevational distribution of the biota. We felt that if we could obtain reasonably detailed knowledge of the species composition of the beetle fauna along an elevational transect, we might be able to provide a basis for paleoclimatic analyses. In this respect, our study is the first of its kind involving beetles, and although the results are preliminary, we believe that our assumption was at least partially correct.

THE COLEOPTERA OF THE LAKE REGION

The earliest description of occasional specimens collected from southern South America were made during voyages of exploration to the Pacific. Later, in 1834 and again in 1835, Darwin made small collections of beetles from the forests on the Isla de Chiloe and at Valdivia. His specimens were later designated as types and are preserved in excellent condition in the British Museum. The most comprehensive early treatment of the fauna, with descriptions by Solier and Blanchard, was published in the 1849 and 1851 volumes of Gay's "Historia fisica y politica de Chile". German colonization in the mid-nineteenth century made the Lake Region accessible to travel and in succeeding years the area was visited by several Chilean and European coleopterists.

After more than a century of study, the systematics of the fauna are reasonably well known. Little interest, however, has ever been paid to the distribution and ecology of the fauna. To improve this situation, the Lake Region Coleoptera fauna was sampled during the austral summers of 1977-78 and 1978-79, and beetles were collected from locations in the Cordillera de la Costa, the Valle Longitudinal, and the Cordillera de los Andes, between latitudes $40°$ and $43°S$ (Fig. 1). The Parque Nacional de Puyehue ($40°S$, $72°W$) was chosen as the location for an elevational transect because the vegetation is well known, sites above 500 m are relatively undisturbed, and sites above tree-line are accessible.

The region receives most of its weather from cyclonic storms that are generated at the boundary of the subantarctic and subtropical air masses. Moisture carried by strong westerly winds, "the roaring forties", results in 2000 mm of rainfall annually in the lowlands and 5000 mm at tree-line (Muñoz 1980). Temperatures are modified by the cold, offshore Peru current and westerly winds. At the Parque Nacional de Puyehue mean January (summer) temperatures are $14°C$ in the lowlands, and $11°C$ at tree-line (Muñoz 1980).

The vegetation of the Parque Nacional de Puyehue has been studied and described by Heusser (1966, 1974, 1981), Muñoz (1980), and Veblen et al. (1977, 1979, 1980). Heusser (1974) described five vegetational zones (Fig. 2), that range from lowland deciduous forest at 200 m to alpine tundra at 1300 m. Veblen et al. (1977) proposed a slightly different arrangement with eight vegetational types (Fig. 2). The boundaries between major vegetational types, however, were placed at similar elevations to those indicated by Heusser (1974). Diversity in trees and density of undergrowth decrease with elevation, with the result that the subalpine forests are completely different in appearance from those of the low montane zone. In the upper Subantarctic Deciduous Forest, there are three species of Nothofagus, while in the lower Valdivian Rain Forest, there are numerous genera of tall evergreen flowering trees. The lower forests with lianas, large ferns, patches of the impenetrable cane Chusquea, abundant shrubs, and numerous species of mosses and lichens have an exotic appearance.

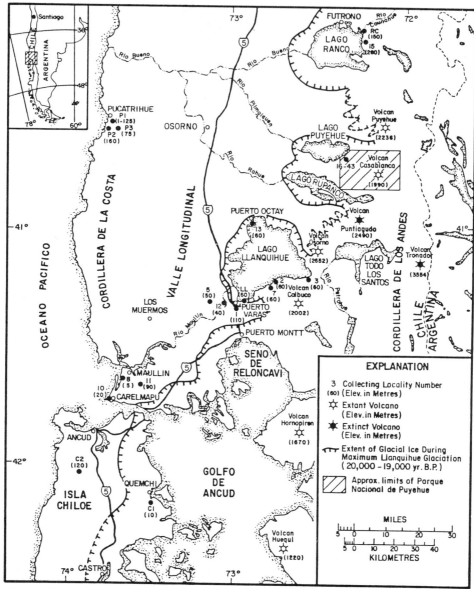

Figure 1. Map of the southern Chilean Lake Region showing location and elevation of beetle fauna collecting sites, location of the Parque Nacional de Puyehue, extent of glacial ice during the Last Glacial Maximum and location of extent and inactive volcanoes. Extent of glacial ice compiled from Mercer (1976), Heusser and Flint (1977) and Porter (1981). Information on volcanoes from Casertano (1963).

Beetles were collected from sites located at approximately 100 m steps from the shores of Lago Puyehue (200 m) to the vegetationless ash slopes (1450 m) above tree-line. Similar sampling techniques were employed at each location and where possible, Coleoptera from aquatic, water-marginal, and forest habitats were collected. The objective at each locality was to sample as many habitats as possible.

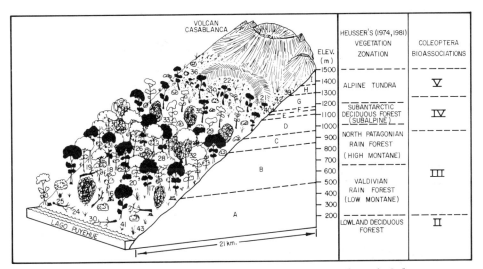

Figure 2. Schematic representation of vegetational and Coleoptera associations from an elevational transect in the Parque Nacional de Puyehue, Osorno Province, Chile (Fig. 1). Vegetational zonation indicated by letters is adapted from Veblen et al. (1977): (A) Disturbed forest, (B) Dombeyi-Eucryphietum forest (Coigue-Ulmo forest), (C) Laurelio-Weinmannietum forest (North Patagonia Teniu forest), (D) Mixed Nothofagus betuloides and N. pumilio forest, (E) Nothofagus pumilio forest, (F) Nothofagus krummholz, (G) Scrub-grassland, (H) Fellfield, (I) Bare scoria. Vegetation zonation indicated in column from Heusser (1974, 1981). Numbers on diagram are beetle fauna collecting localities. Coleoptera Bioassociations II-V are from this study.

Ecological information, such as host plant information, density and type of vegetation, was recorded for each habitat.

More than 9000 individuals were collected and in the taxonomic study that followed, these were assigned to 484 taxa in 47 families. The families especially well represented are the Carabidae (62 taxa), Staphylinidae (54 taxa), Scarabaeidae (29 taxa) and Curculionidae (59 taxa). In order to determine if there are patterns in the distribution of the Coleoptera, the various collecting locations were compared with one another on the basis of their taxa. Several similarity co-efficients were calculated but the one chosen to represent the data because it emphasizes similarity, was the Dice coefficient (C_D) where $C_D = 2C/(N1 + N2)$; C = number of shared taxa; N1, N2 = numbers of taxa present in the first and second samples respectively. In order to construct a dendrogram and trellis diagram, the similarity coefficients were ordered using a WPGMA clustering technique (Fig. 3).

The results indicate five discrete clusters that are referred to as Bioassociations I-V. Bioassociations II-V are represented in the Parque Nacional de Puyehue (Fig. 2) and their boundaries coincide approximately with the boundaries of the vegetational zones: Bioassociation II represents the low-elevation disturbed-vegetation beetle fauna and Bioassociation V the tundra beetle fauna. To a large extent, the results of the cluster analysis confirm what was expected, namely that the beetle fauna would be zoned. The method, however, provides an independent and unbiased way of assigning divisions to the zonation.

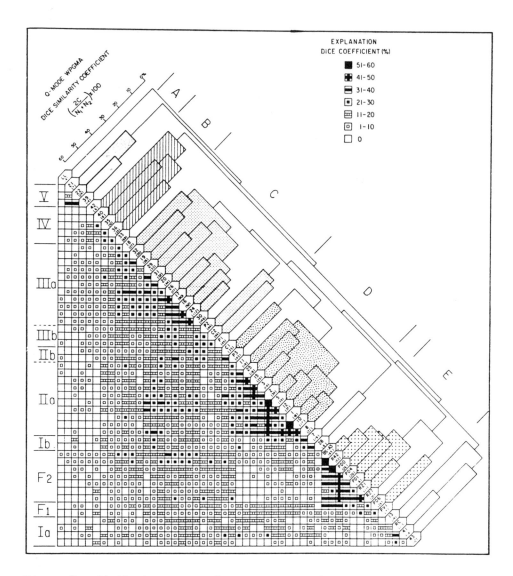

Figure 3. Similarity matrix (trellis diagram) and binary tree (dendrogram) showing patterns of similarity among 41 beetle fauna collecting sites and 8 fossil assemblages. Large numbers along diagonal are collecting localities, small numbers in parentheses are taxa collected. Clusters are: A, Bioassociation V; B, Bioassociation IV; C, Bioassociation III; D, Bioassociation II; E, Rio Caunahue fossil assemblages (PD6, PD4, R42, PD1, PD0, R21 - Fig. 5).

The diversity of the fauna (Fig. 4), if the low elevation disturbed ground species are excluded, decreases with elevation. This was also expected and probably results from a decrease in the diversity of habitats. The percentage of taxa unique to each zone is highest in Bioassociation III (56 %) and Bioassociation V (48 %) which correspond to montane forest and alpine tundra habitats, respectively.

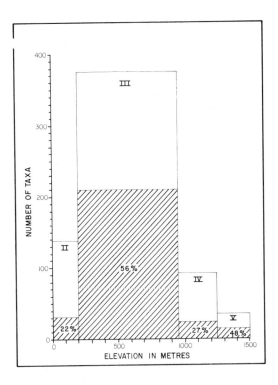

Figure 4. Diversity diagram showing the number of taxa occurring in each bioassociation (II-V) and the number and percent of taxa restricted to each bioassociation.

This demonstrates that the bioassociations can be defined qualitatively as well as quantitatively.

The close parallelism between Coleoptera Bioassociations and vegetational zones (Fig. 2) might suggest that the faunal zonation was controlled by the distribution of the flora. In addition, many species of beetles are dependent on plants, as a source of either food or shelter. However, there are as many beetles in the fauna that do not have a specific dependence on plants, and at any rate, dependence is not unidirectional. For example, many species of plants and trees, particularly in Chilean rain forests, owe their existence to insect pollination. The interdependence that is implied suggests that other factors, especially climatic ones, are responsible for both the floral and faunal zonation, and that the distribution developed mutually.

FOSSIL ASSEMBLAGES IN THE LAKE REGION

Paleoclimatic inferences from fossil assemblages are facilitated by the discovery of modern analogues. Several fossil assemblages have now been examined from low elevation sites in the Valle Longitudinal. Those that pre-date the last major ice advance into the lowlands are different from those that colonized the region following deglaciation. Assemblages from Rupanco, 24 000 BP and 28 490 BP, Bella Vista Park, Puerto Varas, 15 715 BP, and Dalcahue, Chiloe, 14 970 BP contain rela-

tively few species and are dominated by individuals of the weevil
Listroderes dentipennis Gmn. The low diversity suggests that the fauna
is from a cold climate, an inference supported by the palynological
data of Heusser et al. (1981).

The absence of an analogue for this fauna in the Lake Region reveals
an obvious weakness in our method for interpreting paleoclimates. The
assumption that knowledge of the elevational distribution of the fauna
in the Lake Region would provide an adequate base for interpretation
was clearly incorrect for these earlier faunas. The possibility exists
that significant latitudinal shifts in the distribution of the fauna
may have occurred. The fauna that colonized the region following
deglaciation of the lowlands was much more diverse. The difference in
faunas may represent the change from a glacial to an interglacial
climate.

RIO CAUNAHUE ASSEMBLAGE

The Rio Caunahue sections, cutbanks on the west side of the river,
expose one of the most complete Post-glacial sections in southern
Chile. The Rio Caunahue enters Lago Ranco east of Futrono, and the
sections are located approximately 2 km upstream from where Route T-55
crosses the bridge over the river. Approximately 13 m of laminated
silts and clays with ash layers and organic horizons are exposed below
a prominent terrace. The lower 4 m of silts and clays are inorganic
and overlie a coarse gravel consisting of angular and subrounded boulders of igneous lithologies. A date of 12 810 BP (Mercer pers. comm.,
1983) from near the base of the organic part of the section replaces an
earlier date of 14 635 BP (Hoganson and Ashworth 1982).

A tentative history for the site is suggested by stratigraphy and
geomorphological relationships: After glacial retreat at about 14 000
BP, downslope movement of screes and till formed the basal valley
fill. A narrow arm of Lago Ranco filling the lower stretches of the
Caunahue Valley came into existence as ice retreated from the main lake
basin. The lake was presumably at the level of the high terrace at
this time. The earliest sediments formed in a lake that had no or only
sparse marginal vegetation. Colonization of the lake shores is marked
by the appearance of organics in the section at about 13 900 BP.
Silts, clays and organic detritus transported to the lake by the river,
occasionally supplemented with volcanic ashes from different sources,
accumulated in the lake until at least 4525 BP. During the Late Holocene, a morainic dam on the west side of Lago Ranco was breached. As a
result, water levels lowered, and the Rio Caunahue started to downcut
in the soft lake sediments. As erosion continued, a sediment-filled
gorge was flushed of its fill and the river abandoned its western
channel for lower exits to the lake.

The early Post-glacial interval of the Rio Caunahue section, 13 000 -
10 000 BP, is represented by the lower 3.4 m of organic-bearing silts
and clays (Fig. 5). The contact with the coarse gravel is about 4 m
lower in section than the organic horizon dated at 12 810 BP. The 4 m
of intervening laminated silts and clays are completely inorganic. A
prominent 4 cm band of white ash that occurs in the sequence enables
the comparison of the section from which beetles were studied with the
pollen profile described by Heusser (1981). Sediment samples for the
Coleoptera study were collected in 4 kg, 10 cm blocks through the part
of the section dated 12 000 - 10 000 BP, and from thin layers of con-

centrated organic debris in the interval 12 810 - 12 000 BP (Fig. 5).
Several thousand well-preserved beetle fragments, mostly disarticulated
heads, pronota and elytra, were recovered from the sediments by wet
sieving (300 micron mesh) and kerosene (paraffin) floatation following
procedures described by Ashworth (1979). The fossils, representing at
least 179 taxa in 37 families (Table 1), were identified by comparison
with modern specimens from our own reference collection, the
collections of the USNM (Smithsonian Institution), and from the private
collections of systematic experts. The fossils are part of the
collection of the North Dakota State University Fossil Beetle Laboratory.

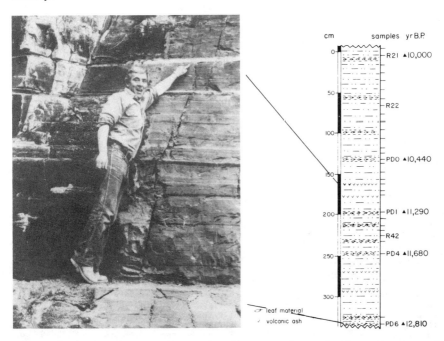

Figure 5. Lithology, radiocarbon dates and sample locations at the Rio
Caunahue site. The photograph is adjacent to the section that was
sampled. Ashworth is standing on the horizon dated at 12 810 BP and is
pointing to a prominent ash bed. The dark bands are the leaf layers.

PALEOCLIMATE IMPLIED BY THE RIO CAUNAHUE COLEOPTERA

The Post-glacial environments of the Rio Caunahue Valley can be inferred from an integration of the habitat data of the species represented by fossils. The majority of the taxa are forest inhabitants, a few are aquatic, and a few are restricted to water-marginal habitats.

Aquatic taxa are rare and, with the exception of the lowest interval, dominated by running-water species, the elmids. The presence of running-water species, and the virtual absence of standing-water species in a depositional environment that was clearly lacustrine from the laminated sediments, implies transport and accumulation of the fossils in relatively deep, quiet water.

Shoreline taxa are represented by carabids such as Bembidion and Agonum and various staphylinids, including Nomimoceras marginicollis. Some species have preferences for open, sandy spots, others for wet,

Table 1. Fossil taxa from the Rio Caunahue site. Systematic order follows Blackwelder (1944-47). Numbers in columns refer to minimum numbers of individuals. Age of the sample intervals is in thousands of years BP. A) 10-11; B) 11-12; C) 12-13. The elevational range of existing species in the Parque Nacional de Puyehue is shown by * : below 650 m; ** : up to 1000 m; *** : above 1000 m; no asterisk: throughout elevational range; and + : not collected during the modern survey.

COLEOPTERA TAXA	A	B	C	COLEOPTERA TAXA	A	B	C
TRACHYPACHIDAE				Stenus chilensis Ben.+	1		
Systolosoma brevis Sol.	1			Stenus spp.	2	3	1
CARABIDAE				Baryopsis araucanus C.& S.*	1		1
Ceroglossus valdivia (Hope)		1		Cheilocolpus sp. 3			1
Ceroglossus sp.		1		Cheilocolpus spp.	1		
Creobius eydouxi Guér.	1	1	1	Loncovilius (Lienturius) sp.**	3	2	
Bembidion cf. posticalis Gmn.*			1	Quediinae gen. indet.		1	1
Bembidion setiventre Neg.*	4	2	1	cf. Bolitobius asperipennis			
Bembidion sp. 5*		4	1	C.& S.+		2	
Bembidion sp. 8*		1	1	Leucotachinus luteonitens			
Bembidion spp.		2	3	(F.& G.)*	6	22	11
Aemalodera cf. centromaculata				Nomimoceras marginicollis			
Sol.**			1	(Sol.)**	2	1	7
Aemalodera spp.		2	1	Aleocharinae sp. 3*		1	2
Trechisibus nigripennis Group				Aleocharinae sp. 8*		1	
Sol.	2	2	2	Aleocharinae sp. 9*	2	1	3
Trechisibus sp.		1		Aleocharinae sp. 10*			1
Trechinotus striatulus M.& N.*		2	1	Aleocharinae group 1	1	1	1
Trechini gen. indet.	1	5		Aleocharinae gen. indet.	8 6	60	31
Trirammatus (Feroniomorpha)				Staphylinidae gen. indet.	9	10	4
sp. 1*	1			**PSELAPHIDAE**			
Trirammatus sp.*			1	Dalminiomus araucanus Jean.*		1	
Parhypates (sensu stricto) sp.	1	4		Dalminiomus spp.*	2 1	7	2
Metius sp.**			1	Achillia spp.	2 9	38	17
Abropus carnifex Fabr.		1	1	Tyropsis spp.**	8	25	13
Agonum sp. 1*		1		Pselaphidae gen. indet.	2	3	1
Agonum sp. 4*		1	1	**LUCANIDAE**			
Agonum spp.	1	3	1	cf. Sclerognathus femoralis			
Pelmatellus (sensu lato) sp.1	1			Geur.*		1	
Bradycellus (Goniocellus) sp.			1	Sclerognathus sp.		1	
Bradycellus (Stenocellus) sp.*	?	2		**SCARABAEIDAE**			
Bradycellus sp.	?		1	Sericoides chlorosticta Blanch.**			1
Carabidae gen. indet.	2	4	1	Sericoides viridis Sol.*	5	7	2
DYTISCIDAE				Sericoides sp. 4			1
Lancetes sp.**		1		Sericoides sp. 5*		1	
HYDROPHILIDAE				Sericoides sp. 10*	1		?
Hydrochus stolpi Gmn.*			1	Sericoides spp.	3	5	4
Enochrus vicinus So.			1	Scarabaeidae gen. indet.	1	2	1
Enochrus sp.			1	**HELODIDAE**			
PTILIIDAE				cf. Microcara sp.	?	10	2
Ptiliidae gen. indet.			1	Prionocyphon sp.		1	
LEIODIDAE				Helodidae gen. indet.	1	4	
Dasypelates sp.1**	7	11	4	**ELMIDAE**			
Dasypelates spp.**	?	2	?	Austrolimnius chiloensis			
Eunemadus chilensis Ptvn.	1	4	?	(Champ.)*	1	1	
Colon sp.*		1		Austrolimnius sp.*	3	2	
Leiodidae gen. indet.	1	3	1	Stethelmis sp.	3	1	
SCYDMAENIDAE				**ELATERIDAE**			
Euconnus spp.*	?	3	2	Semiotus luteipennis Guer.		1	
STAPHYLINIDAE				Deromecus sp. 1*			1
Pseudopsis cf. adustipennis				cf. Deromecus spp.		2	1
F.& G.+	1	1	1	cf. Medonia sp.*		1	
cf. Pseudopsis spp.+	3	3	1	Negastrius sp.*	1		2
Glypholoma pustuliferum Jean.			2	Elateridae gen. indet.	3	11	4
Omaliopsis spp.	?	2	1	**LAMPYRIDAE**			
Omaliinae gen. indet.	2	3	2	Pyractonema nigripennis			
Carpelimus spp.+	8	6	5	group Sol.		1	
Thinodromus sp. 1*	1			**DERONDONTIDAE**			
Thinodromus sp. 3*		1	1	Nothoderodontus dentatus			
Thinodromus sp. 4*	6	3	2	Lawr.+	4	6	5
Thinodromus spp.*	7	12	9	**CANTHARIDAE**			
cf. Oxytelinae gen. indet.+			1	Hyponotum cf. violaceipenne			
				(Pic)***		1	

Table 1 (cont.).

COLEOPTERA TAXA	Sample Intervals		
	A	B	C
Hyponotum spp.	6	14	11
Micronotum nodicorne (Sol.)	2	2	1
Plectocephalon testaceum (Pic)	1	1	1
cf. Cantharidae gen. indet.	1		
ANOBIIDAE			
Caenocara spp.*	1	10	2
Strichtoptyohus cf. brevicollis (Sol.)*		1	
Strichtoptyohus sp.*		2	
BOSTRICHIDAE			
Bostrichidae gen. indet.+		3	
TROGOSITIDAE			
cf. Diontolobus sp. 1**	3	5	3
cf. Diontolobus sp. 2**	4	6	
cf. Diontolobus sp.**	1		
MELYRIDAE			
Dasytes haemorrhoidalis Sol.**	5	3	1
Melyridae gen. indet.**	2		
NITIDULIDAE			
Brachypterus n. sp.*	1		?
Perilopsis flava Rtr.*	25	61	21
Cryptarcha sp. 1*		1	
Cybocephalus sp.+	1	1	?
Nitidulidae gen. indet.	3	4	5
RHIZOPHAGIDAE			
cf. Rhizophagidae gen. indet.+	2	1	
CRYPTOPHAGIDAE			
Pseudochrodes suturalis Rttr.*	2	1	1
Cryptophagidae gen. indet.	1		1
COCCINELLIDAE			
Rhizobius chilianus Mader*		1	
Orynipus spp.**	1	2	
Adalia kuscheli Mader*		1	
Adalia spp.*		3	1
Sarapidus cf. australis Gord.*	1	5	
Strictospilus darwini Brths.**	16	42	5
LATHRIDIIDAE			
Aridius heteronotus (Belon)	1	1	
Lathridiidae gen. indet.	11	9	3
COLYDIIDAE			
cf. Colydiidae gen. indet.+	5	2	1
SALPINGIDAE			
Cycloderus rubricollis Sol.*	1	1	
Vincenzellus sp.		1	
MELANDRYIDAE			
Orchesia sp. 1	14	44	17
Orchesia sp. 2*	2	3	2
MORDELLIDAE			
Mordellidae cf. sp. 3*			1
Mordellidae gen. indet.*	5	5	1
CERAMBYCIDAE			
Hoplonotus spinifer Blanch.*	1		
CRYSOMELIDAE			
Pachybrachis sp.*	1		
Strichosa eburata Blanch.**		1	
Altica sp. 1			1
Altica sp. 2***		1	
Alticinae sp. 1	2	1	
Alticinae gen. indet.		1	1
cf. Crepidodera sp. 1*	5	1	1
cf. Crepidodera sp.*	1		
Chrysomelidae gen. indet.	?	1	2
NEMONYCHITOMACER			
Rhynchitomacer flavus Voss+	8	21	2
Rhynchitomacer fuscus Kusch.+		1	3
ANTHRIBIDAE			
Ormiscus parvulus (Blanch.)*		1	
Anthribidae gen. indet.	2	9	3

COLEOPTERA TAXA	Sample Intervals		
	A	B	C
SCOLYTIDAE			
Pityophthorus sp.+	1		
Amphicranus sp.+	1		
cf. Corthylus sp.+	1		
Gnathotrupes cf. sextuberculatus Sch.+		1	
Gnathotrupes spp.*	10	6	5
Phloeotribus cf. spinipennis Sch.+	3	15	4
Scolytidae gen. indet.*	1	5	2
ATTELABIDAE			
Eugnamptoplesius violaceipennis (F.& G.)*	1	2	1
Minurus testaceous Wtfh.*	1	2	1
CURCULIONIDAE			2
Nototactus angustirostris Kusch.**	1	3	2
Polydrusus nothofagi Kusch.*	1	1	1
Dasydema hirtella Blanch.	4	3	1
Listroderes dentipennis Gmn.			1
Tartarisus signatipennis (Blanch.)**		3	
Nothofagobius brevirostris Kusch.		1	
Nothofaginus lineaticollis Kusch.			1
Neopsilorhinus sp.*			1
Rhopalomerus tenuirostris (Blanch.)**	3	1	
cf. Omoides sp.**			1
Epaetius carinulatus Kusch.*	1		
Notiodes sp.+	2	4	2
Aoratolcus estriatus Kusch.+	1	1	
Wittmerius longirostris Kusch.+	2	20	5
Erirrhininae gen. indet.	17	41	10
Apion spp.	?	7	8
Allomagdalis cryptonyx Kusch.+		1	
Berberidicola crenulatus (Blanch.)**		1	1
Berberidicola exaratus (Blanch.)**		2	1
Psepholax dentipes (Blanch.)*		1	
Lophocephala fasciolata Blanch.+	1		
Cryptorhynchinae gen. indet.	5	11	2
Pentarthrum castaneum (Blanch.)+	2	4	5
Curculionidae gen. indet.	3	6	7

swampy patches. The helodids, certain coccinellids and the weevil Notiodes occur on emergent vegetation.

Forest associated beetles dominate the assemblages. Members of forest floor communities occur throughout the section and are represented by large, flightless carabids such as Ceroglossus and Creobius, leiodids such as Dasypelates, numerous staphylinids, pselaphids, scydmaenids, elaterids and apterous leaf litter weevils of the subfamily Cryptorhynchinae.

Species that are associated with understorey shrubs such as Fuchsia, Berberis and the bamboo, Chusquea, are abundant. Some of the species are phytophagous, others predaceous and some general scavengers. In this group is the carabid Abropus, the abundant scarab Sericoides, the nitidulid Perilopsis flava, and the weevil Berberidicola. Species of coccinellids, trogositids, attelabids, melandryids, mordellids and cantharids are also found on understorey vegetation.

Weevils that are associated with trees occur throughout the assemblages. Notatactus angustirostris, Polydrusus nothofagi, Nothofagobius brevirostris, Nothofaginus lineaticollis, Wittmerius longirostris, Allomagdalis cryptonyx occur on species of the southern beech, Nothofagus. Species that are associated with decaying wood are represented by bostrichids, lucanids and scolytids and those associated with fungus by anobiids and the derodontid, Nothoderodontus dentatus.

The Caunahue assemblages imply that the depositional environment from 12 810 - 10 000 BP was a lake with forested margins. The climate that prevailed during this time can be inferred by comparing the species composition of the fossil assemblages to their existing elevational distribution in the Parque Nacional de Puyehue. Of the 179 fossil taxa, 24 were not found in the survey of the modern fauna, 84 occur at elevations below 1 000 m and are restricted to Bioassociations II and III, only two are restricted to Bioassocation IV, and 69 occur in Bioassociations II-IV. Individuals of the latter group are generally more abundant at lower elevations.

Dice coefficients for similarity were calculated for selected fossil samples spanning the interval from 12 810 BP to 10 000 BP. They were then compared, using WPGMA clustering techniques, to the modern samples collected in the Parque Nacional de Puyehue. The fossil samples form a discrete cluster which shows the greatest similarity with the samples of Bioassociation III, the fauna of the rain forest. These results were obtained by reading the matrix of the trellis diagram (Fig. 3).

Both types of analyses, one using the ranges of individual taxa and the other a less biased mathematical comparison of groups of taxa, indicate that the climate of the early Post-glacial in the Rio Caunahue Valley was relatively constant and similar to that of the present. This inference is supported by Heusser's (1981) pollen study of the Rio Caunahue site.

The Rio Caunahue pollen spectrum, which appears to span the entire early Post-glacial interval, was divided into three zones. The disappearance of the pollen of Podocarpus andinus and a slight decrease in the pollen of Graminae appear to be the reasons for defining a lower zone (RC-3), and a slight increase in the pollen of Weinmannia appears to be the principal reason for defining an upper zone (RC-1). These changes in pollen could imply a slight amelioration of climate during the early Post-glacial interval, but mostly they imply a constant climate.

The conclusion implied by both beetles and pollen is that the climate of the early Post-glacial in the Rio Caunahue Valley was relatively constant and not too different from that of the present.

DISCUSSION

Heusser and Streeter (1980) derived regression equations that related pollen taxa from surface samples to temperature and precipitation. The equations were applied to the fossil pollen from a lake core at Alerce, in southern Chile. The derived paleotemperature curve shows that during the early Post-glacial, temperatures ameliorated by as much as $10\,^\circ C$, reaching their maximum at 11 300 BP. The warm peak was followed by a cold phase that lasted until 9500 BP. The lowest temperatures occurred about 10 200 BP. Thus Post-glacial warming was interrupted by a significant cold phase. The difference in calculated temperatures between the warm and cold peaks was considerably greater than had previously been reported by Heusser (1966, 1974). Precipitation during the cold phase was calculated to be 2000 mm higher than it is at present. Heusser and Streeter (1980), however, state that this figure may be exaggerated. The Alerce site is 142 km south of the Rio Caunahue site and both sites are at low elevations in the Valle Longitudinal. The climate at any particular time would be expected to be similar at the two locations and certainly would not be expected to be as different as it appears to have been in the interval 11 000 - 9500 BP.

In attempting to resolve the disparity between the interpretations, two positions can be taken. The first is that the paleoclimatic inferences for both locations are correct. The second is that one of the interpretations is incorrect. The possible premise that both interpretations are incorrect will not be considered in this discussion. In considering the first position, there appear to be two possible scenarios. The first scenario is that warm local conditions existed in a regionally cold climate, and the second, its corollary, is that locally cold conditions existed in a regionally warm climate. Modification of climate occurs frequently but the possible causes in this situation, either proximity to a glacier or to hot springs, could not account for the magnitude of the differences.

Having ruled out the first premise as unlikely, we are left with the less popular choice that one of the interpretations is questionable. At this point it is difficult to have an unbiased appraisal of the situation. At both sites, however, there are conditions that could mask or possibly make paleoclimatic interpretation difficult.

It is conceivable that most of the fossils in the Caunahue sediments were transported to the lake by the river. The fossil assemblages are a mixed bag of habitat specialists that, potentially at least, could incorporate Coleoptera from different climatic regimes. The observation is, however, that the assemblages are completely dominated by low-elevation taxa. Mixing on the scale that could affect a paleoclimatic interpretation did not occur. It is worth noting, however, that if there had been a problem with contamination, it could only have been by the addition of species from higher elevations.

It could also be argued that the relatively constant nature in the composition of the Rio Caunahue beetle assemblages might result from all the taxa having very broad tolerances to climatic factors. In other words, climatic changes would be masked because of the insensitivity of the indicators. The tremendous changes in the fauna that occur through a $3\,^\circ C$ difference in mean January temperatures in the Parque Nacional de Puyehue suggest, however, that the beetles are very sensitive to change. There is no doubt that the lowering of temperatures proposed by Heusser and Streeter (1980) would have caused the extinction of the Coleoptera species represented by fossils in the Caunahue assemblages.

The quantitative paleoclimatic data of Heusser and Streeter (1980) were derived from pollen diagrams published earlier by Heusser (1966). A comparison of the paleoclimatic curves for the interval 11 000 - 9500 BP with the actual pollen diagrams leaves us with the following question: How can such small changes in the pollen profile imply such major changes in climate? The pollen changes entail a slight increase in Nothofagus dombeyi type, a slight increase in Podocarpus nubigenus, a larger increase in Myrtaceae, a decrease in Tepualia, and a reduction in Weinmannia.

Birks (1981) discusses the pros and cons of the multivariate approach to palynology at length and concludes that one of the major assumptions of the method, that plant distributions are in equilibrium with climate, has been invalidated by several recent studies. Heusser and Streeter (1980) do not indicate that the changes in the Alerce pollen profile could be caused by any factors other than climatic ones. Disease, succession and a differential migration ability are all known to affect changes in pollen profiles, yet their possible effects at the Alerce site are not discussed. It is a matter of speculation whether a literal interpretation of pollen-climate transfer function data is possible if changes induced by non-climatic factors are not filtered out.

Evidence from other sources can invariably be found to support a particular argument. Thus Heusser and Streeter (1980) cite evidence from Peru, New Guinea, New Zealand and South Georgia that supports a climatic deterioration between 11 000 and 10 000 BP. In support of the argument for relatively constant conditions during the early Post-glacial we cite the pollen studies of Markgraf (1979) for adjoining parts of Argentina and the glacial geological studies of Mercer (1976, 1983) for the Lake Region of Patagonia.

SUMMARY

Both beetle and pollen evidence from the Rio Caunahue site indicate that the climate of the Chilean Lake Region during the early Post-glacial, 13 000 - 10 000 BP was relatively constant, with temperatures similar to those of today.

The evidence for a major climatic deterioration in southern Chile equivalent in age to the Younger Dryas needs to be re-evaluated in the light of the evidence from Rio Caunahue.

In order to resolve the problem of conflicting evidence, a site satisfying the requirements of a combined pollen, macroscopic plant, and beetle study needs to be located in the western parts of the Valle Longitudinal, away from the influence of glaciers.

ACKNOWLEDGEMENTS

This study was supported by National Science Foundation grants ATM 77-14683 to Ohio State University and ATM 78-20372 to North Dakota State University. We are especially grateful to Howard Mooers, John Mercer, and to Nano, Tito and Maria of the Aguilar Gomez family of Puerto Varas. Coleopterists who assisted with identifications are too numerous to mention but Guillermo Kuschel, Charlie O'Brien and the staff of the Smithsonian Institution and the Agriculture Canada insect collections deserve our special thanks.

REFERENCES

Ashworth AC 1979. Quaternary Coleoptera studies in North America: past and present. In: TL Erwin, GE Ball and R Whitehead (eds.), Carabid Beetles. Dr W Junk, The Hague. p. 395-504.

Birks HJB 1981. The use of pollen analysis in the reconstruction of past climates: a review. In: TML Wigley, MJ Ingram and G Farmer (eds.), Climate and History. Cambridge University Press, Cambridge. p. 111-138.

Blackwelder RE 1944-47. Checklist of the coleopterous insects of Mexico, Central America, the West Indies and South America, Parts 1-5. Smithsonian Inst., U. S. Nat. Mus. Bull. 185:1-925.

Casertano L 1963. Catalogue of the active volcanoes and solfatara fields of the Chilean continent. In: International Volcanological Association catalogue of the active volcanoes of the world, Part XV.

Coope GR 1967. The value of Quaternary insect faunas in the interpretation of ancient ecology and climate. In: EJ Cushing and HE Wright Jr. (eds.), Quaternary Paleoecology. Yale University Press, New Haven. p. 359-380.

Coope GR 1970. Interpretations of Quaternary insect fossils. Ann. Rev. Entomol. 15:97-120.

Coope GR 1975. Climatic fluctuations in northwest Europe since the Last Interglacial, indicated by fossil assemblages of Coleoptera. In: AE Wright and F Moseley (eds.), Ice Ages: Ancient and Modern. Seel House Press, Liverpool. p. 153-168.

Coope GR 1977. Fossil coleopteran assemblages as sensitive indicators of climatic changes during the Devensian (last) cold stage. Phil. Trans. R. Soc. Lond. B. 280:313-340.

Coope GR 1978. Constancy of insect species versus inconstancy of Quaternary environments. In: LA Mound and N Waloff (eds.), Diversity of Insect Faunas. Blackwell Scientific Publications, Oxford. p. 176-186.

Coope GR 1979. Late Cenozoic fossil Coleoptera: Evolution, Biogeography and Ecology. Ann. Rev. ecol. Syst. 10:247-267.

Coope GR 1981. Episodes of local extinction of insect species during the Quaternary as indicators of climatic change. In: J Neal and J Flenley (eds.), The Quaternary of Britain. Pergamon Press, Oxford. p. 216-221.

Coope GR and Brophy JA 1972. Late glacial environmental changes indicated by coleopteran succession from North Wales. Boreas 1:97-142.

Eldridge N and Gould SJ 1972. Punctuated equilibria: an alternative view to phyletic gradualism. In: TJM Schopf (ed.), Models in Paleobiology. p. 82-115.

Heusser CJ 1966. Late Pleistocene pollen diagrams from the Province of Llanquihue, southern Chile. Proc. Amer. phil. Soc. 110:269-305.

Heusser CJ 1974. Vegetation and climate of the southern Chilean lake district during and since the last interglaciation. Quat. Res. 4:290-315.

Heusser CJ 1981. Palynology of the last Interglacial-Glacial cycle in midlatitudes of southern Chile. Quat. Res. 16:293-321.

Heusser CJ and Flint RF 1977. Quaternary glaciations and environment of northern Isla Chiloe, Chile. Geology 5:305-308.

Heusser CJ and Streeter SS 1980. A temperature and precipitation record of the past 16 000 years in southern Chile. Science 210:1345-1347.

Heusser CJ, Streeter SS and Stuiver M 1981. Temperature and precipitation record in southern Chile extended to ∼43 yr ago. Nature 294:65-67.

Hoganson JW and Ashworth AC 1982. The Late-glacial climate of the Chilean Lake Region implied by fossil beetles. Proc. Third North Am. paleont. Conv. 1:251-256.

Laugenie C and Mercer JH 1973. Southern Chile: chronology of the last glaciation. In: Abstracts, IX INQUA Congress, Christchurch. p. 202-203.

Lindroth CH 1956. Movements and changes of area at the climatic limit of terrestial animal species. In: KG Wingstrand (ed.), Bertil Hamström, Lund. p. 226-230.

Markgraf V 1979. Paleoclimatic reconstruction of the last 15 000 years in subantarctic and temperate regions of Argentina. IV Symposium des Palynologues de Langue Francaise: Palynologie et Climats, Paris.

Markgraf V and Bradbury JP 1982. Holocene climatic history of South America. Striae 16:40-45.

Mercer JH 1972. Chilean glacial chronology 20 000 - 11 000 carbon-14 years ago: some global comparisons. Science 176:1118-1120.

Mercer JH 1976. Glacial history of southernmost South America. Quat. Res. 6:125-166.

Mercer JH 1982. Holocene glacier variations in southern South America. Striae 18:35-40.

Mercer JH 1984. Simultaneous and near-equal climatic change in the two hemispheres in last glacial and in last glacial and interglacial times: implications for Milankovitch theory and future CO_2-induced warming. In: Climate processes and climate sensitivity. AGU geophys. mon. ser. (in press).

Mercer JH, this volume. Late Cainozoic glacial variations in South America south of the equator.

Morgan AV, Morgan Anne, Ashworth AC and Matthews JV Jr. 1983. Late Wisconsin insects: a review of fossil beetles in North America. In: HE Wright (ed.), Late Quaternary environments of the United States, Univ. Minnesota Press, Minneapolis. p. 354-363.

Muñoz SM 1980. Flora del Parque Nacional Puyehue. Editorial Universitaria, S.A. Santiago, Chile: 1-557.

Porter SC 1981. Pleistocene glaciation in the southern Lake District of Chile. Quat. Res. 16:263-292.

Veblen TT and Ashton DR 1979. Successional pattern above timberline in south-central Chile. Vegetation 40:39-47.

Veblen TT, Ashton DR, Schlegel FM and Veblen AT 1977. Plant succession in a timberline depressed by vulcanism in south-central Chile. J. Biogeogr. 4:275-294.

Veblen TT, Schlegel FM and Escobar RB 1980. Structure and dynamics of old-growth Nothofagus forests in the Valdivian Andes, Chile. J. Ecol. 68:1-31.

Villagran C 1980. Vegetationsgeschichte und pflanzensoziologische Untersuchungen im Vincente Pérez Rosalew Nationalpark (Chile). Diss. Botanicae 54:1-165.

Late Glacial glaciation and the development of climate in southern South America

W.LAUER & P.FRANKENBERG
Universität Bonn, Germany

ABSTRACT. Evidence for an abrupt glacial advance in both the SW Altiplano of Bolivia and in southern Chile at about 13 000 BP suggests a cold set-back in the Southern Hemisphere during a generally humid period. Circulation models are presented to explain this phenomena.

INTRODUCTION

It can be proved that reglaciation in the mountains of South America took place during the Late Glacial phase. This apparently short glaciation process set in relatively uniformly between approximately 14 000 and 11 000 BP. The event is confirmed by moraines at the edge of the Altiplano in Bolivia, Columbia, and Mexico as well as in the extratropical lake area of southern Chile.

This brief study deals with the Late Glacial development of the Bolivian Altiplano and of the lake area of southern Chile. It can be shown that the Late Glacial advances of glaciers occurred almost simultaneously and that they can be assigned to a certain zonation of climatic conditions. Therefore, an attempt is made to find a circulation model by means of which the almost simultaneous advance in different climatic zones at about or after 13 000 1 000 BP can be explained.

Since the two areas under discussion now have different types of climate (Bolivia is located in the arid marginal tropics and southern Chile in the fully humid ectropics), it is probably that the phases of glacier advance did not occur at exactly the same time, but only when the hydro-thermic conditions favoured such a quasi-simultaneous glacier advance in the respective topographical situations.

BOLIVIA

Recently it was possible to identify evidence of a Late Glacial glaciation in the Bolivian East Andes (Lauer 1977) and to relate this to a renewed peak water level of Lake Tauca on the Bolivian Altiplano (Servant and Fontes 1978). Very distinct and younger glacial valleys were cut into the morphological systems of the older phase of the most recent glacial period in the form of tongue moraines, whose moraine edges are located at an elevation between 4300 and 4500 m altitude in the flat foreland, but extend down to 3800 m in the individual valleys (Fig. 1).

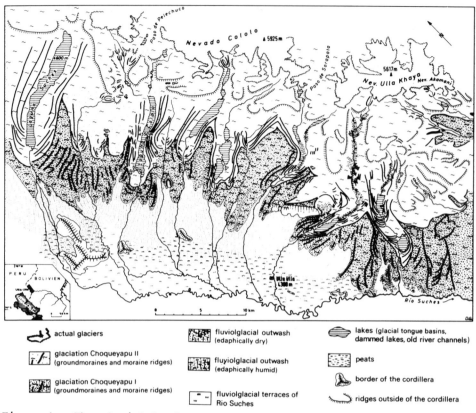

Figure 1. The glacial landscape of the Apolobamba-foreland, Bolivia (Orig. W Lauer).

The young morphological shape shows extremely fresh erosion and accumulation forms. The long lateral moraines bear deep gullies, channels and scores. They mostly accommodate tongues or secondary lakes or large moorlike areas. The strong rinsing did not permit the development of a front moraine but formed erratic fluvioglacial deposits. These forms are certainly the result of a rapidly advancing and equally rapidly deteriorating glacier system, which in the course of a short and cold but mainly more humid period created sharp forms. These are forms of so-called "warm-glaciers", whose formation is interpreted as evidence of a humid climate.

Pollen analyses and C14 dates from the Apolobamba-foreland obtained by Kurt Graf (1981) prove that the more recent glaciation complex was deposited before 11 000 BP, thus fixing the terminus ante quem. Since it is possible at the same time to assign the young and fresh phase of glacier advance to the Tauca-maximal stage of the glacial highland-lake, as was proved by Servant and Fontes (1978), it would be reasonable to assume that the glacier advance took place within a short and humid period between 13 000 and 11 500 BP. This Late Glacial phase is called Choqueyapu II (Troll and Finsterwalder 1935; Servant and Fontes 1978).

From these new findings one may draw the conclusion that no extensive glaciation of the foreland took place during the Last Glacial Maximum between 20 000 and 14 000 BP, since the very dry and cold climate of this phase did not permit the formation of long glacier tongues. Thus, it is proposed that long glacier tongues could only develop in the arid regions of the marginal tropics, if a temperature decrease was accompanied by a humid phase.

Kessler (1963) has proved that the high water level of Lake Minchin, which corresponds to the older complex of the moraines of the Last Glacial period, could only occur at a temperature of 6° C lower than today's if the water budget was balanced by an increase of the relative humidity or precipitation. Kessler (this volume) has calculated a precipitation increase of approximately 40 % for the water level of Lake Tauca as compared with that of today, if the temperature decrease was 3° C.

Thus, while the older phases (Choqueyapu I) probably belong to the period between 40 000 and 30 000 BP (or 70 000 BP?), the above-mentioned tongue basins, which are distinctly shaped and in most cases filled with water, must be assigned to a Late Glacial glacier advance, which occurred between 13 000 and 11 000 BP (Choqueyapu II).

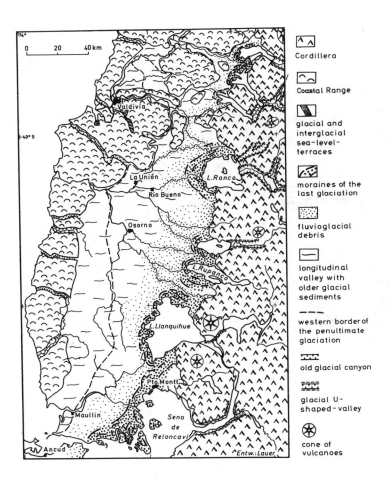

Figure 2. Glacial landscapes of the Chilean lake district.

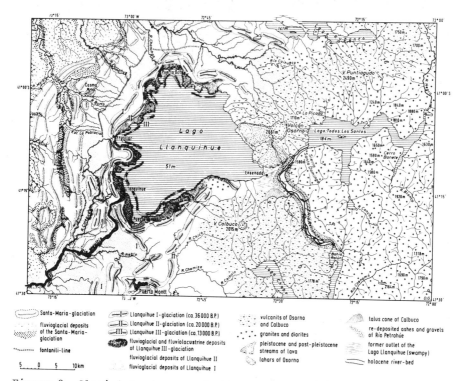

Figure 3. Glacial landscape of the Llanquihue lake region (Orig. W Lauer).

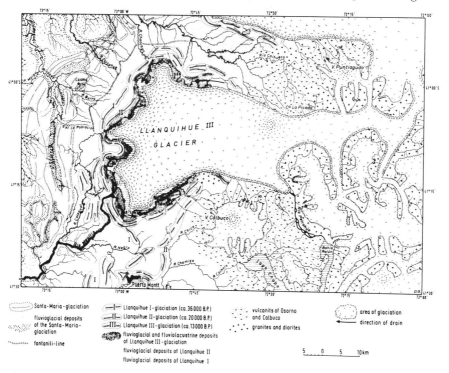

Figure 4. The stage of Llanquihue III glaciation (Orig. W Lauer).

SOUTHERN CHILE

The lake district of southern Chile today is a humid region. Recent studies by Mercer, Porter, Heusser and Laugenie have shown that the three glacier stages on the western edge of the southern Chilean lakes that were mapped by Lauer in 1957/58 (unpubl. ms.) belong to the Last Glacial period (Fig. 2). At Lake Llanquihue, two of the front moraines rest directly on the western edge of the lake, while a younger stage is located under water off the west bank of the lake (Fig. 3).

The moraines have been re-evaluated by Mercer (1976): Llanquihue I was probably deposited around 70 000 BP (Mercer) or between 40 000 BP and 30 000 BP (Porter), Llanquihue II between 20 000 BP and 19 000 BP, and Llanquihue III between 14 000 BP and 12 000 BP. The phase of the most recent glacier advance started around 15 700 BP and culminated at 13 000 BP (13 145 BP according to Mercer and Heusser).

At present the banks of the lake show a centripetally draining groove extending from north to south, which joins with the upper terrace of the present lake outlet (Rio Maullin), as was also confirmed recently by Porter (1981) (Fig. 4). Thus the water level of the lake must have risen more than 80 m. The ice-dammed lake drained via the Maullin system. The glacier then rapidly withdrew, leaving Lake Llanquihue completely free of ice at the beginning of the Holocene. However, the water level of the lake fluctuated several times, particularly as a result of the volcanic activities of the Calbuco and Osorno.

The level of the ice-dammed lake fell more and more, causing the lake to drain via Rio Petrohué before the ice had reached the rear area of the present lake. The water level probably decreased catastrophically as a result of the sudden formation of the new outlet at the rear end of the lake, probably to a level below the present mark, viz. to 30 m altitude (Fig. 5).

Underwater observations have shown that the lake is enclosed almost completely by a very wide terrace. The drainage groove to the Rio Petrohué could be found in the southeast corner of the lake, in the course of groundwater drillings carried out for a forestry company in 1954. The groove is located at approximately this depth. Thus, it is quite probable that the lake drained through this groove for a longer period during the Holocene. Volcanic activities of the Calbuco and Osorno resulted in the refilling of the groove during the Holocene and caused the lake to rise to the present level of 51 m.

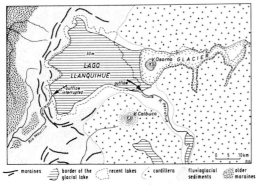

Figure 5. The Petrohué-stage of the Holocene drainage of Lake Llanquihue (Orig. W Lauer).

On the basis of the glaciomorphological findings from Bolivia and southern Chile, it would be appropriate to look for a circulation pattern that could have caused the relative synchronous events in different parts of the Andes.

We propose that a more humid climate than that of today should be assumed for the Late Glacial short-phased glacier advances, not only on the Bolivian Altiplano but also in the lake area of southern Chile. Since the Late Glacial glacier advance on the Altiplano occurred almost simultaneously with the increase of the lake water level, and since pollen diagrams of southern Chile, which is very humid anyhow, also suggest that the glacier advances were linked to an increase in humidity, it may be assumed that the circulation pattern for the Late Glacial phase between 13 000 and 11 000 BP changed in favour of a slight increase in precipitation, which accompanied the decrease in temperature, thus initiating the advance of the glaciers in both Bolivia and southern Chile.

For a better understanding of the atmospheric circulation at the time of the Late Glacial glacier advances around 13 000 BP it would seem to be necessary to proceed from the circulation conditions prevailing at the peak of the Last Glacial period (approximately 18 000 BP). This high glacial period was not only colder but also generally drier than today (see Gates 1976). Figure 6 shows the glaciation boundaries for the Antarctic as they existed in summer and winter 18 000 years ago and the present pack-ice and ice-shelf boundaries. At 18 000 BP the summer glaciation of the Antarctic amounted to ten times the area of today, the winter glaciation being only twice as large. Consequently the excessive expansion of the ice during summer was the decisive factor. The largest reduction of the ocean temperature was reflected in the South Atlantic by a far-reaching northward shift of the Arctic front, while further north, the water temperatures of the Atlantic remained almost unchanged as compared with those of today (CLIMAP 1976).

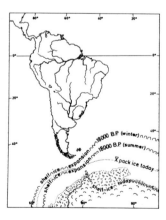

Figure 6. Modelling of the cover in the Antarctic 18 000 BP and at present (after different authors).

On the Pacific side of South America the water temperatures near the Chile and Peru Currents were strongly reduced by a maximum of 7.1 C, even west of the Galapagos Islands. This development was accompanied by an increase in the upwelling of the cold deep water due to the intensified Hadley cell (trade wind circulation) which was the result of an increase of the meridional temperature gradient between pole and equator.

Compared with today, the ectropic circulation elements shifted northward by 3-5 degrees latitude due to the extensive ocean glaciation of the Antarctic (Morley and Hays 1979; Heusser 1981; CLIMAP 1976). Of decisive importance for the circulation pattern at that time was probably the fact that at least during winter (July) the subtropic high pressure cell was located over the continent, whereas it is situated bicellularly over the Pacific and Atlantic oceans today (CLIMAP 1976). The circulation pattern outlined in Fig. 7a created an essential regional differentiation as a result of a general temperature and humidity reduction. The intensified upwelling off the west coast of South America led to a considerable cooling of the ocean surface in that area.

The strongly reduced ocean temperatures resulted in a sharp decline of the evaporation and, thus, a remarkable decrease of the precipitable water in the atmosphere, particularly in the areas of the westerlies which were shifted northward by five degrees latitude as compared with today. In the area of the luff-effects of the westwinds, a minor glaciation was nevertheless possible. Despite the very strong temperature depression, only short-block glaciers were formed at that time due to the aridity in the Andes of Bolivia which, like the entire area of central South America, were mainly under the influence of high pressure conditions, causing a strong reduction in precipitation.

During the high glacial period in Patagonia the intensification of the westwind drift led to an increase of the lee-effects, i.e. aridity. At approximately 18 000 BP, except for some luff areas in the Andes, much more arid conditions than today must have prevailed south of 20° S. For the more innertropic parts of South America the slight shift of the Atlantic-equatorial divergence suggests that the ITC, located in a balanced position between the Arctic and Antarctic glaciation, was not significantly different from its present position.

The outlined circulation pattern of the Antarctic Glacial Maximum about 18 000 BP, as has been explained repeatedly by other authors, will not be dealt with in detail. It probably prevailed with its dry and cool conditions until 17 000 BP (Heusser). A decisive improvement of climate took place after 15 000 BP (Lorius et al. 1979), corresponding to the Varas Interstadial in Chile and the Bölling Interstadial in Europe (see also Porter 1981). At that time the Andean glaciers of Chile had already melted down to 50 % of their high glacial state at 20 000 BP. This was accompanied by a rapid increase of the ocean temperatures (Emiliani 1955). According to Mercer (1978) the oceans of the Southern Hemisphere warmed up relatively quickly as a result of the small land masses and the insignificant terrestrial and oceanic glaciation after the solar radiation term, as defined by Milankovic, showed a positive trend again after 16 000 BP.

The Southern Hemisphere was able to convert the increased radiation yield more quickly into a warming of the ocean than the Northern Hemisphere. According to Hays (1978) the time-lag of the Northern Hemisphere as compared with the ocean temperatures of the Southern Hemisphere amounted to approximately 3000 years.

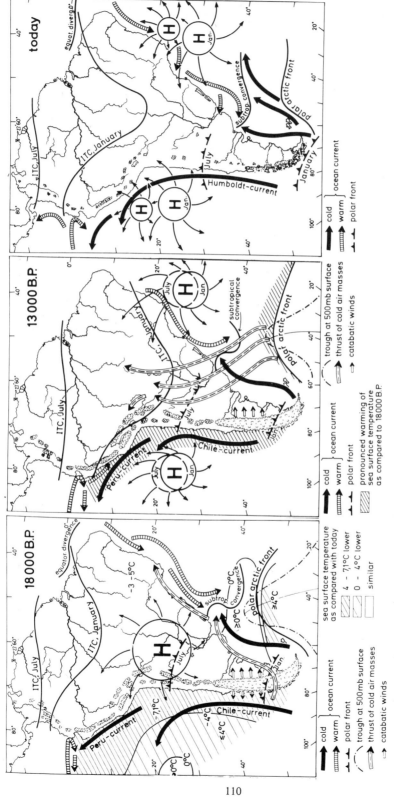

Figures 7. Modelling of the atmospheric circulation at about 18 000 BP (7a), at about 13 000 BP (7b), and at present (7c) (after different authors).

The initially mainly summer-melting processes of the west Antarctic oceanic ice shield resulted in an increase in sea level, which caused the ice of the west Antarctic to lose its ground contact, thus intensifying the melting processes. Ice surges occurred as described by Wilson (1978). The melting of the calved ice masses absorbed enormous quantities of energy from the atmosphere and from ocean waters, causing a further cold period between 13 500 BP and 12 500 BP, particularly in the coastal areas of southern South America, despite increased solar radiation. However, the water temperatures remained clearly above the level of the Glacial Maximum. This resulted in a considerable increase of evaporation (Henning and Flohn 1980). Due to the reduced south pole/equator thermal gradient the atmospheric circulation decreased, which in turn reduced upwelling off the west coast of South America and finally led to a significant increase of precipitation. This fact and the short cold set-back permitted the glaciers of the cordillera in Chile to advance abruptly again (Llanquihue III). In other words, this renewed advance was the result of the deglaciation of the Antarctic.

The proposed development of the climate in South America after the Glacial Maximum is supported by an isotope climatic record of the Dome-C ice core in East Antarctica by Lorius et al. (1979), which shows evidence of a sharp temperature rise after 15 000 BP. The record is in good agreement with Vostok and Byrd records of Antarctica (Dodson 1977; Webster and Streten 1978) and even with pollen analyses from New Guinea. This cold set-back coincides quite well with the glacier advances described above. The circulation pattern proposed for South America at 13 000 BP (Fig. 7b) suggests that the glaciers of Bolivia and southern Chile advanced abruptly despite a general improvement of climate. The decisive factor under already more humid climatic conditions was a short cold set-back due to the melting processes in the Antarctic. Perhaps the end of a glacial period will always show a cold set-back because of the melting processes.

The extensive melting of the Antarctic sea ice and relatively stable inland ice conditions in the Northern Hemisphere led to quasi-unipolar type of glaciation with a distinct southward shift particularly of the equatorial circulation elements and, probably, the subtropic high pressure cells. The asymmetry mainly occurred during summer, since the summer sea ice of the western Antarctic had disappeared. In addition, one must assume that the Northern and Southern Hemispheres showed different circulation intensities. The atmospheric circulation was stronger in the Northern Hemisphere and weakened already in the Southern Hemisphere (Kelvin-Bergeron-Effect). Thus, the elements of the ITC were shifted signficantly southward (Fig. 7b), particularly as a result of the dominance of the NE trade winds. Consequently, the Bolivian NE highland received more summer rain over a longer period than today. The Apolobamba Cordillera at the edge of the highland towards the Amazon lowland probably had a two-peak rainy season at that time.

A second decisive factor was the fact that the high pressure areas were located to the south and away from the continent over the Atlantic and Pacific oceans. Thus, cold air masses were able to advance northward into the interior of South America particularly during winter, after they had absorbed sufficient moisture over the relatively warm South Atlantic ocean. As a consequence the winter rainfall also increased considerably in comparison with the high glacial period. This applied to both Patagonia and the SW-Altiplano of Bolivia. One must assume that the winter aridity was lessened during this period, which contributed to an advance of the glaciers.

In southern Chile the humidiy-bearing westerly winds contributed to a renewed glacier build-up as a result of the retarded circulation and decreased upwelling. However, the cold set-back originated in the melting areas of the South Atlantic and South Pacific where the largest energy consumption occurred. This led to the situation that the glaciers in Chile advanced again after 15 000 BP but that the main phase of advance in Bolivia did not start until approximately 13 000 BP. In Peru the Late Glacial glaciation is dated at 11 000 BP (Mercer and Palacios 1977).

After the sea ice had largely melted in the Southern Hemisphere and the ice conditions had reached a new state of equilibrium, the radiation surplus was converted into sensible heat. In South America the hypsithermal was reached as early as 9000 BP (Heusser and Streeter 1980). During the Post-glacial period the Southern Hemisphere thus had a 3000-year lead in comparison with the climatic development of the Northern Hemisphere, where vast areas of the inland were still covered by ice at 9000 BP. However, the developments in the Southern Hemisphere are more or less synchronous and only differ as regards their position in the hygrothermically differentiated zones, as produced by the circulation pattern.

REFERENCES

CLIMAP Project Members 1976. The Surface of the Ice-Age earth. Quantitative geologic evidence is used to reconstruct boundary conditions for the climate 18 000 years ago. Science 191:1131-1137.

Drewry DJ 1978. Aspects of the early evolution of the West Antarctic ice. In: EM van Zinderen Bakker (ed.), Antarctic glacial history and world palaeoenvironments. Balkema, Rotterdam. p. 25-32.

Dodson JC 1977. Late Quaternary palaeoecology of Wyrie Swamp, southeastern South Australia. Quat. Res. 8:97-114.

Emiliani C 1955. Pleistocene temperatures. J. Geol. 63:538-578.

Frenzel H 1982. Forschungen zur Geographie des Eiszeitalters (Pleistozän) und der Nacheiszeit (Holozän). Jb. Akad. Mainz: 167-170.

Flohn H 1980. Geophysikalische Grundlagen einer anthropogenen Klimamodifikation. Veröff. Jungius-Ges. d. Wiss. Hamburg 44: 195-218.

Gates WL 1976. Modelling the Ice-Age climate. The July climate of 18 000 years ago has been simulated with a global atmospheric model. Science 191:1138-1144.

Graf K 1981. Palynological investigations of two post-glacial peat bogs near the boundary of Bolivia and Peru. J. Biogeogr. 8:353-368.

Hays JD 1978. A review of the Late Quaternary climatic history of Antarctic Seas. In: EM van Zinderen Bakker (ed.), Antarctic glacial history and world palaeoenvironments. Balkema, Rotterdam. p.57-71.

Henning I and Henning D 1980. Kontinent-Karten der potentiellen Landverdunstung, berechnet mit dem Penman-Ansatz. Met. Rdsch. 33:18-30.

Henning D and Flohn H 1980. Some aspects of evaporation and sensible heat flux of the tropical Atlantic. Beitr. Physik der Atmosphäre 53, Nr. 3:430-441.

Heusser CJ 1966. Late-Pleistocene pollen diagrams from the province of Llanquihue, southern Chile. Proc. Am. phil. Soc. 110, No. 4: 269-305.

Heusser CJ 1981. Palynology of the Last Interglacial-Glacial cycle in midlatitudes of southern Chile. Quat. Res. 16:293-321.

Heusser CJ 1983. Quaternary pollen record from Laguna de Tagua Tagua, Chile. Science 219:1429-1432.

Heusser CJ and Streeter SS 1980. A temperature and precipitation record of the past 16 000 years in southern Chile. Science 210:1345-1347.

Heusser CJ, Streeter SS and Stuiver M 1981. Temperature and precipitation record in southern Chile extended to ∼43 000 years ago. Nature 294:65-67.

Imbrie J, Van Douk J and Kipp NG 1973. Palaeoclimatic investigation of a Late Pleistocene Caribbean deep-sea core: comparison of isotope and faunal methods. Quat. Res. 3:10-38.

Kennett JP 1978. Cainocoic evolution of circumantarctic palaeoceanography. In: EM van Zinderen Bakker (ed.), Antarctic glacial history and world palaeoenvironments. Balkema, Rotterdam. p. 41-56.

Kessler A 1963. Über Klima und Wasserhaushalt des Altiplano (Bolivien, Perú) während des Hochstandes der letzten Vereisung. Erdkunde 17:165-173.

Kessler A, this volume. The Palaeohydrology of the Late Pleistocene Lake Tauca on the Southern Altiplano (Bolivia) and recent climatic fluctuations.

Lamb HH 1966. The Changing Climate, London. p. 150-156.

Lamb HH and Woodroffe A 1970. Atmospheric circulation during the Last Ice Age. Quat. Res. 1:29-58.

Lauer W 1968. Die Glaziallandschaft des südchilenischen Seengebietes. Acta Geographica 20:215-236.

Lauer W 1979. Im Vorland der Apolobamba-Kordillere. Physisch-geographische Beobachtungen auf einer kurzen Studienreise nach Bolivien. In: R Hartmann und U Oberem (eds.), Estudios Americanistas II. p. 9-15.

Lauer W. Berichte der Kommission für Erdwissenschaftliche Forschung. Jb. Akad. Mainz, 1977:125-127; 1981:136-138; 1982: 158-161.

Laugenie CA and Mercer JH 1973. Southern Chile: A chronology of the Last Glaciation. In: IX INQUA Congress Abstr., Christchurch, NZ. p. 202-203.

Lohmann GP 1978. Response of the Deep Sea to Ice Ages. Oceanus 21: 58-64.

Lorius C, Merlivat L, Jouzel J and Purchet M 1979. A 30 000-yr isotope climatic record from Antarctic ice. Nature 280:644-648.

Mercer JH 1976. Glacial history of southernmost South America. Quat. Res. 6:125-166.

Mercer JH 1978. Glacial development and temperature trends in the Antarctic and in South America. In: EM van Zinderen Bakker (ed.), Antarctic glacial history and world palaeoenvironments. Balkema, Rotterdam. p. 73-93.

Mercer JH and Palacios O 1977. Radiocarbon dating of the last glaciation in Perú. Geology 5:600-604.

Molina-Cruz A 1977. The relation of the southern trade winds to upwelling processes during the last 75 000 years. Quat. Res. 8: 324-338.

Morley JJ and Hays JD 1979. Comparison of glacial and inter-glacial oceanographic conditions in the South Atlantic from variations in calcium carbonate and radiolarian distributions. Quat. Res. 12: 396-408.

Nogami M 1982. Circulación atmosférica durante la última época glacial en los Andes. Atmospheric circulation pattern during the last glaciation over the Andes mountains. Revta Geografia Norte Grande, Pontificia Universidad Católica de Chile 9:41-48.

Porter StC 1981. Pleistocene glaciation in the southern Lake District of Chile. Quat. Res. 16:263-292.

Rayner JN and Howarth DA 1979. Antarctic Sea Ice: 1972-1975. Geogr. Rev. 4:202-223.

Servant M and Fontes J-Ch 1978. Les lacs Quaternaires des hauts plateaux des Andes Boliviennes. Premières interprétations paléoclimatiques. Cah. ORSTOM, sér. géol, vol. X, no. 1:9-23.

Stewart RW 1978. The role of Sea Ice in Climate. Oceanus 21:47-57.

Troll C and Finsterwalder R 1935. Die Karten der Cordillera Real und des Talkessels von La Paz (Bolivien) und die Diluvialgeschichte der zentralen Anden. Petrol. Mitt. 81:393-399; 445-455.

Webster PJ and Streten NA 1978. Late Quarternary Ice Age climates of tropical Australasia. Interpretations and reconstructions. 10:279-309.

Wilson AT 1978. Past surges in the West Antarctic ice sheet and their climatological significance. In: EM van Zinderen Bakker (ed.), Antarctic glacial history and world palaeoenvironments. Balkema, Rotterdam. p. 33-39.

The palaeohydrology of the Late Pleistocene Lake Tauca on the Bolivian Altiplano and recent climatic fluctuations

ALBRECHT KESSLER
Universität Freiburg, Germany

ABSTRACT. Applying the water budget to the palaeohydrology of the Late Pleistocene Lake Tauca (12 500 - 11 000 BP) on the Altiplano (South American Andes), several cases are discussed to show how the climatic conditions must have changed in comparison with the recent climate in order to make the formation of the palaeolake possible. It is shown that precipitation must have increased by at least 30 % in the entire Altiplano-basin in comparison with that of today. The example of the recent water budget fluctuations of Lake Titicaca shows that the conditions during the Tauca-period can best be explained by a southward shift of the atmospheric circulation belts in the area of the Altiplano.

INTRODUCTION

The Peruvian-Bolivian Altiplano is a closed topographic depression (Fig. 1). Since the northern part is higher and wetter, it is drained from north to south by the Desaguadero river which flows into a closed lake, the Lake Poopo. Today the southern Altiplano also includes the Salars of Coipasa and Uyuni.

The investigations by Servant and Fontes (1978) showed that three lacustrine transgressions have occurred since the Middle Pleistocene, which united the three basins to form a large lake. The Palaeolake Tauca, the last one, was formed in the Late Pleistocene between 12 500 and 11 000 BP. It covered 43 000 km^2 and had a maximal depth of 60 m. By about 10 000 BP Lake Tauca had been reduced to approximately the size of Lake Poopo. During the Holocene there were only small fluctuations of the lake levels. Kessler (1966) reported on recent variations and the formation of a new lake (Uru-Uru) on the southern Altiplano.

Mercer and Palacios (1977) were able to prove that the Last Glaciation in the Cordillera de Vilcanota and the Quelccaya ice cap in the area of the northern Altiplano culminated some time between about 28 000 and 14 000 BP. A rather minor readvance of the Quelccaya ice cap was in progress about 11 500 BP and culminated about 11 000 BP (Huancane II moraine). By 10 000 it was hardly, if at all, larger than it is today. The snow line depression during the Huancane II readvance was probably not more than 300 m.

Evidence of a Late Pleistocene glacier advance has also been found in other parts of the Altiplano catchment area (Cordillera de Apolobamba,

Lauer 1982; Cordillera Real, Nogami 1970). The Huancane II readvance can also be compared with the oxygen isotope profile observed in the Dome C ice core of east Antarctica (Lorius et al. 1979). The ice core chronology indicates a relatively cold period from 11 000 to 13 000 BP.

The following is a reconstruction of the climate during the Huancane II readvance using the water budget equation for Lake Tauca. This method has already been tested on numerous closed lakes (e.g. Bobek 1937; Leopold 1951; Snyder and Langbein 1962; Kessler 1963; Grove and Pullan 1963; Haude 1969; Galloway 1970; Butzer et al. 1973; Kutzbach 1980).

Since Lake Tauca already existed during the Huancane II readvance, its presence during this period cannot be explained by an additional runoff from melting glaciers of the basin. Its origin must be due to changes in evaporation and/or precipitation. In any case it is remarkable that during the glacial retreat from position Huancane II to position Huancane I between 11 000 BP and 10 000 BP Lake Tauca dried out despite being additionally fed by melting glaciers during this period.

QUANTITATIVE ESTIMATES OF PAST EVAPORATION AND PRECIPITATION OF THE PALAEOLAKE TAUCA-PERIOD

In order to calculate the water budget of Palaeolake Tauca, the Altiplano was divided into several areas (Fig. 1 and Table 1). The figures for the recent situation are from Kessler (1963, 1970) and Kessler and Monheim (1968).

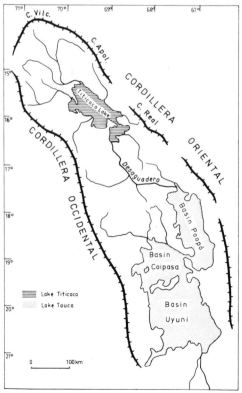

Figure 1. Late Pleistocene Lake Tauca and the Altiplano (Peru, Bolivia) according to Servant and Fontes (1978).

Table 1. Hydrological and climatological data of the Altiplano.

	Area (km^2)	Recent prec. (m/year)	Recent runoff coefficient
Area of northern Lake Tauca	22 700	0.31	
Area of southern Lake Tauca	20 300	0.18	
Catchment area of Lake Tauca without catchment area of Desaguadero river	73 500	0.22	0.07
Catchment area of Desaguadero river without catchment area of Lake Titicaca	30 250	0.45	0.07
Area of Lake Titicaca (Late Pleistocene)	9 700	0.93	
Catchment area of Lake Titicaca (Late Pleistocene)	48 400	0.73	0.21

The mean annual water budget of a closed lake at equilibrium can be expressed by the equation

$$P_L + R_C - E_L = 0$$

where P_L is the annual precipitation onto the lake, R_C is the mean total annual runoff from the catchment area and E_L is the mean annual lake evaporation. In the various case studies the increase in runoff from the catchment area which occurred together with a decrease in evapotranspiration due to temperature depression, was calculated with the water budget equation of the catchment area:

$$R_C = P_C - E_C$$

where P_C is the annual precipitation onto the catchment area, E_C is the annual evapotranspiration from the catchment area.

According to a diagram by Wundt (1953) the increase in runoff when the precipitation onto the catchment area increases implies an increase in relative humidity (RH) of the air in the basin (Table 2).

Table 2. The change in the runoff coefficient $K = R_c/P_c$ dependent on precipitation onto the basin P_c (according to Wundt 1953).

Pc(cm/year)	20	30	40	50	60	70	80	90	100	110	120	130
K	0.100	0.155	0.210	0.263	0.317	0.363	0.410	0.450	0.490	0.530	0.570	0.600

Evaporation from the lake and its change dependent on the surface temperature and relative humidity of the air was calculated with an aerodynamic bulk formula:

$$E_L = c\, f(u)(Q-q) = c\, f(u)(1-RH)Q$$

where c is a constant, f(u) is a function of the wind velocity, Q is the vapour pressure at the water surface and q is the vapour pressure of air. The decrease in evapotranspiration of the catchment area with decrease in temperature was calculated using the same factor as the evaporation from the lake surface.

For estimating the recent lake evaporation of the Altiplano the value for Lake Titicaca was used, namely $E_L = 1.48$ (m per year). But since the southern Altiplano is drier, E_L is probably somewhat higher. For this reason the following calculations represent the minimal conditions for an increase in precipitation. The maximal temperature depression for the Huancane II readvance (= 3 °C) was calculated from the snow line depression (300 m) and the temperature gradient of the air near the ground (0.8-0.9 °C/100 m, according to Kessler 1963).

RESULTS OF THE CALCULATIONS

The following individual cases will demonstrate the sensitivity of the model. Table 3 shows what percentage of the annual water supply would be represented by the precipitation onto Lake Tauca (A), the runoff from the catchment without discharge from Lake Titicaca (B), and the discharge from Lake Titicaca (C) in each case.

Case 1: $\Delta P = 0$, $\Delta E = 0$, $\Delta T = 0$.

What would have happened if the Palaeolake Tauca had been subject to today's climatic conditions, that is, if changes in precipitation, evaporation and temperature had not occurred? Our calculations show that the lake would have dried up in about 50 years. In this example the precipitation onto the Lake Tauca makes up 72% (Table 3).

Case 2: $\Delta P \neq 0$, $\Delta E = 0$.

The second question is: By how much must the precipitation in the entire basin be increased in comparison with today's conditions for the Palaeolake Tauca to exist? The annual precipitation amount would have to be increased by 85% all over the catchment area. As Table 3 shows, the discharge from Lake Titicaca would then supply 50% of the water for Lake Tauca.

Case 3: $\Delta E \neq 0$, $\Delta T \neq 0$, $\Delta P = 0$.

As already mentioned, a temperature decrease of 3 °C can be deduced from the snow line depression during the Huancane II readvance. This indicates a decrease in evaporation of 18% in comparison with the situation today. It can also be assumed that precipitation and relative humidity have not changed. This suggests that the existence of the Palaeolake Tauca was not secured. It would disappear in about 100 years. The water supply of the Palaeolake Tauca would be relatively equally distributed among all three components.

Case 4: $\Delta E \neq 0$, $\Delta T \neq 0$, $\Delta P \neq 0$.

This is based on the same conditions as in case 3. The deficit is compensated by an increase in precipitation. In the case of a decrease in evaporation by 18% due to the temperature, precipitation in the entire basin must be increased by 40% in order to guarantee the existence of the Palaeolake Tauca.

Case 5: $\Delta E \neq 0$, $\Delta T \neq 0$, $\Delta P \neq 0$, $\Delta RH \neq 0$ (Lake).

In calculating case 4 we take into account that the precipitation rise causes an increase in the relative humidity of the air close to the ground (RH) in the catchment area. In order to heighten the minimal conditions for a possible increase in precipitation it will also be assumed that the relative humidity over the sea surfaces increased by 5% compared with today's conditions. Taking the temperature change into consideration this would mean a decrease in E_L by a total of 27%. The necessary increase in precipitation (ΔP) would then still be as much as 30%.

Table 3. Mean annual water supply for the Palaeolake Tauca in percent.

A Precipitation onto Lake Tauca
B Runoff from the catchment without discharge from Lake Titicaca
C Discharge of Lake Titicaca

	A	B	C
Case 1	72	14	14
Case 2	32	18	50
Case 3	39	26	35
Case 4	29	25	46
Case 5	30	24	46

The results can be summarized as follows: The existence of the Palaeolake Tauca cannot be explained without assuming an increase in precipitation of at least 30% in the entire basin. Although the discharge from Lake Titicaca is of secondary importance for the water budget of the southern Altiplano today, it played a very important role during the Late Pleistocene for the Palaeolake Tauca (Table 3, case 5).

REFLECTIONS ABOUT THE PALAEOCLIMATE

It has been pointed out repeatedly (e.g. Heine 1977; Hays 1978; Flohn 1981; Mercer 1978; Salinger 1981; Street-Perrott and Roberts 1983) that during the period from 18 000 BP and 9400 BP (Southern Hemisphere thermal maximum) there was an unequal rise in temperature in the two hemispheres. This indicates that at least for a time there was a southward shift of the circulation belts.

The following climatic conditions exist on the Altiplano today (see Fig. 2). Tropical easterlies reach the basin during the rainy season from November to April. During the dry period from May to September/October the area is influenced by the SE-Pacific anticyclone. Westerly winds prevail all over the troposphere. The axis of the subtropical jet stream shifts northwards to a latitude of about 20° to 25°S. The position in summer time is at about 30° to 35°S. During an occasional southward shift in comparison with today's position of the equatorial trough and the subtropical anticyclone belts, the summer rainy period in the Altiplano would have been lengthened during the Late Pleistocene. This assumption would best explain the Late Pleistocene climatic conditions.

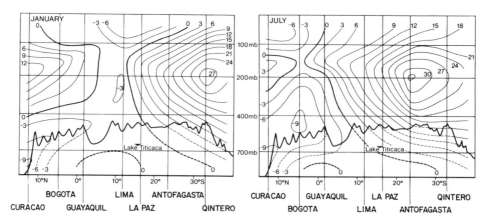

Figure 2. Mean cross-section of the zonal wind above the west of South America, with the peak line of the Andes. Zonal component of the resultant wind in m/s, positive values denote eastward flow.

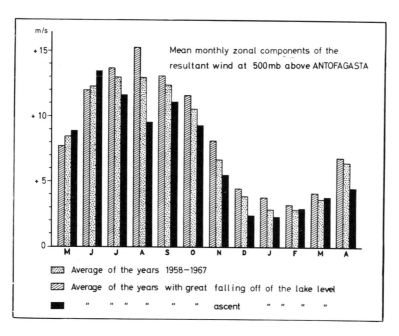

Figure 3. Mean monthly components of the resultant wind at 500 mb above Antofagasta.

RECENT CLIMATIC FLUCTUATIONS

The recent climatic variations on the Altiplano can best be illustrated using the fluctuations in the level of Lake Titicaca. The annual fluctuations of about 70 cm are superimposed by long-term alterations of about 5 m. Examination of the water budget (Kessler 1974) showed that the long-term fluctuations of the level are due mainly to changes in precipitation onto the lake and the catchment area. Fluctuations in outflow therefore be regarded as a good indicator for the changes in precipitation on the Altiplano.

In order to clarify the connection between recent circulation anomalies and fluctuations of precipitation, the monthly mean values of the zonal wind component at 500 mb above Antofagasta were examined from May 1958 to April 1967 and correlated with the level fluctuations of Lake Titicaca. As Fig. 3 shows, high westerly wind components resulted when the lake level decreased and low components resulted when the lake level increased (with the exception of May and June). In other words, during the increase of the belt of the westerlies and its displacement to the north, a deficit of the amount of precipitation occurred, and vice versa. These observations suggest that a similar shift of the circulation belts occurred in the Late Pleistocene during the Huancane II readvance and the formation of Lake Tauca.

In this context it should be mentioned that the considerable rise in the level of Lake Titicaca in 1973-1976 (Kessler 1981) occurred together with an increase in the snow cover on the land surfaces of the Northern Hemisphere and with a decrease in the sea ice cover on the Southern Hemisphere (Kukla 1981). This global asymmetry of the snow and ice cover and the simultaneous rise of the water level in Lake Titicaca can be seen qualitatively as a recent analogy to the Late Pleistocene Lake Tauca period.

REFERENCES

Bobek H 1937. Die Rolle der Eiszeit in Nordwestiran. Z. Gletscherkunde 25:130-183.
Butzer KW, Fock GJ, Stuckenrath R and Zilch A 1973. Palaeohydrology of Late Pleistocene lake Alexandersfontein, Kimberley, South Africa. Nature 243:328-330.
Flohn H 1981. Tropical climate variations during Late Pleistocene and early Holocene. In: A Berger (ed.), Climatic variations and variability. Nato Adv. Study Institutes Series, Ser. C, vol. 72. p. 233-242.
Galloway RW 1970. The full-glacial climate in the southwestern United States. Ann. Ass. Am. Geographers 60:245-256.
Graf K 1981. Palynological investigations of two post-glacial peat bogs near the boundary of Bolivia and Peru. J. Biogeogr. 8:353-368.
Grove AT and Pullan RA 1963. Some aspects of the Pleistocene palaeogeography of the Chad Basin. In: FC Howell and F Bourliere (eds.), African Ecology and Human Evolution, No. 36. Adeline, Chicago. p. 230-245.
Hastenrath S 1967. Observations on the snow line in the Peruvian Andes. J. Glaciol. 6:541-550.
Haude W 1969. Erfordern die Hochstände des Toten Meeres die Annahme von Pluvial-Zeiten während des Pleistozäns? Met. Rdsch. 22:29-40.
Hays JD 1978. A review of the late quaternary climatic history of Antarctic seas. In: EM van Zinderen Bakker (ed.), Antarctic glacial history and world palaeoenvironments. Balkema, Rotterdam. p. 57-71.
Heine K 1977. Beobachtungen und Überlegungen zur eiszeitlichen Depres= sion von Schneegrenze und Strukturbodengrenze in den Tropen und Sub= tropen. Erdkunde, Arch.f.wiss.Geographie 31:161-178.

Kessler A 1963. Climate and hydrology of the Altiplano (Bolivia, Peru) during the climax of the last glaciation (Germ.). Erdkunde, Arch.f. wiss.Geographie 17:165-173.

Kessler A 1966. Recent changes in the course of the Desaguadero and the origin of lake Uru-Uru, Bolivian Altiplano (Germ.). Erdkunde, Arch.f. wiss.Geographie 20:194-204.

Kessler A 1970. On the annual variation of the potential evaporation in the basin of lake Titicaca (Germ.). Arch. Met. Geoph. Biokl., Ser. B 18:239-252.

Kessler A 1974. Atmospheric circulation anomalies and level fluctuations of lake Titicaca (Germ.). Bonner Met. Abh. 17:361-372.

Kessler A 1981. Fluctuations of the water budget on the Altiplano and of the atmospheric circulation (Germ.). Aachener Geographische Arbeiten 14:111-122.

Kessler A and Monheim F 1968. The water budget of lake Titicaca according to new measurements (Germ.). Erdkunde, Arch.f.wiss.Geographie 22:275-283.

Kukla G 1981. Climatic role of snow covers. In: I Allison (ed.), Sea level, ice and climatic change. IAHS Publ. no. 131. p. 79-107.

Kutzbach JE 1980. Estimates of past climate at paleolake Chad, North Africa, based on a hydrological and energy-balanced model. Quat. Res. 14:210-223.

Lauer W 1982. Die jungglaziale Vorlandvergletscherung am Fuße der Apolobamba Kordillere. Vortrag auf der DEUQUA-Tagung in Zürich, 16.8.1982 (manuscript 7 pp.).

Leopold LB 1951. Pleistocene climate in New Mexico. Am. J. Sci. 249:152-168.

Lorius C, Merlivat L, Jouzel J and Pourchet M 1979. A 30 000 year isotope climatic record from Antarctic ice. Nature 280:644-648.

Mercer JH 1978. Glacial development and temperature trends in the Antarctic and in South America. In: EM van Zinderen Bakker (ed.), Antarctic glacial history and world palaeoenvironments. Balkema, Rotterdam. p. 73-93.

Mercer JH and Palacios O 1977. Radiocarbon dating of the last glaciation in Peru. Geology 5:600-604.

Nogami M 1970. El retroceso de los glaciares en la cordillera real, Bolivia. Geogr. Rev. Japan 43:338-346.

Salinger MJ 1981. Palaeoclimates north and south. Nature 291:106-107.

Servant M and Fontes J-C 1978. Les lacs quaternaires des hauts plateaux des Andes Boliviennes, premières interprétations paléoclimatiques. Cah. O.R.S.T.O.M. Sér. Géol. 10:9-23.

Snyder CT and Langbein WB 1962. The Pleistocene lake in Spring valley, Nevada, and its climatic implications. J. geophys. Res. 67:2385-2394.

Street-Perrott FA and Roberts N 1983. Fluctuations in closed-basin lakes as an indicator of past atmospheric circulation patterns. In: FA Street-Perrott et al. (eds.), Variations in the global water budget. D Reidel Publ. Company. p. 331-345.

Wundt W 1953. Gewässerkunde. Springer, Berlin/Göttingen/Heidelberg. 320 pp.

Late Quaternary climatic changes in the basin of Lake Valencia, Venezuela, and their significance for regional paleoclimates

L.PEETERS
Vrije Universiteit, Brussel, Belgium

ABSTRACT. A tentative correlation of the results for the lake basin obtained by fieldwork, archeological research and a detailed study of deep core samples from the lake bottom allows the reconstruction of a sequence of paleoclimates. Semi-arid climatic conditions existed at the end of the Pleistocene, and during the Holocene there was a general trend towards the present humid tropical climate, interrupted by short spells of less humid conditions. A similar succession of paleoclimates is also found in several areas of Atlantic tropical South America.

INTRODUCTION

At about 100 km W of Caracas, straddling the states of Carabobo and Aragua, lies the lake of Valencia (Table 1). The lake is located in a tectonic depression between the Cordillera de la Costa and the Serrania del Interior and occupies an area of about 370 km^2. During the Late Quaternary several climatic changes occurred in the lake basin and the sequence of those changes is represented in Table 1.

SOURCE MATERIAL

The tentative correlations shown in Table 1 are based on the results of geomorphological fieldwork, archeological research and the detailed study of samples from a deep core in the lake. Fieldwork on geomorphology revealed the existence of former lacustrine deposits. It has been proposed that these transgressions and regressions were due to climatic changes. Moreover, it appears that all transgressions – except the oldest one – culminated at about 427 m, which represents the natural level of overflow of the lake during historical times. This points to the fact that Holocene faulting of the lake basin (Schubert and Laredo 1979) was not strong enough to influence the hydrological balance of the lake during this period.

Archeological research provided C14-dates and the description of the sites allows correlation with the fieldwork on geomorphology. Analysis of the deep core samples produced very detailed information on lithological changes and on the paleoenvironment, the latter confirming the link between changes in the lake level and climatic changes. Moreover, most of the C14-dates in Table 1 are derived from this study (see also Schubert 1980).

DISCUSSION

Evidence for the existence of a humid tropical climate during the period of the initial lake Valencia I is based on the presence of important linear erosion. V-shaped valleys were formed and filled up afterwards by Pleistocene sediments of considerable thickness. According to V Lopez (1942), the solid bedrock of rio Cabriales is hidden by 160 m of gravel, sand and clay. Consequently, the period of linear erosion in a humid tropical climate belongs to the end of the Tertiary.

Table 1.

LAKE	AGE	C14 DATE	CLIMATE	GEOMORPHOLOGY	\multicolumn{3}{c}{EVOLUTION OF LAKE VALENCIA DURING THE QUATERNARY}
					MAR CARIBE / CORDILLERA DE LA COSTA / LAGO DE VALENCIA / SERRANIA DEL INTERIOR / • CARACAS		
Valencia I	end Tertiary		humid tropical	V-shaped valleys linear erosion			
Valencia II	Pleistocene		predominant semi-arid	filling up of V-shaped valleys - aggradation	ARCHEOLOGY		
	end Pleistocene	12.930 BP 10.500 BP	semi-arid	plains-pediplanation			
Changes in lake level of Valencia II: transgression VII-0 regression ?	Holocene	between 10.500 BP and 8000 BP	alternation of humid tropical (transgression) and less humid (regression)	filling up of river beds in lower course (transgression) - erosion in lower course of tributaries (regression)			
transgression VII-1 regression		8000 BP 7.130 BP					
transgression VII-2 regression		6.000 BP 5.480 BP					
transgression VII-3 regression		between 5.480 BP and 4.400 BP 4.400 BP		probably exoreic (transgression) - endoreic (regression)	human bones Guacara - potsherds El Roble	Meso-indian epoch (partim) Neo-indian epoch (partim) (periods II-III-IV)
transgression Valencia III (427 m)	partly historical	post 4400 BP till shortly after 920-940 A.D.	humid tropical	filling up of river beds in lower course - exoreic	urns El Roble La Mata Tocoron	
historical lowering of lake level	from XVIII th century on			canyons in lower course of tributaries - lake terraces - endoreic		

The description of the paleoenvironment during phase IV (Salgado-Labouriau 1980) and the period of zone 1 (Platt Badbury et al. 1981) corresponds quite well to what has been supposed formerly about lake Valencia II (Peeters 1968), that is, a marsh environment bordered by an open xeric vegetation on mostly bare soil, the climate being much drier than that of today. Such a small, shallow lake existed during the upper part of the Pleistocene as phase IV and, according to Salgado-Labouriau (1980), represents the end of the Pleistocene.

During the paleoenvironment of zone 2 the lake attained sufficient depth to have seasonal thermal stratification and was at least 20 m deep in the centre of the basin shortly after the beginning of this period (10 500 BP) (Platt Badbury et al. 1981). According to the depth of the deep core sediments this corresponds to a lake level of at least

381 m at the beginning of this period. As the oldest lacustrine deposits (V II-0) encountered in drillings close to the lake shore are located at 397 m (Peeters 1973), a rise of level over 16 m during this period is not unlikely and the lacustrine deposits of transgression V II-0 could have originated within the paleoenvironment of zone 2. The characteristics of this period offer indications of a climatic change which could signify a regression. Hence, it is not excluded that alluvial deposits stopped the sedimentation of Planorbis clay at the place of the drilling, which means that the transgression V II-0 could be considered as the initial stage of transgression V II-1.

The climax of the trend towards freshness and high lake levels occurred at 8000 BP (Platt Badbury et al. 1981) and is correlated in Table 1 with deposits of transgression V II-1, which probably reached the level of overflow at 427 m. The existence of stromatolite beds at 421 m, dated 7130 BP and the increase of clastic material in the deep core sediments (Platt Badbury et al. 1981) point to a regression. Transgression V II-2 is supposed to have occurred at 6000 BP because at this time the important input of clastic material in the deep core sediments came to an end (Platt Badbury et al. 1981). Stromatolite beds at 418 m (5480 BP) (Platt Badbury et al. 1981) testify to a regression. According to field observations, this regression was succeeded by transgression V II-3. When comparing geomorphological fieldwork (Peeters 1971) with the results of excavations at Morro Guacara by H Peñalver (1971), it appears that the human skeleton, dated 4400 BP, belongs to a layer of sandy clay and grit, resting upon deposits of V II-3 covered by deposits of lake Valencia III. Hence, the transgression of V II-3 occurred between 5480 and 4400 BP, the latter representing the age of a regression.

Consequently, the final transgression (transgression V III) started after 4400 BP. As parts of the archeological site of La Mata are dated AD 920-940 (Cruxent and Rouse 1958) and as this site was probably abandoned due to the rising lake level (Peeters 1968), lake Valencia III must have reached the natural level of overflow at 427 m shortly after AD 920-940 and became exoreic. This represents the situation at the time of the arrival of the Spaniards at the beginning of the 16th century.

Anthropic activities signifying the historical lowering of the lake level are mentioned by A Bockh (1956). The different historical lake levels are recorded by A Jelambi (1970).

FINAL REMARKS

Predominantly semi-arid climatic conditions existed in the lake basin during the Pleistocene. Pleistocene semi-arid climatic conditions are also reported from the Amazon basin (Journaux 1975) and in southern Brazil (Klammer 1981, 1982) and Uruguay (Prost 1977) as far as the first part of the Quaternary is concerned. At the end of the Pleistocene the climate was much drier than that of today and during the Holocene there was a general trend towards the present humid tropical climate, interrupted by short spells of less humid conditions. Tentative correlations between the lake basin and other parts of the humid tropics (Peeters 1970, 1971; Platt Badbury et al. 1981; Salgado-Labourian 1980) show that such a sequence of climatic changes during the Late Quaternary also occurred in Columbia, the Rupununi savanna of Guyana, the Andes of Venezuela and Brazil (see also Zonneveld 1975;

Van der Hammen 1974; Damuth and Fairbridge 1970).

Difficulties arise when trying to correlate the Holocene C14-dates of lake Valencia with results obtained outside the lake basin. One must bear in mind that climatic changes of the Holocene occurred only during very short periods and the response of different geographical environments was not always the same.

Coastal areas are expected to react promptly to an increase in humidity due to a transgression. But predominantly continental areas, protected against marine influences by high mountains and lying in a Föhn zone (such as lake Valencia), probably do not. Comparing the results of lake Valencia with those obtained for lakes in the Columbian llanos (Van der Hammen 1974) the following C14-dates could be considered as representing a more general trend: the less humid period of 5480 BP to 4400 BP and the increase in humidity after 4400 BP. But dates for the drier period of 2200 BP in Columbia are lacking for the lake Valencia basin.

Nevertheless, the results obtained in the lake basin are of general interest as far as the Atlantic area of tropical South America is concerned.

REFERENCES

Bockh A 1956. El desecamiento del lago de Valencia. Caracas, Fundac. Mendoza. 256 p.

Cruxent JM and Rouse I 1958. An archeological chronology of Venezuela, vol. 1. Soc. sc., Monogr. 6, Pan Am. Univ.

Damuth JE and Fairbridge RW 1970. Equatorial Atlantic deep-sea arkasic sands and ice-age aridity in tropical South America. Bull. geol. Soc. Am. 81:189-206.

Jelambi O 1970. Grafico de la variacion del nivel del lago de Valencia segun datos recopilados. El Lago, año 3, 21:267.

Journaux A 1975. Géomorphologie des bordures de l'Amazonie brésilienne. Bull. Ass. géogr. franc. no 422-423:5-19.

Klammer G 1981. Landforms, cyclic erosion and deposition and Late Cenozoic changes in climate in southern Brazil. Z. Geomorph. N.F. 25, 2:146-165.

Klammer G 1982. Die Paläowüste der Pantanal von Mato Grosso und die pleistozäne Klimageschichte der brasilianischen Randtropen. Z. Geomorph. N.F. 26, 4:393-416.

Lopez VM 1942. Geologia del valle de Valencia. Rev. de Fomento 45-46: 1-28.

Peeters L 1968. Origen y evolucion de la cuenca del lago de Valencia, Venezuela. Caracas 66 p.

Peeters L 1970. Les relations entre l'évolution du lac de Valencia (Vénézuela) et les paléoclimats du Quaternaire. Rev. Géogr. phys. et Géol. dyn. (2), 12, 2:157-160.

Peeters L 1971. Nuevos datos acerca de la evolucion de la cuenca del lago de Valencia (Venezuela) durante el Pleistoceno superior y el Holoceno. Caracas 38 p.

Peeters L 1973. Evolucion de la cuenca del lago de Valencia de acuerdo a resultados de perforaciones. El Lago, año 7, 41:861-874.

Peñalver H and De Chacin AR 1968. Informe preliminar de El Roble-Investigacion arqueologica n° 3. Boll. del Inst. de Antrop. e Hist. del Est. Carabobo 2:16-25.

Peñalver H 1971. Excavaciones en el Morro Guacara - Informe preliminar. Boll. del Inst. de Antrop. e Hist. del Est. Carabobo 3-4:11-15.

Platt Badbury J, Leyden B, Salgado-Labouriau M, Lewis WM, Schubert C, Binford MW, Frey DG, Whitehead DR and Weibezahn FW 1981. Late Quaternary environmental history of lake Valencia, Venezuela. Science 214: 1299-1305.

Prost Th 1977. Oscillations climatiques et formations continentales des bords du rio de la Plata en Uruguay et leurs rapports avec les oscillations du niveau marin. Rech franç. sur le Quat., INQUA 1977, supplém. Bull. Ass. franç. Et. du Quat. 50:273-288.

Salgado-Labouriau M 1980. A pollen diagram of the Pleistocene-Holocene boundary of lake Valencia, Venezuela. Rev. Paleob. and Palynol. 30: 297-312.

Schubert C and Laredo M 1979. Late Pleistocene faulting in lake Valencia basin, north-central Venezuela. Geology 7:289-292.

Schubert C 1980. Contribution to the paleolimnology of lake Valencia, Venezuela : seismic stratigraphy. Catena 7: 275-292.

Van der Hammen T 1974. The Pleistocene changes of vegetation and climate in tropical South America. J. Biogeogr. 1:3-26.

Zonneveld JIS 1975. Some problems of tropical geomorphology. Z. f. Geomorph. N.F. 19, 4:377-392.

Australasia

New Zealand climate: The last 5 million years

M.J. SALINGER
New Zealand Meteorological Service, Wellington

ABSTRACT. The interaction of the high axial ranges of New Zealand with the humid westerly circulation, to produce quite distinct regional climates, makes the country an ideal platform from which to monitor past climates. Variations from the present circulation can be readily detected by a change in the spatial pattern of climate; this spatial sensitivity is used to reconstruct larger-scale palaeocirculation patterns on a several-thousand-year time scale.

This paper reviews and discusses the implications of the climate record of the last 5 million years. The climate has varied from warm temperate at the beginning of the Pliocene to sub-Antarctic during the Pleistocene glaciations. Significantly, the timing of recent variations leads similar fluctuations in the Northern Hemisphere by several millennia. Reconstructions of past circulation are described for the Late Pleistocene and Holocene only because data are inadequate before 20 000 BP. These suggest enhanced mid-latitude westerly circulation during the Last Glacial Maximum. Decreased westerly circulation and enhanced tropical Walker Circulation are indicated for the period of Holocene maximum warming. Importantly, the reconstructions confirm results from numerical modelling studies and indicate possible future climate scenarios.

1. INTRODUCTION

New Zealand, a narrow country occupying the temperate zone (34° to 45°S) in the largely oceanic Southern Hemisphere (Fig. 1), is a key area for indicating regional and global palaeoclimatic trends. Apart from Tasmania and Patagonia, New Zealand is one of the few locations in the southern mid-latitudes from which detailed Late Cenozoic climatic data are available. It is one of the few areas of the globe where the abundance of palaeoenvironmental evidence from some time intervals allows reconstruction of palaeocirculation patterns.

Abundant New Zealand palaeoenvironmental data have been obtained from well correlated fossiliferous Late Cenozoic marine sequences (Hornibrook 1971; Fleming 1975). Plant fossil evidence is contained in a number of Cenozoic terrestrial, fresh-water and marine deposits (Couper and McQueen 1954; McQueen et al. 1968; Moar 1973; Mildenhall 1973, 1975a, 1975b, 1977 and 1980; Mildenhall and Pocknall 1983). Additionally, there are many areas of South Island where evidence of past cold

Figure 1(a). Location diagram.

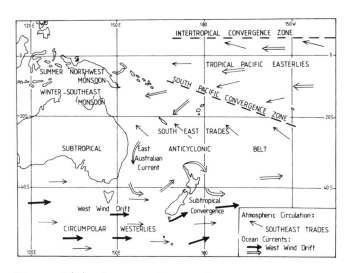

Figure 1(b). Present-day southwest Pacific atmospheric and oceanic circulation.

climate is recorded in prominent glacial deposits (Gage 1958; Suggate 1965; Suggate and Moar 1970; Soons and Gullentops 1973; Chinn 1983; Mabin 1983). Hence there is abundant evidence available from New Zealand to enable an important contribution to the understanding of Late Cenozoic climate.

This paper reviews the significant climatic trends recorded from New Zealand for the Pliocene to the present day. Only since the Last Glacial Maximum (at approximately 26 000 - 22 000 BP) are the data detailed enough to permit a reconstruction of palaeocirculation patterns. These are described for 18 000, 9000 and 6000 BP using fossil pollen data.

2. PRESENT-DAY CLIMATIC SETTING

Southern New Zealand (Fig. 1) is today located in the circumpolar westerly vortex, while the north of the country protrudes into the subtropical ridge of migratory anticyclones that move east across the Pacific (Steiner 1980). Both islands have high southwest/northeast trending axial ranges that significantly obstruct the humid westerly circulation (Watts 1947; Maunder 1971) and give rise to a variety of local climates with contrasting temperature and rainfall patterns. High topography means that both local and regional variations in the general circulation of the atmosphere are strongly reflected by changes in the spatial climatic patterns (Salinger 1980a, 1980b). Figures 2a and 2b which show isolines of equal degree of correlation relate the degree of zonality or meridionality of airflow to rainfall patterns. In Fig. 2a, $Z1$ is a measure of the zonal westerly flow over New Zealand (Trenberth 1976a). When westerly flow is strong, rainfall is increased in the west, along the south coast of South Island and in the central North Island mountains and decreased on the east coast of North Island. Conversely, when westerly flow is weak (i.e. enhanced easterly flow) the opposite pattern occurs. In Fig. 2b, $M1$ is a measure of the meridional southerly flow in the region (Trenberth 1976a). Enhanced southerly flow ($M1$ positive) yields increased rainfall in the south of South Island and reduced rainfall in the north of both islands. Reduced southerly flow ($M1$ negative) gives the reverse rainfall pattern.

The sensitivity of New Zealand climate to circulation variations on a larger scale is further illustrated by correlations with the Southern Oscillation. This quasi-periodic variation in pressure gradient across the tropical Pacific determines the strength of the surface easterlies of the Walker Circulation (Walker and Bliss 1932, 1936). The strength of the Southern Oscillation and Walker Circulation can be quantified by a number of closely related indices (Troup 1965; Kidson 1975; Wright 1975; Trenberth 1976b). The correlation of the Southern Oscillation Index (SOI), using the Wright's index, with New Zealand temperatures is statistically highly significant. Figure 2c shows that annual rainfall is significantly correlated with the SOI. The correlations are sharply defined regionally with high positive correlations in the northeast and east of North Island and north of South Island. In the south of South Island higher rainfalls are associated with a low SOI. Such a sensitivity of temperature and rainfall to present local and regional circulation changes can be used to advantage in reconstructing palaeocirculation patterns. In the regions indicated in Figs. 2a-2c as the most sensitive, details of zonal flow, meridional flow and the Walker Circulation can be inferred directly.

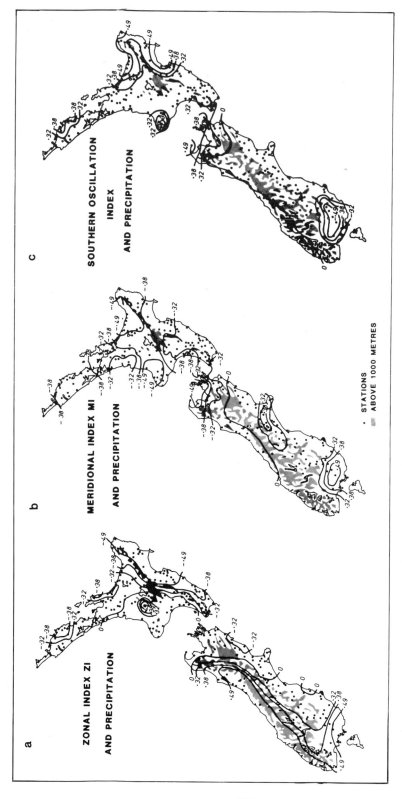

Figure 2(a). Correlation isopleths of Trenberth's (1976a) zonal index Z1 with annual rainfall, 1951-1975.
Figure 2(b). Correlation isopleths of Trenberth's (1976a) meridional index M1 with annual rainfall, 1951-1975.
Figure 2(c). Correlation isopleths of the Southern Oscillation Index (Troup 1965) with annual rainfall, 1951-1975.

Variations in the regional atmospheric circulation also directly affect the main oceanic surface currents. In the New Zealand region these include the East Australian Current and the West Wind Drift. The former, a mild eastern boundary current, travels across the Tasman Sea toward nothern New Zealand, then flows southward along the east coast of North Island. The cool circumpolar West Wind Drift circulates around the southern oceans and sweeps northeastwards up the east coast of South Island. The two currents meet to the east of central New Zealand in the subtropical convergence. The latitudinal position of the subtropical convergence can vary quite markedly (Brodie 1960) due to fluctuations in the westerlies and in the subtropical circulation features such as the Walker Circulation. Long-term changes in the position of the subtropical convergence will be reflected in the nature of marine fossils. New Zealand is therefore ideally located to detect climatic changes in the southwest Pacific.

New Zealand began to take its present shape less than 6 m.y. ago and the axial ranges only attained their present altitudes 0.5 m.y. ago (during the Kaikoura Orogeny). Hence during the Pliocene and Early Pleistocene the present land area was either submerged or topographically lower and present relationships of rainfall distribution with zonal and meridional flow cannot be used to reconstruct past circulation patterns.

3. PALAEOENVIRONMENTAL DATA SOURCES

Late Cenozoic palaeoclimatic data are inferred from the following fossil groups: diatoms, calcareous nannoplankton, planktonic and shallow water foraminifera, molluscs, pollen and spores. The range of environments, from warm temperate to alpine with high to low rainfall, provides habitats for a wide spectrum of plants from grassland to rain forest. Data on former fauna and flora distributions are contained in marine and terrestrial sequences. The glaciers, which came into existence during the Late Cenozoic mountain building period, have left a record of Pliocene and Pleistocene cold events in their deposits. Summaries of marine evidence in Devereux (1967) and Hornibrook (1971) and terrestrial evidence in Mildenhall (1980) and Mildenhall and Pocknall (1983) cover the Pliocene. Early Pleistocene events have been reviewed by Couper and McQueen (1954), Fleming (1975), Mildenhall (1980), Pillans et al. (1982), and Beu and Edwards (1983). Useful summaries of the abundant Late Pleistocene and Holocene evidence of glacial episodes (Burrows 1979; Pillans et al. 1982; Chinn 1983) and vegetational history (Moar 1973a; McGlone 1983) are contained in the CLIMANZ proceedings (1983).

4. PAST CLIMATES

The Pliocene (c. 5-2 m.y. BP)

During the Pliocene marine temperatures show a decline from 15 to 12 °C at 41 °S (Devereux 1967), comparable to the 3 °C decline recorded in sub-Antarctic sea surface temperatures for the same period (Shackleton and Kennett 1975). Planktonic foraminiferal assemblages are similar to those of today (Hornibrook 1971). In the Early Pliocene the northward migration of cool water molluscs reached the southeast of North Island

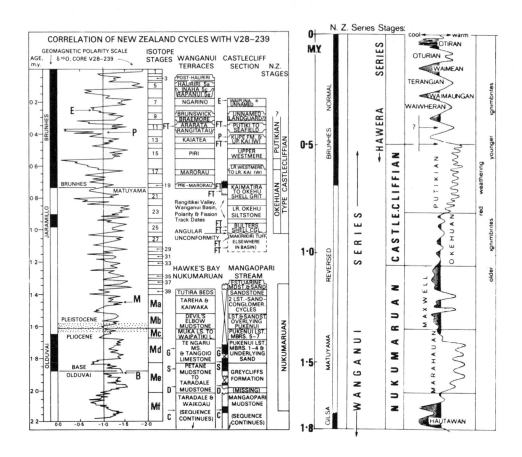

Figure 3(a). A synthesis of New Zealand Pleistocene and Late Pliocene glacio-eustatic cycles from Beu and Edwards (1983) with kind permission. Cycles are correlated with the oxygen isotope curve for core V28-239 (Shackleton and Opdyke 1976; fitted to a year scale and additional stages designated by Gardner 1982), with addition of the Pliocene/Pleistocene boundary (Backman et al. 1983; Tauxe et al. 1983), correlation lines at Brunhes/Matuyama and base Olduvai, and positions of FAD Emiliania huxleyi ("E") and LAD Pseudoemiliania lacunosa ("P") (Gartner 1977; Thierstein et al. 1977). Other symbols: B = LAD Discoaster brouweri; C = FAD Globorotalia crassula; D = top of the zone of overlap of dextrally coiling Globorotalia crassaformis with G. crassula; FT = positions of fission track dates; G = FAD Gephyrocapsa sinuousa; M = LAD Calcidiscus macintyrei; S = LAD Struthiolaria (Pelicaria) acuminata.
(Note : LAD = last appearance datum; FAD = first appearance datum.)

Figure 3(b). A tentative summary of the New Zealand Quaternary stages after Fleming (1975) with kind permission. The base of the Castlecliffian is now considered by Boellstorff and Te Punga (1977) and Pillans (1983) to be as young as 340 000 BP.

into Hawke's Bay (Hornibrook 1975). The Pliocene palynofloras record the extinction of many species adapted to warm temperate, humid environments and the evolution of taxa adapted to montane and cool to cold temperate environments which prevailed from the Late Miocene. Most notable is the decline in the Nothofagus brassi group and the rise to dominance of the Nothofagus fusca group (McQueen et al. 1968; Mildenhall and Harris 1970; Mildenhall 1980; Mildenhall and Pocknall 1983). During the Early Pliocene the Nothofagus brassi group occurred in Northern New Zealand with temperate pollen types such as Dacrydium cupressinum and the Nothofagus fusca group dominant further south. This period was the last time that palm pollen was important as a major constituent of the coastal flora (Mildenhall 1980). At one site in the north of South Island vegetation was reduced to grassland: at central New Zealand sites samples show a range of environments, although the podocarp/Nothofagus fusca group and associated cool temperate taxa were widespread (Mildenhall 1980). The last major occurrence of the Nothofagus brassi group was in the Late Pliocene but it is sporadically present throughout the Quaternary (Mildenhall, pers. comm.). At the present altitudinal limit of the Nothofagus brassi group in New Guinea annual mean temperatures average 13 °C (Sluiter, pers. comm.); hence northern New Zealand temperatures must have declined below 13 °C for the taxon to be eliminated or else it became too dry and windy. All evidence indicates that the Pliocene had climates fluctuating rapidly from warm to cold temperate (with glaciations) relative to the present, the complexities of which are yet to be understood.

The Pleistocene excluding the Last Glacial cycle (2 m.y. - 70 000 BP)

Climatic trends for the Pleistocene are summarized in Fig. 3a from Beu and Edwards (1983). Figure 3b also shows Pleistocene stages and fluctuations, as inferred tentatively by Fleming (1975). The Early Pleistocene had numerous glacio-eustatic cycles in each stage of which the high and low points are stadials and interstadials respectively (Beu and Edwards 1983). The Late Pleistocene, covered by the Hawera Series time-stratigraphic unit, had at least three major glaciations and three major interglacial intervals according to South Island glacial evidence (Table 1).

The continuous deposition of thick Wanganui series terrestrial and marine sediments (Te Punga 1952; Fleming 1953) provides a good record of Early Pleistocene glacial events. Cool temperatures during the Early Pleistocene are indicated by cold water molluscs in marine sediments (Fleming 1953). There is a wide range of fossil plant indicators of climate from arid Acacia-Myrtaceae assemblages in the north (Mildenhall 1975a, 1975b), cool temperate Nothofagus/podocarp forest in the west of South Island and an absence of trees in the south. Couper and McQueen (1954) estimate that vegetation belts were lower by 800 m, equivalent to the temperature decline of up to 4.5 °C below the present. The Porika Glaciation has been tentatively assigned to the Hautawan stage (Mildenhall and Suggate 1981). The occurrence of cool moist pollen assemblages of Cyperaceae, Gramineae and Cyathea in the west of North Island and cool indicators in Wanganui sediments (Fleming 1953) suggests that cool climates continued into the Okehuan Stage (Mildenhall 1975c).

Late Pleistocene trends are summarized in Table 1 and include at least three glacials and three interglacials. This table is an interpretation of the available data from the Late Pleistocene and Holocene. It will clearly need to be revised as more information becomes available on the number and timing of glacials and stadials and transalpine correlations. Fission track dating (Seward 1974; Boellstorff and Te Punga 1977; Pillans 1983) suggests an age of 340 000 BP at the Hawera-Castlecliffian boundary for the beginning of warmer conditions during the Waiwheran, the earliest Late Pleistocene interglacial. The interglacial is represented by weathering of Waimakariri Valley glacial deposits (Gage 1958; Suggate 1965) and may correlate with the Braemore marine terraces dated at 340 000 BP in south Taranaki (Pillans 1983).

Colder temperatures during the Waimaungan glacial are represented by Hohonu glacial deposits in north Westland (Suggate 1965) and some of the Avoca glacial deposits in the Waimakariri Valley (Gage 1958). Warmer temperatures of the subsequent Terangian interglacial are indicated by weathering of Waimakariri Valley glacial deposits (Gage 1958) and in south Taranaki by the Ngarino marine terraces dated to 210 000 BP (Pillans 1983). Suggate (1965) has identified the next period of cold climate as the Waimean glacial and assigned Kumara 1 deposits in north Westland, Woodstock deposits in the Waimakariri Valley and Woodlands deposits in the Rakaia Valley to this event. However, recent evidence (Pillans et al. 1982) suggests some of these deposits to be younger and to include advances in the early Otiran glacial and Oturi interglacial (Table 1).

The Oturi interglacial, the last before the present cycle, possibly includes three identifiable periods of milder climate. There are three south Taranaki terraces, the Rapanui, Inaha and Hauriri, dated at 140 000 to 120 000, 100 000 and 80 000 BP respectively (Pillans 1983). Of the two South Island west coast shorelines, the Awatuna 2 has been assigned an age of about 80 000 BP (Moar and Suggate 1979). Oturi vegetational history in the central North Island shows a change from sub-alpine shrubland/grassland to Libocedrus - Dacrydium cupressinum dominant forest at about 80 000 BP (McGlone and Topping 1983). These trends correspond to a change from Nothofagus dominant forest in the early Oturi to Dacrydium cupressinum dominant forest, then a return to Nothofagus dominant forest at the end of the Oturi on the west coast of South Island (Moar and Suggate 1979). Mean temperatures are estimated at $1°$ to $2°C$ below present-day values (McGlone and Topping 1983). Climatic deterioration marking the beginning of the Otiran glacial stage occurred about 70 000 BP.

The Otiran Glaciation (70 000 - c. 13 000 BP)

Abundant palaeoenvironmental evidence exists for the Otiran (Table 1). The first glacial advance of the Otiran is represented by the Tui Creek glacial episodes in the Rakaia Valley and possible Kumara 1 advances in Westland (Soons and Gullentops 1973; Pillans et al. 1982). Colder climates are indicated by greater advances in this stade than in subsequent ones. Vegetation in the central North Island deteriorated from conifer hardwood forest to sparse grassland/shrubland with much eroding ground (McGlone and Topping 1983). Temperatures estimated from palaeosnowlines (Chinn 1983) and speleothems (Hendy and Wilson 1968) indicate a depression of 4 to $5°C$ compared with present-day values (Fig. 4a).

There is some evidence of an important interstade before 30 000 BP.

A change from sparse grassland/shrubland to subalpine shrubland/grassland and some Nothofagus forest from 42 000 until about 30 000 BP is indicated by central North Island pollen data (McGlone and Topping 1983). The latter part of the interstade is recorded on the west coast of South Island by replacement of Dacrydium cupressinum dominant forest from 30 000 BP wih the Nothofagus fusca group and eventually with Nothofagus menziesii at Hokitika (Moar and Suggate 1973) and a minor spread of Nothofagus before 31 000 BP followed by an immediate decline at Westport (Moar and Suggate 1979). Speleothem data (Hendy and Wilson 1968) confirm the existence of this interstade and McGlone (1983) estimates temperatures of $3\,^\circ C$ below present values.

The largest glacial advances in the late Otiran (Table 1) are represented by Kumara 2 (1) - Blackwater 1 and 2 - Bayfield 1 and 2 glacial episodes (Soons and Burrows 1978; Pillans et al. 1982). There is some uncertainty in dating the commencement of this advance but the presence of large glaciers in the Lake Wakatipu system in Otago around 26 000 BP (Bell 1982) suggests cooling before this date, a conclusion supported by the speleothem curve (Fig. 4). Evidence from Canterbury (Soons and

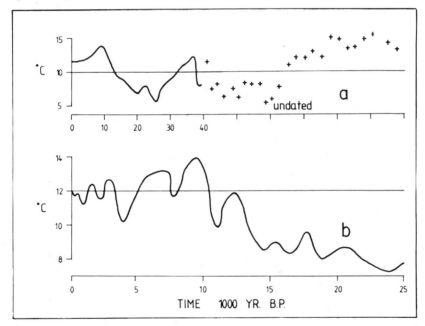

Figure 4(a). Oxygen isotope palaeotemperatures. From Hendy and Wilson (1968) with kind permission.

Figure 4(b). Reconstructed temperature trends for the last 25 000 years inferred from pollen, glacial and speleothem data.

Burrows 1978) indicates the end of this episode by an ice retreat of the Bayfield 1 and 2 advances before 22 000 BP, coeval with suggested pauses in Westland aggradation at 22 300 and 20 200 BP (Moar and Suggate 1973). Pollen sequences from the Waikato Basin, Volcanic Plateau and north Westland show progressive decline from forest/scrubland, or forest/grassland at 27 000 BP to open grassland by 20 000 BP

Table 1. Correlation of New Zealand Late Pleistocene glacial advances and interglacials.

Stages	N Westland Glacials	West Coast Cliffs	Upper Buller Glacials	Waimakariri River Glacials	Rakaia River Glacials	Wanganui Taranaki Interglacials	Date
				ARANUIAN			
OTIRA GLACIATION	Kumara 3 Late Kumara 2₂			Poulter 2 Poulter 1	Acheron 2 Acheron 1		13 000 15 000 17 000
				Minor interstadial			18 000
	Kumara 2₂			Blackwater 3	Bayfield 3		19 000
				Minor interstadial			22 000
	Kumara 2₁		Black Hill	Blackwater 2 Blackwater 1	Bayfield 2 Bayfield 1		27 000 c. 40 000
				Important interstadial			
	Kumara 1			Otarama	Tui Creek 3 Tui Creek 2 Tui Creek 1		
		Awatuna 2				Hauriri	70 000 80 000
	?			OTURI INTERGLACIAL		Inaha	100 000
		Awatuna 1				Rapanui	140 000 –
							120 000
WAIMEA GLACIATION	Kumara 1		Tophouse	Woodstock	Woodlands		
		Karoro ?					
				TERANGI INTERGLACIAL		Ngarino	210 000
WAIMAUNGA GLACIATION	Hohonu	Albion 3	Kikiwa	Avoca			
				WAIWHERA INTERGLACIAL		Brunswick Braemore	310 000 340 000

(McGlone 1983). In nearly all pollen spectra podocarps and Nothofagus make up a small component of the total pollen, suggesting that forest was not eliminated but restricted to favourable sites. Palaeosnowline and pollen evidence give a temperature depression of 4 to 5 °C below present-day values during the late Otiran maximum cooling episode (Chinn 1983; McGlone 1983).

The break in cold conditions around 22 000 BP, also indicated by replacement of subalpine grassland with Nothofagus menziesii forest from 21 900 until 19 200 BP (Harris et al. 1983) at Wellington, ended with Kumara 2(2) - Bayfield 3 - Blackwater 3 advances (Suggate and Moar 1970); Soons and Burrows 1978; Moar 1980). Minor recessions between Kumara 2(2) advances coeval with Bayfield 3 - Blackwater 3 retreat indicate a brief interstade from 17 000 to 18 000 BP (Chinn 1983). The final Kumara 2(2) advance (Suggate 1965; Moar 1980 and Mabin 1983) is dated around 17 000 to 16 000 BP. Throughout the late Otiran scrub in the west and grassland in the East, with many lowland forest species surviving in favourable sites (McGlone 1983), indicate a not exceptionally cold but a much drier climate in inland and eastern areas with more extremes of temperature.

The final Otiran stade is represented by the Kumara 3 - Poulter 2 - Acheron 2 advances from 15 000 to about 13 000 BP (Gage 1958; Soons and Gullentops 1973; Mabin 1983). Glacier expansion was much less than before 22 000 BP with palaeosnowline estimates of a 3° depression in mean temperature compared to the present (Chinn 1983).

The Aranuian (c. 13 000 BP to Present)

The beginning of the Aranuian (Table 2) was heralded by rapid glacier retreat and ice diminishing to volumes close to that of present-day amounts by 12 000 BP (Chinn 1983). Glacial activity was restricted to smaller advances in the Rakaia Valley and on the west coast of South Island which are dated between 11 500 and 11 000 BP. The response of vegetation to warming was rapid; Podocarpus spicatus dominant podocarp/hardwood forest had occupied nearly all central and northern North Island lowland and lower montane sites by 13 000 BP (Harris 1963; McGlone and Topping 1977). Dacrydium cupressinum dominant podocarp/hardwood forest was established at Westport by 12 000 BP (Moar and Suggate 1979). In other South Island areas except the dry interior, scrub/grassland replaced grassland. Transition from scrub and grassland to forest was very rapid, taking only about 300 years (McGlone 1983). Forests grew under a cooler climate than at present and many areas were drier though not as dry as during the late Otiran. Chinn (1983) estimates a mean temperature depression of 2.5 °C for the glacial advance between 11 500 and 11 000 BP (Fig. 4b).

Chinn (1983) has summarized events in the Waimakariri Valley. Two glacier advances dated at 9500 and 8000 BP are recorded although these are minor compared with those of the Otiran (Lintott and Burrows 1973; Burrows 1979; Chinn 1981). After 8000 BP there is no evidence of any advance until 4000 BP. Between 10 000 and 8000 BP New Zealand had a more complete forest cover than at any other time since the Otiran. The South Island had a near simultaneous spread of podocarp forest, Dacrydium cupressinum dominant in the west and Podocarpus spicatus dominant in the east, onto lowland regions previously occupied by scrub and grassland (McIntyre and McKellar 1970; Moar 1971). In southern and eastern areas of the North Island, forest developed where it was previ-

Table 2. Tentative correlation of the New Zealand glacial events (from Chinn 1983).

Years BP	Waimakariri Valley	Westland	Rakaia Valley	Cameron Valley	Mt Cook	Ben Ohau Range
0	Barker		Whitcombe	Arrowsmith	Mt Cook	Dun Fiunary
430						Whale Stm
1 000	Omalley		Lyell		Foliage Hill	Jacks Stm
2 500						
4 000	Arthurs Pass		Meins Knob	Marquee	Hermitage	Ferintosh
8 000	McGrath II		Jagged Stm	Wildman II		
9 500	McGrath I		Lake Stm	Wildman I		Birch Hill
11 000						
11 500		Waiho Loop	Acheron III			

ously absent. In the older northern forests the dominant podocarp changed from Podocarpus spicatus to Dacrydium cupressinum (McGlone 1983). Widespread occurrences of frost and drought sensitive species such as Ascarina lucida (McGlone and Moar 1977) and Tetrapathaea tetrandra (Mildenhall 1979) suggest that this period was the wettest in northern and eastern areas of both islands, and the mildest and most equable of the Aranuian. Speleothem palaeotemperatures (Hendy and Wilson 1968) support the conclusions based on pollen, providing temperature estimates of between 1 and 2 $^\circ$C above modern values. A slight change is indicated after 7500 BP when upland South Island areas and Central Otago became forested (McGlone 1983).

Speleothem palaeotemperatures (Fig. 4a) indicate a decline after 7000 BP coeval with small glacial events in the Southern Alps (Burrows 1979) which are dated at about 4000, 1000 and 430 BP (Chinn 1981). From 5000 BP Dacrydium cupressinum became less common in North Island forests, Ascarina lucida declined dramatically (McGlone and Moar 1977; Mildenhall 1977) and Nothofagus first spread in the south of North Island and throughout South Island. Many areas of podocarp/hardwood forests were replaced by mixed Nothofagus/podocarp forests. Dacrydium cupressinum with Nothofagus menziesii replaced Podocarpus spicatus in coastal Southland and Otago (McGlone 1983). The spread of Nothofagus forest ceased between 3000 to 2000 BP, suggesting a new forest equilibrium with respect to podocarp forest. The vegetational and speleothem evidence indicates the development of a cooler but less equable climate after 5000 BP. A return to drier conditions occurred in many areas but especially in the north and east of both islands.

5. PALAEOCIRCULATION PATTERNS

Fossil pollen data are detailed enough to allow discrimination of local rainfall relative to modern values after 20 000 BP. From the climatic relationships discussed in Section 2 qualitative rainfall reconstructions are presented for 18 000, 9000 and 6000 BP (Figs. 5a-5c). In the Northern Hemisphere 18 000 BP was the period of maximum ice volume for the Last Glacial advance while 6000 BP was the warmest period in the Holocene. 9000 BP is selected because it was the warmest period of the Holocene in New Zealand.

At 18 000 BP, in North Island sites (Fig. 5a) from Waikato Basin, Tongariro, Taranaki, Lake Poukawa and Wellington, scrub and grassland tended to dominate (McIntyre 1970; McGlone and Topping 1977; McGlone 1980b). Charcoal occurs in the deposits and is especially abundant at the Taranaki and Lake Poukawa sites, indicating that the vegetation was subjected to frequent fire. Some Nothofagus menziesii forest did occupy the higher areas bordering the Waikato Basin and Wellington. In South Island at Hokitika (Moar and Suggate 1973) there is evidence of forest dominated by Nothofagus menziesii and at Westport (Moar and Suggate 1979) scrub and grassland dominated. Present-day Nothofagus menziesii distribution and early Post-glacial occurrences at Milford Sound and Lake Manapouri (Johnson 1978; Wardle and McKellar 1978) indicate that Nothofagus menziesii was present in the west and south of South Island. In other parts forest was restricted and grassland was widespread.

Figure 5a illustrates the climatic inferences from the proxy data. The data support a colder climate which was considerably drier, especially in the north and east of North Island. Additional North Island evidence comes from the widespread occurrence of loess and sand dunes, and from Lake Rotorua which reached its lowest recorded level at this time (Kennedy et al. 1978). South Island data support an intensification of east-west moisture gradients. The vegetation indicates a cool, moist climate in the west and south but very dry conditions in the east. All palaeoenvironmental evidence points to a climatic regime associated with a strengthened westerly circulation (Salinger 1980a), still wet in the west and south of South Island but drier elsewhere. This conclusion is supported by the more easterly zonal distribution of quartz grains off the east coast of New Zealand (Thiede 1979). Similarly, the stronger westerly and southwesterly circulation is also associated with higher rainfall in southern New Zealand and dry conditions in the north and east. This pattern accords with a weaker Pacific Walker Circulation (Kidson 1975; Gordon, pers. comm.).

Between 10 000 and 8000 BP all North Island evidence indicates a milder and more equable climate (Fig. 5b) than the present. Pollen spectra from Hamilton, Hauraki Plains, Mt Egmont, Tongariro, the Ruahine Ranges and Wellington indicate a dominance of Dacrydium cupressinum and contain the environmentally sensitive Ascarina lucida and Dodonaea viscosa and other indicator species (Harris 1951 and 1958; McGlone and Topping 1977; McGlone 1980a). Dacrydium cupressinum, Ascarina and Dodonaea have a decided preference for sites with high rainfall and no soil moisture deficit at any time of the year, and are intolerant of extremes in temperature. South Island data show more spatial distribution. Occurrence of Dacrydium cupressinum forest containing Ascarina lucida in the Upper Maruia Valley and at Hokitika, Springs Junction, Bell Hill and Fox Glacier suggests a warmer climate (Moar 1971; Moar and Suggate 1979). At the same time Dodonaea viscosa occurred in Nelson (Dodson 1978), suggesting a more equable climate. Evidence from Tophouse, Cass, Timaru and near Christchurch suggests a wetter climate with less year-to-year rainfall variability (Moar 1971;

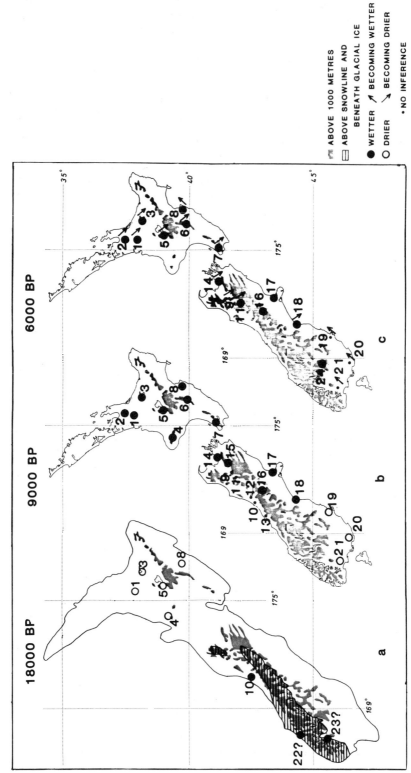

Figure 5. Reconstructed moisture conditions from pollen data relative to the present for (a) 18 000 BP, (b) 9000 BP, and (c) 6000 BP. Sites: 1. Hamilton, 2. Hauraki Plains, 3. Lake Rotorua, 4. Mt Egmont, 5. Tongariro, 6. Ruahine Ranges, 7. Wellington, 8. Lake Poukawa, 9. Upper Maruia Valley, 10. Hokitika, 11. Springs Junction, 12. Bell Hill, 13. Fox Glacier, 14. Nelson, 15. Tophouse, 16. Cass, 17. Christchurch, 18. Timaru, 19. Swampy Summit, 20. and 21. Coastal Southland, 22. Milford Sound, 23. Lake Manapouri, 24. Old Man Range.

Lintott and Burrows 1973). Dated charcoal remains from Canterbury support this conclusion (Cox and Mead 1963). Only in the south do pollen data suggest drier conditions. At Swampy Summit and in coastal Southland sites (McIntyre and McKellar 1970; McGlone 1980b) the forest was dominated by Podocarpus spicatus, which indicates a warmer but drier climate than the present. Palaeoclimatic inferences at 9000 BP are summarized in Fig. 5b. The climatic pattern of high rainfall in the north and east of North Island and the north of South Island with drier conditions in the south of South Island is associated with a stronger Walker Circulation and weaker westerly circulation over New Zealand, the opposite of the 18 000 BP inferences.

The vegetational evidence of 6000 BP suggests a weakening of the 9000 BP pattern (Fig. 5c). Pollen spectra indicate that conditions were milder than at present but not as warm as at 9000 BP. Most North Island sites were wetter than at present although not as wet as at 9000 BP, indicating a drying trend from after 9000 BP. Conversely, areas in the south of South Island exhibit a rainfall increase when Podocarpus spicatus dominant forest was replaced by Dacrydium cupressinum dominant forest (McIntyre and McKellar 1970; McGlone 1980b). Such a climatic pattern suggests a slight weakening of the Walker Circulation and strengthening of the westerly circulation over the New Zealand region.

6. DISCUSSION AND CONCLUSIONS

The data presented show that the New Zealand region experienced a wide range of climates from warm temperate conditions at the beginning of the Pliocene to sub-Antarctic cold during the Pleistocene glaciations. Although temperatures are estimated to have decreased up to 5 °C on 20th century values during the Pleistocene, geomorphic and fossil plant evidence confirms that the climate remained oceanic in character. The abundance of fluvioglacial deposits as a result of running water rather than periglacial landforms (Gage 1965) indicates that winters of continuous frost and snow were rare. The survival of drought sensitive species throughout the Pleistocene and their rapid spread in the Holocene indicate that these species survived in favoured coastal refugia (Wardle 1963). Hence the New Zealand Pleistocene was a period of colder climate with mean temperatures several degrees below modern-day values. Warm and cold sequences of weather would still have occurred even during winter, unlike in Europe and North America, where winters were of continuous cold weather and permanent snow and ice.

Late Pleistocene and Holocene changes in the southwest Pacific occurred several millennia before those in the Northern Hemisphere. Evidence suggests a Last Glacial Maximum in New Zealand between 26 000 and 22 000 BP in contrast with a 18 000 BP maximum in the Northern Hemisphere. The rapid warming observed after the last Otiran glacial episode is in advance of northern latitudes. Finally, the maximum Holocene warming around 9000 BP is 3000 years before maximum warmth is observed in Europe and North America. These climatic trends support the discovery that Southern Hemisphere sea surface temperatures usually lead Northern Hemisphere changes by about 3000 years (Hays 1977). One explanation of the hemispheric asynchroneity in temperature trends is the rapid freezing and melting of Antarctic sea ice. The northern continental ice sheets respond slowly in comparison. The faster response of Antarctic sea ice may cause southern mid-latitude temperatures to cool or warm more rapidly than equivalent northern latitude temperatures. Certainly the rapid present-day seasonal fluctuations in the sea ice area (Streten and Pike 1980) support such a hypothesis. These pertinent facts should be carefully considered before correlations of Northern Hemisphere events are applied to the Southern Hemisphere; the two need not vary synchronously, an assumption that is so often made.

The palaeoclimatic reconstructions from fossil pollen data at 18 000, 9000 and 6000 BP illustrate how the determination of local climates permits the reconstruction of regional circulation trends. The strong influence of topography is the prime factor responsible in the distinct responses of rainfall to variations in the circulation. 18 000 and 9000 BP represent the two extremes of cold and warm temperature in the Late Quaternary. The results suggest that the regional westerly circulation is stronger in periods of colder climate and weaker in periods of warmer climate. Indeed, available data at 9000 BP are consistent with a strengthened tropical Walker Circulation. The pattern at 6000 BP, after a slight cooling had occurred, shows a weakening of the 9000 BP pattern.

The palaeocirculation reconstructions are consistent with numerical modelling studies using glacial and hypsithermal age boundary conditions (Williams et al. 1974; Gates 1976; Manabe and Hahn 1977; Kutzbach 1981). The models show a strengthened mid-latitude westerly circulation under a cold climatic regime and a weaker mid-latitude westerly circulation under a warm climatic regime. The reconstructions and evidence reviewed here indicate that New Zealand, in a location open to both mid-latitude and sub-tropical atmospheric centres of action, is an ideal platform from which to monitor past climates. These influences varied in importance during the Late Cenozoic, causing New Zealand to have a wide range of climates from warm temperate to sub-Antarctic. This range of past climates indicates possible scenarios which future decision makers may need to consider.

ACKNOWLEDGEMENTS

The author is very appreciative of the assistance of AG Beu, TJH Chinn, MJ Crozier, AR Edwards, JW Kidson, MCG Mabin, MS McGlone, WJ Maunder, DC Mildenhall, B Mullen, DT Pocknall, RB Pillans and JM Soons for fruitful discussions, ideas and suggestions towards improving the manuscript.

REFERENCES

Backman J, Shackleton NJ and Tauxe L 1983. Quantitative nannofossil correlation to open ocean deep-sea sections from Plio-Pleistocene boundary at Vrica, Italy. Nature 304: 156-158.

Bell DH 1982. Geomorphic evolution of a valley system: The Kawarau Valley, Central Otago. In: JM Soons and MJ Selby (eds.), Landforms of New Zealand. Longman Paul, Auckland. p. 7-41.

Beu AG and Edwards AR, in press. New Zealand Pleistocene and late Pliocene glacio-eustatic cycles. Palaeogeogr. Palaeoclimatol. Palaeoecol.

Beu AG and Hornibrook NdeB 1983. Guide book for tour B1 Neogene geology - North Island east coast basin. 15th Pacific Science Congress, Dunedin. 28pp.

Boellstorff JD and Te Punga MT 1977. Fission track ages and correlation of Middle and Lower Pleistocene sequences from Nebraska and New Zealand. N.Z.J. Geol. Geophys. 20: 47-58.

Brodie JW 1960. Coastal surface currents around New Zealand. N.Z.J. Geol. Geophys. 3: 235-252.

Burrows CJ 1979. A chronology for cool-climate episodes in the Southern Hemisphere 12000 - 1000 years BP. Palaeogeogr. Palaeoclimatol. Palaeoecol. 27: 287-347.

Chinn TJH 1981. Use of rock weathering-rind thickness for Holocene absolute age-dating in New Zealand. Arc. Alp. Res. 13: 33-45.

Chinn TJH 1983. New Zealand glacial record, 10 to 15000 BP. Proc. 1st CLIMANZ Conf. Aust. Acad. Sci. Canberra. p. 71-72.

CLIMANZ 1983. Proc. 1st CLIMANZ Conf. Aust. Acad. Sci. Canberra. 149pp. + diagr.

Couper RA and McQueen DR 1954. Pliocene and Pleistocene plant fossils of New Zealand and their climatic interpretation. N.Z.J. Sci. Technol. 35B: 398-420.

Cox JE and Mead CB 1963. Soil evidence relating to post-glacial and climate on the Canterbury Plains. N.Z. Ecol. Soc. Proc. 10: 28-38.

Devereux I 1967. Oxygen isotope palaeotemperature measurement on New Zealand Tertiary fossils. N.Z.J. Sci. 10: 988-1011.

Dodson JR 1978. A vegetational history from north-east Nelson. N.Z.J. Bot. 16: 371-378.

Fleming CA 1944. Molluscan evidence of Pliocene climatic change in New Zealand. Trans. R. Soc. N.Z. 74: 207-220.

Fleming CA 1953. The geology of the Wanganui subdivision. N.Z. geol. Surv. Bull. 52.

Fleming CA 1975. The quaternary record of New Zealand and Australia. In: RP Suggate and MM Cresswell (eds.), Quaternary Studies. R. Soc. N.Z., Wellington. pp. 1-20.

Gage M 1958. Late Pleistocene glaciations of the Waimakariri Valley, Canterbury, New Zealand. N.Z.J. Geol. Geophys. 1: 123-155.

Gage M 1965. Some characteristics of Pleistocene cold climates in New Zealand. Trans. R. Soc. N.Z. 3: 11-21.

Gage M and Suggate RP 1958. Glacial chronology of the New Zealand Pleistocene. Bull. geol. Soc. Amer. 69: 589-598.

Gardner JV 1982. High-resolution carbonate and organic-carbon stratigraphies for the late Neogene and Quaternary from the western Caribbean and eastern equatorial Pacific. In: WL Prell, JV Gardner et al. (eds.), Initial reports of the Deep Sea Drilling Project 68. p. 347-364.

Gartner S 1977. Calcareous nannofossil biostratigraphy and revised Zonation of the Pleistocene. Mar. Micropaleont. 2: 1-25.

Gates WL 1976. The numerical simulation of ice-age climate with a global general circulation model. J. Atm. Sci. 33:1844-1873.

Harris WF 1951. New Zealand plants and their story: clues to the past. N. Z. Sci. Rev. 9:5-9.

Harris WF 1959. Pollen bearing deposits and forest history. In: The ecology of the Hutt Valley. N.Z. Ecol. Soc. Proc. 6:46-48.

Harris WF 1963. Paleo-ecological evidence from pollen and spores. N. Z. Ecol. Soc. Proc. 10:38-44.

Harris WF, Mildenhall DC and McQueen DR 1983. Past and present effects of the last glacial episode (Otiran) in the Cook Strait area. Proc. 1st CLIMANZ Conf. Aust. Acad. Sci. Canberra.

Hays JD 1977. The late Quaternary history of Antarctic seas. In: Abstracts X INQUA Congress, Birmingham. Geoabstracts, Norwich. p.200.

Hendy CH and Wilson AT 1968. Palaeoclimatic data from speleothems. Nature 219:48-51.

Hornibrook NdeB 1971. New Zealand Tertiary climate. N. Z. geol. Surv. Rep. No. 47. 19pp.

Hornibrook NdeB 1975. Globorotalia trunatuliniodes and the Pliocene-Pleistocene boundary in northern Hawke's Bay, New Zealand. In: Y Takayanagi and T Saito (eds.), Progress in micropalaeontology, Micropaleontology Press, New York. p.83-102.

Johnson PN 1978. Holocene plant remains from the shores of Lake Manapouri, New Zealand. N. Z. J. Bot. 16:141-145.

Kennedy NM, Pullar WA and Pain CF 1978. Late Quaternary land surfaces and geomorphic changes in the Rotorua Basin, North Island, New Zealand. N. Z. J. Sci. 21:249-264.

Kidson JM 1975. Tropical eigenvector analysis and the Southern Oscillation. Mon. Weath. Rev. 103:187-196.

Kutzbach JE 1981. Monsoon climate of early Holocene: climate experiment using the earth's orbital parameters for 9000 years ago. Science 214: 59-61.

Lintott WH and Burrows CJ 1973. A pollen diagram and macro-fossils from Kettlehole Bog, Cass, South Island, New Zealand. N. Z. J. Bot. 11:269-282.

Mabin MCG 1983. Late Otiran sedimentation and glacial chronology in the Warwick Valley, southeast Nelson. N. Z. J. Geol. Geophys., in press.

McGlone MS 1980a. Late Quaternary vegetation history of central North Island, New Zealand. Ph.D thesis, Univ. Canterbury, Christchurch.

McGlone MS 1980b. Post-glacial vegetation change in coastal Southland. N. Z. J. Ecol. 3:153-154.

McGlone MS 1983. New Zealand vegetation and climate since the last glaciation: 32+/-KA, 25-20KA, 18+/-2KA, 15-10KA, 7+/-2KA. In: JMA Chappell and A grindrod (eds.), Proc. 1st CLIMANZ Conf. Aust. Acad. Science. ANU, Canberra. p.16-17. 38-39. 63-66. 81-82. 102-103.

McGlone MS and Moar NT 1977. The Ascarina decline and post-glacial climatic change in New Zealand. N. Z. J. Bot. 15:485-489.

McGlone MS and Topping WW 1977. Aranuian (post-glacial) pollen diagrams from the Tongariro region, North Island, New Zealand. N. Z. J. Bot. 15:749-760.

McGlone MS and Topping WW 1983. Late Quaternary vegetation, Tongariro region, central North Island, New Zealand. N. Z. J. Bot. 21:53-76.

McIntyre DJ 1970. Appendix 1. Pollen analyses from the Lindale section, Paraparaumu. In: CA Flemming (ed.), Radiocarbon dating and pollen analysis from the Otiran periglacial fans in western Wellington. Trans. R. Soc. N. Z. Earth Sciences 7:197-208.

McIntyre DJ and McKellar IC 1970. A radiocarbon dated post glacial pollen profile from Swampy Hill, Dunedin, New Zealand. N. Z. J. Geol. Geophys. 13:346-349.

McQueen DR, Mildenhall DC and Bell CJE 1968. Palaeobotanical evidence for changes in the Tertiary climates of New Zealand. Tuatara 16:49-56.

Manabe S and Hahn DG 1977. Simulation of the tropical climate of an ice-age. J. Geophys. Res. 82:3889-3911.

Maunder WJ 1971. The climate of New Zealand - physical and dynamical features. In: J Gentilli (ed.), World Survey of Climatology 13:213-227.

Mildenhall DC 1973. Vegetational changes in the New Zealand Quaternary based on pollen studies, with brief comment on the Pliocene background. In: GD Mansergh (ed.), The New Zealand Quaternary. An Introduction, IX INQUA Congress, Christchurch. p.20-32.

Mildenhall DC 1975a. Palynology of the Acacia-bearing beds in the Komako district, Pohangina Valley, North Island, New Zealand. N. Z. J. Geol. Geophys. 18:209-228.

Mildenhall DC 1975b. Lower Pleistocene palynomorphs from the Ohuka Carbonaceous Sandstone, south-west Auckland, New Zealand. N. Z. J. Geol. Geophys. 18:675-681.

Mildenhall DC 1975c. New fossil spore from the Pakihikura pumice (Okehuan, Quaternary). Rangitikeu Valley, New Zealand. N. Z. J. Bot. 20:242-245.

Mildenhall DC 1977. Appendix 2. Hautawan and presumed Hautawan palynomorphs from northern Hawke's Bay. N. Z. J. Geol. Geophys. 20:242-245.

Mildenhall DC 1979. Holocene pollen diagrams from Pauatahanui Inlet, Porirua, New Zealand. N. Z. J. Geol. Geophys. 22:585-591.

Mildenhall DC 1980. New Zealand late Cretaceous and Cenozoic plant biogeography: A contribution. Palaeogeogr. Palaeoclimatol. Palaeoecol. 31:197-233.

Mildenhall DC and Harris WF 1970. A cool climate po-len assemblage from the type Waipipian (middle Pliocene) of New Zealand. N. Z. J. Geol. Geophys. 13:586-591.

Mildenhall DC and Pocknall DT, this volume. Palaeobotanical evidence for changes in Miocene and Pliocene climates in New Zealand.

Mildenhall DC and Suggate RP 1981. Palynology and age of the Tadmor group (late Miocene-Pliocene) and Porika Formation (early Pleistocene), South Island, New Zealand. N. Z. J. Geol. Geophys. 24:515-528.

Moar NT 1971. Contributions to the Quaternary history of the New Zealand flora, 6. Aranuian pollen diagrams from Canterbury, Nelson and North Westland, South Island. N. Z. J. Bot. 9:80-145.

Moar NT 1973a. Late Pleistocene vegetation and environment in southern New Zealand. In: Palaeoecol. Africa 8:179-198.

Moar NT 1973b. Contributions to the Quaternary history of the New Zealand flora, 7. Two Aranuian pollen diagrams from central South Island. N. Z. J. Bot. 11:291-304.

Moar NT 1980. Late Otiran and early Aranuian grassland in central South Island. N. Z. J. Ecol. 3:4-12.

Moar NT and Suggate RP 1973. Pollen analysis of late Otiran and Aranuian sediments at Blue Spur Road (S51), North Westland. N. Z. Geol. Geophys. 16:333-344.

Moar NT and Suggate RP 1979. Contributions to the Quaternary history of the New Zealand flora, 8. Interglacial and glacial vegetation in the Westport district, South Island. N. Z. J. Bot. 17:361-387.

Pillans RB 1983. Upper Quaternary marine terrace chronology and deformation, South Taranaki, New Zealand. Geol. 11:292-297.

Pillans RB, Pullar WA, Selby MJ and Soons JM 1982. The age and the development of the New Zealand landscape. In: JM Soons and MJ Selby (eds.), Landforms of New Zealand. Longman Paul, Auckland. p.15-44.

Salinger MJ 1980a. New Zealand climate: 1. Precipitation patterns. Mon. Weath. Rev. 108:1892-1904.

Salinger MJ 1980b. New Zealand climate: 2. Temperature patterns. Mon. Weath. Rev. 108:1905-1912.

Salinger MJ 1982. New Zealand climate: Scenarios for a warm high-CO_2 world. Weather and Climate 2:9-15.

Seward D 1974. Age of New Zealand Pleistocene substages by fission track dating of glass shards from tephra horizons. Earth Planet. Sc. Lett. 24:242-248.

Shackleton NJ and Kennett JP 1975. Palaeotemperature history of the Cenozoic and the initiation of Antarctic glaciation: oxygen isotope and carbon isotope analysis in DSDP sites 277,279,281. In: JP Kennett, RE Houtz et al. (eds.), Initial reports of the Deep Sea Drilling Project 29. US Govt Printing Office, Washington, DC. p.743-755.

Shackleton NJ and Opdyke ND 1976. Oxygen-isotope and paleomagnetic stratigraphy of Pacific core V28-239, late Pliocene to latest Pleistocene. Geol. Soc. Am. Mem. 145: 449-464.

Soons JM 1981. CLIMANZ: A venture in Australia-New Zealand co-operation in the field of climate change. Proc. 11th N. Z. geog. Conf. 201-203.

Soons JM and Burrows CJ 1978. Dates for Otiran deposits, including plant microfossils and macrofossils, from Rakaia Valley. N. Z. J. Geol. Geophys. 21:607-615.

Soons JM and Gullentops FW 1973. Glacial advances in the Rakaia Valley, New Zealand. N. Z. J. Geol. Geophys. 16:425-438.

Steiner JT 1980. The climate of the south-west Pacific region. N. Z. met. Serv. misc. Publs, No. 166. 35pp.

Streten NA and Pike DJ 1980. Characteristics of the broad-scale Antarctic sea ice extent and the associated atmospheric circulation 1972-1977. Arch. Met. Geoph. Biokl. Ser. A. 29:279-299.

Suggate RP 1965. Late Pleistocene geology of the northern part of the South Island, New Zealand. N. Z. geol. Surv. Bull. 77.

Suggate RP and Moar NT 1970. Revision of the chronology of the late Otira glacial. N. Z. J. Geol. Geophys. 13:742-746.

Tauxe L, Opdyke ND, Pasini G and Elmi C 1983. Age of the Plio-Pleistocene boundary in the Vrica section, southern Italy. Nature 304: 125-129.

Te Punga MT 1952. The geology of the Rangitikei Valley. Mem. N. Z. Geol. Surv. 8.

Thiede J 1979. Wind regimes over the late Quaternary southwest Pacific Ocean. Geol. 7:259-262.

Thierstein HR, Gitzenauer KR, Molfino B and Shackleton NJ 1977. Global synchroneity of late Quaternary cocclith datum levels: validation by oxygen isotopes. Geol. 5: 400-404.

Trenberth KE 1976a. Fluctuations and trends in indices of the Southern Hemisphere circulation. Q. J. R. met. Soc. 102:65-75.

Trenberth KE 1976b. Spatial and temporal variations in the Southern Oscillation. Q. J. R. met. Soc. 102:639-653.

Troup AJ 1965. The Southern Oscillation. Q. J. R. met. Soc. 91:490-506.

Walker GT and Bliss EW 1932. World Weather V. Mem. R. met. Soc. 4:53-84.

Walker GT and Bliss EW 1932. World Weather VI. Mem. R. met. Soc. 4:N39.

Wardle P 1963. Evolution and distribution of the New Zealand flora as affected by Quaternary climates. N. Z. J. Bot. 1:3-17.

Wardle P and McKellar MH 1978. Nothofagus menziesii leaves dated at 7490 yr BP in till-like sediments at Milford Sound, New Zealand. N. Z. J. Bot. 16:153-157.

Watts IEM 1947. The relation of New Zealand weather and climate: An analysis of the westerlies. N. Z. Geogr. 3:115-129.

Williams J, Barry RG and Washington WM 1974. Simulation of the atmospheric circulation using NCAR global circulation model with ice age boundary conditions. J. appl. Met. 13:305-317.

Wright PB 1975. An index of the Southern Oscillation. Climatic Research Unit Publ., CRURP4. 22pp.

The changing face of the evidence: An examination of proxy data for climate change during the Late Pleistocene in New Zealand

J.M. SOONS
University of Canterbury, Christchurch, New Zealand

ABSTRACT. The evidence for changing climate during the Late Pleistocene depends on the interpretation of a variety of sedimentological, geomorphological and palynological data, in a framework determined by the scientific conventions of a particular time. Over time, this framework changes, but frequently the interpretation is not adjusted, or is adjusted only partially. Consequently, the basis for climatic reconstructions may be inadequate and long distance correlations falsified. This paper examines some of the changes through which New Zealand proxy data have passed in the last twenty-five years, and reviews the time scale of Late Glacial events on South Island.

INTRODUCTION

The problems inherent in proxy data for climate change - usually the only data available - are generally well-known, and have been discussed by a number of writers (e.g. Frenzel 1973). Nevertheless, it may not come amiss to review some of these problems in the context of a specific area, when that area is the subject of an attempt to model the changes of climate which it has experienced. Salinger has presented a comprehensive review of the current state of the art of proxy climate data for New Zealand (Salinger, this volume). This paper traces some of the historical changes that have led to the development of the present picture, insofar as it is based on glacial advances, and examines some of the unresolved problems, for one or two specific cases, that may well result in further modifications of existing views.

It may be useful first to review the processes which provide data for a climatic reconstruction. Initially, a field worker - geologist, botanist, geomorphologist or whatever - finds a site of interest and describes it, with appropriate analyses according to the type of material present and the expertise of the worker. An interpretation is then made. This is the first major step in departure from the purely factual. A radiometric date - or dates - for the site, may or may not be obtained. Age and significance are then reviewed on the basis of prevailing Quaternary concepts, whether global or regional. As an example, a site at which plant remains are found will be the subject of careful study, but will inevitably progress from a careful description and analysis to a series of interpretations as to the environmental conditions obtaining at the time the deposit was formed, the probable climate of the time, and its relationship to other, more or less distant sites. A radiometric date may permit the site to be slotted into an existing chronology - or may be rejected on the grounds that it does not fit such a chronology. If no date is available, the site may be allocated a place in the chronological framework on the basis of a range of subjective judgements, but all with a tendency to reinforce existing dogma. All through this process a series of interpretations is being made, some of which may not be apparent to later users of material, and not all of which may be revised as new concepts and information become available.

In New Zealand, as in most other places, the data on which climate interpretations are based have been obtained over a time period during which some fundamental changes took place in concepts of the nature of Quaternary events. There is no reason to suppose that this process has been completed.

DEVELOPMENT OF A NEW ZEALAND GLACIAL SEQUENCE

Until about twenty-five years ago a two-glaciation model of Quaternary events seems to have been widely accepted in New Zealand, although a monoglacial hypothesis was strongly advocated by at least one worker. The modern developments begin, however, in 1958 when three interglacials were recognized (Gage 1958). These separated four advances, or series of advances, on a variety of grounds, including degree and depth of dissection, and weathering, as well as spatial and stratigraphic position. Gage and Suggate (1958) compared glacial events on both sides of the Southern Alps and regrouped the advances into two glaciations, eliminating two of the interglacials recognized by Gage (Table 1).

Table 1. New Zealand glacial sequence after Gage and Suggate (1958).

Stage	Advances	
	Westland	Canterbury
Otiran	Kumara 3	Poulter
	Kumara 2	Blackwater *
	Kumara 1	Otarama *
	Hohonu	Woodstock
	Interglacial	
Waimaungan	Waimaunga	Avoca

*Interglacials indentified by Gage (1958).

In 1961 this sequence was modified, adding a new glaciation and reinstating one of the 'lost' interglacials (Gage 1961) (Table 2). Together with the (much earlier) Ross Glaciation this revision brought the New Zealand Quaternary sequence into line with the Alpine four-glaciation model, as it was apparently intended to do. The revival of the Otarama/Woodstock interglacial was not the result of new information, since the floristic evidence on which it was based was already

Table 2. New Zealand glacial sequence after Gage (1961).

Stage	Advances	
	Westland	Canterbury
Otiran	Kumara 3 Kumara 2 (2 advances) Kumara 1	Poulter Blackwater Otarama
	Interglacial	
Waimaungan	Hohonu	Woodstock
	Interglacial	
Porikan	Porika	Avoca

known and described by Gage (1958). The addition of the Porikan to the sequence did, however, depend on new evidence.

A further modification of the sequence was published in 1965. The addition of another glacial to the Late Pleistocene sequence (Table 3) was based on information from a new site, Sunday Creek, which yielded evidence of an interglacial to glacial vegetation change, and lay beyond the effective range of C14-dating (Suggate 1965; Dickson 1972).

Table 3. New Zealand glacial sequence after Suggate (1965).

Glaciation	Interglacial	Advances	
		Westland	Canterbury
Otira		Kumara 3 Kumara 2 $^{2}_{1}$	Poulter Blackwater Otarama
Waimea	Oturi	Kumara 1	Woodstock
Waimaunga	Terangi Waiwhero	Hohonu Porika	Avoca (part) Avoca (part)

The Sunday Creek deposits lie below those of the older of the two Kumara 2 advances, and thus provided a basis for transferring the Kumara 1 advance out of the Otira Glaciation to the new Waimea Glaciation. Further support was given by the identification of two episodes of high sea-level during the Oturi Interglacial.

This revision broke previously accepted correlations across the Southern Alps and has been the cause of considerable debate (e.g. Soons 1966; Gage and Soons 1973). No new information affecting the Canterbury sequence was offered to support the change, and Gage's (1958) suggestion of an Otarama/Blackwater interglacial (Table 1), which might have been considered comparable to the new status of the Kumara 1/Kumara 2 interval, was apparently seen as lacking substance.

Both the Blackwater and Otarama advances and their correlatives in Canterbury were multiple events, and it is always possible that new information may result in a re-assessment of trans-Alpine correlations.

A consequence of the 1965 revision was that it stimulated renewed investigation into the Westland area, from which derives much of the evidence on which the glacial/interglacial sequence is based. The results of this work still give only an incomplete picture of Late Pleistocene events and have yet to be fully incorporated into a standard sequence. Preliminary revisions have been suggested (e.g. Pillans 1983) and a much more complex story than was envisaged in 1965 seems certain to emerge. Much of the Westland story is based on relationships of marine deposits and raised shorelines, regarded as evidence of high interglacial sea-levels, and gravel deposits and aggradation surfaces equated with periods of glacial advance. Preservation of these at several levels is related to ongoing tectonic uplift. Additional marine deposits have been identified, while two, or perhaps three surfaces appear now to be included in that mapped as Kumara 1 in 1965 (Suggate, pers. comm.). There is also evidence that rates of uplift in the area are spatially more variable than was previously suspected. This will have its effect on revision of the glacial/interglacial sequence, since some correlations are essentially based on altitude.

From the point of view of reconstructions of patterns of climate change, the series of modifications of the New Zealand glacial sequence would have been reflected in rather different climatic interpretations, depending on precisely when such interpretations were undertaken. Correlations with other areas and reconstruction of large-scale trends based on present published material are likely to require revision as results of ongoing work are incorporated into the literature.

LATE GLACIAL EVENTS IN THE SOUTH ISLAND

A very real problem underlying the changes in the sequence is the absence of reliable dating. This continues to be a problem even in the time range for which 'good' C14 dates may be expected. As an example we can consider the late glacial/post-glacial transition, and trace the evolution of ideas about the sequence of advances and retreats. Here the real difficulty is not usually the dates themselves, but the interpretation.

The first published attempts to date late glacial events (Gage 1958; Gage and Suggate 1958) largely avoided discussion of absolute dates. In 1961 Gage mentioned that a late Otiran advance had occurred before 15 000 BP, citing (in a footnote) a line of argument developed by Suggate in 1965 (Gage 1961). This was essentially that all the main glacial lake basins of the South Island were "last occupied by ice during the same period of ice advance that formed narrow inner moraines close to, or little separated from, the wider moraine belts of the main last glaciation advance" (Suggate 1965). One such inner moraine was dated as somewhat older than 15 000 BP by McKellar, who regarded it as a correlative of the Kumara 2 advance, while equating it at the same time with the Poulter advance in the Waimakariri Valley (McKellar 1960). This and a limited number of radiometric dates ranging from about 16 500 to 14 000 BP formed the basis for suggesting that the Kumara 3 advance, now recognized as a double event, occurred between 18 000 and 16 000 BP and between 14 500 and 14 000 BP (Suggate 1965). Of these supporting dates one of 16 600 BP could, as Suggate pointed out, be linked only uncertainly with the nearest moraines, some 15 km up-valley; another of 14 100 BP does not directly date a glacial event, but rather accumulation resulting from local fan deposition, under cool climate conditions. A date of 14 800 BP is derived from material underlying an outwash surface traceable to a correlative of the first Kumara 3 advance, and thus predates the event (Mabin 1983).

In 1970 Suggate and Moar revised the 1965 datings, in the light of new information which placed the end of the Kumara 2_2 aggradation at some time after 18 600 BP. They suggested that the early Kumara 3 advance took place between 17 000 and 16 000 BP, and the younger between 14 500 and 14 000 BP (Table 4).

Table 4. Late Glacial stages in Westland after Suggate and Moar (1970).

Date (BP)				Advances
14 000			major retreat	
14 000	–	14 500		Younger Kumara 3
14 500	–	16 000	recession	
16 000	–	17 000		Older Kumara 3
17 000	–	18 000	recession	
18 000	–	22 300		Younger Kumara 2

Later work suggests that they probably took place even later (Table 5). Soons and Burrows (1978) were able to show that a correlative of the Kumara 2_2 advance in the Rakaia Valley peaked later than 19 200 BP, while the type Kumara 2_2 moraine was formed at, or slightly later than, 18 000 BP (Moar 1980). Since it is clear on both sides of the Southern Alps that a substantial retreat of ice, and some modification of landforms happened between the Kumara 2_2 and Kumara 3 advances, it is unlikely that the Kumara 3 advances were earlier than 16 000 BP: the date of 16 600 BP cited by Suggate (1965) and predating a period of aggradation may be relevant in this context.

Table 5. New Zealand Late Glacial stages, this paper.

11 500	–	13 000		Younger Kumara 3
			minor recession	
13 500	–	15 000		Older Kumara 3
			recession	
17 000	–	18 000		Younger Kumara 2

The end of the Kumara 3 advances is not well-defined. It seems probable, however, that ice was present in significant amounts in the major valleys later than the date of 14 000 BP suggested by Suggate and Moar (1970). Mabin (1983) has suggested a later date on the grounds that a date of 14 800 BP comes from a horizon stratigraphically below outwash of an early Kumara 3 advance; ice was present in the Hope Valley, damming a lake in a tributary valley at about 13 300 BP (Clayton 1968); in the Lake Pukaki basin a date of 13 900 BP comes from a site alongside a lateral moraine (Mansergh, pers. comm.), and another of 13 500 BP from a kame terrace (Moar 1980). In South Westland ice was still present, although probably retreating, in the Paringa Valley at 13 400 BP (Suggate 1968). A date of 11 900 BP from the outlet area of Lake Pukaki suggests that ice was still present in the basin at about 12 000 years (Mansergh, pers. comm.); in the Rakaia Valley a small lake was dammed by ice at about 11 600 BP. This lake was drained, and a minor advance followed at a slightly later date, 60 km down-valley from the present glaciers (Fig. 1).

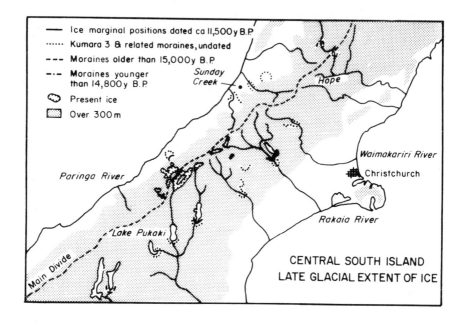

Figure 1. Late Glacial ice margins in central South Island of New Zealand.

West of the divide, the striking Waiho Loop moraine of the Franz Josef Glacier is now reliably dated at about 11 000 BP (Wardle 1978).

Whether all, or any, of these dates refer strictly to correlatives of the Kumara 3 advances is perhaps a matter for debate, and is likely to remain such until either the type Kumara 3 moraines in the Taramakau Valley or the Poulter moraines in the Waimakariri Valley are dated. Nevertheless, they demonstrate persistence of large ice masses in the New Zealand mountain areas until a relatively late date. This is consistent with a rather delayed pattern of forest re-establishment in the South Island, which has been the subject of comment by several workers (e.g. Burrows 1974; Moar 1980). A period of relatively cool climate after 14 000 BP until 12 000 - 11 000 BP seems to be indicated, during which glaciers, at least on the eastern side of the divide, slowly wasted from their Kumara 3 maximum.

The dates suggested in Table 5 for events in the Late Glacial period do not conform closely to the vegetation changes which can be identified for the same time-span (Moar 1980 and pers. comm.). At most sites for which pollen information is available from 25 000 BP to 11 000 BP the vegetation does not record the fluctuations found in the record of glacial events. This may be because the sites investigated are for the most part not close to former positions of ice, and may also reflect a lack of sensitivity of grass and shrubland vegetation to the factors which controlled glacier variation. An important control on glacier behaviour is the conditions in the névé field feeding the glacier; these may be rather different from those affecting the behaviour of plants growing two or three thousand metres lower down and - in many cases - in a coastal environment. A comparison of the sequence of events based on glacier behaviour with that based on vegetation pattern changes may thus be valuable in that they may monitor different climatic elements. In a period of generally depressed temperatures, glaciers may advance or recede in response to precipitation changes, while deterioration and recovery of shrub and tree species may reflect changes in seasonality as well as overall cooling and warming. It is thus important to establish changes in individual elements of proxy data, as well as to integrate all information in order to reconstruct an overall picture of climate change.

CONCLUSIONS

In reviewing the changes which have been made to both long- and short-term sequences of glacial events in New Zealand, the problems of correlations with other parts of the world are highlighted, and hence the difficulties in reconstructing climate patterns and global trends. The process traced is typical. Firstly workers identify the broad picture, and then, having established a framework, are able to work out detail with more confidence. The absolute time framework has undergone a similar series of changes. As more dates become available, and as the details of the stratigraphy and morphology are elaborated, new dates result in adjustment to the early, relatively rough chronology. Some of these later changes, however, may require that substantial alterations be made to the picture of the sequence of climate changes.

There is no reason to suppose that the process of the elaboration of chronology has been completed. New Zealand has relatively few workers in the Quaternary field, and there are still areas of the country which are remote, or where access is by no means easy. The climatologist therefore needs to develop climatic reconstructions which are sufficiently flexible to be able to accommodate future modifications. He may also be able to play a more active role by indicating what constitute probable effects, which may be used to monitor some of the proxy data. There is considerable scope for active co-operation in this area.

REFERENCES

Burrows CJ 1974. The Otiran/Aranuian boundary: reply. N.Z. J. Geol. Geophys. 3:742-6.
Burrows CJ 1980. Radiocarbon dates for post-Otiran glacial activity in the Mount Cook region, New Zealand. N.Z. J. Geol. Geophys. 23:239-48.
Clayton L 1968. Late Pleistocene glaciations of the Waiau valleys, North Canterbury. N.Z. J. Geol. Geophys. 11:753-67.
Dickson M 1972. Palynology of a Late Oturi interglacial and early Otira glacial sequence from Sunday Creek (S51), Westland, New Zealand. N.Z. J. Geol. Geophys. 15:590-8.
Frenzel B 1973. Climatic Fluctuations of the Ice Age. English transl. AEM Nairn, Case Western Reserve Univ. Press, Cleveland. 306 pp.

Gage M 1958. Late Pleistocene glaciations of the Waimakariri Valley, Canterbury, New Zealand. N.Z. J. Geol. Geophys. 1:123-55.

Gage M 1961. New Zealand glaciations and the duration of the Pleistocene. J. Glaciol. 3:940-3.

Gage M and Suggate RP 1958. Glacial chronology of the New Zealand Pleistocene. Bull. geological Soc. Am. 69:589-98.

Gage M and Soons JM 1973. Early Otiran glacial chronology - a re-examination. Ninth Congress, INQUA, Abstr., Christchurch 2-10 Dec. 1973.

Mabin MCG 1983. The Late Pleistocene glacial sequence in the middle Maruia valley, southeast Nelson, New Zealand. N.Z. J. Geol. Geophys. 26:85-96.

McKellar IC 1960. Pleistocene deposits of the Upper Clutha Valley, Otago, New Zealand. N.Z. J. Geol. Geophys. 3:432-460.

Moar NT 1980. Late Otiran and early Aranuian grassland in central South Island. N.Z. J. Ecology 3:4-12.

Pillans B 1983. Upper Quaternary marine terrace chronology and deformation, South Taranaki, New Zealand. Geology 11:292-297.

Salinger MS, this vol. New Zealand Climate: the last 5 million years.

Soons JM 1966. A review of 'Late Pleistocene geology of the northern part of the South Island', by RP Suggate. N.Z. Geographer 22:100-2.

Soons JM and Burrows CJ 1978. Dates for Otiran deposits, including plant microfossils and macrofossils, from Rakaia Valley. N.Z. J. Geol. Geophys. 21:607-615.

Suggate RP 1965. Late Pleistocene geology of the northern part of the South Island, New Zealand. N.Z. geol. Survey, Bull. n.s. 77, Govt Printer, Wellington.

Suggate RP 1968. The Paringa Formation, Westland, New Zealand. N.Z. J. Geol. Geophys. 11:345-355.

Suggate RP and Moar NT 1970. Revision of the chronology of the late Otira Glacial. N.Z. J. Geol. Geophys. 3:742-6.

Wardle P 1978. Further radiocarbon dates from Westland National Park and the Omoeroa River mouth, New Zealand. N.Z. J. Botany 16:147-52.

Palaeobotanical evidence for changes in Miocene and Pliocene climates in New Zealand

D.C.MILDENHALL & D.T.POCKNALL
New Zealand Geological Survey, Lower Hutt

ABSTRACT. Pollen analysis of numerous long sedimentary sequences ranging in age through Miocene and Pliocene shows that New Zealand was covered in a mosaic of vegetational types influenced by latitude, altitude, climate, tectonism and its oceanic position. During this time the New Zealand vegetation changed from a predominantly warm temperate Nothofagus brassi (beech) forest to a predominantly cool temperate N. fusca/podocarp forest disturbed by periodic glacial conditions during which shrubland and grassland expanded and forest was preserved only locally. During the same interval New Zealand moved c. 10° further north to the present day latitude (34°-46°S) and the Kaikoura Orogeny commenced to form a mountainous landmass, especially in South Island.

The Early to Middle Miocene vegetation evolved from residual taxa which remained after an Oligocene marine transgression had reduced New Zealand's land area to an archipelago. Some taxa also arrived via long distance dispersal. A rich floral diversity included rare subtropical elements and a large number of taxa now restricted to Australia, New Caledonia and/or Papua - New Guinea (e.g. brassi beech, Casuarina, Microcachrys/Microstrobos, Acacia, Melaleuca, Acmena, Eucalyptus, etc.). The Late Miocene vegetation was regionally differentiated with brassi beech forest in the north and northwest, Podocarpus/Dacrydium cupressinum forest in the southwest and fusca beech forest in the southeast. These forest types are interpreted as representing wet humid westerly conditions in the north, cooler wet westerly conditions in the south and drier conditions in the southeast in the lee of mountains which rose during the Late Miocene.

The Pliocene vegetation differed from that of the Miocene with extinction of plants adapted to warm temperate humid environments and evolution of plants adapted to montane and cool to cold temperate environments. Plant diversity declined although some taxa, notably Coprosma, Hebe, Dracophyllum and many Compositae evolved rapidly into new ecological niches. Climatic conditions varied from warm to cold temperate as a result of glaciations which fluctuated in intensity throughout the period.

INTRODUCTION

The evidence for Miocene/Pliocene climates presented in this paper is derived entirely from palynological analysis of many sections located

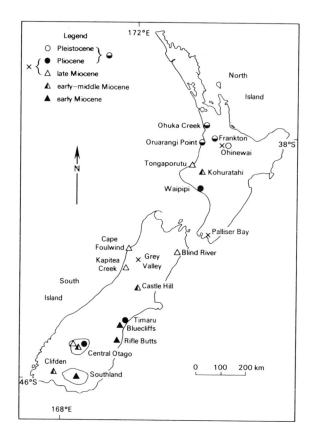

Figure 1. Localities and geological age of plant microfossils.

throughout New Zealand. Couper (1960) and Mildenhall (1980) have previously attempted brief syntheses based on the then current data, and numerous biogeographic papers have summarized previously published palaeobotanical data (see Fleming 1979). Most of the localities upon which the data presented below are based are given in Fig. 1. Figure 2 shows the stratigraphic range of the better dated sequences.

New Zealand is today, and was during the Miocene, a long narrow country straddling the boundary between the Pacific and Indian plates. During the Miocene the land mass extended between latitudes $53°S$ and $33°S$. The beginning of the Miocene saw New Zealand emerging from an Oligocene marine transgression which resulted in peneplanation, particularly in the south of South Island and in the north of North Island. Through the Miocene and Pliocene New Zealand moved into its current latitudinal range ($46°S$ and $34°S$), contracted in length and entered a phase of mountain building which continues to the present day. These changes of latitude, altitude, climate and land area resulted in marked vegetational changes showing distinct latitudinal (and ?altitudinal) differences. (See Suggate et al. 1978:738-740 for palaeogeographic maps of the Late Cenozoic.)

Vegetational history as portrayed by palynology is limited by difficulties in pollen identification. Many dispersed pollen types can be identified only to genus or family, e.g. Coprosma, Dracophyllum, Hebe

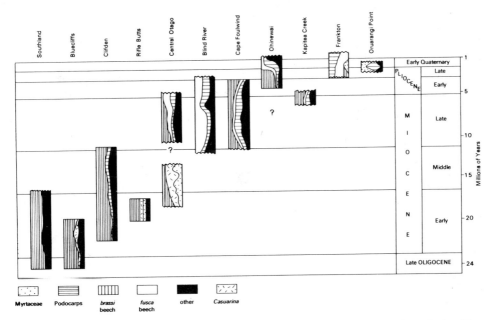

Figure 2. Stratigraphic range of important sections mentioned in the text. Saw-toothed ends to some columns indicate uncertainty of age.

and Compositae. This is unfortunate because some modern species which would be useful palaeoecological indicators have restricted distributions. Also, any individual pollen assemblage may represent a number of habitats and associations; taxa vary in their pollen production so that the amount of pollen recovered is not commensurate in importance with that of its parent plant; and the depositional environment may size grade and sort pollen and differentially preserve the more resistant pollen and spores.

Bearing these problems in mind the Miocene-Early Pleistocene vegetation can be studied and changes correlated with climatic and edaphic changes. In this summary paper we have avoided using local stage names and instead use Early to Middle and Late Miocene, Early and Late Pliocene and Early Pleistocene. No evidence will be presented to substantiate the assessment of ages of the sequences used in the following discussion; however, reference can be made to Suggate et al. (1978) where some of this information is available.

Throughout the paper we use Nothofagus brassi, N. fusca and N. menziesii to refer to the pollen types, not to the existence of these three species.

PALYNOLOGY AND CLIMATIC CHANGE

Early to Middle Miocene

Most sequences studied come from terrestrial deposits in South Island although a few well-dated samples from predominantly marine sequences in North Island have also been studied.

New Zealand coastal vegetation in the period was dominated by Nothofagus brassi beech, predominantly Nothofagidites cranwellae (Couper),

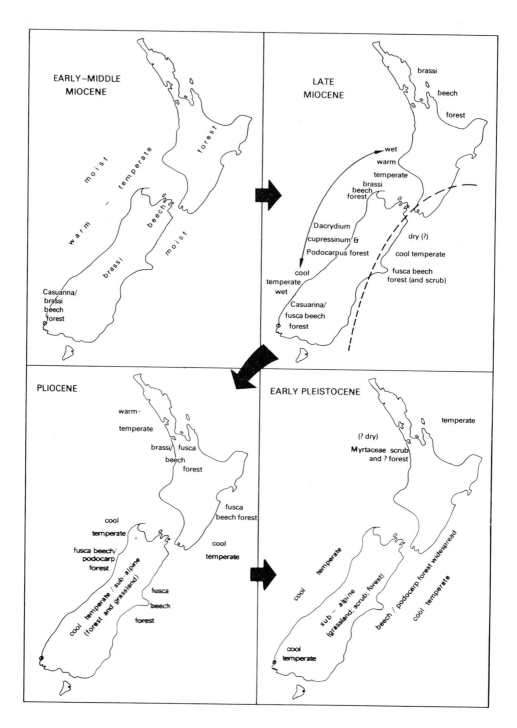

Figure 3. Summary of Miocene to early Pleistocene terrestrial paleoenvironments.

but with lower frequencies of N. falcatus (Cookson), N. emarcidus (Cookson) and N. matauraensis (Couper). Casuarina (Haloragacidites harrisii (Couper)) was also common. Inland the vegetation was more varied according to environment, e.g. lacustrine beds, swamps, levees, or flood plains.

In Southland the Early Miocene Gore Lignite Measures accumulated in a deltaic environment in which a considerable thickness of swampy vegetation was deposited. Sediments consisting primarily of sandstone, mudstone, conglomerate and laterally extensive or sporadic lignite seams represent the Proteacidites isopogiformis and Tricolpites lapispinosus Zones (Pocknall and Mildenhall, in press). During periods of peat swamp formation forests were restricted to levees along streams and rivers although some tree species existed on better drained regions of the swamps (Pocknall 1982a). The dominant forest elements were brassi beech species with a variety of understorey plants belonging to many different plant families. Podocarps were commonly emergent above the beech canopy. Occasionally Myrtaceae, Chloranthaceae (Clavatipollenites ascarinoides McIntyre), Liliaceae (Liliacidites spp.), Palmae (Monosulcites otagoensis Couper) and Podocarpaceae account for up to 10 % of total pollen and must have been locally important elements in the flora. Components of the swamp vegetation include Gleicheniaceae, Restionaceae, Sparganiaceae and Agavaceae, the last-mentioned family being represented by an extinct species of Phormium. The swamps and water courses may have been bordered by Casuarina but it is difficult to ascertain the part Casuarina played since it is extinct in New Zealand and occupies a variety of habitats in other southeast Pacific areas including Australia.

Palynological analyses indicate a predominantly cool to warm temperate climate with high humidity. Forest dominated by brassi beech occurs today in lower montane areas of Papua-New Guinea and New Caledonia under conditions of high rainfall distributed evenly throughout the year. Both the forest and swamp vegetation of Southland had taxa in common with the present floras of New Caledonia, North Queensland and Papua-New Guinea, the southeastern coast of Australia, the offshore islands of New Zealand and mainland New Zealand. The complexity of Miocene vegetation in Southland is clearly apparent.

Other sequences from the southeast and southwest of South Island differ only in detail. Coastal sequences tend to have more pteridophyte spores preserved but this is probably the result of the depositional environment concentrating them at the expense of the more fragile pollen types.

In the southwest assemblages from an Early to Middle Miocene marine sequence from Clifden show little difference from the Gore Lignite Measures except for an increase in pteridophyte spores and a slightly higher percentage of fusca beech pollen in some samples. Assemblages are quite rich in variety and probably represent optimum diversity in the Tertiary.

Sequences of similar age have been studied from Bluecliffs, Rifle Butts and Gimmerburn (Central Otago). All are similar in overall diversity and dominance of brassi beech, indicating a similar environment of warm temperate conditions. At Bluecliffs (Pocknall 1982b) other taxa of importance include Casuarina, fusca beech, Palmae (M. otagoensis), gymnosperms (notably Dacrydium cupressinum, Phyllocladus, and Microstrobus/Microcachrys) and Pteridophyta. At Rifle Butts (Pocknall 1981), dated as Early Miocene by foraminifera, all beech groups, Casuarina, Myrtaceae, Dacrydium cupressinum and Pteridophyta are well represented. At Gimmerburn diverse assemblages are dominated by Myrtaceae, brassi beech (Nothofagidites cranwellae) and Casuarina, with Liliaceae, Araucariaceae, Epacridaceae, Dacrydium cupressinum, fusca beech and Restionaceae common components in most samples. The presence of key taxa suggests that this sequence is consistent with the Proteacidites isopogiformis Zone described from Southland.

Early(?) to Middle Miocene sequences from Central Otago show a very diverse range of assemblages from brassi beech dominant (Nevis Basin) to Sparganiaceae dominant (Kawarau River). Lacustrine and fluviatile sediments consisting of sand, siltstone, minor quartzose conglomerate and lignite were deposited in one or several of the large basins. At Kawarau River samples are dominated by brassi beech, Casuarina, Myrtaceae (including Eucalyptus and Acmena), Mallotus/Macaranga, Palmae, Liliaceae, Sparganiaceae, Podocarpus, and Cyatheaceae. The climate is interpreted as warm temperate with plentiful rainfall distributed evenly throughout the year. Occasional drier periods may have occurred to allow Mallotus or Macaranga shrublands to exist, although the abundance of freshwater colonial algae (Botryococcus and Pediastrum) and pollen from semi-aquatic herbs suggests that there was abundant available moisture most of the time. The vegetation is interpreted as representing a lake or river margin community of Casuarina, Palmae and Liliaceae with a flood-plain and levee community of Myrtaceae, Fagaceae (mainly brassi beech) and emergent Podocarpaceae (mainly Podocarpus and Phyllocladus), Araucariaceae, Dacrydium cupressinum, and fusca beech were common locally. Unlike coastal sequences, pollen of the Amaryllidaceae (Monogemmites gemmatus (Couper)) and Chloranthaceae (Clavatipollenites ascarinoides) are rare. The presence of Tricolpites latispinosus McIntyre indicates that this sequence may fall within the Tricolpites latispinosus Zone (Pocknall and Mildenhall, in press); it may, however, also be slightly younger.

A further sequence in Central Otago at Vinegar Hill may be similarly interpreted. Here nearshore lacustrine sediments, marginal to a fluvio-deltaic hinterland, contain terrestrial vertebrates (Douglas et al. 1981) and spore-pollen assemblages dominated by brassi beech (Nothofagidites cranwellae), Casuarina, Myrtaceae, Podocarpaceae and Mallotus/Macaranga.

A transitional palynoflora between Early and Late Miocene is found in the Castle Hill Basin, Canterbury. Here the main elements in the spore-pollen assemblages are Casuarina, brassi beech (N. cranwellae), Myrtaceae, fusca beech, menziesii beech and to a lesser extent Compositae (Tubuliflorae). Spore-pollen assemblages dominated by Casuarina and brassi beech are not found in Late Miocene sediments in South Island and Compositae pollen is rare in Early and even Middle Miocene sediments.

In North Island the only Early to Middle Miocene sequence studied in detail is at Kohuratahi (Taranaki). Assemblages are generally dominated by brassi beech (N. cranwellae) and Casuarina although near the top of the sequence and throughout the Early Miocene podocarps (particularly Podocarpus species and Dacrydium cupressinum) are more abundant. In the Middle Miocene, Myrtaceae and Nothofagidites spinosus (Couper) dominate with N. cranwellae and Casuarina less important. Palaeoenvironmental interpretation of this sequence accords well with that envisaged for the southern regions of South Island where during the Early to Middle Miocene predominantly westerly winds (see Kemp 1978) brought and distributed rainfall evenly throughout the year.

Late Miocene

During the Late Miocene brassi beech was dominant only in the north of North Island although it remained consistently present throughout New Zealand. Percentages of Casuarina fluctuated widely while fusca beech and podocarps increased in importance. Regional differentiation became increasingly marked, and species diversity was lower.

In South Canterbury and Central Otago two palynologically dated Late Miocene sequences are dominated by fusca beech and Casuarina. In South Canterbury brassi beech, Phyllocladus, Podocarpus, Dacrydium cupressinum, Myrtaceae and Compositae (Tubuliflorae) are also present and indicate cool temperate climatic conditions. In Central Otago, at Gimmerburn, plants growing near the site of deposition are locally overrepresented and include Chenopodiaceae, Restionaceae and Cyperaceae. Pollen and spores from other plants tolerant of wet or semi-aquatic environments are very common, especially Dacrydinum bidwillii/biforme type, Haloragis, Sparganiaceae and Sphagnum. The colonial freshwater algae Pediastrum and Botryococcus which inhabit ponds and lakes are abundant. Brassi beech, menziesii beech, Phyllocladus, Podocarpus and Dacrydium cupressinum are present in small numbers and Compositae (Tubuliflorae) increase in abundance throughout. The floras are indicative of moist cool temperate climatic conditions.

A further Central Otago sequence from near St Bathans may range in age up into the Early Pliocene. Here the dominant taxa include fusca beech, Podocarpus, brassi beech (mostly N. cranwellae), Dacrydium cupressinum, Compositae (Tubuliflorae), Malvaceae, Casuarina and Phyllocladus in decreasing order of importance. This type of assemblage, lacking Palmae and Liliaceae and with low species diversity, represents forest growing under a cool temperate and moist climatic regime. The high percentage (6 %) of Compositae suggests that upland sites with cooler climatic conditions were not far removed from the site of deposition.

To the northeast at Blind River a thick sequence of marine sands, silts, shell beds and conglomerate yield spore-pollen assemblages sparse in abundance and diversity due to a New Zealand-wide decrease in species diversity, the distance from shore, and the rapid deposition of sediment diluting spore-pollen concentration. The dominant elements in the assemblages vary from fusca beech to Compositae (Tubuliflorae). The change from brassi beech in the south in the late Middle Miocene to fusca beech at the base of the Late Miocene is dramatic and represents a change to a drier climate on the east coast of New Zealand, and possibly colder, as a result of the uplift of mountain ranges (Kaikoura Orogeny) which commenced about 15 m.y. ago. The Late Miocene Compositae shrub assemblage may result from very cool conditions or be a rain shadow effect. The spore-pollen assemblages represent a cool temperate to temperate coastal beech/podocarp forest in which Proteaceae, Chloranthaceae, Palmae and Pteridophyta were rare or absent. These contrast with assemblages to the north and west and suggest a prevailing westerly air flow at the time. Pollen of Chenopodiaceae, Caryophyllaceae, Compositae (Tubuliflorae), Rubiaceae, Malvaceae, and Gramineae are as common as could be expected in a coastal environment. Two key taxa for correlation occur in the sequence, Randia (Triporopollenites bellus Partridge), making its earliest New Zealand appearance, and Mallotus/Macaranga (Nyssapollenites endobalteus McIntyre), making its last New Zealand appearance. Randia is regarded as a sub-tropical forest margin taxon, not found south of latitude $30°20'S$ in Australia today and the

Mallotus/Macaranga complex is not found south of latitude 32°10'S in Australia today (Martin 1978). Climatic conditions warm enough to support these taxa would be temperate, or warm temperate.

At the same latitude (44°- 42°S) on the west coast similar sediments to those at Blind River occur at Cape Foulwind. These middle to outer shelf, predominantly blue-grey micaceous muddy siltstone and fine sandstone, contain assemblages sparse in pollen with many grains etched and dark in colour. Water sorting has concentrated the more resistant spores, especially Cyathea. The assemblages are dominated by pollen of brassi beech, Metrosideros, and Dacrydium cupressinum with Chloranthaceae, Palmae, Proteaceae and Pteridophyta common. All assemblages are typical of a coastal rain forest growing in cool to warm temperate conditions with high humidity. Because of its position on the west coast of New Zealand this locality must have received moisture from prevailing westerlies enabling the development and maintenance of cool to warm temperate brassi beech forest. Key taxa are rare but include Randia, Mallotus/Macaranga and Micrantheum (M. spinyspora Martin), a genus typical of temperate eucalypt-sclerophyll formations in Australia, and found associated with assemblages dominated by Myrtaceae and/or Casuarina in the Australian Plio-Pleistocene (Martin 1974). At Cape Foulwind it is associated with Metrosideros pollen and indicates temperate conditions.

To the south at Kapitea Creek, preservation, spore-pollen abundance, water sorting, and sedimentology are similar to those found at Cape Foulwind. All samples are dominated by Cyathea, brassi beech (N. cranwellae), Podocarpus and Dacrydium cupressinun consistent with the climate and vegetation assessment for Cape Foulwind. More warm climate indicators are found at Kapitea Creek than at Cape Foulwind. These include Casuarina, Chloranthaceae, Palmae, Proteaceae, Pteridophyta and the tribe Cupanieae of the Sapindaceae. The Cupanieae are generally common in humid warm temperate assemblages and do not reach south of latitude 32°20' in Australia today (Martin 1978).

In the extreme south of North Island a marine sequence in Palliser Bay, with sediments similar to those in the south, yields assemblages domninated by cool temperate taxa, viz. fusca beech, with smaller percentages of Dacrydium cupressinum, Podocarpus, brassi beech, Compositae (Tubuliforae) and Phyllocladus. The sequence is most like that at Blind River indicating that moisture laden westerly winds did not affect the area in the Late Miocene. A number of key taxa for correlation first appear including Anthocerotales (hornworts), Microstrobos (Podosporites erugatus Mildenhall), and Salicornia australis and Stellaria (herbaceous Chenopodiaceae and Caryophyllaceae respectively).

At Tongaporutu in western North Island a marine sequence contains diverse assemblages dominated by brassi beech, fusca beech, Dacrydium cupressinum and Podocarpus, but in many samples poor preservation meant percentages could not be accurately assessed. Liliaceae, Palmae, Amaryllidaceae, Phyllocladus, Dacrycarpus dacrydioides, menziesii beech, Compositae (Tubuliflorae), Chloranthaceae and Pteridophyta are common. Pollen of Myrtaceae is sparse, possibly through water-sorting or poor preservation. The assemblages are consistent with rain forest growing under moist, temperate to warm temperate conditions. All the key taxa mentioned above occur in this sequence as well as Aquifoliaceae (Ilexpollenites), and Acacia.

Two other localities, north of Tongaporutu, have been examined. Both sequences are dominated by brassi beech (N. cranwellae) with varying frequencies of fusca beech, Podocarpus, Dacrydium cupressinum and

Dacrycarpus dacrydioides. The assemblages are all consistent with derivation from a moist warm temperate rain forest.

During the Late Miocene marked differences occurred in the regional dominance of tree species. If the relative percentages of brassi beech and Dacrydium cupressinum versus fusca beech, menziesii beech and Podocarpus represent humid versus drier, cooler conditions then the New Zealand climate fluctuated markedly throughout the Late Miocene. On the west coast of South Island moist cool to warm temperate conditions gave rise to a forest dominated by Dacrydium cupressinum and brassi beech while on the east coast a ?dry temperate to cool temperate forest dominated by Podocarpus and fusca beech existed. This cool temperate forest also occurred in the south of North Island. In the north and northwest of North Island brassi beech dominated with podocarps less important. This was probably a reflection of slightly higher temperatures with a gradual migration of podocarps north as temperatures fell during the Late Miocene and Early Pliocene. The predominantly westerly winds would have brought rain to the west coast and north of North Island but would have had less effect on the eastern coasts because of an inferred rain shadow effect caused by uplift of the main axial ranges. At the same time as the mountains were uplifted, ocean currents were bringing cool water to the east coast and warm water to the west coast, enhancing the rain shadow effect; if, however, in the east conditions were too dry beech/podocarp forest would probably not have developed.

Pliocene

During the Pliocene climatic changes increased in magnitude and frequency and the vegetation reflects these changes. Many sub-tropical and humid warm temperate elements became extinct or were substantially reduced in distribution. The Kaikoura Orogeny increased in intensity and many thick conglomerates were deposited, particularly in central New Zealand. Depending on local variations in climate, edaphic factors, latitude, altitude and tectonism, the vegetation was dominated by brassi beech, fusca beech, Phyllocladus, Podocarpus, Dacrydium cupressinum, Myrtaceae, Umbelliferae, Gramineae, Cyperaceae, Restionaceae, Compositae (Tubuliflorae) and Rubiaceae. Regional differentiation was even more marked, however, and palynological dating is more uncertain as it is difficult to determine which period of relative cold or warmth is being dealth with unless there is good stratigraphic and radiometric dating control. Sequences of pollen bearing sediment are found only in offshore localities which usually contain water-sorted, poorly preserved, and often reworked assemblages, and in thick peat deposits where the regional vegetation is masked by local herbaceous taxa making pollen of key regional taxa difficult to find. Most other terrestral sequences are either discontinuous or are thick non-fossiliferous gravel deposits.

A large number of Pliocene samples have been collected from both North and South Islands, but few sequences have been studied. This summary will deal only with a few well-dated samples and sequences.

The earliest Pliocene is represented by a few samples from North Island dominated by brassi beech (N. cranwellae), Dacrydium cupressinum, Amaryllidaceae and fusca beech. Cool temperate or temperate floral elements dominate in the south and warm humid elements in the north and west. A number of warm temperate or sub-tropical taxa still persist indicating a climate similar to that of the Late Miocene.

During the Middle and Late Pliocene, deteriorating climatic conditions took toll of the warm temperate taxa and a large number of species disappeared from New Zealand. These were replaced by cool temperate taxa and there followed a large radiation of pre-existing taxa into new ecological niches of which Coprosma, Hebe, Dracophyllum, Aciphylla, Celmisia, Senecio, Olearia and Ranunculus are just a few. Taxa to disappear include Bombax (Bombacaceae), Randia (Rubiaceae), Ephedra, Joinvillea, and the fossil genus Lymingtonia (Nyctaginaceae). Taxa to appear at this time or slightly earlier include Nothofagidites longispina (Couper), a brassi beech grain with long processes restricted to the Pliocene in the north of North Island, Microstrobos, Colobanthus, Wahlenbergia, Pimelea, Tupeia antarctica, Bulbinella hookeri, Calystegia, Nertera and Stellaria (see Mildenhall (1980) for notes on other modern genera appearing at this time).

In South Island Middle to Late Pliocene assemblages are dominated by fusca beech and various podocarps (including the extinct Microcachrys). Middle Pliocene sequences from Grey Valley (Mildenhall 1978b) show a wide range of dominants reflecting fluctuations in rainfall and temperature. Gramineae, Umbelliferae, Cyperaceae, fusca beech, Dacrydium cupressinum and Myrtaceae dominate and Beauprea (Proteaceae), Sapindaceae, Zygogynum or Exospermum (Winteraceae), Polygalaceae, brassi beech and Randia are consistently present in temperate assemblages but in very small numbers. Climatic conditions ranged from very cold (sub-alpine; Gramineae/Umbelliferae dominant assemblage) to cool temperate (beech/podocarp dominant assemblage) but no evidence of dry conditions exists. All other South Island samples show evidence of wet/cold or wet/temperate conditions but the sequence of events is obscured by poor dating. At Timaru, however, a well-dated Middle Pliocene sequence (Vella 1977) is dominated entirely by fusca beech (Gair 1961).

In North Island numerous isolated samples and a number of sequences through Middle and Late Pliocene sediments indicate the same range of climatic conditions but the vegetation apparently did not react to quite the same marked degree as that in the south. For example a Middle Pliocene sequence at Waipipi (Mildenhall 1978a) shows the existence of forest conditions with fusca beech, Podocarpus, Dacrydium cupressinum and brassi beech in decreasing order of abundance. Conditions were moist, cool temperate, and somewhat like that of Southland at the present day (Mildenhall and Harris 1970) except for the continued presence of warm temperate elements. A similar sequence occurs at Palliser Bay but here brassi beech is less evident and shrub pollen types are more common. This suggests that a marked latitudinal temperature gradient existed during Pliocene time with cool temperate taxa more common in the south and warm temperate taxa more common in the north. Brassi beech pollen, however, is never dominant in Late Pliocene assemblages (contra Mildenhall 1973: 22) always being overshadowed by pollen of fusca beech and podocarps. Other sequences examined from North Island all exhibit fusca beech dominant assemblages with fluctuating podocarp and brassi beech pollen frequencies presumably in response to the climatic changes that caused marked movements in South Island vegetation.

Pliocene/Pleistocene boundary

Three sequences from South Auckland region crossing the Pliocene-Pleistocene boundary have been recently examined. At Ohinewai, near Huntly, a sequence of terrestrial mudstones with intercalated peats and pumiceous sands and silts, shows a gradual change from a brassi beech dominant forest in the Late Miocene and Early Pliocene to a fusca beech dominant forest in the Late Pliocene and to podocarp dominant forest in the Early Pleistocene. The brassi beech forest represents warm temperate humid conditions while a gradual decrease in diversity up the sequence represents the effects of a cooling climate. A number of taxa restricted to the Miocene in South Island became extinct through this part of the sequence. As brassi beech gave way to fusca beech, herbaceous and shrubby taxa also increased as conditions remained moist but became temperate to cool temperate. The extinctions continued and a number of taxa appeared for the first time: Lycopodium australianum, Compositae (Liguliflorae), and Proteacidites franktonensis Couper; some seem to have distinct local ranges useful for correlation. The Pliocene-Pleistocene boundary is tentatively placed at the disappearance of brassi beech, Microcachrys and Beauprea (Couper and Harris 1960) and the appearance of the taxa listed above. In the upper part of the sequence podocarps gradually gain ascendency.

Climatic conditions are difficult to assess at Ohinewai because the regional pollen rain was low with respect to Restionaceae which locally inhabited the acid peat swamp and is over-represented. Cool temperatures are inferred as pollen assemblages from the topmost part of the sequence are dominated first by Gramineae, Dacrydium bidwillii/biforme type, and Compositae (Tubuliflorae) representing cool, relatively dry conditions; then Quintinia, Dracophyllum, Phyllocladus, Podocarpus and Dacrydium cupressinum, representing mild wetter situations and finally by Dacrydium cupressinum, D. bidwillii/biforme type, Compositae (Tubuliflorae), Cyperaceae, Gramineae and Restionaceae representing wetter (?)cooler conditions.

Nearby at Frankton a sequence, previously analysed by Couper and Harris (1960), has been re-examined. This apparently continuous sequence of pumiceous sands and clays and intercalated peats and peaty sands is dominated by Restionaceae pollen. The Late Pliocene-Early Pleistocene sediments are highly productive and, disregarding the local Restionaceae pollen, give a good record of regional vegetation. The sequence begins with brassi and fusca beech in approximately equal proportions followed by an interval representing cooler conditions where podocarps, including Phyllocladus become common. Brassi beech recovered slightly immediately below the Pliocene-Pleistocene boundary, but then finally declined. Fusca beech and podocarps (including Dacrydium cupressinum and Phyllocladus) dominate. the Late Pliocene was warm temperate and moist, the Early Pleistocene temperate to cool temperate and moist with brief cooler intervals represented by sharp but brief rises in the Phyllocladus curve. In the latest Pliocene Rugulatisporites micraulaxus Partridge, Tetracolporites ixerboides Couper, Lymingtonia cenozoica Pocknall and Mildenhall, Tricolpites latispinosus McIntyre, Polycolpites reticulatus Couper (Sterculiaceae) and Luminidites reticulatus (Couper) become extinct. In the earliest Pleistocene Microstrobos, Harrisipollenites annulatus Mildenhall and Crosbie, Dictyophyllidites arcuatus Pocknall and Mildenhall, Proteacidites franktonensis and Beauprea become extinct, while Acacia, "Assamiapollenites" incognitus Pocknall and Mildenhall and Peromonolites problematicus Couper are locally restricted to the Early Pleistocene.

A complex sequence of marine and non-marine carbonaceous shales, muds and palaeosols near Oruarangi Point in western North Island is regarded as crossing the Pliocene-Pleistocene boundary. The sequence is coeval with those exposed at nearby localities at Ohuka described by Couper and McQueen (1954) and Mildenhall (1975). The sequence is dominated by Myrtaceae pollen with brassi beech pollen present only in one sample. Most of the key taxa listed from Frankton are present in the sequence but the position of the boundary is uncertain. Further, more detailed, stratigraphy including fission track dating is necessary to determine the relationship between the coastal Myrtaceae (including Eucalyptus) dominated assemblages and the inland beech/podocarp dominated assemblages. The prevailing westerly winds and coastal dunes may have created a specialized open coastal forest/scrub formation in which Metrosideros, Eucalyptus, Leptospermum, Acacia, menziesii beeech, Proteaceae, Chloranthaceae, Rubiaceae, Dodonaea viscosa and Podocarpus were prominent. Conditions may have been dry and windy resulting in high evapotranspiration.

CONCLUSION

During the Early to Middle Miocene brassi beech forest covered New Zealand and climatic conditions were humid and warm temperate. In the Late Miocene the vegetation pattern changed to one where brassi beech was more common in the north and west of New Zealand with fusca beech more common in the south and east. Podocarps and other cool temperate taxa show a pattern of gradual northward migration. During the Pliocene a temperature gradient south to north is revealed by the continued dominance of brassi beech in the north, its scarcity in the south and the late extinction in the north of many taxa that had previously become extinct in South Island. The Early Pleistocene is marked by cool climate conditions with windy coastal environments. Rainfall was generally high and evenly spread throughout the Miocene although a rainshadow effect in the lee of the mountains formed by the Kaikoura Orogeny is in evidence in the Late Miocene and Pliocene in the Canterbury and southern North Island areas. Figure 3 summarizes these conclusions.

ACKNOWLEDGEMENTS

We gratefully acknowledge the critical appraisal of the manuscript by JI Raine (New Zealand Geological Survey).

REFERENCES

Couper RA 1980. New Zealand Mesozoic and Cainozoic plant microfossils. N.Z. Geol. Surv. Pal. Bull. p.32. 87 pp.
Couper RA and Harris WF 1960. Pliocene and Pleistocene plant microfossils from drillholes near Frankton, New Zealand. N.S.J. Geol. Geophys. 3:15-22.
Couper RA and McQueen DR 1954. Pliocene and Pleistocene plant fossils of New Zealand and their climatic interpretation. N.Z.J. Sci. Technol. B 35:398-420.
Douglas BJ, Lindquist JK, Fordyce RE and Campbell JD 1981. Early Miocene terrestrial vertebrates from Central Otago. N.Z. Geol. Soc. Newsletter no. 53:17.

Fleming CA 1979. The geological history of New Zealand and its life. Auckland Univ. Press.
Gair HS 1961. Drillhole evidence of the Pliocene-Pleistocene boundary at Timaru, South Canterbury. N. Z. J. Geol. Geophys. 4:89-97.
Kemp EM 1978. Tertiary climatic evolution and vegetation history in the southeast Indian Ocean region. Palaeogeogr. Palaeoclimatol. Palaeoecol. 24:169-208.
Martin HA 1974. The identification of some Tertiary pollen belonging to the family Euphorbiaceae. Aust. J. Bot. 22:271-291.
Martin HA 1978. Evolution of the Australian flora and vegetation through the Tertiary: evidence from pollen. Alcheringa 2:181-202.
Mildenhall DC 1973. Vegetational changes in the New Zealand Quaternary based on pollen studies, with brtief comment on the Pliocene background. In: GD Mansergh (ed.), The New Zealand Quaternary: an introduction. IX INQUA Congress, Christchurch, New Zealand. p. 20-32.
Mildenhall DC 1975. Lower Pleistocene palynomorphs form the Ohuka Carbonaceous Sandstone, Southwest Auckland, New Zealand. N. Z. J. Geol. Geophys. 18:675-681.
Mildenhall DC 1978a. Palynology of the Waipipian and Hautawan Stages (Pliocene and Pleistocene), Wanganui, New Zealand. Note, N. Z. J. Geol. Geophys. 21:775-777.
Mildenhall DC 1978b. Palynomorphs from Miocene-Pliocene sediments, Grey Valley (K31 - metric), South Island, New Zealand. N. Z. Geol. Surv. Rep. PAL. 24. 18 pp.
Mildenhall DC 1980. New Zealand late Cretaceous and Cenozoic plant biogeography: a contribution. Palaeogeogr. Palaeoclimatol. Palaeoecol. 31:197-233.
Mildenhall DC and Harris WF 1970. A cool climate pollen assemblage from the type Waipipian (Middle Pliocene) of New Zealand. N. Z. J. Geol. Geophys. 13:586-591.
Pocknall DT 1981. Pollen and spores from the Rifle Butts Formation (Altonian, Lower Miocene), Otago, New Zealand. N. Z. Geol. Surv. Rep. PAL. 40. 15 pp.
Pocknall DT 1982a. Early Miocene vegetation at Kapuka, Southland, New Zealand: a study based on pollen analysis of a coal seam. N. Z. Geol. Surv. Rep. PAL. 48. 15 pp.
Pocknall DT 1982b. Palynology of the Bluecliffs Siltstone (early Miocene), Otaio River, South Canterbury, New Zealand. N. Z. Geol. Surv. Rep. PAL. 55. 24 pp.
Pocknall DT and Mildenhall DC, in press. Late Oligocene - Early Miocene spores and pollen from Southland, New Zealand. N. Z. Geol. Surv. Pal. Bull. 51.
Suggate RP, Stevens GR and Te Punga MT (eds.) 1978. The Geology of New Zealand. Govt. Printer, Wellington. 2 vols. 820 pp.
Vella P 1977. Report to INQUA Subcommission on the Neogene - Quaternary boundary. The boundary in New Zealand. In: Neogene - Quaternary boundary, Proc. 2nd Sympos., Bologna 1975, Giornale di Geol. (2) 41 (1-2), Bologna. p. 347-358.

Holocene climates of the Vestfold Hills, Antarctica, and Macquarie Island

JOHN PICKARD
Antarctic Division, Kingston, Tasmania

P.M. SELKIRK
Macquarie University, North Ryde, Australia

D.R. SELKIRK
University of Sydney, Australia

ABSTRACT. The Vestfold Hills (68°35'S 78°00'E) are a 400 km² ice-free oasis on the coast of East Antarctica. Mean daily temperatures are >0°C for <2 months annually. The terminal Pleistocene ice sheet advance (Vestfold Glaciation) covered the hills. Subsequent Holocene ice recession at 0.7-3.1 m per year and isostatic uplift cut off marine inlets and exposed Holocene marine fossil beds below terraces around lakes which became saline. Wind direction has remained constant for the past 4000 years. Ablation rates of the ice sheet and a nearby glacier are consistent with the overall rate of Holocene ice sheet retreat. Rates of surface lowering of an ice-cored moraine are consistent with the postulated age of the Chelnok Glaciation which formed the moraine. Thus the fossil and geomorphic evidence shows no substantial climatic change in the Holocene.

Subantarctic Macquarie Island (54°30'S 158°57'E) supports fjaeldmark, herbfield and grassland vegetation with c. 40 vascular and c. 110 bryophyte species. Its hyperoceanic climate has a mean annual temperature of 4.5°C and rainfall of 926 mm. The Antarctic Convergence lies south of the island but during the Last Glacial Maximum lay north of it. Sea temperatures were then 2°C lower and land temperatures 3-6°C lower than at present. The extent of glaciation on the island is uncertain. Palynological studies of peat from three sites show that plant remains had begun accumulating by 9500 BP, and suggest that there was no major climatic fluctuation during the Holocene on Macquarie Island.

The apparent lack of major climatic change during the Holocene in both the Vestfold Hills and on Macquarie Island is similar to interpretations from elsewhere in Antarctica, Marion Island and South Georgia. However, it differs from interpretations of events on Kerguelen. Possible reasons for the apparently constant climate are that it was only relatively constant or, that the ecological amplitudes of the fossil species are so wide that minor climatic changes are not reflected.

INTRODUCTION

There is evidence from New Zealand and South America that the Last Glacial period ended in the Southern Hemisphere about 14 000 BP (Burrows 1979). The period from the end of the glacial to c. 8000 BP can be regarded as transitional between glacial and the fully interglacial conditions of the present. Evidence from several subantarctic islands and Antarctica corroborates this view. Recent geomorphological studies in the Vestfold Hills, Antarctica (68°30'S 78°00'E) (Figs. 1 & 2) and subantarctic Macquarie Island (54°30'S 158°57'E) (Figs. 1 & 3) provide data which add to the growing body of palaeoclimatic information.

The Vestfold Hills are a 400 km^2 coastal oasis of low rocky hills. The area was completely covered with at least 1000 m of ice by the Vestfold Glaciation at the Last Glacial Maximum, but subsequent retreat has exposed the hills and hundreds of lakes (Adamson and Pickard 1983). Many of the lakes are now hypersaline remnants of marine inlets cut off from the sea by isostatic uplift. Holocene marine fossils are abundant below terraces above these lakes. Adamson and Pickard (1983) interpreted the fossils as indicating local events within overall global changes and concluded that it was very difficult to separate local and global climatic effects. In contrast, Zhang et al. (1983) adopted a global view, correlating events in the Vestfolds with events at McMurdo Sound, Antarctica and in East China. They also concluded that at least one series of moraines coincided with the Neoglaciation of 3000-2000 BP .

Macquarie Island is a rectangular island with steep scarps c. 250 m high on all margins. The plateau surface undulates with numerous lakes in shallow valleys. The island is tectonically active (Selkirk et al. 1983). Wireless Hill at the northern extremity of the island has risen c. 80 m in the last 5500 years. The glacial history of the island is debated (see below).

In this paper available evidence from both the Vestfold Hills and Macquarie Island is examined and the most conservative interpretation of the Holocene climate consistent with the evidence is presented.

VESTFOLD HILLS

The present climate of the Vestfold Hills is typical of the antarctic coast but it is ameliorated by the large oasis of exposed rock. Burton and Campbell (1980) summarized available data from Davis Station for the period 1957-1975. The two extreme months of June and December show a considerable range in solar radiation, temperature and wind. Mean daily temperatures are $>0°C$ for <2 months annually.

Fossil evidence of Holocene climate

The available fossil evidence from marine, terrestrial and freshwater organisms gives little indication of change in climate over the past 8000 years. Extant molluscs and polychaete worms flourished from at least 8000 BP (Fig. 4, Table 1). All these species (8 bivalves, 5 gastropods and 3 polychaetes) currently have circumantarctic distributions and wide depth ranges (0-1500 m), typical of the antarctic benthic fauna (Yegorova 1973).

Setty et al. (1980) recorded 14 planktonic and 42 benthic foraminifera from terraces at Deep Lake (Fig. 2). Marine molluscs and a serpulid worm on the terraces are C14-dated at c. 5000 BP (Table 2). All

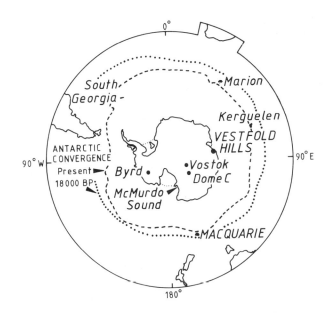

Figure 1. Map of Antarctica and adjacent oceans showing locations of Vestfold Hills, Macquarie Island and other places mentioned in the text. Positions of Antarctic Convergence (dashed: present, dotted: 18 000 BP) from Hays (1978, 1983).

the forams are extant in various sectors of the ocean around Antarctica.

The terrestrial and freshwater moss Bryum algens Card. and associated algae grew in Clear Lake c. 8000 years ago (Pickard and Seppelt, in press). This moss currently occurs around the continent in both lake and terrestrial habitats (Seppelt 1983; Kasper et al. 1983).

Several species of cyanobacteria formed papery stromatolites up to 0.15 m thick at Lake Watts. Dating of top and bottom of these stromatolites shows that they grew continuously from 2700 to 1800 BP (Table 1). Today these species form felts and stromatolites in lakes and ponds of varying salinity both in the Vestfolds and elsewhere in Antarctica (Parker et al. 1981). Diatoms associated with the Lake Watts stromatolites are widespread freshwater species.

Geomorphic evidence of Holocene Climate

Several lines of geomorphic evidence also indicate little change in climate during the Holocene. Pickard (1982) analysed directions of wind-eroded grooves and ventifact faces in both gneiss country rock and dolerite dykes. He showed that the directions of these features coincided with the present mean wind direction of c. 67° true and has for the past 4000 years.

The continental ice sheet retreated from the Vestfolds at an average rate of 1-2 m per year (Adamson and Pickard 1983) but with wide local variations (0.7-3.1 m per year). Measured ablation rates of 0.3-0.4 m

Table 1. Summary of ages and present distribution of fossil organisms from the Vestfold Hills.

Organism #	Ages*	Distribution in Vestfold Hills	Present distribution ecology†
MOLLUSCA: BIVALVIA Laternula elliptica (King and Broderip)	7370±95 to 2410±90	Below terraces in Death Valley, Watts System and elsewhere	Circumantarctic, Scotia Arc. to South Georgia, Kerguelen. Depth range 1-472 m, most live specimens from <100 m, probably commonest <20 m.
Limatula (Antarctolima) hodgsoni (Smith)	6500±105	"	Circumantarctic, Scotia Arc. 6-695 m, occasionally 1180 m.
Adamussium colbecki (Smith)	7305±130	"	Circumantarctic, Scotia Arc. 4-1308 m, most abundant 23-183 m.
ANNELIDA: POLYCHAETA Serpula narconensis (Baird)	6150±95 to 5340±90	"	Circumantarctic, Scotia Arc. Kerguelen. At low tide mark.
CYANOBACTERIA Several spp. including Phormidium frigidum, Nostoc puctiforme, and Tribonema sp.	top: 1890±50 (SUA2023) bottom: 2720±60 (SUA2022)	Below terraces, unconformably overlies beds of Serpula	Widely distributed in antarctic lakes of varying salinity.
MUSCI (Mosses) Bryum algens Card.	8355±250 to 7310±150	Thin bands in sand above lake	Circumantarctic. Terrestrial and in fresh lakes to at least 30 m depth.

\#Voucher specimens of dated material are lodged in the National Museum of New Zealand (molluscs), the Australian Museum (serpulids) and Antarctic Division (mosses).
*Except where laboratory numbers are given, dates are quoted from Adamson and Pickard (1983). Ages are given in conventional radiocarbon years. Ages of marine organisms are not corrected for the antarctic marine reservoir effect of c. 1300 years. See also Fig. 4.
†Compiled from data in the literature and information supplied by RK Dell (National Museum of New Zealand), PA Broady (University of Melbourne) and HA ten Hove (Rijksuniversiteit, Utrecht).

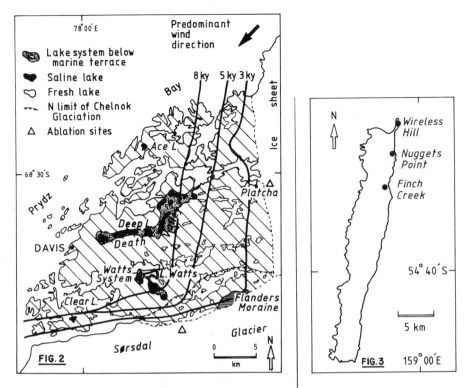

Figure 2. Map of Vestfold Hills, Antarctica showing places and features mentioned in text and position of the ice sheet at various times during the Holocene (from Adamson and Pickard 1983).

Figure 3. Map of Macquarie Island showing location of peat profiles.

per year on the ice sheet at Platcha and on the Sørsdal Glacier (Fig. 2) (Pickard, unpubl. data) are consistent with the overall rates of retreat during the Holocene.

Flanders Moraine is an ice-cored moraine formed by stagnation of ice which surged north from the Sørsdal Glacier some 1000-2000 years ago (Adamson and Pickard 1983). Present rates of surface lowering (0.05 m per year: Pickard in press) suggest a minimum age of 1200 years for the Chelnok Glaciation. As this age is consistent with the age estimate based on the ice sheet retreat mentioned above, it tentatively supports the idea of little climatic change. However, the Chelnok Glaciation itself may well have been the consequence of a warmer period with increased precipitation on the ice sheet. Simultaneous rapid melt of the margin of the glacier would leave an unstable boundary when the higher ice in the glacier finally debouched from the ice sheet.

Water levels of various lakes are currently reflecting short-term changes (HR Burton, pers. comm.). The level of Lake Watts rose c. 3 m from 1948-1980, Deep Lake rose 2 m from 1963-83 and Ace Lake rose 1 m from 1978-82. The cause of these changes in water balance is unknown: it could be either an increase in precipitation or a decrease in evaporation, or both. The fossil stromatolites at Lake Watts show that the current water level is the highest for the last 1800 years.

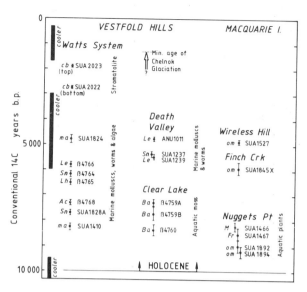

Figure 4. Summary of some radiocarbon dates from three localities in the Vestfold Hills and three sites on Macquarie Island. Dates are given in conventional radiocarbon years. See Table 1 and Adamson and Pickard (1983) for details of Vestfold dates, and Table 2 for Macquarie dates. Also shown are cold periods suggested by Burrows (1979) for the Southern Hemisphere using data in the Byrd and Vostok ice cores.

Abbreviations: Ac: Adamussium colbecki, Ba: Bryum algens, Fr: organic matter including Fissidens rigidulus, Le: Laternula elliptica, Lh: Limatula hodgsonii, M: organic matter including Myriophyllum sp., Sn: Serpula narconensis, cb: cyanobacteria, ma: marine algae, om: organic matter.

MACQUARIE ISLAND

Present environment and former glaciation
Subantarctic Macquarie Island now supports a vegetation of fjaeldmark, herbfield, grassland and mire with a flora of c. 40 vascular and c. 110 bryophyte species. The climate is hyperoceanic with little variation in temperature, rainfall, sunshine days and probably radiation throughout the year (Seppelt 1980). The mean annual temperature is 4.5 C and the rainfall is 926 mm per year.

The Antarctic Convergence now lies to the south of the island but at the Last Glacial Maximum 18 000 BP it was north of the island (Fig. 1). At that time sea tempertures were 2 C lower than at present (Hays 1978, 1983), and temperatures over the island were 3-6 C lower (Colhoun and Goede 1974). The island was glaciated at this time but the extent of this glaciation is uncertain. It seems clear, however, that only limited glaciation was involved and that some parts of the island were snow- and ice-free (Colhoun and Goede 1974; Löffler and Sullivan 1980). Colhoun and Goede (1974) postulated 40 % coverage by permanent ice or snow during the Last Glacial period, Löffler and Sullivan (1980) somewhat more. Either estimate leaves ample ice-free areas as refugia for vegetation. The estimates of temperature depression during this period (Hays 1983; Colhoun and Goede 1974) do not preclude this possibility.

Palynological evidence of Holocene climates

Two of us (PMS and DRS) have examined three peat profiles on the island (Selkirk and Selkirk 1982, Selkirk et al. 1983). The three profiles provide an overlapping record of Holocene vegetation on the Island. Old lake deposits at Nuggets Point (Fig. 3) now exposed by cliff retreat, have pollen of Stilbocarpa, Pleurophyllum and Poaceae in 9800 and 10 200 year old sediments (Table 2, Fig. 4). 9000-odd year old macrofossil remains of the aquatic angiosperm Myriophyllum and the mosses Fissidens rigidulus and Drepanocladus aduncus occur in sediments from the same profile (Selkirk and Selkirk 1982). The base of a peat profile from a ridge near Finch Creek is c. 6790 years old; and a peat profile from Wireless Hill has a basal age of c. 5500 BP (Selkirk et al. 1983). Detailed examination of the Wireless Hill profile and preliminary examination of the Finch Creek and Nuggets Point profiles show no marked changes in pollen composition which could be attributed to climatic change.

Table 2. Radiocarbon dates from Macquarie Island.

Lab No.	Conv. C14-age (yrs BP± std dev)	Calibr. age* (yrs BP ± 95 % confid. inverval)	Site and depth below surface (cm)
SUA-1894	9400±220	10200±440	Nuggets Pt, 440
SUA-1892	9000±250	9800±500	Nuggets Pt, 240
SUA-1466	8230±240	9030±480	Nuggets Pt, 339
SUA-1467	8560±200	9360±400	Nuggets Pt, 147
SUA-1845X	5930±240	6790±410	Finch Crk, 190
SUA-1527	4880±90	5580±260	Wireless Hill, 394

* Calibrated following method of Klein et al. (1982) for ages less than 8000 radiocarbon years. Dates of 8000-10 000 BP are calibrated by adding 800 years to the conventional date and doubling the standard deviation (Barbetti 1980 and pers. comm.).

COMPARISON WITH OTHER ANTARCTIC AND SUBANTARCTIC AREAS

Comparison with continental Antarctica is difficult. Most research has concentrated on Late Pleistocene events (Hendy et al. 1979; Stuiver et al. 1981). Well-dated Holocene sequences are uncommon. Ice cores from Byrd, Vostok (Burrows 1979) and Dome C (Lorius et al. 1979) indicate several cooler periods in the Holocene (Fig. 4). Beget (1983) found world-wide evidence of cooler periods c. 2000 and 8000 BP. The Vestfold Hills data do not yet unequivocally support these interpretations.

The conclusions from Macquarie Island are consistent with findings from other subantarctic islands. Like Macquarie, Marion Island and the Kerguelen Islands (Fig. 1) were partially glaciated (Hall 1980; Young and Schofield 1973a), though recent findings by Hall (pers. comm.) suggest that extensive glaciation during the Last Glacial may have occurred. Palynological and geomorphic data indicate the start of warming on Marion c. 14 000 BP and on Kerguelen c. 12 000 BP. At least some vegetation survived the Last Glacial Maximum on both islands in ice-free refugia close to sea-level (Schalke and van Zinderen Bakker 1971; Hall 1980; Young and Schofield 1973a, b). South Georgia was fully glaciated (Clapperton et al. 1978) but glacial retreat c. 10 000 BP allowed establishment of vegetation by c. 9500 BP. The plants either dispersed from refugia now below sea-level (Barrow 1978) or immigrated from nearby South America (Smith 1981). Pollen records suggest little change in the last 11 000-12 000 years on Marion Island (Schalke and van Zinderen Bakker 1971) and in the last 9500 years on South Georgia (Barrow 1983). On Kerguelen however Young and Schofield (1973b) interpret evidence for a cool period 5000-5400 BP, and Roche-Bellair (1976a, b) describes several climatic fluctuations during the past 5000 years.

CONCLUSIONS

Our fossil and geomorphic evidence supported by extensive radiocarbon dating is remarkably consistent. Taken at its face value it indicates little change in climate over most of the Holocene. However, the ecological amplitudes of the fossil species are so wide that minor climate changes need not have been recorded. For example, Bryum algens, like other antarctic mosses (Rastorfer 1970), probably has photosynthetic and growth optima at c. $20\,^{\circ}C$. Thus a warming of even 5 $^{\circ}C$ in mean annual temperature would have little effect, as this species is primarily limited by characteristics of the substrate and not climate.

The Chelnok Glaciation may reflect the global warming recorded in the Byrd, Vostok and Dome C ice cores (Burrows 1979; Lorius et al. 1979), the Neoglacial cooling (Zang et al. 1983; Beget 1983), or a purely local event. As the question remains open, we conclude that there is scant evidence of changes in temperature, wind and probably solar radiation in the Vestfold Hills during the Holocene. Similarly, on Macquarie Island, changes in the pollen record of each species appear to represent localized vegetation changes for reasons other than climatic fluctuations (Selkirk et al. 1983).

ACKNOWLEDGEMENTS

We thank the Antarctic Division of the Australian Department of Science and Technology for financial and logistic support when we visited Davis (JP 1978-79, 1979-81) and Macquarie Island (PMS 1970-80, 1981) with the Australian National Antarctic Research Expeditions. Radiocarbon dating in the Vestfold Hills was supported by an Australian Research Grants Scheme (ARGS) grant to DA Adamson (Macquarie University), and on Macquarie Island by an ARGS grant to PMS and DRS. JP received financial support from a Commonwealth Post Graduate Research Award.

REFERENCES

Adamson D and Pickard J 1983. Late Quaternary ice movement across the Vestfold Hills, East Antarctica. In: RL Oliver, PR James and JB Jago (eds.), Antarctic earth science. Aust. Acad. Sci., Canberra. p. 465-469.

Barbetti M 1980. Geomagnetic strength over the last 50 000 years and changes in atmospheric C-14 concentrations: emerging trends. Radiocarbon 20:191-199.

Barrow CJ 1978. Postglacial pollen diagrams from South Georgia (sub-Antarctic) and West Falkland Island (South Atlantic). J. Biogeogr. 5:251-274.

Barrow CJ 1983. Palynological studies in South Georgia: III. Three profiles from near King Edward Cove, Cumberland East Bay. Br. Antarctic Surv. Bull. 58:43-60.

Barrow CJ and Smith RIL 1983. Palynological studies in South Georgia: II. Two profiles in Sphagnum Valley, Cumberland West Bay. Br. Antarctic Surv. Bull. 58:15-42.

Beget JE 1983. Radiocarbon-dated evidence of world-wide early Holocene climatic change. Geology 11:389-393.

Burrows CJ 1979. A chronology for cool-climate episodes in the Southern Hemisphere 12 000 - 1000 yr B.P. Palaeogeogr. Palaeoclimatol. Palaeoecol. 27:287-347.

Burton HR and Campbell PJ 1980. The climate of the Vestfold Hills, Davis Station, Antarctica, with a note on its effect on the hydrology of hypersaline Deep Lake. Aust. natn Antarctic Res. Expeds scient. Rep. Ser. D 129:1-50.

Clapperton CM, Sugden DE, Birnie RU, Hanson JD and Thom G 1978. Glacier fluctuations in South Georgia and comparison with other island groups in the Scotia Sea. In: Van Zinderen Bakker EM (ed.), Antarctic glacial history and world palaeoenvironments. Balkema, Rotterdam. p.95-104.

Colhoun EA and Goede A 1974. A reconnaissance survey of the glaciation of Macquarie Island. Pap. and Proc. R. Soc. Tasmania 108:1-19.

Hall K 1980. Late glacial ice cover and palaeotemperatures on sub-Antarctic Marion Island. Palaeogeogr. Palaeoclimatol. Palaeoecol. 29:243-259.

Hays JD 1978. A review of the Late Quaternary climatic history of Antarctic Seas. In: Van Zinderen Bakker EM (ed.), Antarctic glacial history and world palaeoenvironments. Balkema, Rotterdam. p.57-71.

Hays JD 1983. Evolution of the Southern Ocean 18 ± 2 ka. In: JMA Chappell and A Grindrod (eds.), Proc. First CLIMANZ. Dep. Biogeogr. Geomorph. Res. School Pacific Stud., ANU, Canberra. p.55.

Hendy CH, Healy TR, Rayner EM, Shaw J and Wilson AT 1979. Late Pleistocene glacial chronology of the Taylor Valley, Antarctica, and the global climate. Quat. Res. 11:172-184.

Kaspar M, Simmons GM, Parker BC, Seaburg KG, Wharton RA and Smith RIL 1983. Bryum Hedw. collected from Lake Vanda, Antarctica. The Bryologist 85:424-430.

Klein J, Lerman JC, Damon PE and Ralph EK 1982. Calibration of radiocarbon dates: tables based on consensus data of the workshop on calibrating the radiocarbon timescale. Radiocarbon 24:103-150.

Löffler E and Sullivan M 1980. The extent of previous glaciation on Macquarie Island. Search 11:246-247.

Lorius C, Merlivat L, Jouzel J and Pourchet M 1979. A 30 000-yr isotope climatic record from Antarctic ice. Nature 280:644-648.

Parker BC, Simmons GM, Love FG, Wharton RA and Seaburg KG 1981. Modern stromatolites in antarctic dry valley lakes. Bioscience 31:656-661.

Pickard J 1982. Holocene winds of the Vestfold Hills, Antarctica. New Zealand J. Geol. Geophys. 25:353-358.

Pickard J, in press. Surface lowering of ice-cored moraine by wandering lakes. J. Glaciol. 29.

Pickard J and Seppelt RD, in press. Holocene occurrence of the moss Bryum algens Card. in the Vestfold Hills, Antarctica. J. Bryology.

Rastorfer JR 1970. Effects of light intensity and temperature on photosynthesis and respiration of two east Antarctic mosses, Bryum argenteum and Bryum antarcticum. The Bryologist 73:544-556.

Roche-Bellair N 1976a. L'holocene de l'anse Betsy (Île Kerguelen). C. r. Acad. Sciences Paris Ser. D 282:1347-1348.

Roche-Bellair N 1976b. Les variation climatiques de l'holocene supérieur des Îles Kerguelen: d'après la coupe d'une tourbire de la plaine de Dante (côte meridionale). C. r. Acad. Sciences Paris Ser. D 282:1257-1260.

Schalke HJWG and van Zinderen Bakker EM 1971. History of the vegetation in Marion and Prince Edward Islands. In: EM van Zinderen Bakker, JM Winterbottom and RA Dyer (eds.), Marion and Prince Edward Islands. Balkema, Rotterdam. p.89-99.

Selkirk PM and Selkirk DR 1982. Late Quaternary mosses from Macquarie Island. J. Hattori Botan. Lab. 52:167-169.

Selkirk DR and Selkirk PM 1983. Preliminary report on some peats from Macquarie Island 7+2 ka. In: JMA Chappell and A Grindrod (eds.), Proc. First CLIMANZ. Dep. Biogeogr. Geomorph., Res. School Pacific Stud., ANU, Canberra. p.115-117.

Selkirk DR, Selkirk PM and Griffin K 1983. Palynological evidence of Holocene environmental change and uplift on Wireless Hill, Macquarie Island. Proc. Linnean Soc. New South Wales 107:1-17.

Seppelt RD 1980. A synoptic moss flora of Macquarie Island. Antarctic Div. tech. Memorandum 93:1-8.

Seppelt RD 1983. The status of the antarctic moss Bryum korotkevicziae. Lindbergia 9:21-26.

Setty MGAP, Williams R and Kerry KR 1980. Foraminifera from Deep Lake terraces, Vestfold Hills, Antarctica. J. Foraminiferal Res. 10:303-312.

Smith RIL 1981. Types of peat and peat-forming vegetation on South Georgia. Br. Antarctic Surv. Bull. 53:119-139.

Stuiver M, Denton GH, Hughes TJ and Fastook JL 1981. History of the marine ice sheet in West Antarctica during the last glaciation: a working hypothesis. In: GH Denton and TJ Hughes (eds.), The last great ice sheets. Wiley-Interscience, New York. p.319-436.

Yegorova EN 1973. Some features of the distribution and ecology of Davis Sea mollusks. Inf. Bull. Soviet Antarctic Expeds 84:344-348.

Young SB and Schofield EK 1973a. Palynological evidence for the late glacial occurrence of Pringlea and Lyallia on Kerguelen Islands. Rhodora 75:239-247.

Young SB and Schofield EK 1973b. Pollen evidence for late Quaternary climate changes on Kerguelen Islands. Nature 245:311-312.

Zhang Q, Xie Y and Li Y 1983. A preliminary study on the evolution of the post late Pleistocene Vestfold Hills environment, East Antarctica. In: NL Oliver, PR James and JB Jago (eds.), Antarctic earth science. Aust. Acad. Sci., Canberra. p.473-477.

Glacial age environments of inland Australia

J.M.BOWLER
Australian National University, Canberra

R.J.WASSON
CSIRO, Canberra, Australia

ABSTRACT. The Australian continent preserves a record of changes in climatic zones ranging from summer monsoonal regions into the mid-latitude westerly winter rainfall zone. The absence of major orographic belts reduces the effect of local climatic complications. Similarly, the relatively small area of glacial ice had little effect on climatic modifications. Thus the climatic pattern involves migration of the ITCZ and procession of subtropical high pressure systems that are relatively predictable today and which may be reconstructed with some degree of confidence in the past.

Analysis of the array of fluvial, aeolian and lacustrine sediments and landforms of the Australian interior yields an account of environmental changes through the period of the Last Glacial cycle. Within the range of radiocarbon dating, this records two major events: a long period during which lakes in southern Australia were greatly expanded, followed by drying and accentuated aridity. Evidence from a closely dated sequence in the Willandra Lakes suggests that the expanded lacustral phase began about 50 000 BP, with an oscillation at 36 000 BP, associated with a brief phase of dune instability. Lake level oscillations of diminishing amplitude continued and came to an end about 20 000 BP, giving way to a period of extensive dune building. At this time the arid zone, as indicated by expanded areas of dune activity, extended polewards, producing longitudinal dunes even into northeastern Tasmania. Events in the monsoon rainfall zone, although not securely dated, occurred in the same sequence as those in the south. There expanded lakes, as typified at Lake Woods, were followed by drying and dune reactivation in low latitudes.

Glacial Age environments were characterized by increased wind strength although wind direction remained approximately similar to that of today, with the exception of the westerlies on the southern margin of the high pressure belt. Strengthening of the westerlies produced west-to-east dune alignments where south-westerly sand-shifting resultants apply today.

Within the Australian arid and semi-arid region, high water-tables associated with the early humid phase exerted a profound influence on the processes, composition and forms of the subsequent dunes which developed during the more arid phase. Thus saline pelletal clays originating from lake floors produced lunette dunes and deflation from salinized swales resulted in extensive pelletal clay deposits being transported into longitudinal dunes.

Reconstruction of environments of the Last Glacial Maximum suggests that higher wind speeds and amplified advective heat transfer in Glacial Age summers resulted in evaporation rates higher than those of today. These were associated with colder Glacial Age winters, the accentuated seasonality producing conditions inimical to vigorous plant growth and favouring high run-off. These conditions resulted in a range of hydrologic and sedimentary phenomena that persisted in the rivers, lakes and dunes across the Australian landscape until the onset of conditions, about 13 000 BP, more akin to those of today.

INTRODUCTION

Located in the middle to low latitudes, spanning the range of $10°$ to $44°$ S, the Australian continent, like southern Africa, straddles climatic zones from the Tropic of Capricorn through the subtropical high pressure belt to the zone of southern westerlies. The effects of this range of climatic zones is imprinted on the landscape, on its sediments and soils; it finds expression through the diversity of its various land forms.

While the direct effects of glaciation on the Australian continent remain relatively slight, with the exception of Tasmania which experienced substantial glaciation, the indirect effects as measured by changes in water balance and variations in intensity and frequency of aeolian events in the dry interior have been substantial. Indeed, such was the diversity and extent of these effects in the legacy of rivers, lakes and dunes that few elements of the landscape can be said to have escaped the influence of changes that accompanied the Last Glacial to Post-glacial cycle spannig the past 120 000 years.

We here review some of the expressions of hydrologic and climatic changes of Late Quaternary age as recorded particularly in the dry continental interior. We shall explore the interconnection between mountains and plain and between lake and dunes. In so doing we shall attempt to define causal links between stream flow, lake levels and some of the processes that have contributed to the diverse array of aeolian landforms that help characterise inland Australia today.

Others have reviewed Late Quaternary environments of the Australian continent, notably Bowler et al. (1976) and Hope (1983). More recently a comprehensive treatment of the Australasian region has become available through the CLIMANZ Project, an international programme aimed at a synthesis of the Late Glacial environments of Australia, New Zealand, Papua New Guinea and the Southern Ocean. The proceedings of the first CLIMANZ Conference (Chappel and Grindrod 1983) provided ample testimony for the diverse range of evidence now available for Glacial Age environmental changes through this expansive region of the Southern Hemisphere.

In Australia, as in many other low latitude desert regions, the contribution of glacial processes to our understanding of Quaternary environments is now subordinate to those reconstructions based on sedimentation sequences which are often far removed from glacial or periglacial areas. Thus, in this paper, the evidence from rivers, lakes and dunes is more substantial and consistently better dated than that from the high altitude phenomena. Whilst this is partly a reflection of personal research interests, it is also consistent with the relative amount of information now coming forward from the inland and more arid environments, many of which seem to have been neglected in the history of Pleistocene stratigraphic studies both in Australia and elsewhere.

REGIONAL SETTING

The Australian continent, lying astride the zone of sub-tropical high pressure cells, combines uniformly low relief with relative climatic

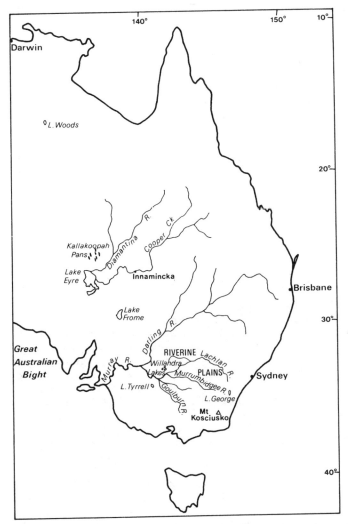

Figure 1. Locality map showing sites referred to in the text.

uniformity over large regions. The absence of high orographic barriers reduces the components of local geographic variability in circulation patterns, thereby reducing the complications that must be considered in palaeoclimatic reconstructions.

On the mainland, solid precipitation is restricted to the southeastern highlands where Mt. Kosciusko reaches the maximum elevation of the continent, 2228 m (Fig. 1). The low drainage divide that runs the length of the east coast diverts streams inland contributing to the continent's major drainage basin, the Murray-Darling system. With the exception of this basin, the entire southern and central inland regions of the continent lack any permanent streams. Over large areas, ephemeral internal drainage systems feed into saline pans of various sizes ranging from L. Eyre at the one end of the scale to numerous pans less than 1 km in diameter at the other.

The northern half of the continent receives its rainfall from the summer monsoonal system while south of latitude 30-33° the winter westerlies remain dominant. Clear skies and low humidity contribute to very high levels of surface readiation, especially in summer, within the latitudinal range 22-27° on the southern margin of the monsoonal rainfall zone.

Continentality is not extreme, and the influence of the surrounding maritime environment extends deeply into the innermost regions of the continent. Air masses that bring rain to the Alice Springs region often have their origins in tropical cyclonic depressions. Additionally, whilst winter diurnal temperature may fall below freezing, the combination of low relief, low altitude and relatively high humidity prevents development of very low temperature extremes. Thus the Australian continent nowhere experiences either the high degree of seasonal contrast, found for instance in the landlocked deserts of Asia, or the altitudinal gradients of South America. In contrast to Saharan Africa, Australia lacks the hyperarid environments even in the driest regions near L. Eyre and the Simpson Desert where rainfall, although highly irregular, is sufficient to maintain a vegetative cover on most desert dunes.

GLACIAL AND PERIGLACIAL ENVIRONMENTS

The actual extent and number of glacial episodes experienced by the highlands of southeastern Australia still remain a matter of controversy. It is apparent that the Australian continent experienced only restricted glaciation during the last cold phase. A small ice cap on the summit peaks of the Snowy Mountains was estimated by Galloway (1963) to have covered a maximum of only 50 km^2. Periglacial processes on the other hand had widespread effects on slopes down to and below 1000 m as summarized in earlier reviews (Bowler et al. 1976).

The most important evidence of extensive glaciation is limited to Tasmania where Colhoun and co-workers have dated the period of maximum glacial advance (the Margaret Glaciation) in an important sequence on the West Coast Range. Here 3 m of proglacial gravels and silts overlie a buried soil of an earlier glacial outwash phase. A radiocarbon date from wood 10 cm above the soil, in sediment representing the maximum advance of the last glaciation, provides a date of 18 800 ± 500 BP (Colhoun 1983).

Although the age of recession is not well defined, the appearance of peats at 15 000 BP on summit slopes on the mainland at Mt. Kosciusko (Fig. 1) (Costin 1972) indicates retreat of ice and return to warmer conditions there by that time. Similarly in Tasmania, ice caps and valley glaciers had disappeared by 13 000 BP and the highest cirques were evacuated before 10 000 BP (Calhoun 1983). Like the disappearance of the ice, the actual age of onset of glacial and periglacial events is not well dated. The best available evidence still comes from the Toolong Range where Caine and Jennings (1968) dated a Nothofagus stump below a block stream at 34 000 BP, suggesting that periglacial processes were in action soon after that date. Thus we assume that by 32 000

BP periglacial processes were active in the slopes of the southeastern highlands.

On the continental mainland, the greatest influence of glacial and periglacial processes in the southeast lay in the hydrologic influence they exerted on rivers draining the highland slopes. Changes in run-off, seasonality and many aspects of the flow regimes were transmitted by the rivers through the Riverine Plain to the inland semi-arid dunefields of northern Victoria, western New South Wales and South Australia.

RIVERS OF THE EAST

The rivers draining the slopes of southeastern Australia link the area affected by glacial and periglacial processes with the more arid inland plains (Fig. 1). The only region of the entire southern half of the continent to possess even moderate water resources is that of the southeast centred on the Murray Basin. Composed of two major geomorphic provinces, the Mallee dunefield to the west and the Riverine Plain to the east, the Murray Basin is traversed by four of Australia's main rivers draining the southeastern montane slopes, the Lachlan, Murrumbidgee, Murray and Goulburn rivers. Additionally, the Darling rising in the Queensland summer rainfall zone, enters the basin from the north to join the Murray in the centre of the sandy Mallee country. Thus four of the five rivers have their origin in the southeastern highlands, the regions most affected by cold climate processes today and most sensitive to Glacial Age changes in hydrologic regimes.

The alluvial sediments of rivers draining the southeastern highlands preserve a long history of both morphologic and sedimentologic changes in the channel regimes. Whilst evidence is available from coastal terrace studies (Walker 1962; Warner 1972; Bowler 1970) the most comprehensive data are available from the inland plains. There the rivers of the Murray Basin have yielded most data, especially those on the Riverine Plain. This region has been the site of intensive studies for many years. Indeed the pioneering work of Butler (1950, 1958) in establishing his soil-stratigraphic sequence for this region led a succession of others to conduct geomorphic and stratigraphic studies in the same region (Pels 1964; Langford-Smith 1960; Schumm 1968; Bowler 1978a). These combine to present a comprehensive picture of Late Glacial changes in the fluvial regimes of this extensive region.

In summary, the fluvial systems of the last 40 000 years available for radiocarbon dating, record three major hydrologic phases (Fig. 2). Studied in detail in the Goulburn-Murray system of northern Victoria, equivalent phases can be identified in the channel records of the Murrumbidgee and other rivers of the region. In the Goulburn Valley near Shepparton, Bowler identified two early phases in which channel and meander morphologies were both an order of magnitude larger than those of the modern streams. Named the Tallygaroopna and Kotupna phases respectively, from oldest to youngest, they are spatially separated, the Kotupna having deviated from its ancestral Tallygaroopna course. Radiocarbon dates at the point of separation (Bowler 1978a) place this event at 25-26 000 BP, although in the light of information on sample contamination this date must now be seen as a minimum only. Both early phases of large channel and meander morphology possess sand dunes on their northern or northeastern margins, source bordering dunes that were developed by deflation from exposed point-bars by prevailing westerly winds during low flow stages.

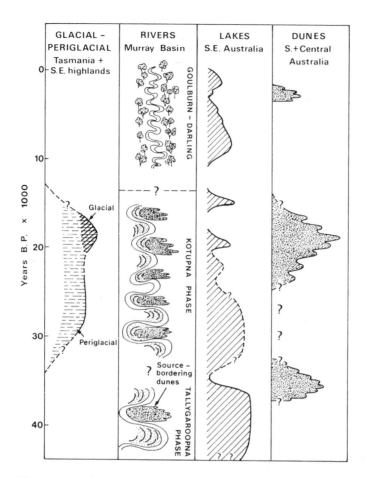

Figure 2. Correlation diagram showing patterns of change registered in southeastern and central Australia through the past 40 000 years. Uncertainties in dating beyond 30 000 BP limit reliability of correlations in the earlier part of the time scale.

The Kotupna phase is expressed morphologically as a low-lying belt 3-5 km wide with ancient meander scrolls, point-bar traces and source bordering dunes within which narrow, sinuous channels of the present rivers often follow meandering courses inherited from the past regime.

Radiocarbon dates from sandy point-bars of the Kotupna system indicate its continuation until just after 15 000 BP, while finer sediments of the silty-clay channels of the present-day suspended load regime date from 10 000 BP, until recent. Thus the transition from the large channel morphology with sandy sediments to the small, highly sinuous suspended load channels of the modern regime is placed between about 11 and 14 000 BP. The actual age and mechanism of the transition are not clear although, as will be apparent later, there is an important correlation with hydrologic changes documented from elsewhere in the system.

The differences between the ancient and modern channels have been interpreted as evidence of much greater discharges in past regimes (Schumm 1968; Bowler 1978a). The association of large channels with sandy sediments dated to within the period of periglacial processes in the highlands led Bowler to propose a causal relationship between plains and uplands. Periglacial activity would have reduced vegetation cover, amplified run-off and contributed large quantities of coarse sediments into channel systems. However accurate this estimate may be, it does not explain the entire evidence.

In 1977 a stratigraphic section provided by a gas pipeline trench across the Darling system near Tilpa furnished Bowler et al. (1978) with an opportunity to make a comparative study of a fluvial system whose catchments, extending well into Queensland, would have remained relatively unaffected by periglacial activity.

The modern Darling is a small, sinuous channel characterised by a silty clay suspended load regime. But like the Murray-Goulburn system, the Darling also possess an ancestral phase of channels and point-bars dramatically different from those of today. The pipeline trench, passing across a section with ancient channel traces and point-bar sediments, provided a sequence of radiocarbon dates. These confirm that the transition from the wide channel, large meander wavelength phase with sandy sediments to the modern small wavelength regime occurred some time about 12 000 BP, suggesting a remarkable similarity between both the morphologic sequences and the age of the transition between them, a similarity that has its parallel in the events recorded in the Goulburn-Murray systems further south. But in this case, the controlling factors were almost entirely independent of periglacial catchment controls.

The convergent evolution of form and the similarities of the ages of transitions between different channel regimes has now been explored in the Cooper Creek system of the desert interior (Wasson 1983).

Cooper Creek is a major stream rising in the summer rainfall zone of central Queensland and terminating in L. Eyre (Fig. 1). Major changes have occurred in both channel morphology and sediment load of this river, and have been dated in the area downstream of Innamincka (Wasson 1983). Prior to 23 000 BP, sand was deposited over large areas, probably by braided channels. From 23 000 to 12 000 BP large meandering channels developed and carried a mixed load, depositing muddy fine sand overbank, thereby burying the sandy deposits of the putative braided channels. The veneer of finer sediment was the source for clay pellets and quartz particles blown into longitudinal and transverse dunes on the floodflats. At 13 000-12 000 BP low sinuosity suspended load channels began to bury the now relict meandering channels with muddy sediments, a process that is continuing today. A similar sequence of morphologic changes is evident in the Diamantina River but this has not been dated (Fig. 1).

While the nature of the morphologic and sedimentologic changes in the Goulburn, Murray, Darling and Cooper streams has not been identical, the major changes occur in the same direction. That is, the streams of the Pre-glacial Maximum and Glacial Maximum periods were much more vigorous than their Post-glacial (<12 000-15 000 BP) counterparts. The decreased bedload component in each stream can be ascribed to Post-glacial revegetation of their catchment areas.

We believe the explanation for the convergent evolution of channel morphologies and sediment regimes across different climatic gradients on the inland plains lies in some common factor other than changes in catchment regimes alone. The answer is to be found in the regional occurrence of high groundwater conditions inherited from the earlier humid environments which will be discussed later. During periods of high groundwater, flow in the streams would have been accompanied by high pore water pressures, leading to extensive bank failure and thus contributing a very rapid meander enlargement, while the development of wide channels provided a continuous source of sandy bedload materials. These characteristics are common to the Goulburn, Murray, Darling and Cooper systems in the period from about 15 000 to 30 000 BP.

INLAND LAKES AND DUNES

The Mallee Story

The streams carrying large quantities of water to the inland plains and especially to the Murray Basin supplied many inland lakes. Scattered throughout the Mallee dunefield and the Riverine Plains are innumerable lake basins and pans in various stages of hydrologic development. Although many remain dry or are in a saline playa phase today, all preserve legacies of past episodes of either high water levels or phases of deflation from lake floors or even both. High water levels are recorded by the presence of strandlines, beaches or low salinity faunas. Deflationary episodes are indicated mainly by the presence of almost ubiquitous transverse dunes on the eastern margins. Often composed of large percentages of clay with gypsum, they follow closely the crescentic basin outline producing the now well known lunette dune type. The frequent occurrence of these clay-rich dunes across the southern half of the continent, and the sensitive hydrologic conditions necessary for their formation, together have provided one of the main lines of evidence permitting us to reconstruct both ages and processes of basin and dunefield evolution. The Willandra Lakes (Fig. 1) remain one of the most instructive regions. Here the presence of major hydrologic changes ranging from widespread lacustral events to arid phase drying of lakes, evaporite production and dune building was first established in a time controlled context (Bowler 1971). It is appropriate here to review this evidence briefly (Fig. 2).

Willandra Lakes

In the Willandra Lakes which are fed by a distributary of the Lachlan River, a stratigraphic succession spanning the Last Glacial cycle has been reconstructed from drilling lake floor sediments and from detailed analysis of shoreline stratigraphy. Both basin and strandline sediments of Last Glacial Age are separated from sediments of an earlier glacial cycle by a major palaeosol, the Golgol soil (Bowler 1971). This represents a long period of non-deposition. From the end of the penultimate glaciation about 130 000 BP the system remained dry as it is today, permitting pedogenesis to develop over both lake floor and shorelines dunes. By extrapolating radiocarbon dates within the basin sediments, the onset of the last lacustral phase is located at near 50 000 BP.

There followed a long period during which fresh water filled the basins to overflow levels, whilst well sorted gravels and quartz sand beaches developed on the eastern (downwind) margins under the influence of strong wave action.

An oscillation at 36 000 BP interrupted the long lacustral episode. It is marked by a weak palaeosol with human occupation as is indicated by burnt shell middens. Associated with this event there is the first appearance in the shorelines sequence of saline aeolian clays which require both low water levels and increasing salinity in the basin for their formation. This registers the first appearance of water deficit conditions and implies a fall below the outlet level in the interconnected system. This change was previously dated at about 25 000 BP (Bowler 1971, 1978b), but recent evaluation of the reliability of some radiocarbon dates obtained, casts doubt on the validity of organic dates in the time range greater than 25 000 years. Events in this part of the time scale are thus being re-evaluated (Bowler et al., in prep.). The opportunity now offers itself in the Willandra sequence for cross checking the reliability of dates determined on different materials; specifically the comparison of synchronous pairs of unionid shells and organic samples, fractionated organic samples and, finally, the crucial test of stratigraphic consistency. Without being able to detail here the arguments advanced by Bowler et al. (in prep.), we may summarize their conclusions. Consistent evidence suggests that all organic samples older than 20 000 to 25 000 BP are subject to substantial and irregular patterns of contamination by younger organic complexes. This makes them appear too young. In some cases, the errors involve differences of up to 10 000 apparent years between dates on shell and organic samples from the same midden. In all tests of consistency, the unionid shells provide more reliable results in this part of the time scale near the limits of radiocarbon detection.

This new evidence requires modification of previously published reconstructions in this part of the time scale (Bowler 1978b). Moreover, the availability of organic materials in the semi-arid Willandra area with its relatively frequent occurrences of organic concentrates in Aboriginal fire places would, by most current standards, be regarded as providing good materials for dating. The new evidence calls into question not only some of the older dates from this sequence, but raises serious doubts about the validity of comparable evidence from other regions. It highlights the difficulty, and urges caution, in correlating any events even on a local, let alone an intercontinental, basis for that part of the time scale in the 25 000-40 000 BP range unless thorough independent checks are available. Indeed the inconsistencies only appear in the Willandra sequence because there are so many dates (more than 120) from the same stratigraphic sequence in a restricted region. Such controls are all too rarely available so that many cases of similar contamination will inevitably go undetected.

Passing to the younger and more relaible dated part of the system, from about 22 000 BP onwards, the saline and often gypseous pelletal clays increase in the aeolian lakeshore components, indicating a progressive decrease in surface water. This trend was interrupted by a brief return to deeper water conditions about 20 000 BP and again about 18 500 BP. The drying trend culminated in a phase of maximum dune construction within the interval 17 500 to 16 000 BP, a period defined by many radiocarbon dates within the uppermost aeolian unit (Zanci unit of Bowler 1971) on the shoreline of L. Mungo and other lakes in the system.

Following the long period of deep water environments, the basin sediment response was influenced largely by changes in the groundwater regime. While the lakes had been full, they acted largely as groundwater recharge systems. With the onset of a surface water deficit, the piezometric gradients were reversed, saline groundwater drained back into the drying basins, maintaining the supply of salts and helping to perpetuate the seasonal efflorescence of halite causing pelletization and deflation of clay aggregates and associated gypsum crystals from the basin floors.

The dunes ceased migration as brief pulses of fresh water returned about 15 000 BP, indicated by many occurrences of unionid shells especially in lakes highest in the catchment. The water available was insufficient to reactivate all basins. However, by this time, the water-table had already fallen to a level beyond which the basins returned once more to a recharge state. Consequently there was no further build-up of salts and no reactivation of clay pellet formation during the 15 000 BP brief spell of freshwater environments. Elsewhere in the Murray Basin, radiocarbon dates from L. Albacutya, Tyrrell and the Darling River lakes point towards a regional expression of those events identified in the Willandra system (Fig. 2).

Lake Frome and Lake Eyre

Studies currently in progress at L. Frome and L. Eyre in connection with the ANU SLEADS Project (Salt Lakes, Evaporites and Aeolian Deposits - ANU project concentrating on coring depositional centres within the arid zone of the continent) point to the existence there of a sequence of hydrologic changes which reproduce in kind if not in exact age, the sequence of hydrologic changes documented from the Willandra Lakes. Thus at Frome, a long period of relatively deep low salinity environments indicated by a 20 metre shoreline on the western side of the lake and by the existence of Chara oogonids, Corbiculina and other fauna was terminated by drying and a phase of deflation. At this time major gypseous transverse dunes were constructed in a manner directly analogous and apparently synchronous with that described for the Willandra system (Bowler 1976; Callen et al. 1983). Preliminary dates from basin sediments indicate the presence there of a major disconformity between 25 000 and 16 000 BP equivalent to the aeolian deflationary phase.

Whilst the L. Eyre stratigraphic sequence reproduces the deep water lacustral phase followed by a corresponding deflationary episode, dates are not yet available to provide accurate chronologic comparisons with events at L. Frome and the Willandra Lakes.

DESERT SEQUENCE, DUNE BUILDING

Thus far we have identified intensified phases of aeolian processes in conjunction with the drying of lakes. At Willandra, construction of extensive transverse dunes between the interval 25 000-15 000 BP coincided with the re-activation of quartz dunes of the longitudinal and sub-parabolic types throughout the region. Some such dunes advanced onto the lake floors and, in places, were trimmed back by episodic return of surface water to the basins. The advance of others across drying lake floors, as in the Chibnalwood lakes, coincided with deflation from basin centres and the construction of transverse clay and gypsum dunes on eastern margins.

At L. Frome, construction of gypsum lunettes and islands within the playa are attributed to the interval between about 23 000-15 000 BP suggesting a regional control by the same climatic variables that affected the Willandra system.

Within the Australian desert dunefields we recognize a distribution between dunes that are essentially quartzose and those that consist of quartz grains and clay pellets (Bowler and Magee 1978; Wasson 1983, and in press). Thin section analysis of the short clay-rich longitudinal dunes of the semi-arid Mallee suggested the presence of clays of originally pelletal form, implying that their formation was by processes similar to those that produce clay lunettes. These require saline efflorescence with the breakdown of clays into pelletal form, circumstances favoured by high water-tables. Dunes of the Mallee type have accreted layer upon layer in a conformable sequence and they have not migrated significantly downwind.

The mineralogic evidence with pelletal clay fabrics, together with internal structural features, suggest a common factor in the processes controlling both the origins of the transverse gypseous clay lunettes and the clay-rich longitudinal forms with which they are often associated in the same region. Whilst the former were constructed from the products of deflation from saline basin floors, the longitudinal forms involve deflation from salinized swales, removing clays and associated salts into adjacent ridges. Identification of the role of salts in the formation of dunes of different types explains the apparent anomaly in what seemed to have been synchronous age relationships between dunes of different form and composition. There now appears to be a common denominator in the formation of both, namely the presence of salts in adjacent depressions largely controlled by high water-tables.

Later investigations in the Strzelecki and Simpson dunefields (Wasson 1983, and in press) show that pellet-rich longitudinal dunes occur in two geomorphic settings. The first is in an area of mud-rich alluviation by Cooper Creek and its distributaries. Here the dunes are long and have accreted layer upon layer vertically but they have also migrated appreciably downwind. Low angle beds within these dunes are pellet-rich, while avalanche sets contain less pellets. The phase of muddy alluviation between 23 000 and 12 000 BP (see above) was coeval with the construction of these pelletal dunes - with deflation controlled by groundwater.

The second geomorphic setting occurs in the Simpson dunefield, between L. Eyre and the northern boundary of the Kallakoopah pans (Wasson 1983), where groundwater controlled deflation has cut pans into pre-dune alluvium and lacustrine deposits. The longitudinal dunes here accreted vertically and migrated downwind. This setting is similar to the Mallee insofar as the dunes received sediment from pre-existing deposits, rather than from coevally deposited sediments as in the Strzelecki dunefield. The dunes in both the Strzelecki and Simpson are, however, different from the Mallee type. The short Mallee dunes are richer in clay than the others, and so their downwind migration has been severely limited by the hygroscopic nature of the clay pellets. Moreover, during the glacial dune building episode the Strzelecki and Simpson dunefields were probably drier than the Mallee, allowing the dunes to be more mobile.

The red quartzose dunes of the Strzelecki and Simpson dunefields lie on slightly higher surfaces than do the pale pellet-rich dunes. The red dunes show no signs of groundwater controlled deflation, and were nourished by the deflation of old alluvium and lake deposits. Some of the longitudinal dunes of the Malllee are essentially quartzose, but most quartzose dunes in this area are sub-parabolic.

Limited stratigraphic evidence shows that the quartzose and pellet-rich dunes in the Mallee, Strzelecki and Simpson were accumulating at much the same time, although dune construction in the Strzelecki continued later than in the Mallee. The conditions that produced pellets required strong evaporation to concentrate salts, strong winds to erode the salinised sediments, and low precipitation at least in the Strzelecki and Simpson to allow pellet migration. These climatic conditions were also ideal for the mobilisation of quartzose sediments, so that both dune types accumulated at much the same time.

FLINDERS RANGES REGION

In northern South Australia, important evidence is available from the head of Spencer Gulf and from the nearby Flinders Ranges. In this semi-arid region, GE Williams (1973) established a chronologically controlled sequence of depositional events related to past climates. A phase of extensive fan building, the Pooraka Formation, was interpreted as due to cold arid environments with periods of intense rainfall contributing to the stripping of bare slopes. Local dune building argues for aridity.

Radiocarbon dates of this event fall near and, in one case, beyond the limits of radiocarbon detection; probably >40 000 BP. Similarly, carbonate concretions from soils formed within the Pooraka Formation provide apparent ages from 20 300 to 35 000 years. In view of the long period involving soil atmosphere exchange in the formation of carbonates, the soil must be seen as forming before 30 000 BP with subsequent exchange providing age estimates up to 20 000 years. Indeed one date is as young as 5800 BP (GE Williams 1973, Table 2).

Given the environmental reconstruction proposed by GE Williams, the unit may well represent events equivalent to the low sea level cold environments of oxygen isotope stage 4 (approx. 60 000-70 000 BP).

At nearby Dempseys Lagoon at the head of Spencer Gulf, DLG Williams (1982, 1983) has reported a succession of aeolian deposits separated by palaeosols. The youngest of these rests on marine transgressive sediments with the Last Interglacial about 124 000 BP. Radiocarbon dates on emu egg shells indicate a phase of aeolian activity before 36 000 BP, an event that may well correlate with the cold arid fan building described by GE Williams on the Flinders slopes nearby.

NORTHERN AUSTRALIAN EVIDENCE

Evidence presented so far is drawn from central and southern Australia. With the exception of L. Eyre and, to a lesser extent, L. Frome, it is a reflection of Glacial Age environments operating within the zone of the westerlies, the winter rainfall zone. Evidence from the summer monsoonal region remains sparse. To some extent L. Eyre with its catchment area (including Cooper Creek) extending into central and northern Queensland will assist in redressing this imbalance.

In northern Australia, deep vertical moisture penetration and strong oxidation combine to remove most organic carbon from Quaternary sediments, imposing severe constraints on the application of radiocarbon dating. However, within these limitations some observations are pertinent:

1. Many basins preserve evidence of expanded shorelines which, by reason of the weak soils developed on them, are considered to be of Late Quaternary age.

2. This expanded lacustral episode was followed by a contraction of water during which longitudinal dunes often transgressed across drainage lines or onto basin floors.

3. Following dune transgression, the forms became stabilized and water returned to the lakes, in some cases flooding dune corridors.

This succession is represented in the sequence recorded at L. Woods (Fig. 3). Here a major Late Quaternary expanded lake phase is outlined by large strandline deposits. These define a former lake of 5470 km^2, compared to 423 km^2, the area of the present average flood-out. The strandlines, consisting of red quartz sands, rise to 12 m above the lake floor.

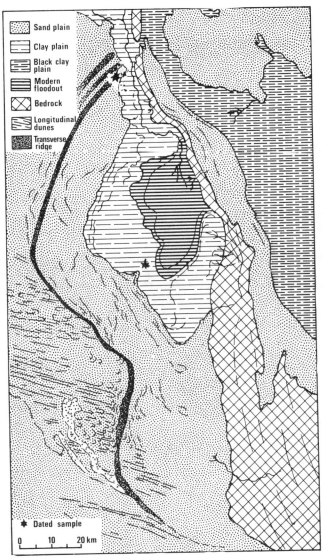

Figure 3. Geomorphic map of Lake Woods, Northern Territory (Lat. 18° S), showing extent of late Pleistocene expanded lake defined by the lakeshore transverse ridge compared to restricted area of present flood-out. Last transverse ridge post-dates longitudinal desert dunes to its west, with limited development of longitudinals in south within confines of ancient lake demonstrating reactivation of dunefields after contraction from maximum lacustral phase, dated as older than 23 000 BP.

In 1967, extensive fieldwork in the area (JME.) failed to reveal any organic matter suitable for dating. However, an ostracod limestone cemented by partly recrystallised calcite, in the high strandline deposit, provided an apparent radiocarbon age of 22 600 BP (Jones and Bowler 1980), a minimum estimate for the high water phase. A second date (Fig. 3) relates to a freshwater mussel (unionid) shell in growth position on the lake floor lying near the limit of flooding in historic times. This date (800 BP) emphasises the difference between the maximum events of the past 1000 years and those of the major lacustral environments of Late Quaternary time.

Thus the actual nature and sequence of events, although as yet undated in northern Australia, parallels the sequence dated from the south. The actual time-relationships, whether in phase or to some degree out of phase, remain to be determined. Glacial age expansion of desert environments on the equatorial side of the high pressure belt may well have been synchronous with that established here for the southern margin. However, the synchrony claimed (MAJ Williams 1975) has not yet been established positively in Australia. In either case, the similarity of magnitude and sequence of changes between the southern and northern portions of the continent is beyond doubt.

GROUNDWATER, SALTS AND INHERITANCE

In the previous discussion, many effects have been attributed to hydrologic phases of a transitional nature, especially those which accompanied the change from humid lacustral conditions with abundant water in the landscape to dry regimes in which aeolian processes became dominant. We recognize here principles of fundamental importance in the evolution of the Australian arid regions, the relevance of which has largely escaped attention. The principles involved may be stated succinctly:

1. The response of the landscape to any particular hydrologic change is conditioned by the rate, direction and magnitude of that change, and,
2. The landscape response is determined as much by the pre-existing conditions as by the nature of the new mix of parameters.

In other words, the landscape has a memory. This notion can be illustrated (Fig. 4) by reference to the Glacial Age landscapes of Australia in which the aeolian/arid period (D in Fig. 4) followed the lacustral/humid phase (A), so that the former has a memory of the latter. Changes between climatic stages follow pathways defined by the directions shown. The phases of this unidirectional change can be summarized as follows:

1. A to B: The onset of surface water deficit conditions will produce an immediate response registered as a fall in lake levels. Continuation of this trend results in an increase in salinity with changes in lake faunas, for the example from fresh to brackish water mollusca. In the L. Eyre catchment area streams became less vigorous.

2. B to C: Further development of the drying trend results in the deposition of early evaporites, especially gypsum producing the laminite clay-gypsum couplets characteristic of this stage in sediments of many Australian salt lakes. With progressive fall in surface water levels, regional and local water-tables begin to readjust to the new regime. Piezometric gradients are partially reversed, permitting discharge into lake basins and producing extensive areas of groundwater outcrop in now saline playa environments.

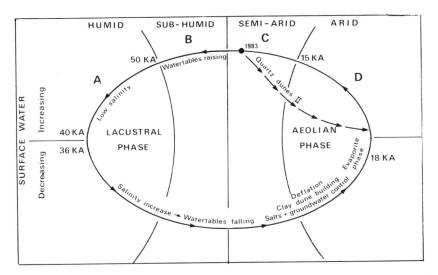

Figure 4. Diagram to illustrate unidirectional nature of changes characteristic of Late Glacial environments documented from the Australian record. The principle of inheritance in landscape evolution demonstrated here provides a key to understanding past changes, present dynamics and future predictions.

3. C to D: Efflorescence of salt breaks surface clay into pelletal aggregates preparing the surface for deflation by late summer winds. Low lying depressions and dune swales, so affected, become major sources of materials suitable for dune building. These processes reach maximum expression with dune building stage D.

4. D to C: With relaxation of hydrologic stress, involving increased humidity, more vegetation can grow on dunes which are then stabilized; pedogenetic effects extend across the landscape and salts, previously concentrated near the surface or redistributed by deflation, are leached from upper soil zones. Modern or Holocene regimes may be depicted as lying near stage C.

5. C to B: A further increase in humidity will result in reactivation of many lakes which remain dry today. This will involve a phase of groundwater recharge with consequent rise in water-tables; it would set the scene for a repetition of the cycle.

This process involves a number of irreversible components and produces an asymmetric pattern of change with a number of important consequences. If in 1984 we were to enter a period of increased aridity such as experienced in a succession of drought years, the responses of much of the landscape would be quite different from those that accompanied the onset of Glacial Age aridity, largely controlled by inherited effects (memory) of high salinity, the legacy of long lacustral environments. A modern increase in aridity would result in remobilization of dunes already in existence, especially by activating the already relatively unstable quartz dunes rather than the production of any new sets of gypsum or clay-rich dunes. This response, illustrated by the pathway (labelled II in Fig. 4), represents a minor oscillation within the larger cycle but the nature of the landscape response remains quite different in both cases. Late Holocene dune reactivation is a good example of this (Thom and Wasson 1982).

Thus to understand any single hydrologic stage within any part of a Glacial-Interglacial cycle, it is essential that we understand and clarify those events which preceded it and which, to a large extent, determined it. This deterministic view of Quaternary environments has important implications both for interpreting the past as, indeed, for predicting the future. It means that in large areas present aridity cannot be used as a model against which to test theories of Glacial Age events. Conversely, prediction of future environmental changes cannot be gauged by identifying the effects of comparable changes during Glacial cycles.

These same principles, developed here for the interaction of lacustrine and aeolian processes, can be extended to the evolution of fluvial systems of comparable age. The transition from large, high discharge bed-load channels of the Riverine Plains and L. Eyre catchment, to small, deep and sinuous suspended load systems, and the inherited influence of glacial systems, finds comparable expression in the preservation of underfit streams.

PALAEOHYDROLOGY AND CLIMATE

One of the most significant features that emerges from the Australian evidence involves the widespread extent of surface waters in the middle of the Last Glacial cycle. A major problem surrounds the question, how much water was involved and what hydrologic and climatic factors controlled it?

Bowler (1981) related climatic variables to the ratio between basin catchment and lake surface area (Ac/Al). From the ratio between areas of expanded lacustral water bodies (mega-lakes), and of the contracted size, the run-off must have been more than double that of the present hydrologic regime. Of the various factors involved - precipitation, evaporation and run-off - the most easily changed is run-off. Increase of rainfall intensity and variations in seasonality, especially if associated with changes in plant cover, may have resulted in large increases in run-off without major changes in mean annual values of precipitation or evaporation. Thus the filling of the lakes in the southeast, the Willandra system, L. George and L. Tyrrell was certainly affected by increased run-off from the highlands, where in some areas periglacial diminution of vegetation would certainly have assisted. However, the filling of the lakes, estimated to date from about 60 000--55 000 BP, seems much earlier than the onset of major cold episodes in the mountains as indicated by the development of blockstreams about 35 000 BP (Caine and Jennings 1968). Moreover, evidence cited earlier from the Darling River and Cooper Creek, whose catchment areas were unaffected by periglacial processes, indicates the persistence there of high water-tables following the lacustral environments, events unconnected with cold climate processes. Similarly, the filling of L. Eyre and L. Frome required large changes other than those associated with periglacial conditions. Therefore it is hard to escape the conclusion that substantial changes in precipitation were involved.

From the widespread distribution of lacustral environments, these seem to have involved:

a) an increase in summer rainfall to have maintained L. Eyre as a deep and relatively freshwater body, and,

b) increased run-off from the southern highlands and the Flinders Ranges to maintain levels in the Willandra, Tyrrell, George and Frome systems.

The actual climatic system that produced such excess surface water, apparently in both summer and winter rainfall regions, remains to be explored further as does evidence for synchrony of responses in different climatic zones.

Perhaps the most important stage in the Quaternary evolution of the Australian landscape, in terms of regional extent and the degree of modification involved, was that which accompanied peak aridity. Measured in terms of widespread simultaneous drying of lakes and the efficacy of aeolian processes, this is dated to between 25 000-16 000 BP with dune building continuing until 12 000 BP in more arid regions.

Throughout central Australia, the vast majority of desert dunes is now fixed. The conditions that prevailed during the last phase of mobilization require explanation. In modern southern Australia, the high velocity winds responsible for both sand movement and dust transport are associated with the easterly travelling frontal systems that sweep across the southern portion of the continent. The steep pressure gradients that precede the fronts generate strong northerly winds. These advectively transport large quantities of heat from the continental interior often bringing summer heat-wave conditions to the towns and cities of the southeast. A typical example of a summer frontal system that transported large quantities of dust to southeastern Australia occurred in January 1983. One most significant effect of such winds produced in this way is a great increase in evaporation and transpiration rates in their paths. These effects are evident in a comparison of evaporation rates for 1982 with those of previous years during which the frontal winds were weaker (Fig. 5).

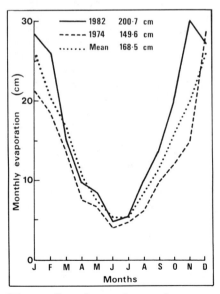

Figure 5. Comparison of evaporation variations for a wet year (1974) compared to a drought year (1982), and annual mean values at Canberra airport.

In Glacial times, the effects of increased pressure gradients would certainly have accentuated velocities of northerly winds. Combined with reduced cloud cover and high summer insolation, such conditions may actually have resulted in an increase in summer evaporation rates above today's values, rather than glacially reduced evaporation as many previous workers have assumed. Moreover, as stated elsewhere (Bowler 1975), the trajectory of the Glacial Age hot summer winds with advective heat transfer amplified by increased velocity, increased turbulence and a vertical heat flux greater than today, would have limited the snow and ice fields of the southeastern highlands which would have lain in their direct path. Like the disappearance of permanent ice, the reduction in lake levels probably involved two factors which reinforced each other:

a) reduced run-off consequent on a reduction in actual precipitation below that of the preceding lacustral phase, and,

b) increased summer evaporation rates resulting from combined effects of higher wind speeds and increased advective heat flow into and out of the regions.

GLACIAL AGE WIND REGIMES

The desert dunes of Australia are arranged in a continent-wide anti-clockwise whorl (Fig. 6), their directions of elongation judged from the orientation of y-junctions between longitudinal dunes (Jennings 1968). The dunes have moved towards the east in the southern part of the continent, essentially towards the north in the Simpson and Strzelecki dunefields, and towards the west in the northern part of the whorl. In general, these directions agree with the present prevailing winds: westerly in the zone of travelling anticyclones, easterly in the 'trades' and southern in the zone between.

Combining the calculations of the resultant direction of winds with velocities greater than the impact threshold of sand movement (Brookfield 1970 a,b; Fryberger 1979; Sprigg 1982) with data recently analysed for Moomba in the Strzelecki dunefield, we can gain some idea of the relationship between modern potential sand drift and dune orientation. It is now clear (Wasson and Hyde 1983) that longitudinal dunes form where there is only a moderate amount of sand under a wind regime that has a distribution with two modes which form either an acute or obtuse angle (Fryberger 1979). The resultant lies between the modes and is approximately parallel to the dunes. The degree of parallelism is a statistical function of the standard deviation of dune orientation and the seasonal variability of the resultant.

From the resultant drift potentials calculated by the authors cited above, there is little obvious divergence between longitudinal dune orientation and sand moving winds in the Great Sandy, Tanami, Simpson and northern Strzelecki (Fig. 6). In the southern Strzelecki, east of L. Frome, Sprigg (1982) demonstrated a southwesterly resultant which lies at about 45° to the trend of the longitudinal dunes but is normal to small transverse dunes being formed on the edge of the lake. The resultant is also at 45° to late Holocene longitudinal dunes on the western side of the lake. Further south, in the Murray Mallee at Wanbi, Sprigg (1982) has recorded a similar divergence. The modern resultant is south-westerly while the longitudinal dunes are east-west. At Coobowie, on Eyre Peninsula, the modern resultant is south-southwesterly while a little north on the Peninsula and on adjacent Kangaroo Island, longitudinal dunes are oriented towards east-south-

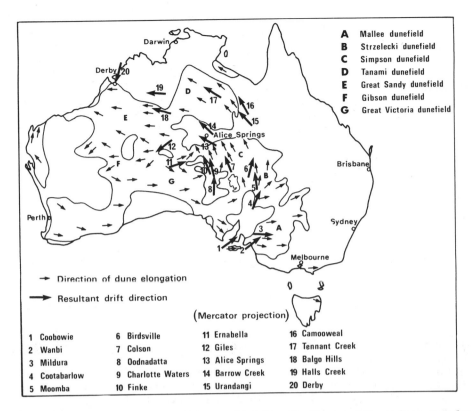

Figure 6. Mean trends of Australia desert dunes with modern resultant drift directions.

east. Sea breezes obviously play some role in this area, but are insufficient to explain such a marked divergence. Further east at Mildura the resultant (using the technique of Fryberger 1979) is within $10°$ of both the orientation of modern longitudinal dunes at L. Mungo and the fossil Mallee-type longitudinal dunes between Mildura and the Willandra Lakes.

The majority of these analyses demonstrate that Glacial Age dune-building resultants and modern resultants are remarkably similar. Divergence occurs only at Coobowie, Wanbi and L. Frome. The greater role of westerly and north-westerly winds of Glacial Age summers (postulated by Bowler 1975) can be explained by two hypotheses. Firstly, the altered long-wave patterns suggested by Webster and Streten (1978) would have had most influence in the present southern coastal region east of the Australian Bight, where the divergence noted above occurs. The long-wave patterns postulated by Webster and Streten is not consistent with the palaeoclimatic history of New Zealand (J Salinger, pers. comm.) so a second hypothesis is needed to explain the pattern of modern and Glacial Age dune building winds. The enhanced circulation of the Glacial resulted from a steeper meridional atmospheric temperature gradient. The steepening of the pressure gradients in the anticyclone belt may have increased the role of pre-frontal north-westerlies. On the available evidence we cannot chose between these hypotheses. Furthermore, if a shift in the anticyclone belt did occur, the statistical relationship between average dune trend and modern drift resultants does not allow close resolution of the change. Ash and Wasson (1983) showed that over much of the Australian desert dunefield the vegetation cover is insufficient to stop major sand movement. The limiting factor appears to be wind; that is, the modern environment is far less windy than the glacial environment. A minimum increase in windiness (expressed as percentage of days experiencing winds above the sand moving threshold) by 20-30 % would mobilize most of the dunes in Australia. Such an increase probably would have overcome the resistance to movement offered by clay pellets in dune swales and on lake surfaces.

Evidence from beyond the Australian continent supports the proposed increase in windiness. The increased meridional temperature gradient of the Last Glacial Maximum is thought to have increased the energy cycle of the atmosphere, thus increasing wind speeds by 17 % in the southern hemisphere (Newell et al. 1981) irrespective of season. Streten (1973) has correlated the present equatorward location of the ice margin with the strength of the present Southern Ocean surface westerlies, thereby showing that the equatorward position of the Antarctic sea ice at the Glacial Maximum (CLIMAP 1976) must have resulted in stronger westerlies. As pointed out by Webster and Streten (1978) from conservation of momentum arguments, strong westerlies are associated with strong tropical easterlies. Dust in Antarctic ice has been used by Petit et al. (1981) to indicate a 50-80 % increase in wind speed during the Glacial Maximum. Similar results come from the northern hemisphere (Parkin and Shackleton 1973; Sarnthein et al. 1981).

The Glacial Age westerly wind climate of the Australian interior was more rigorous than that of the Post-glacial and therefore a more important component in evaporation during both winter and summer. In the north the tropical easterlies were probably stronger and so we suggest that dune building was roughly synchronous across the continent. This hypothesis is supported by the likely reduction of vegetation cover caused by reduced monsoon vigour (Manabe and Hahn 1977). But vegetation clearly played a part in the dunefields, for dunes in the wettest parts of the quartzose dunefields rarely contain high-angle beds. This indicates that these low rounded dunes were never highly mobile, in turn showing that vegetation was a significant modulator of sand transport. Without investigation of pollen from lakes in interior Australia, we cannot describe the vegetation of this dry, windy time.

DISCUSSION AND CORRELATION

The sequence described here from glacial, fluvial, lacustrine and aeolian environments may be summarised diagrammatically in a chronologically controlled context as in Fig. 2. The latter part of the Last Glacial cycle exhibits a number of substantial environmental changes, with processes operating across the landscape at levels of intensity rarely experienced today. Thus while glacial and periglacial conditions were active in the highlands from at least about 30 000 BP, there was an associated increase in run-off in the rivers draining towards the now arid interior.

The early Tallygaroopna phase of enlarged channel dimensions may correlate with the early part of the Mungo lacustral. At this time numerous small basins across the interior of the continent experienced a major phase of greatly expanded surface waters. This prolonged condition would certainly have resulted in extensive groundwater recharge producing a hydrologic status that, in turn, affected the channels supplying surface waters to the system. During seasonal low stage flows channel banks failed at rates much in excess of those in the present regime in which rivers generally lie above the levels of surrounding water-tables. As described earlier, this condition is seen as the main factor in controlling the behaviour of channels from the Cooper, Diamantina and Darling in the north to the Murrumbidgee, Murray and Goulburn in the south.

About 36 000 to 25 000 BP the whole system underwent drastic change. Although water-tables remained high enough to perpetuate the large meander geometry, the lakes began to experience surface water deficit. As the landscape changed from one of abundant surface waters to a drier regime, salinity increased, enabling salts to concentrate in soils and sediments of waterlogged depressions, thus preparing the ground for deflation by winds already strengthened by pressure gradients steepened by changes in the thermal regime of the Southern Ocean and increased atmospheric meridional temperature gradients.

The period from about 20 000 to 16 000 BP is seen as that in which both transverse and longitudinal dunes were most active. New dunes were created, old systems were reactivated. As previously emphasised (Bowler 1978) there are two stages of evidence from southeastern Australia, spanning the middle of the Last Glacial cycle from about 50 000 to 15 000 BP; a major humid interval with water bodies much expanded beyond those of today and an arid phase in which dune building and therefore desert conditions expanded on the southern side of the sub-tropical high pressure belt with longitudinal dunes now identified even into Tasmania (Bowden 1983). A comparable expansion on the tropical side of the desert, although not yet securely dated, almost certainly relates to this period.

Thus the landscape was subjected to successive wet and dry episodes. Indeed it is this succession which we believe is of major importance in determining the process response to conditions that prevailed during the Last Glacial Maximum. We shall return to this point below.

Glacial-Age Hydrology

One of the major anomalies of mid-latitude deserts concerns the drying of lakes at the time of Glacial Maximum, a period of assumed temperature depression. In approaching this problem many workers have taken the view that the period of glacial temperature lowering followed the modern pattern of seasonal temperature change but was shifted to a lower mean temperature. It is often assumed that an evaporation-temperature correlation can be used to reconstruct evaporative loss under past climatic conditions. This assumption is demonstrably false.

Evaporation (E) is particularly susceptible to change in radiant energy received at the earth's surface, relative humidity and wind speed. But temperature (T) also is largely a function of radiant energy. Therefore, whilst both T and E are dependent on radiation, E is controlled by additional variables, namely wind speed and humidity.

The complexity of the relationship may be gauged by examining evaporation-temperature relationships in different parts of the present climatic system. In parts of central Asia such as in the Qaidam Basin, Qinghai Province of China, where mean annual temperatures are only 3-5° C, evaporation may exceed 3 metres, a value much higher than that experienced in many regions with higher mean annual temperatures. In these circumstances, conditions of very low humidity and high wind speeds combine with high summer radiant energy to produce high evaporation rates even while mean annual temperature may remain very low. The correlation of evaporation with temperature is therefore an invalid one.

Glacial Age environments in central Australia were characterised by long dry periods, high wind speeds and high summer radiant energy, conditions which find their analogs in the drought periods we still experience in such regions today. At 20 000 BP the amount of radiant energy at latitudes 30-40° S was essentially the same as today. Under conditions of reduced cloud cover associated with long dry periods, surface energy may even have been enhanced.

In considering the magnitude and direction of Glacial Age changes, they may be expected to have followed the same trends as those observed during wet and drought periods today. The pattern reconstructed for Glacial Age conditions in central Australia in many ways resembles the droughts experienced in recent times (Fig. 5). During Glacial summers, the deserts, far removed from the modifying influence of nearby glacial or periglacial environments, responded directly to incoming solar energy, wind regimes and local humidity levels. These conditions postulated to have existed during Glacial times are exactly those that favour increased rather than decreased summer evaporation rates.

In Fig. 5 annual pan evaporation measurements are presented from Canberra for a wet year, 1974 and a year of drought, 1982. The 1982 evaporation loss was 25 % above that of the humid year. Of the additional 500 mm recorded in the drought year, 70 % of the increase occurred during the 4 months November to February.

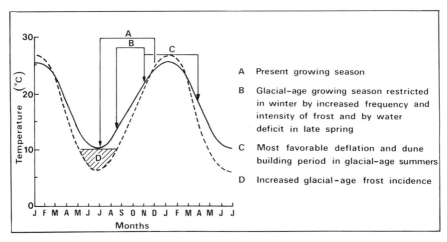

Figure 7. Modern mean monthly temperatures for Mildura, southeastern Australia, compared to postulated full-glacial temperature curve (dashed line). A winter depression of up to 6° C compares to a summer increase in temperatures due to increased advective heat transfer from the continental interior.

During Glacial Age regimes, we therefore postulate that the combination of low humidity, high wind speeds, increased pressure gradients and high radiant summer energy would have produced evaporative losses higher by a factor of up to 20 % above conditions of today even if precipitation remained near present levels. Any reduction of precipitation would serve to accentuate conditions of aridity without necessarily being everywhere the basic cause.

Whilst the summers were hot, dry and windy, the northerly migration of the high pressure belts in winter would have permitted cold Antarctic and Southern Ocean air masses to penetrate deeply into the continental interior, increasing the incidence of frosts and solid precipitation on uplands and generally producing conditions inimical to plant regeneration and growth. These conditions are summarised in Fig. 7. Under these conditions of increased continentality or amplified seasonal contrasts the growing season would have been considerably shortened by increased frost incidence in spring and by the onset of conditions of severe water stress in summer. Combined with strong winds, such circumstances would have imposed severe stress on plants, especially on eucalypt woodlands sensitive as they are to frost incidence. It is conditions such as these, combined with strong winds, which helped reduce vegetation cover over large parts of the landscape and which assisted the reactivation of ancient dunefields and the generation of new ones.

REFERENCES

Ash JE and Wasson RJ 1983. Vegetation and sand mobility in the Australian desert dunefield. Z. Geomorph. N.F., Suppl. Bd. 45:7-25.

Berger AL 1978. Long-term variations of caloric insolation resulting from the Earth's orbital elements. Quat. Res. 9:139-167.

Bowden AR 1983. Relict terrestrial dunes: legacies of a former climate in coastal northeastern Tasmania. Z. Geomorph. N.F., Suppl. Bd. 45: 153-174.

Bowler JM 1970. Alluvial terraces in the Maribyrnong Valley near Keilor, Victoria. Mem. natn. Mus. Vict. 30:15-58.

Bowler JM 1971. Pleistocene salinities and climatic change: Evidence from lakes and lunettes in southeastern Australia. In: DJ Mulvaney and J Golson (eds.), Aboriginal Man and Environment in Australia. ANU Press, Canberra. p.47-65.

Bowler JM 1973. Clay dunes: their occurrence, formation and environmental significance. Earth-sc. Rev. 9:315-338.

Bowler JM 1975. Deglacial events in southern Australia: their age, nature, and palaeoclimatic significance. Bull. R. Soc. N.Z. 13:75-82.

Bowler JM 1976. Aridity in Australia: age, origins and expression in aeolian landforms and sediments. Earth-sc. Rev. 12:279-310.

Bowler JM 1978a. Quaternary climates and tectonics in the evolution of the Riverine Plain, southeastern Australia. In: JL Davies and MAJ Williams (eds.), Landform evolution in Australasia. ANU Press, Canberra. p. 70-112.

Bowler JM 1978b. Glacial age aeolian events at high and low latitudes: a Southern Hemisphere perspective. In: EM van Zinderen Bakker (ed.), Antarctic Glacial History and World Palaeoenvironments. Balkema, Rotterdam. p. 149-172.

Bowler JM and Magee JW 1978. Geomorphology of the Mallee region in semi-arid northern Victoria and western New South Wales. Proc. R. Soc. Vict. 90:5-26.

Bowler JM and Polach HA 1971. Radiocarbon analyses of soil carbonates: an evaluation from palaeosols in southeastern Australia. In: D Yaalon (ed.), Palaeopedology - Origin, Nature and Dating of Palaeosols. Int. Soc. Soil Sci. and Israel Univ. Press, Jerusalem. p. 97-108.

Bowler JM 1981. Australian salt lakes: a palaeohydrological approach. Hydrobiologia 82: 431-444.

Bowler JM, Stockton E and Walker M 1978. Quaternary stratigraphy of the Darling River near tilpa, New South Wales. Proc. R. Soc. Vict. 90:78-88.

Bowler JM, Hope GS, Jennings JN, Singh G and Walker D 1976. Late Quaternary climates of Australia and New Guinea. Quat. Res. 6: 359-394.

Bowler JM, McBryde I and Head J., in prep. Radiocarbon dates from shell and organic carbon, Willandra Lakes.

Brookfield M 1970. Dune trends and wind regime in Central Australia. Z. Geomorph., Suppl. Bd. 10:121-153.

Brookfield M 1970b. Winds of Arid Australia. CSIRO Div. Land Res., Tech. Pap. 30. 59 pp.

Butler BE 1950. A theory of prior streams as a causal factor of soil occurrence in the Riverine plain of south-eastern Australia. Aust. J. Agric. Res. 1:231-52.

Butler BE 1958. Depositional systems of the Riverine Plain of south-eastern Australia in relation to soils. Soil Publ. CSIRO Aust. No. 10.

Caine N and Jennings JN 1968. Some blockstreams of the Toolong Range Kosciusko State Park, New South Wales. J. and Proc. R. Soc. NSW 101: 93-103.

Callen RA, Wasson RJ and Gillespie R 1983. Reliability of radiocarbon dating of pedogenic carbonate in the Australian arid zone. Sedim. Geol. 35:1-14.

Chappell JMA and A Grindrod (eds.), 1983. CLIMANZ: a symposium of results and discussions concerned with Late Quaternary climatic history of Australia, New Zealand and surrounding seas. Dept. Biogeog. and Geomorph. ANU, Canberra. 2 vols.

Churchward HM 1961. Soil studies at Swan Hill, Victoria, Australia. 1. Soil layering. J. Soil Sci. 12:73-86.

CLIMAP project members 1976. The surface of the Ice-Age. Earth Science 191:1131-1137.

Colhoun EA 1983. The Climate of Tasmania 18 $\pm$ 2 ka. In: JMA Chappell and A Grindrod (eds.), CLIMANZ. Proc. 1st Climanz Conf., Feb. 1981. ANU, Canberra.

Conacher AJ 1971. The significance of vegetation, fire and man in the stabilization of sand dunes near the Warburton Ranges, central Australia. Earth-sc. J. 5:92-94.

Costin AB 1972. Carbon-14 dates from the Snowy Mountains area, southeastern Australia and their interpretation. Quat. Res. 2:579-590.

Fryberger S 1979. Dune forms and wind regime. In: ED McKee (ed.), A Study of Global Sand Seas.U.S. Geol. Surv. Prof. Pap. 1052. p. 137-170.

Galloway RW 1963. Glaciation in the Snowy Mountains: A reappraisal. Proc. Linnean Soc. N.S.W. 88:180-198.

Hills ES 1939. The Physiography of north-western Victoria. Proc. R. Soc. Vict. 51:293-320.

Hope GS 1983. Australian Environmental Changes: timing, directions, magnitudes and rates. In: PS Martin and RG Klein (eds.), Pleistocene extinctions: the search for causes. Arizona Press, Tucson Univ.

Hyde R and Wasson RJ, in press. Radiative and meteorological control on the movement of sand at Lake Mungo, N.S.W. Australia. In: ME Bookfield and TS Ahlbrandt (eds.), Aeolian Sediment and Processes.

Jennings JN 1968. A revised map of the desert dunes of Australia. Aust. Geographer 10:408-409.

Jennings JN 1975. Desert dunes and estuarine fill in the Fitzroy estuary (north-western Australia). Catena 2:215-262.

Jones R and Bowler JM 1980. Struggle for the savanna: Northern Australia in ecological and prehistoric perspective. Research School of Pacific Studies, ANU, Canberra. p. 3-31.

Langford-Smith T 1960. The dead river systems of the Murrumbidgee. Geog. Rev. 50:368-389.

Macumber PG 1970. Lunette initiation in the Kerang district. Mining & Geol. J. Vic. 6:16-18.

Manabe S and Hahn DG 1977. Simulation of the tropical climate of an ice age. J. Geophys. Res. 82 (27):3889-3911.

Newell RE, Gould-Stewarts S and Chung JC 1981. A possible interpretation of paleoclimate reconstructions for 18 000 BP for the region 60°N to 60°S, 60°W to 100°E. In: JA Coetzee and EM van Zinderen Bakker (eds.), Palaeoecology of Africa. Balkema, Rotterdam. p. 1-20.

Parkin DW and Shackleton NJ 1973. Trade wind and temperature correlation down a deep sea core of the Saharan Coast. Nature 245:445-457.

Pels S 1964. The present and ancestral Murray River system. Aust. geograph. Stud. 2:111-119.

Penman HL 1963. Vegetation and Hydrology. Techn. Commun. No. 53 Commonwealth Bureau of Soils (Farnham Royal). 124pp.

Petit J-R A, Briat M and Royer A 1981. Ice age aerosol content from East Antarctic ice core samples and past wind strength. Nature 293: 391-394.

Sarnthein M 1978. Sand deserts during glacial maximum and climatic optimum. Nature 272:43-46.

Sarnthein M, Tetzlaff G, Koopmann B, Walter K and Pflaumann U 1981. Glacial and interglacial wind regimes over the eastern subtropical Atlantic and north-west Africa. Nature 293:193-196.

Scheidegger AE 1983. Instability principle in geomorphic equilibrium. Z. Geomorph. N.F. 27:1-19.

Schumm SA 1968. River adjustment to altered hydrologic regimen - Murrumbidgee River and palaeochannels, Australia. U.S. geol. Surv. Prof. Pap. 598.

Sprigg RC 1982. Alternating wind cycles of the Quaternary era and their influences on aeolian sedimentation in and around the dune deserts of southeastern Australia. In: RJ Wasson (ed.), Quaternary Dust Mantles of China, New Zealand and Australia. Dept. of Biogeog. and Geomorph., ANU, Canberra. p. 211-240.

Streten NA 1973. Satellite observations of the summer decay of the Antarctic sea ice. Arch. Meteorologie, Geophysik & Bioklimatologie, Serie A; Meteorologie und Geophysik 22:119-134.

Thom BG and Wasson RJ (eds.), 1982. Holocene Research in Australia. Dept. Geogr., Univ. of New South Wales, Duntroon.

Twidale CR 1972. Evolution of sand dunes in the Simpson Desert, central Australia. Trans. Inst. Br. Geogr. 56:77-109.

Walker PH 1962. Terrace chronology and soil formation on the south coast of New South Wales. J. Soil Sci. 13:178-186.

Warner RF 1972. River terrace types in the coastal valleys of New South Wales. Aust. Geographer 12:1-22.

Wasson RJ 1976. Holocene aeolian landforms in the Belarabon area, S.W. of Cobar, N.S.W. J. and Proc. R. Soc. N.S.W. 109:91-101.

Wasson RJ 1983. The Cainozoic history of the Strzelecki and Simpson dunefields (Australia), and the origin of the desert dunes. Z. Geomorph. N.F., Suppl. Bd. 45:85-115.

Wasson RJ and Hyde R 1983. Factors determining desert dune type. Nature 304:337-339.

Wasson RJ, in press. Dune sediment types, sand colour, sediment provenance, and hydrology in the Strzelecki-Simpson dunefield, Australia. In: ME Brookfield and TS Ahlbrandt (eds.), Eolian Sediments and Processes.

Wasson RJ, Rajaguru SN, Misra VN, Agrawal DP, Dhir RP, Singhvi AK and Kameswara Rao K 1983. Geomorphology, late Quaternary stratigraphy and palaeoclimatology of the Thar dunefield. Z. Geomorph. N.F., Suppl. Bd. 45:117-152.

Webster PJ and Streten NA 1978. Quaternary Ice Age Climates of Tropical Australasia: Interpretations and reconstructions. Quat. Res. 3:279-399.

Williams DLG 1982. Multiple episodes of Pleistocene dune-building at the head of Spencer Gulf, South Australia. Search 13:88-90.

Williams DLG 1983. Depositional environments in the Flinders and Mt. Lofty Ranges of South Australia 32 $\pm$ 5 ka. In: JMA Chappell and A Grindrod (eds.), CLIMANZ etc.

Williams GE 1973. Late Quaternary piedmont sedimentation, soil formation and paleoclimate in arid South Australia. Z. Geomorph. N.F. 17:102-125.

Williams MAJ 1975. Late Pleistocene tropical aridity: synchronous in both hemispheres. Nature 253:617-618.

Evolution of Australian landscapes and the physical environment of aboriginal man

J.B.FIRMAN
Geological Survey, Parkside, Australia

ABSTRACT. The record of landscape evolution in Australia provides a useful framework for archaeological and anthropolgial studies. Episodic changes in climate, materials, landform, flora and fauna alternated with the development of palaeosols during times of relative stability. Aboriginal man inherited abundant relict features and materials from past landscapes. He also witnessed profound changes in climate and sea level. Differences in original materials and landform and the rate at which processes operated produced considerable diversity in the modern landscape. Aboriginal tribes came to occupy separate regions. Differences in settlement, economy, sociology and culture may all be traced in part to environmental constraints.

INTRODUCTION

In this text the evolution of the South Australian landscape is set down in relation to the evolution of man. Information is presented that derives from regional stratigraphic and other studies of climate, earth materials, topography, and flora and fauna in the past. That is, the study is palaeogeographic in so far as it deals with the description and distribution of environmental elements or factors, and evolutionary in that it deals with changes from one landscape to another through time. Because of the importance of relict features and materials of value to aboriginal man, emphasis is placed upon their time of origin. The aim of the study is to arrive at an explanation of the role of aboriginal man in terms of the ecological systems in which he lived. For older systems this can be only a partial explanation of the climatic and physical constraints on aboriginal man. It is only in the modern landscape that matters of structure, function and the interrelationships of man with other living organisms can be fully studied in ecological terms.
 The evolutionary approach facilitates the study of environmental changes through time. For example, climates have undergone dramatic changes; materials of value to man are formed at a particular time in a particular place, and they may not be accessible for use until revealed by later erosion; certain terrains such as the great dune fields have not always been a feature of the landscape - contrary to popular notions of the great antiquity of all Australian landforms - and soils and their associated flora and fauna have undergone profound changes.

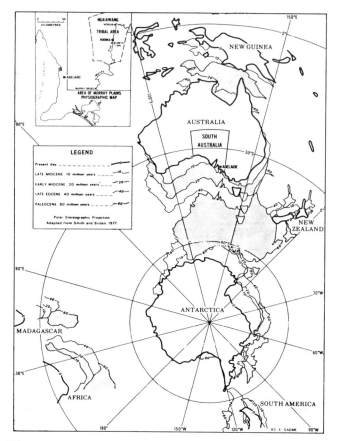

Figure 1. Movement of landmasses during the Cainozoic. Dates in m.y. BP.

Examples are given from the Adelaide region and from the western margin of the Murray Basin including the gorge tract of the Murray River with its important archaeological sites, which was occupied until the coming of Europeans by the Ngaiwang people and their Pleistocene predecessors.

LANDSCAPES OF THE PAST

Interaction of the physiographic factors, rock type and structure, climate, landform, soil and organisms (including man and vegetation) has produced characteristic landscapes at different times in the past (Firman 1981). Many of these landscapes were formed following continental break-up about 55 m.y. ago (see Fig. 1 adapted from Smith and Briden 1977). Because palaeosols are excellent indicators of past events, landscapes have been described according to the palaeosols developed on them.

In Australia, the interplay between tectonics, erosion and deposition has produced toposequences with older sediments and associated palaeosols high in the landscape on the flanks of the ranges and younger sediments and palaeosols low in the landscape within the younger basins. Toposequences in the present landscape record the relationship between soil and geological history in the past.

The record of sedimentation, palaeosol development, and floral evolution (Fig. 2) is taken from Firman (1983). The climatic curve is interpreted from this information and from the floral and faunal record in Brown et al. (1968). It is assumed that there was a strong continental control of climatic factors for much of the time recorded since separation of Australia and Antarctica, and that there would have been a relative sequence of wet and dry regions from north to south across the continent much as there is today.

Although younger clay palaeosols and carbonate horizons are more common in late Pleistocene landscapes, all the palaeosols appear at ground level as horizons or substrates in great soil group profiles. Many of the "older" or "better differentiated" soil profiles are distinctive assemblages of palaeosols formed at different times in the past and now important features in the modern landscape. Most of the modern soils are characterized by little or no horizon development and by sandy overlays resting upon or developed from older layers, and many of the features in the modern landscape are relicts, that is, they are inherited from older landscapes. The development of most of these features pre-dates the arrival of early man in Australia, but some erosional and depositional features in the present landscape are of more recent origin so that man witnessed their development. The environment in which these younger features were formed relates closely to present landform.

The older weathered and bleached zones

The Playfair weathering zone and the Arckaringa bleached zone are prominent features in Australia (Column 1 on Fig. 2). The Playfair weathering zone is characterized by large irregular, yellow and white patches up to 10 m across. The unit is a discontinuous horizontally-disposed feature which post-dates rocks folded in the Cambrian-Ordovician Delamerian orogeny, and appears to pre-date Carboniferous-Permian sediments.

It is probable that the Playfair weathering zone was developed under warm to tropical conditions of the Permian. Much later, during the Cainozoic, uplift and erosion elevated some of the older basement rocks. Subsequent erosion revealed outcrops of metasedimentary rocks and granites useful as a base for petroglyphs and as a source of material for large stone implements.

The Arckaringa "Palaeosol" is a white bleached zone developed in rocks of late Cretaceous age. It is overlain in many places by sediments of Paleocene-Eocene age. The "palaeosol" was developed at the end of the Cretaceous when a cool wet climate prevailed. Its development occurred at about the same time as the sudden and complete demise of the dinosaurs and many other animals.

Uplift of basin margins and complementary subsidence led to marine transgressions in southern basins during the Oligocene and Miocene. Continental sediments contain fossils of the first known marsupials on the Australian mainland.

Landscapes with ferruginous and siliceous accumulations

Palaeosol-landscapes of the Tertiary are characterized by ferricrete and silcrete pans. There are older pre-Miocene and younger post-Miocene landscapes with similar palaeosols (Column 2 on Fig. 2). The older ferricretes are widespread. There are two kinds of older ferricrete: mottling and iron cement. The mottling is referred to the San Marino Palaeosol and the iron cement to the Yallunda Ferricrete.

Many of the older ferricretes and silcretes, which developed in Paleocene-Eocene sediments, were probably formed during the Oligocene-Miocene. At the end of this time, silcretes were formed in channel gravels shortly after their deposition, possibly in the Miocene. Both ferricretes and silcretes were formed under a subtropical seasonally wet climate. Both the older and younger red iron oxide mottles in old white clays provided a source of ochre for use by modern aboriginals. The marine record suggests the development of a major Antarctic ice cap in the Miocene. Atmospheric circulation continued to intensify and there was a marked decrease in precipitation related to the intense and sudden chilling in Southern Ocean sediments during the latest Miocene.

Vegetation data elsewhere in Australia suggest rain forest cover in the southeast and on the east coast and, in the middle Miocene, the development of grasslands on interfluvial areas in the central part of the continent. Although the continent was far to the north of the Antarctic ice, the disappearance of pollen of the Nothofagus brassi type from regions west of the Great Dividing Range may have coincided with the cooling event in the latest Miocene. Calcicolous vegetation evolved or expanded from formerly restricted habitats to occupy the area of calcareous soils left behind after the retreat of the mid-Tertiary seas from parts of the southern basins.

Younger ferruginous and siliceous materials developed in clastic mantles and sediments following uplift during the Miocene-Pliocene. Many of these palaeosols occur in the Adelaide region. Palaeosols of this kind are thin and discontinuous compared to older ferricretes and silcretes. They include the mottled zone called Ardrossan Palaeosol and the silicified ferruginous detritus called the Karoonda Palaeosol (Fig. 2) which is the youngest palaeosol of this kind. At the time of its development, the elevation of the Mt Lofty Ranges was much lower than today. The ancient surface was ferruginized and later silicified by silica-bearing groundwater which was present throughout the landscape.

Silcretes were much favoured by aboriginal man for the production of flaked tools. Although the older silcretes were removed by later erosion around the margins of the southern Australian Basins, the younger silcrete of the Karoonda Palaeosol is now exposed in the river cliffs, notably at Spring Cart Gully near Berri (east of the map area) where sharp fragments of silica scattered on the ground surface mark a factory site above the Murray River. The Karoonda Palaeosol has been re-worked by ancient rivers in some places to form the Chowilla Sand which can be seen to overlie the eroded surface and underlie the Blanchetown Clay in Murray River cliffs.

In the valley of the Maribyrnong River near Keilor in Victoria there are cemented gravels at the base of the riverine sequence. Pebbles resembling early stone tools occur in the gravels according to Gallus (1976). He places these "artefacts" at about 70 000 BP, but on purely lithostratigraphic grounds the enclosing strata appear to be much older. In South Australia materials of this kind are close to the Pliocene-Pleistocene boundary. If the cemented gravels are artefacts, then man's first entry in Australia could have occurred in the Pliocene-Pleistocene at a time of low sea level.

Landscapes on older structured clays with gypcrete

After the development of the Karoonda Palaeosol, clastic deposition continued in stream channels, lowlands and grabens. These deposits extend seaward of the modern coast and mark a retreat of the sea of unknown extent. The most extensive stratigraphic units are the upper beds of Hindmarsh Clay in St Vincent Basin and Blanchetown Clay in the Murray Basin. An equivalent local unit is the Ngaltinga Clay between Hallett Cove and Sellicks Hill south of Adelaide.

Soil structures, coatings on peds, reddening and other features have been observed in sedimentary clays of the Pleistocene lower sequence in the Adelaide region and in stratigraphically equivalent saprolites. These features mark clay palaeosols, many of which occur as soil horizons or substrates in a variety of soil profiles described in the literature as podsolic soil, solodic soil, black earth and red-brown earth (Column 3 on Fig. 2).

Sedimentation ceased with the development of gypcrete which formed as a groundwater feature on the interfluves and as bedded gypsum in the lowlands. The cessation of sedimentation and development of gypsum deposits and impregnations throughout the province marks the Denison Palaeosurface (Firman 1980) and indicates widespread aridity.

In South Australia much of the early Tertiary flora was destroyed: Nothofagus and Malaysian closed-forest communities no longer exist. The disappearance of pollen of the Nothofagus brassi type west of the Great Dividing Range may have coincided with the cooling event of the latest Miocene. The more humic elements of the Australian flora within the 'sclerophyll' land systems may have been depleted during Pleistocene epochs of aridity. Many forests species of Eucalyptus had apparently had a continuous distribution from south eastern Australian into south Australia; now only distinct remnants of their original distribution remain (Crocker and Wood 1947).

Landscape dominated by calcareous sediments and calcrete

Dissected remnants of the San Marino Palaeosol, the Yallunda Ferricrete, the Clayton River Silcrete, and the Karoonda Palaeosol are found on the flanks of the ranges and capping higher ground on the plains. Sediments younger than the Karoonda Palaeosol are lower in the landscape. This distribution suggests faulting prior to sedimentation in Early Pleistocene.

In the Early Pleistocene Australia was close to the island chain to the north. On the southern coastal margin the ingression of the Pleistocene sea produced shallow marine sequences, including Coomandook Formation (Fig. 2). Aeolian Lower Member Bridgewater Formation of the coastal margin and Telford Gravel on the flanks of the ranges were laid down next. Materials formed about this time in the Ngaiwang tribal area, which contains many important sites of early man, are shown on Fig. 3 which shows assemblages of weathered rocks, sediments and palaeosols characteristic of separate soil-landscapes.

The change in lithology and palaeosols from iron-mottled and silica-cemented sands and gravels of the Pliocene and younger structured clays and sands of the Pliocene-Pleistocene to carbonate-cemented gravels and carbonate and clay sequences of the Early Pleistocene suggests a profound change from seasonally dry to wetter and cooler conditions.

The Bungunnia Limestone, a thin dolomitic micrite containing ostra-

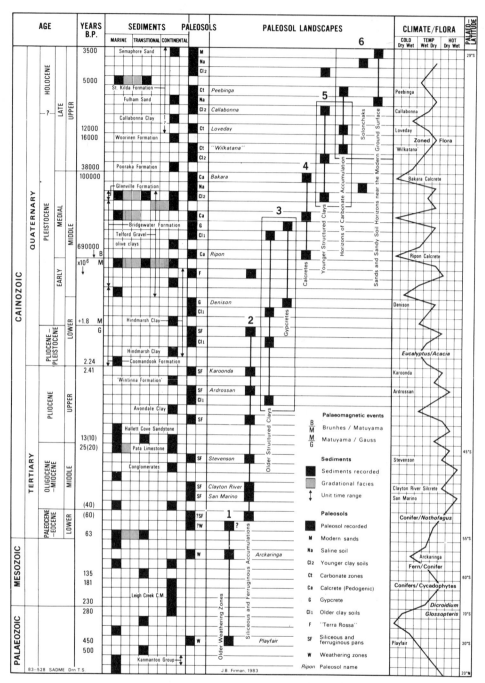

Figure 2. Landscape evolution in the Adelaide region, southern Australia after Firman (1983). Ages on left are based on Kulp in Snelling (1964:30) and (in brackets) Smith and Briden (1977: Fig.34)(Palaeozoic to Upper Tertiary); Singleton et al. (1976)(top of Tertiary); Cook et al. (1977: 86)(Lower/Middle Pleistocene); Gill (1974:211)(top Middle Pleistocene). The blocks show the stratigraphic position of units in well-known sequences; correlations with other units suggest a time range as indicated by arrows.

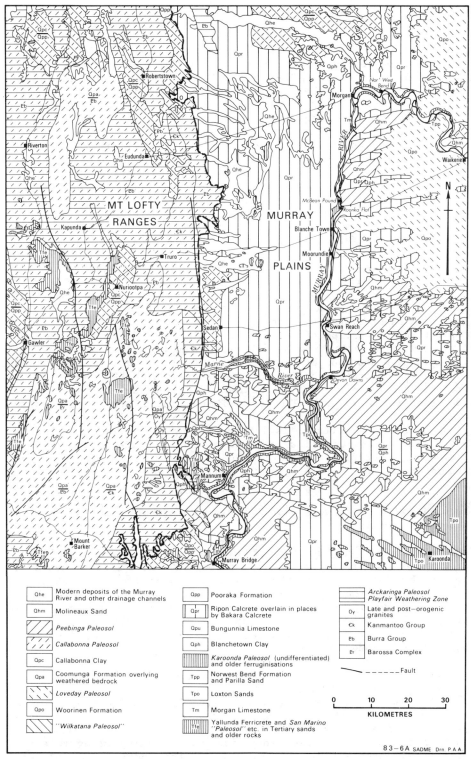

Figure 3. Murray Plains paleosol landscape map.

cods, algae and oolites, overlies the Blanchetown Clay in the Murray Basin. In some places, loess is found between the unit and underlying clay. Bungunnia Limestone is a prominent rock type amongst stone tool assemblages along the gorge tract of the Murray River.

The onset of colder conditions of the Pleistocene was a major event in the changeover from humid rainforest conditions of the Pliocene to the arid conditions of the present, but details of vegetation evolution are too generalized to be matched to the sequence of events interpreted from detailed physical sequences. At times of low sea level during the Medial Pleistocene a Pleistocene periglacial climate prevailed and the landscape was blanketed by calcareous loess and calcrete pans where formed (Column 4 on Fig. 2). Calcretes now occur off-shore on the Sahul Shelf (North Australia) and on the southern continental shelf.

The ancestral Murray River may have occupied an early gorge tract prior to formation of the Ripon Calcrete which drapes ancient stream courses now stranded above river level in the modern gorge.

The Ripon Calcrete is a hard pale brown and pink nodular or massive calcrete with concretionary structures and dark grey clasts produced by reworking of older thin dark grey layers. It occurs as a pedogenic or groundwater calcrete, which developed over calcareous loess and within other deposits during low stands of the sea. Ripon Calcrete is often found amongst artefacts as a hammer stone, and both Ripon and Bakara Calcrete have been used extensively as hearth stones in areas where they occur as erosional remnants.

Uplift and deposition of gravels and colluvial and alluvial sands and clays continued during the Medial Pleistocene. Calcareous loess was also laid down and younger fans of Bakara Calcrete were formed. During a marine incursion the shallow marine-estuarine Glanville Formation was deposited and there was a final period of calcrete development prior to widespread colluvial and alluvial sedimentation of the Late Pleistocene.

Just as the calcareous soils left behind by the retreat of the mid-Tertiary sea probably had a profound effect upon the distribution of calcicolous vegetation, the calcareous soils of this time may also have been important in controlling plant distribution.

Landscape with younger structured clay palaeosols and horizons of carbonate accumulation

Continuing uplift and retreat of the sea led to reworking of older fans, the deposition of valley fill throughout the uplands, the formation of alluvials fans, and the accumulation of sediments in lakes, now of lesser extent than in the earlier Pleistocene. The ancestral Murray was incised through older calcrete pans, beginning on the scarp near Murray Bridge and migrating upstream. Near Roonka the river cut down to about 50 m below cliff top, that is 18 m below present river level and 15 m below modern sea level.

One of the more prominent units on the flanks of the ranges is the Pooraka Formation. In the Ngaiwang tribal area this unit is found as colluvial-alluvial material at the base of older fans on the western margin of the Murray Plains (Cambrai Plateau). The Pooraka Formation has horizons of carbonate nodules characteristic of the "Wilkatana Palaeosol" of Williams (1973). Above the carbonate horizons the reworked top of the Pooraka Formation and adjoining older materials form the Callabonna Clay which marks a wetter period and increased incision.

Soil structures at the top of the Pooraka Formation and Callabonna Clay and manganese coatings on ped faces indicate the development of clay soils. The younger and more prominent is named the Calabonna Palaeosol (Column 5 on Fig. 2). The climate, to judge from the sedimentary evidence, was cool with a higher rainfall than at present.

About 100 000 years ago the numbers of Casuarina trees decreased and eventually gave way to eucalyptus. According to Specht (1972:206) "... a number of very severe arid periods occurred in the Pleistocene to Recent Periods. During the arid periods much of the vegetation died of drought and the soils became unstable and subject to wind erosion and then to water erosion whenever storms occurred". Homo sapiens sapiens had evolved by 50 000 years ago and may have settled in Australia by that time or even earlier.

Sands and sandy soil horizons near modern ground surface

Aggradation of the Murray River valley with the lower beds of valley fill consisting of alluvial clays, silts and coarse sands of the Monoman Formation began with the rise of the Flandrian Sea. The sedimentation was followed by reworking of Woorinen Formation and erosion of previously exposed Tertiary sediments to form the red Bunyip Sand and the yellow Molineaux Sand. The Bunyip Sand occurs at Roonka and near Swan Reach, overlying sediments of an older valley fill probably including the Tartangan Beds C14-dated at 6020 ± 150 BP by Tindale (1957:35) which contain skeletal remains of early man.

The deposits of the Flandrian transgression are distinctive. The most widespread is the St Kilda Formation which contains shallow marine sands, estuarine muds, and shelly deposits of the beach ridges. These deposits intertongue with other sequences, including the gypsiferous Yamba Formation in the Murray Basin and its equivalents on the west coast, dune sands marginal to St Vincent Basin, Monoman and Coonambidgal Formations of the Murray River, and carbonate deposits of the Coorong lagoon and shallow lakes parallel to the coastline in the southeast of south Australia.

The southern dune fields overlie the Late Pleistocene Woorinen Formation and older units. They include an inland belt of reddish Fulham Sand of the western and central basins and Bunyip Sand of the Murray Basin, a southern belt of yellow Molineaux Sand, and a coastal belt of Semaphore Sand. The development of dunes over a wide area indicates an increase in aridity.

The Fulham Sand occurs on the margin of St Vincent Gulf near Adelaide. Here, shelly deposits marking the highest stand of the Flandrian transgression overlie lower-phase Fulham Sand with carbonate pipes and are in turn overlain by younger layers of dune sand. A similar sequence occurs in the Murray River valley, where old aeolian sands with carbonate pipes are apparently younger than the riverine Coonambidgal Formation.

Aeolian gypsum dunes overlying or marginal to lacustrine beds of the Yamba Formation are interbedded with the upper phase of Bunyip Sand and contain a sub-fossil horizon of gypsite.

The Holocene Bunyip Sand and Molineaux Sand were deposited at about the same time in different areas. Two phases of development are seen in the dune fields containing these units. In each case there is an older darker phase in which vertical carbonate pipes developed. Calcareous accumulations of this kind in this stratigraphic position are characteristic of the Peebinga Palaeosol. Downward leaching of carbonates over such a wide area probably marks a period of somewhat higher rainfall.

The deposition of Holocene sediments post-dating the Peebinga Palaeosol was accompanied by minor tectonic activity. There was a decrease in fluvial activity so that modern streams are underfit and lakes are smaller. The thin sediments of this time include talus on the range margins, sandy colluvial, fluvial and alluvial overlays, and the youngest saline crusts and unconsolidated aeolian sands of the modern coast and inland dune fields (Column 6 on Fig. 2).

The following extracts from Specht (1972:205, 217 and 218) are particularly relevant to the composition and distribution of modern flora: "The flora has retained its heritage from the past sub-tropical climate of the Early Tertiary. The dominant species of the 'sclerophyll'. the 'savannah', and the 'mallee' land systems still show a summer-growth rhythm typical of subtropical areas, not Mediterranean climates where the dominant species grow largely during spring (Specht and Rayson 1957; Donald 1958; Holland 1968)." "As the climate deteriorated (during the Late Pleistocene) calcareous coastal species, such as the chenopods, were able to expand and occupy habitats in the drier parts of the 'mallee' and in the arid zone." In the arid zone ..."the flora of the woodlands is chiefly Australian in its composition, the great genera Acacia and Kochia form the bulk of the plant covering.... The flora of the higher rainfall areas of South Australia dominated by 'sclerophyll' communities is centred on the lower portions of Eyre, Yorke, and Fleurieu Peninsulas together with Kangaroo Island."

DISCUSSION

This review has shown that climates of the past had a profound effect upon materials characteristic of particular landscapes and upon associated biota. The formation of successive surfaces and subsequent stream incision produced a diverse heritage of relict materials for use by early man.

Profound climatic changes occurred following the hot dry climate at the end of the Pliocene when the Karoonda Palaeosurface was developed. There were cold dry periods during periglacial conditions in the Medial Pleistocene, cool wet periods during formation of the younger alluvial fans, and hot dry periods after the onset of aridity and during extensive dune building in the Late Pleistocene.

Changes in sea level and tectonic movements caused profound changes in the position of the coastline after the end of the Pliocene. There was an extensive ingression of the sea on the flatter basinal areas of the southern coastal margin in the Early Pleistocene when the upper beds of the Coomandook Formation were laid down. Lesser ingressions occurred during deposition of the Bridgewater Formation and again at the end of the Medial Pleistocene, marked by the Glanville Formation. The last ingression occurred with the rise of the Flandrian Sea and deposition of St Kilda Formation. Between ingressions of the sea there were retreats which exposed the inland margin of the continental shelf and at times established connections with Asia.

Many of the great rivers in Australia, including the Murray River, appear to have been strongly entrenched at least since the beginning of the Medial Pleistocene. Inland of the coast the broader features of the modern landscape - including hills and plains - were present at the end of the Pliocene, although the relief was probably less than it is today. Karst features, lakes and streams have been present as intermittent features in southern Australia since the Early Pleistocene. Calcreted surfaces with calcareous soils were probably much the same in the Medial Pleistocene except that they were probably more extensive and not as eroded and weathered as they are today.

Early man had occupied the continent prior to the rise of the Flandrian Sea when the extensive dune fields of the Woorinen Formation and equivalents were formed. His earliest time of arrival is a matter of conjecture.

Ancient materials were inherited by early man and used to his advantage. In the Ngaiwang area uplift and erosion of the Mt Lofty Ranges revealed the hard Proterozoic and Cambrian rocks from which many of the larger implements such as millstones, pounders, anvils and hammerstones were made. Although the granites provided suitable materials for millstones, finer grained igneous rocks were brought into the area from as far away as the volcanic outcrops in Victoria. Unworked harder rocks from the ranges were used to form ceremonial line and piled stone arrangements. The exposed surfaces of the older rocks formed an excellent base for the petroglyphs.

Ochres were derived from the bleached pre-Permian Playfair weathering zone. The ferruginous mottles provided a ready source of ochre for decrative purposes. From some Tertiary landscapes there was a prolific supply of hard silcretes. Although this substance was missing in the Ngaiwang area, erosion of the Murray River in Late Pleistocene to recent time exposed silcrete near Spring Cart Gully and this became an important factory site.

Along the gorge tract of the Murray River, Bungunnia Limestone was an inferior but commonly used material amongst stone tool assemblages. In the Ngaiwang tribal area, hearthstones and hammerstones were derived from widespread Early to Medial Pleistocene calcretes as well as from Tertiary sediments and older rocks cropping out in the valley of the Murray River itself. There was an inexhaustible supply of flint on the southeast coastal margin.

The accumulation of fire places and stone artefacts in sequence with aeolian and riverine sediments favours correlation of human and landscape evolution. The detailed investigation of this evidence in relation to its regional stratigraphic setting has begun at Roonka, which is the principal locality in the Ngaiwang tribal area.

Some landscapes are characterized by certain physiographic factors or combinations of factors favouring preferential use by man. Important functional differences are reflected in such things as storage and recycling of raw materials, natural routes and trackways, conservation and recreation, and food production. Those things lead to sociological and cultural differences in peoples adapted to particular landscapes. The identification of separate landscapes and the detailed interpretation of physical sequences with regard to their contained faunal, floral and man-made features may lead to time ordering of the evidence, relating the selection of living areas to such things as refinement in manufacture of stone implements, the use of stone in games and as sacred objects, and the use of various materials for ornamentation. The study of landscape evolution in this way in relation to key sites of early man throughout Australia is an important task for the future.

ACKNOWLEDGEMENTS

This report is published with the approval of the Director-General, South Australian Department of Mines and Energy.

REFERENCES

Brown DA, Cample KSW and Crook KAW 1968. The Geological Evolution of Australia and New Zealand. Pergamon Press, Sydney.

Cook PJ, Colwell JB, Firman JB, Lindsay JM, Schwebel DA and Von der Borch CC 1977. The Late Cainozoic sequence of southeast South Australia and Pleistocene sea level changes. BMR. J. Aust. Geol. Geophys. 2:81-80.

Cotton BE (ed.), Aboriginal Man in South and Central Australia 1968. Govt Printer, Adelaide.

Crocker RL and Wood JG 1947. Some historical influences on the development of the South Australian vegetation communities and their bearing on concepts and classification in ecology. Trans. R. Soc. S. Aust. 71:91-136.

Firman JB 1975. Australia - South Australia. In: Rhodes W Fairbridge (ed.), The Encyclopedia of World Regional Geology, Part 1. Western Hemisphere. Encyclopedia of Earth Sciences Series, Vol. III. Dowden, Hutchinson and Ross Inc., Stroundsurg, Pa., U.S.A. p. 61-81.

Firman JB 1980. Regional stratigraphy of the regolith on the southwest margin of the Great Australian Basin Province, South Australia. Flinders Univ. M.Sc. thesis (unpubl.).

Firman JB 1981. Explanatory notes to accompany a physiographic map for use in natural resources management. GDNR thesis, Dept. Natural Resour. Roseworthy Agricult. College, Roseworthy, S. Aust. (unpubl.).

Firman JB 1983. Paleosols in Southern Australia. Proc. XI INQUA Congress, Moscow, 1982 (in press).

Gill ED 1974. Carbon-14 and Uranium/Thorium check on suggested interstadial high sea-level around 30 000 BP. Search 5:211.

Hag Bilal U and Berggren WA 1977. Corrected age of the Pliocene/Pleistocene boundary. Nature 269:483-488.

Playford PG 1979. Floral evolution in the Late Palaeozoic and Early Mesozoic of Australia. Aust. Geol. geol. Soc. Aust. 26:13.

Pretty Graeme L 1977. Archaeology in South Australia. S. Aust. Yb. 1977.

Singleton OP, McDougall J and Mallett CW 1976. The Plio-Pleistocene boundary in southeastern Australia. J. geol. Soc. Aust. 23:299-311.

Smith AG and Briden JC 1977. Mesozoic and Cenozoic Paleocontinental Maps. Cambridge Univ. Press, Cambridge.

Snelling NJ 1964. A review of recent Phanerozoic time-scales. In: Harland et al. (eds.), The Phanerozoic Time-Scale. Geol. Soc., London.

Specht RL 1972. The Vegetation of South Australia. Govt Printer, Adelaide.

Tindale NB 1957. Culture succession in southeastern South Australia from late Pleistocene to the present. Rec. S. Aust. Mus. 13(1):1-49.

Tindale NB 1974. Aboriginal Tribes of Australia. Univ. Calif. Press.

Williams GE 1973. Late Quaternary piedmont sedimentation, soil formation and paleoclimates in South Australia. Z. Geomorph. 17:102-125.

Late Quaternary climatic change
Evidence from a Tasmanian speleothem

A.GOEDE & M.A.HITCHMAN
University of Tasmania, Hobart, Australia

ABSTRACT. This paper reports palaeotemperature studies on two segments of a uniform diameter stalagmite collected from a limestone cave in northern Tasmania. The calcite speleothem has been dated by a combination of C14 and ESR methods and is estimated to have been growing continuously between approximately 12 600 and 2800 BP.

The stalagmite was deposited under conditions of oxygen isotope equilibrium and variations in the O18/O16 ratio are believed to reflect changes in the mean annual temperature at the surface above the site. A profile of 39 $\delta^{18}O_c$ values is presented. A positive relationship between these values and mean annual temperature has been established for the area and this relationship is used to estimate the temperature during deposition.

On this basis the mean annual temperature during the period has varied between approximately 6.2° and 10.6°C compared with a present day value of 9.5°C. Temperatures were frequently higher than at present during the period 12 000 to 9300 BP but consistently lower since then. The lowest temperatures occurred approximately 3800 years ago. There is good evidence from elsewhere in the Southern Hemisphere that terrestrial temperatures were lower than today during the Late Holocene.

INTRODUCTION

The speleothem was collected from a high-level chamber in Lynds Cave, an active stream cave developed in Ordovician limestones. The cave is located in the Mole Creek area of northern Tasmania, Australia (41° 34' 20"S, 146° 13'40"E) (Fig. 1). The cave stream emerges from the base of a limestone cliff on the eastern bank of the Mersey River at an altitude of 300 m above sea level. The elevation of the area above the cave which supplies seepage water to the site is between 400 and 450 m. The area has a mean annual temperature of approximately 9.5°C and a mean annual precipitation of 1500 mm with a pronounced winter maximum. The vegetation is wet sclerophyll forest with an understorey of rainforest species. The cave temperature at the collecting site was found to be 9.5°C ± 1°C.

The material consists of a stalagmite of uniform diameter (code: LY) which was found in a broken condition in two segments. The lower and upper segments have heights of 118 cm and 80 cm respectively. The two do not fit together, which indicates that a piece is missing. A care-

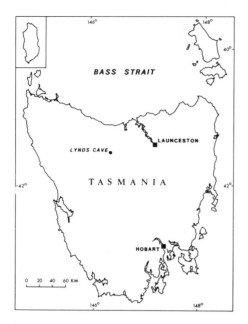

Figure 1. Location map of Lynds Cave.

ful search at the site failed to locate this segment and it is probable that it was either shattered when the stalagmite broke or removed from the cave.

OXYGEN ISOTOPE CONTENT AND TEMPERATURE

The stalagmite was sectioned longitudinally and one half set in plaster. Samples were taken at 50 mm intervals along the core using a 5 mm diameter drill. These were reacted with anhydrous H_3PO_4 under vacuum, using a conventional extraction line, to produce CO_2 gas that was then dried and purified prior to isotope analysis. All isotopic measurements of oxygen and carbon were made using a VG Micromass 602D mass spectrometer. Carbonate values were expressed in ‰ PDB and standardized to New Zealand standards TKL and K2. Water samples are standardized with respect to Vienna SMOW and ANU-Cl and are given relative to SMOW. The subscripts c, p and w denote calcite, precipitation and seepage water respectively.

The modern $\delta^{18}O_p$ and δD_p values and their seasonal variations are estimated for the area as monthly samples of precipitation were collected in 1979 at a site 3 km to the east of Lynds Cave entrance at an elevation of 460 m above sea level (Goede, Green and Harmon 1982). The values ranged from -0.8 to -6.4 ‰ $\delta^{18}O_p$ SMOW and from -2.2 to -37.3 ‰ δD_p SMOW. The weighted mean annual values are -4.82 ‰ and -26.7 ‰ SMOW respectively.

The secular relationship between $\delta^{18}O_c$ and temperature in the area is known from the appropriate altitude from another palaeotemperature study of a uniform diameter stalagmite (code: LT) from Little Trimmer Cave 3 km to the east (Goede, Green and Harmon, unpubl. ms). This 142 cm high stalagmite was deposited between 109 000 and 76 000 yrs BP according to U/Th dating. Eighteen fluid inclusion analyses (δD_w)

were carried out on this specimen in addition to measurements of $\delta^{18}O_c$. A highly significant positive statistical relationship was established between stratigraphically matched values of δD_w and $\delta^{18}O_c$.

$$\delta D_w(SMOW) = 14\delta^{18}O_c + 22 \tag{1}$$

When values of δD_w are known, values of $\delta^{18}O_w$ can be estimated (Schwarcz, Harmon, Thompson and Ford 1976; Harmon, Schwarcz and Ford 1978; Harmon, Schwarcz and O'Neil 1979) using the present-day locally established relationship between them

$$\delta D_p(SMOW) = 7\delta^{18}O_p(SMOW) + 7 \tag{2}$$

Knowing both the $\delta^{18}O_w$ and $\delta^{18}O_c$ values of a number of stratigraphic horizons, the relationship between $\delta^{18}O_c$ values and temperature can be calculated using the formula

$$10^3 \ln \alpha_{c-w} = 2.78 \times 10^6 Ta^{-2} - 3.39 \tag{3}$$

where Ta is the temperature in $^\circ K$ (Harmon, Thompson, Schwarcz and Ford 1978). Using (1), (2) and (3) and assuming a present-day equilibrium value of $\delta^{18}O_c = -4.10$ %. the following relationship is obtained

$$Tc = 4\delta^{18}O_c + 26 \tag{4}$$

where Tc is the temperture in $^\circ C$. This should be regarded as a rough estimate based on the assumption that the slope and deuterium excess values in equation (2) have not deviated from present-day values in the past.

The positive relationship between oxygen isotope composition of calcite and temperature is unusual, but another well documented case of such a relationship has been made for two stalagmites collected from a cave on Vancouver Island (Gascoyne, Schwarcz and Ford 1980; Gascoyne, Ford and Schwarcz 1981). The reason for this phenomenon is believed to be the magnitude of the latitudinal shift in the moisture source areas accompanying climatic change and will be discussed in detail elsewhere (Goede, Green and Harmon, unpubl. ms.).

TEMPERATURE RECORD OF LYNDS CAVE STALAGMITE

Before palaeotemperatures can be determined from the Lynds Cave stalagmite three requirements have to be met:
1. It must be shown that the stalagmite was deposited under conditions of oxygen isotope equilibrium.
2. The isotopic composition of present-day calcite, deposited under conditions of oxygen isotope equilibrium, must be determined.
3. The speleothem must be dated to provide a reliable time scale.

Tests for oxygen isotope equilibrium

Two tests have been suggested to confirm deposition of calcite under equilibrium conditions (Hendy and Wilson 1968).
1. The absence of a strong positive correlation between values of $\delta^{18}O$ and $\delta^{13}C$ of samples taken at intervals along the longitudinal axis. Figure 2 confirms that this is indeed the case.

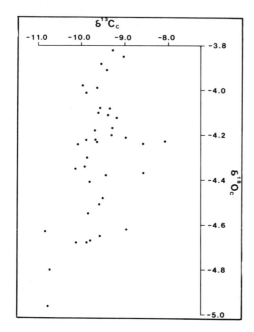

Figure 2. Scatter diagram of thirty-nine $\delta^{18}O_c$ and $\delta^{13}C_c$ values of samples taken along the longitudinal axis of the stalagmite.

2. Samples taken at intervals along a single growth layer outwards from the core should not show a progressive trend towards isotopically heavier values of $\delta^{18}O_c$ which correlate with a similar trend in $\delta^{13}C_c$ values. This was done by taking and analysing sets of seven samples from each of three growth layers A, B and C (Fig. 3). Again the conditions are met except for one unexplained anomalous pair of values for the first point along growth layer B.
Therefore the stalagmite is considered to have been deposited under conditions of oxygen isotope equilibrium and variations in $\delta^{18}O_c$ values should reflect variations in temperature.

Isotopic composition of modern calcite

To determine the oxygen isotope composition of calcium carbonate being deposited at present under conditions of oxygen isotope equilibrium, six actively growing straw stalactite tips were collected in the vicinity of the site. Since straw stalactites are also known to grow rapidly, this material is assumed to have been deposited during the last few years (Harmon, Schwarcz and Ford 1978). Each sample was analysed for $\delta^{18}O$ and $\delta^{13}C$ and the results are shown in Table 1.

Table 1. Isotopic analyses of straw stalactites in Lynds Cave.

Sample No.	$\delta^{18}O_c$ (‰ PDB)	$\delta^{13}C_c$ (‰ PDB)
OX - 21	-4.05	-8.57
OX - 22	-3.76	-6.86
OX - 23	-3.62	-7.19
OX - 24	-3.47	-6.90
OX - 25	-4.16	-7.82
OX - 26	-3.66	-6.72

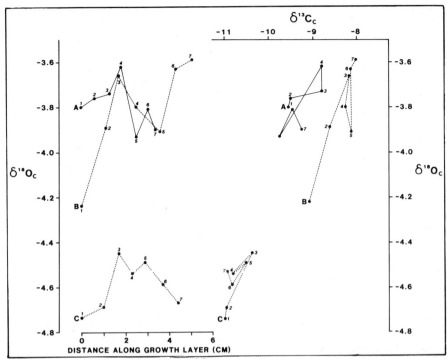

Figure 3. Variation of $\delta^{18}O_c$ with distance along each of three growth layers (on left) and with $\delta^{13}C_c$ for the same growth layers (on right).

The wide range of $\delta^{18}O_c$ values probably reflects a combination of two factors.
1. Values may vary because isotopically heavy summer precipitation may make a more significant seepage contribution to some straw stalactites than to others. In contrast, a fast growing stalagmite such as the one analysed in this paper is likely to obtain a high percentage of its carbonate deposition from seepage derived from winter precipitation under the prevailing climatic conditions.
2. Straw stalactites do not necessarily deposit calcite at oxygen isotope equilibrium. Non-equilibrium deposition is reflected in deposition of calcite that is isotopically heavier with respect to both oxygen and carbon.
The lowest values of $\delta^{18}O_c$ are therefore those which combine a dominance of depositing solutions derived from winter precipitation with deposition at or close to oxygen isotope equilibrium. Lowest values are shown by samples OX-21 and OX-25 which, as expected, also show the lowest values for $\delta^{13}C_c$. The present-day isotopic value of calcite deposited under oxygen isotope equilibrium conditions is believed to be close to the mean of these two samples, that is -4.10 ‰ $\delta^{18}O_c$.

An independent estimate of this value can be made when the isotopic composition of the seepage water and the cave temperature (9.5 °C) are known. The assumption is again made that seepage waters are derived from winter precipitation (weighted mean value $\delta^{18}O_p$ = -5.68 ‰ SMOW). This value is used together with the mean annual cave temperature in equation (3) to obtain a theoretical value for calcite of -4.20 ‰.. This is in close agreement with the previously estimated value of -4.10 ‰..

While this value has been adopted, it must be remembered that it represents an average value for deposition taking place over several years prior to the collection of the material in August 1982. These years may, or may not, have been typical of present-day conditions of deposition. On the other hand, the samples drilled from the stalagmite core represent values that are averaged over a period of tens of years.

ESR DATING

Two methods were used to date the stalagmite: electron spin resonance (ESR) and C14 dating. Details of the ESR method are given in Ikeya (1978) while some of the problems involved in the application of the method are discussed by Wintle in Cook et al. (1982). As an initial step, the intensity of the free radical signal was measured for a large number of samples taken along the centre line of a longitudinal section using a 5 mm diameter drill. The samples were analysed on the JEOL JES-FE3X ESR spectrometer and the intensity values plotted against height (Fig. 4, left). The plot shows a steady increase in the strength of the signal with increasing age. Regression lines are shown for both the lower and upper segments. As a next step seventeen larger samples were taken and each sample subdivided into five sub-samples. Sub-samples were irradiated with incremental doses of γ-rays from a Co60 source. Then sub-samples of each set were measured individually on the ESR spectrometer and the intensity of the signal plotted against the γ-ray doses measured in rad. A regression line was fitted and the equivalent dose (ED) calculated for each sample. The ED values for the samples ranged from 125 to 1519 rad and increased progressively with increasing age.

The ED values represent the equivalent doses of natural radiation the samples have received after deposition of the calcite. These values were plotted against height and regression lines constructed for the relationships for both lower and upper segments of the stalagmite (Fig. 4, right). The scatter of points is much less than that of the plot of signal intensity against height suggesting that some of the scatter in the latter is due to inter-sample variations in sensitivity to radiation damage.

Random measurement errors involved in individual ED determinations are largely eliminated by using the values indicated by the regression lines for age determination at any point. The lines indicate that both the upper and lower segments of the stalagmite experienced constant growth rates if it is assumed that the annual dose rate has remained constant in time. However, the growth rate for the upper segment appears to have been more rapid than for the lower.

The ED can be used to determine the age at any point if the annual dose rate is known since

$$\text{Age (years)} = \frac{\text{Equivalent dose (rad)}}{\text{Annual dose rate (rad/year)}}$$

The annual dose rate consists of two components: the external and internal dose rates. The external dose is made up of radiation received by the sample from the surrounding environment and cosmic radiation. The internal dose rate is due to radioactive impurities within the sample and can be calculated if the concentrations of the elements U, Th and K are known.

The external dose rate was measured by placing a Rad-21 radiation dosimeter at the Lynds Cave site for a fortnight. The amount of radia-

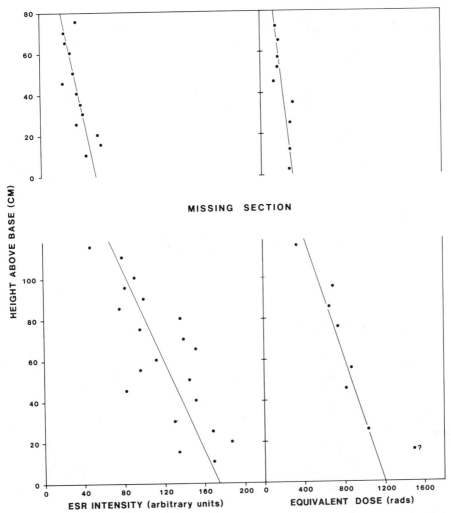

Figure 4. Variations of ESR signal intensity and ED as a function of height above base of stalagmite.

tion measured was equivalent to an external dose rate of 99 millirad/year but accuracy of measurement is low (±20 %). To assess the internal dose rate, a series of sixteen samples was analysed by the CSIRO Division of Energy Chemistry using fluorimetric determinations. All, except the three basal samples, showed uranium concentrations below the detection limit of 0.035 ppm. The three basal samples contained an average of 0.1 ppm. U. Even this amount of uranium will give a dose rate of only 6.4 millirad/year (Aitken 1974). Concentrations of Th and K were not measured but are expected to be very low. The extremely small internal dose rate is very fortunate since it largely eliminates two major problems associated with ESR dating of calcite. They are:
  1. Changes in internal dose rate with time since Th230 formed by the decay of U234 gradually increases the dose (Wintle 1978).
  2. Problems caused within the sample by the concentration of uranium within small inclusions (Walton and Debenham 1980).
Since the external dose rate is small, a crude estimate of the annual

dose rate of 100 millirad/year is assumed. This is at the lower limit of the range of annual dose rates estimated for speleothems by Ikeya (1978). The rate can be tentatively used to calibrate the age range of the stalagmite using the regression lines of ED values against height shown in Fig. 4, right.

Using the assumed dose rate the lowest segment was found to have grown between approximately 12 000 and 4100 yrs BP and the upper segment was deposited between approximately 3100 and 1270 yrs BP. These time ranges must be regarded as crude estimation, especially since ESR dating cannot yet be regarded as completely reliable (Wintle and Huntley 1982). However, it will be shown that ED determinations can be used in combination with C14 dating to provide a more accurate time-scale.

CARBON 14 DATING

Six samples were analysed: the base, middle and top of each of the two segments. Dating was carried out by the National Physical Research Laboratory in Pretoria. The C14 dates obtained are shown in Table 2.

Table 2. Carbon 14 age determination of LY stalagmite in Lynds Cave.

Sample No.	Anal. No.	Medium height above base (cm)	Apparent C14 age (yrs BP)
Lower Segment			
AG32	Pta-2972	2.5	14 500 ± 140
AG33	Pta-3198	57.5	13 000 ± 30
AG34	Pta-2975	112.5	10 300 ± 90
Upper Segment			
AG35	Pta-2976	3.0	8980 ± 90
AG36	Pta-3199	43.0	6400 ± 20
AG37	Pta-2979	78.2	5250 ± 70

The C14 ages cited in Table 2 are subject to a correction due to the reservoir effect. The C13 values of the stalagmite indicate that the apparent initial age could hardly have been less than 1000 years. On the other hand the theoretical maximum for this figure is 5500 years or one half-life of C14 (Hendy 1970). This initial age need not necessarily have remained constant with time, especially if the climate changed significantly. Furthermore, in order to obtain an estimate of the absolute time-scale the dates have to be calibrated to allow for temporal variations in the radiocarbon content of atmospheric carbon dioxide. Calibration tables have been published for conventional radiocarbon ages ranging from 10 to 7240 BP (Klein, Lerman, Damon and Ralph 1982).

We can estimate the reservoir effect from the upper segment of the stalagmite if two basic assumptions are made that during the period of deposition
   1. the reservoir effect remained constant
   2. the annual dose rate remained constant so that ED values bear a linear relationship to time.
The two C14 samples used are AG35 and AG37. The ED values corres-

ponding to the median position of the two C14 samples can be determined from the regression line on Fig. 4, upper right. They are found to be 305 and 131 rad respectively. Therefore the ratio of absolute time between samples AG35 and AG37 should be 2.32.

Let the reservoir effect be x years. Different values of x between the range 1000 to 5500 years are applied to the C14 dates. For each value of x the dates are then converted to the absolute time-scale using the calibration tables and the time ratio is calculated. The value of x which most closely matches the AD ratio of 2.32 was found to be 2160 years and was adopted as the magnitude of the reservoir effect. This value is subtracted from the three C14 dates for the upper segment and they are then calibrated using the calibration tables to provide the time-scale in Fig. 5. The time-scale for the lower segment is constructed on the assumption that the resevoir effect has not changed. The calibration table cannot be used to convert the C14 ages to estimates of absolute time since the ages are beyond the range. Therefore the time-scale for the lower segment must be regarded as considerably less accurate. There appears to be a significant change in the growth rate of the stalagmite between the two segments. The growth rate of the lower segment is calculated as 26.2 cm/1000 year while the upper segment grew at a rate of 20 cm/1000 year. The rates of growth are very rapid compared with four other Tasmanian stalagmites whose growth rates have so far been determined (Goede and Harmon 1983).

PALAEOTEMPERATURE

The isotopic data are plotted against age in Fig. 5. It can be seen that between 12 600 and 9400 BP, values of $\delta^{18}O_c$ ranged between -4.35 and -3.82 ‰.. During this period the average value was slightly above the modern value of -4.10 ‰.. After 9400 BP there was a gradual decline in O18 until a minimum of 4.96 ‰ was reached at about 3800 yrs BP. The record terminates at approximately 2800 BP with O18 values well below present-day levels. The C13 values remain relatively uniform throughout between -8.1 and -10.8 ‰..

The $\delta^{18}O_c$ curve can be interpreted as reflecting variations in palaeotemperature. Equations (2), (3) and (4) are tentatively used to convert oxygen isotope values into estimates of palaeotemperature. The resulting palaeotemperature scale is shown on the top abscissa in Fig. 5. Between 12 600 and 9400 BP temperatures oscillated around the present-day mean annual temperature of 9.5°C but reached a maximum of about 10.6°C at 10 400 BP. Mean annual temperatures remained continuously below the present-day value after 9400 BP and reached a minimum value close to 6.2°C at about 3800 BP. The mean annual temperature appears to have varied by as much as 4.4°C during the growth period of the stalagmite.

Absolute temperature values should be treated with caution due to the assumption made in equation (2). However, relative temperature changes and temperature trends appear to be reliable.

DISCUSSION

The temperature pattern indicated by the oxygen curve suggests an early climatic optimum with temperatures intermittently higher than today between 12 000 and 9000 BP followed by a gradual decline below present-day temperatures through most of the Holocene until about 2800 BP when deposition ceased.

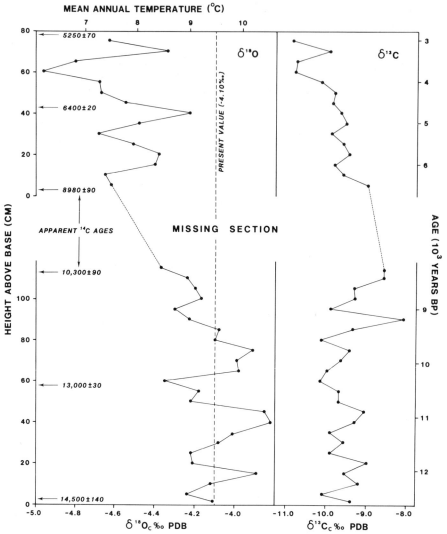

Figure 5. Temporal variations in the values of $\delta^{18}O_c$ amd $\delta^{13}C_c$ with height above base of stalagmite. Mean annual temperature variation with time as well as median positions and apparent ages of six C14 dates are indicated.

The curve should be compared with palynological evidence obtained in southern Tasmania by Macphail (1979) who examined seven Holocene pollen sequences. Each was continuous and extended back at least 10 000 years. At one glacial lake site at 560 m (Lake Vera) Macphail found rapid forest encroachment occurring by 11 500 BP. He deduced that during the Early to Middle Holocene climates were wetter and possibly warmer than at present and suggested a 'climatic optimum' between 8000 and 5000 BP. During this phase cool temperate rainforest dominated by Nothofagus cunninghamii reached its greatest extent. During the Late Holocene there was an expansion of Eucalyptus dominated forests and an increase in shade-intolerant species.

Both isotope and palynological data suggest a change to cooler conditions during the later part of the Holocene. They differ significantly in their timing of the 'climatic optimum'. It must be remembered that while the isotope data reflect predominantly temperature changes, variations in the pollen diagrams are much more sensitive to changes in moisture availabiltiy and its seasonal distribution especially for sites at lower elevations. Detailed comparisons are also not possible at this stage since many of the pollen diagrams are poorly dated. The dating of the lower segment of the Lynds Cave stalagmite is uncertain because it assumes that the reservoir effect of the C14 dates is the same as for the upper segment.

Burrows (1979), working in the South Island of New Zealand (41° - 46°S), produced a radiocarbon dated chronology of glacial events and other cold climate phenomena during the last 12 000 years. He obtained fifteen C14 dates of less than 9000 BP related to glacial advances and other cold climate phenomena. All but three of these dates are younger than 5000 years. This points to a temperature decline during the Holocene.

Williams (1978) summarizes evidence from the Southern Tablelands of New South Wales that hillslopes were unstable and streams were aggrading between 4000 and 1500 BP. As possible causes he suggests lower temperatures, drier and windier conditions and changes in rainfall seasonality.

CONCLUSION

Evidence of Holocene climatic changes from various localities in the Southern Hemisphere appears to be broadly concordant with the trend of palaeotemperatures shown in Fig. 5.

The evidence from the oxygen isotope curve suggests that the mean annual temperature in northern Tasmania during the Holocene may have varied by some 4°C from 1° above to 3° below the present-day value of 9.5°C. The curve also indicates an early 'climatic optimum' between 12 000 and 9000 BP in contrast to palynological evidence which places it between 8000 and 5000 BP. However, in both cases the dating is inadequate.

ACKNOWLEDGEMENTS

We are grateful to the University of Tasmania and the Australian Research Grants Committee for their financial support. Stable isotope analysis and ESR dating were made possible by the facilities provided by the University of Tasmania Central Science Laboratory (CSL). Initiation and development of research into the stable isotope composition of stalagmites owes much to the encouragement, enthusiasm and technical expertise of Dr DC Green of the Tasmanian Department of Mines. Technical assistance with ESR analyses was provided by Mr JC Bignall of the CSL.

The six C14 age determinations were made by Dr JC Vogel of the National Physical Research Laboratory in Pretoria, South Africa. Dr MA Geyh of the Niedersächsisches Landesamt für Bodenforschung, West Germany, suggested the combined use of C14 and ESR analyses to provide a more accurate time-scale. Mr P Pakalns, of the CSIRO Division of Energy Chemistry at Lucas Heights, Australia, was responsible for the fluorimetric determinations of the uranium content of sixteen stalagmite samples. Samples of ANU isotope standards were made available by Mr A Chivas of the School of Earth Sciences at the Australian National University. Our sincere thanks to all of them.

REFERENCES

Aitken MJ 1974. Physics and archaeology. Clarendon Press, Oxford. 291 pp.

Burrows CJ 1979. A chronology for cool-climate episodes in the Southern Hemisphere 12 000-1000 yr B.P. Palaeogeogr. Palaeoclimatol. Palaeoecol. 27:287-347.

Cook J, Stringer CB, Currant AP, Schwarcz HP and Wintle AG 1982. A review of the chronology of the European middle Pleistocene Hominid Record. Yb. Phys. Anthropol. 25:19-65.

Gascoyne M, Ford DC and Schwarcz HP 1981. Late Pleistocene chronology and palaeoclimate of Vancouver Island determined from cave deposits. Can. J. Earth Science 18:1643-1652.

Gascoyne M, Schwarcz HP and Ford DC 1980. A palaeotemperature record for the mid-Wisconsin in Vancouver Island. Nature 285:474-476.

Goede A, Green DC and Harmon RS 1982. Isotopic composition of precipitation, cave drips and actively forming speleothems at three Tasmanian cave sites. Helictite 20:17-27.

Goede A, Green DC and Harmon RS, unpubl. ms. Late Pleistocene palaeotemperature record from a Tasmanian speleothem. 24 pp.

Goede A and Harmon RS 1983. Radiometric dating of Tasmanian speleothems - evidence of cave evolution and climatic change. J. geol. Soc. Aust. 30:89-100.

Harmon RS, Schwarcz HP and Ford DC 1978. Stable isotope geochemistry of speleothems and cave waters from the Flint Ridge-Mammoth cave system, Kentucky: Implications for terrestrial climate change during the period 230 000 to 100 000 years B.P. J. Geol. 86:373-384.

Harmon RS, Schwarcz HP and O'Neil JR 1979. D/H ratios in speleothem fluid inclusions: a guide to variations in the isotopic composition of meteoric precipitation? Earth Planet. Sc. Lett. 42:254-266.

Harmon RS, Thompson P, Schwarcz HP and Ford DC 1978. Late Pleistocene paleoclimates of North America as inferred from stable isotope studies of speleothems. Quat. Res. 9:54-70.

Hendy CH 1970. The use of C14 in the study of cave processes. In: I Olsson (ed.), Radiocarbon variations and absolute chronology. Nobel Symposium 12. p. 419-443.

Hendy CH and Wilson AT 1968. Palaeoclimatic data from speleothems. Nature 219:48-51.

Ikeya M 1978. Electron spin resonance as a method of dating. Archaeometry 20:147-158.

Klein J, Lerman JC, Damon PE and Ralph EK 1982. Calibration of radiocarbon dates. Radiocarbon 24:103-150.

Macphail MK 1979. Vegetation and climates in southern Tasmania since the Last Glaciation. Quat. Res. 11:306-341.

Schwarcz HP, Harmon RS, Thompson P and Ford DC 1976. Stable isotope studies of fluid inclusions in speleothems and their paleoclimatic significance. Geochim. Cosmochim. Acta 40:657-665.

Walton AJ and Debenham NC 1980. Spatial distribution studies of thermoluminescence using a high-gain image intensifier. Nature 284:41-44.

Williams MAJ 1978. Late Holocene hillslope mantles and stream aggradation in the Southern Tablelands, NSW. Search 9:96-97.

Wintle AG 1978. Thermoluminescence dating study of some Quaternary calcite: potential and problems. Can. J. Earth Sci. 15:1977-1986.

Wintle AG and Huntley DJ 1982. Thermoluminescence dating of sediments. Quat. Sci. Rev. 1:31-53.

Southern Africa

SASQUA International Symposium / Swaziland / 29 August - 2 September 1983

Late Quaternary environments in South Africa

K.W.BUTZER
University of Texas, Austin, USA

ABSTRACT. This review discusses a selection of dated local sedimentological sequences from cave, lacustrine, spring, fluvial, and coastal systems in South Africa, drawn primarily from the writer's own research to illustrate trends and changes that relate to time, space, and specific depositional media. It emerges clearly that there have been repeated and significant changes during the Late Quaternary. Provisional correlations of regionally coherent, relative lithostratigraphies for the humid, subhumid and semi-arid areas of the country are presented.

INTRODUCTION

Synthetic analysis of Late Quaternary paleoclimates in southern Africa remains exceedingly difficult. Firstly, chronometric controls are unavailable or unsatisfactory in many sequences where datable materials are absent or prone to substantial errors. This is further complicated for the terminal Pleistocene by repeated, rapid changes in some areas (see Coetzee 1967), so that even ± 5 % error ranges can lead to faulty chronometric controls, allowing little more than broad, inferential correlations between sequences, or to external, stratigraphic standards such as deep-sea oxygen isotope stages.
  Secondly, each category of paleoclimatic evidence has inherent problems and the faunal, palynological, and sedimentological records reflect different first-order controls and may therefore be out of phase and perhaps even appear to be contradictory. Animals of different size and mobility reflect varying scales of habitat (Klein 1980a). Furthermore, they commonly reflect cultural or carnivore selection (Klein 1980b), and they are subject to complex taphonomic processes during and after burial (Gifford 1981). The end-effect is that faunal assemblages are qualitatively and quantitatively biased and difficult to assess as a true reflection of a habitat mosaic. Pollen are also selected in terms of their productivity, dispersal qualities, local preservational media, and differential resistance to oxidation; not subject to the dramatic biotic replacements characteristic of Europe, the resulting pollen assemblages can only be evaluated against standards provided by modern pollen rains that among themselves display subtle and complex variability (Scott 1983). Finally, sediments comprise multiple, event- or process-specific lenses that document depositional microenvironments; these, in turn, reflect the hydrolog-

ical cycle, different degrees of human intervention, the balance of soil formation versus erosion, and the vegetative mat of a medium-scale topographic matrix that in turn forms part of a more complex regional landscape (Butzer 1982: ch. 4). Again there are problems in translating microfacies variability into larger-scale geomorphic and hydrological controls, by means of analog studies.

Thirdly, each category of evidence varies not only in response to the varying amplitudes of different first-order controls, but also with different inherent equilibrium thresholds. In the case of sediments, which are emphasized in this paper, the problem of thresholds is complicated by sediment-transfer systems of different scales, magnitudes of energy flow, and degrees of systemic "openness". Cave microdepositional environments vary significantly within a cave, between two caves in the same macroenvironment, between examples in different landscapes, and between a cave and other point, linear, or areal depositional systems dominated by distinct fluvial, eolian, or gravity processes. Depending on the characteristic geomorphic agents and the system configurations, local sequences of meticulously-reconstructed events may not compare too well within a single macroenvironment, and the discrepancies may be compounded with increasing distance.

These caveats suggest that low-level paleoclimatic interpretation must proceed with caution, and that it should be based on many, internally consistent, local sequences of environmental 'events', preferably inferred from different processes and categories. Such an approach is admittedly antithetical to the CLIMAP precedent of reconstructing whole continental areas for a point in time, based solely on assumptions and simulation (Peterson et al. 1979). The inductive approach is tedious and less tidy, but it is the only scientific method that will ultimately generate the data base for syntheses such as those that CLIMAP has more successfully applied to certain world oceans (see McIntyre et al. 1976).

To facilitate comparison, the data to be outlined for three environmental provinces (Fig. 1) are synthesized in three tables (Figs. 2, 3 and 4). Inferred deep sea O18 isotope stages are indicated on the right-hand margins, as a convenient organizational device for the time-span beyond the range of C14 dating. It is therefore relevant to stress that these charts, however provisional, are designed to provide regionally coherent, relative lithostratigraphies, not to coerce the data into an external, climatostratigraphic framework.

In an overview such as this, it is impossible to describe sites and sections in detail or to argue points of interpretation at length. For the interior sites, including Botswana and Namibia - which are not included here -, more complete evidence with references is marshalled in a monographic paper (Butzer 1984). Many of the geomorphic sequences and lithostratigraphies studied by the writer from south of the Drakensberg and Cape Ranges have been published (Butzer 1973a, 1978, 1979; Butzer and Helgren 1972; Butzer, Beaumont and Vogel 1978). It is for the important coastal sequences at Swartklip, Stanford, and Melkbos that publication is still outstanding, with apologies.

COASTAL SECTOR

The Southwestern Cape. A number of long profiles have been studied in the southwestern Cape Province, along beach cliff exposures, in caves and amid interior eolian terrains indirectly influenced by marine

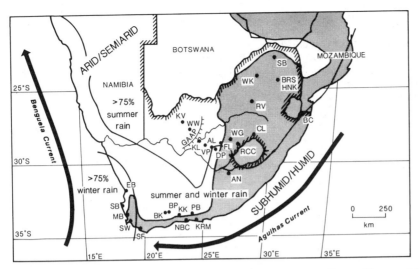

Figure 1. The subhumid and humid sectors of southern African are shaded gray; the area of dominant winter rains is limited to the southwestern Cape, with summer rains dominant in the interior and towards the north (data based on Schultze 1958 and Climate of South Africa 1965). The key sites, discussed from west to east, are indicated by letters, as follows: EB (Elands Bay), SB (Saldanha Bay), MB (Melkbos), SW (Swartklip), SF (Stanford, Die Kelders, Byneskranskop), BK (Buffelskloof), BP Boomplaas), KK (Kangkara), NBC (Nelson Bay Cave), PB (Paardeberg), KRM (Klasies River Mouth), KV (Kathu), WW (Wonderwerk), KL (Klipfontein), AL (Alexandersfontein), VP (Voigtspost), DP (Deelpan), FL (Florisbad and Vlakkraal), WG (Welgevonden), AN (Aliwal North), RCC (Rose Cottage), CL (Clarens), RV (Rietvlei and Moreletta), WK (Wonderkrater), SB (Scot), BRS (Bushman Rock), HNR (Heuningneskrans), BC (Border Cave).

regressions. The criteria include coarser "transgressional" and finer "regressional" sands, colluvia, different types of soils, and frost-weathering phenomena. The chronostratigraphy is primarily based on the major and minor marine cycles of glacial-eustatic origin, interpreted in terms of the deep-sea isotope stratigraphy. Faunal horizons provide general but imprecise guidelines, and there are some artifactual associations and, for the terminal Pleistocene, C14 dates. Only those sites or site complexes with reasonably fixed chronostratigraphies have been selected:

1. Elands Bay, where transgressional sands increase between 20 000 and 11 000 BP in Elands Bay Cave, with complementary data from the surrounding slopes and shore sections and from an interior cave, Diepkloof (Butzer 1979).

2. Melkbosstrand, where several late Middle Pleistocene exposures with fossils and artifacts were studied at Duinefontein on the Escom terrain, continued stratigraphically by an almost complete Upper Pleistocene to Holocene sequence exposed along the beach cliffs just to the north, the Van Riebeeckstrand sector, which was drawn to scale along a 1.5 km stretch (Butzer and WM Michala, unpubl.).

3. Swartklip, in False Bay, where four major profiles, recording the Later Quaternary in detail, were studied between the fossiliferous site and the car park area (Butzer and WM Michala, unpubl.).

4. Stanford, where interior eolianites and related soil developments were delineated for fossiliferous sites on the farms Windheuvel and Linkerhandsgat (Butzer, unpubl.).

These key sites are supplemented by studies of the Elandsfontein dune and soil complex (Butzer 1973b) and several sites among the multi-generational eolianites in Saldanha Bay, specifically Sea Harvest and Hoedjies Punt (Butzer and Michala, unpubl.), which provide floating stratigraphies only.

The stratigraphies themselves are based on (i) macroscopic criteria, such as strata definition, bedding properties, marine and terrestrial shell components, and soil horizonation; (ii) sedimentological criteria, such as Folk (1966) parameters for the fraction coarser than $\phi 4.75$, and their systematic change through long profiles, including upward or downward fining; and (iii) the sedimentological properties of analog samples, chosen in each study area from a range of microenvironment as low- and high-tide beach sands, the crests and swales of various dune types, vleis, and alluvial settings. A discriminant-function program was used to sort out cases representing distinct energy conditions for each area, as based on a total of over 400 samples.

Most of these data are prerequisite to identifying marine transgressive or regressive trends that are primarily of stratigraphic, and only secondarily of environmental interest. In terms of regional climatic implications, for the southwestern Cape, basic concern focuses on the soils, which record a measure of geomorphic stability, colonization by vegetation, and biochemical weathering. Such soils must be interpreted in the context of explicitly xeric, sandy substrates and commonly dynamic geomorphic settings. Location of the shoreline and periodic activation of eolian processes to some extent control whether soil can or cannot form, and whether soils that form will be preserved in the stratigraphic column. But these factors being equal, active pedogenesis indicates less xeric conditions than do active sediment accretion or erosional disconformities.

Four degrees of soil formation can be identified:

1. Calcification, with or without formation of laminar or concretionary "calcrete", represents a pedogenic or diagenic subsoil alteration due in large or main part to intermittent throughflow waters or longer-term seepage. These processes primarily suggest semiarid environments subject to little effective erosion or sedimentation.

2. Humification, without decalcification, produces modest AC-profiles that reflect periods of organic enrichment in microenvironments with a measure of vegetation and geomorphic stability, but little vertical percolation of calcium bicarbonate solutions. Simple humic horizons of this type tend to reflect soil stability and a vegetation mat over sandy substrates in subarid to semiarid environments.

3. Partial decalcification and/or oxidation with rudimentary clay/sesquioxide enrichment are indicated by ACaC- or ABC-profiles. Given longer periods of stability they suggest a semiarid to subhumid environment.

4. Deep and fully decalcified ABC-profiles exhibit varying degrees of iron enrichment in the subsoil, or in the substrate, as a result of throughflow seepage. Since iron mobilization is ineffective with pH values above 5.5, complete prior removal of carbonates is implied. Such soils are anomalous, even in the Pleistocene record of the humid Southern Cape, and reflect a substantially wetter climate.

The relevant information for the southwestern Cape is assembled in Fig. 2. Most prominent in the upper part of the sequence is the Mid-Holocene soil, characterized by humification and decalcification, and increasing in prominence from the drier northwest (Melkbos) to the wetter southeast (Stanford); it suggests subhumid conditions, possibly a little wetter than those prevailing in 1652. The Early Holocene is marked by calcretes, suggesting relatively dry conditions.

Even more than in the case of the Holocene record, surface stability and pedogenesis were the exception rather than the rule during the later Pleistocene. Depending on the degree of geomorphic dynamism, three discrete ACa soils or a single, deeper ABC-profile have a central date of about 21 000 BP. These modest soils suggest increased surface stability and a better groundcover, rather than a genuinely moist climate. The lack of pedogenesis during the preceding span of mid-Upper Pleistocene time is complemented by the mammalian microfaunal evidence from the coastal cave of Die Kelders, which indicates dry and relatively warmer conditions (Avery 1982a).

Several paleosols are found in the time spanned by three high sea levels correlated with deep-sea oxygen isotope stage 5. The most significant falls after the first of these transgressions and, at Elands Bay, predates frost-weathered 'grèzes litées' attributed to substage 5b (Butzer 1979). This cold spell may also be reflected in the abundant, angular spall of the earliest MSA horizons at Die Kelders (Tankard and Schweitzer 1974), in which the microfauna argues for wet and cold conditions (Avery 1982a). Calcretes precede and follow the last of the stage 5 transgressions. However, the most striking, ferruginous paleosols are associated with the fossiliferous and archeological sites at Duinefontein (Klein 1976), which predate all of these transgressions.

The Southern Cape. The Cape South Coast, encompassing the area of the Knysna and Tsitsikama forests, is wetter than the southwestern Cape and lacks a dry summer season. Two important coastal cave sequences that span the Upper Pleistocene and Holocene are available in Nelson Bay Cave (NBC) and Klasies River Mouth (KRM). Adequate dating is available (Fairhall et al. 1976; Singer and Wymer 1982). Surficial deposits and soils have also been studied, providing a potential linkage between cave and external environmental processes (Butzer 1973a, 1978; Butzer and Helgren 1972; Helgren and Butzer 1977).

Chronostratigraphic control for the cave sequences is provided by intercalated beach deposits as well as by fluctuating sand parameters, with additional support from multiple C14 dates. External soil horizons are mainly dated by C14 or with reference to pedogenic or related processes documented within cave sequences. In NBC, iron-rich waters periodically penetrated the cave from acidic soil environments outside, while in KRM periodic decalcification of the calcareous eolianites above the cave resulted in the formation of stalagmites or flowstones inside as a result of abundant, bicarbonate-saturated waters penetrating the fissures of the quartzite cavern. The significance of periodic paleosol horizons is similar in southern and southwestern Cape, because the Holocene forests and woodlands of the former region had, in the past, been periodically invaded by large-scale eolian processes then operating in an essentially treeless environment. But the intensity of weathering recorded by the southern Cape paleosols is more striking, not surprising in view of the more humid nature of the Holocene environment here.

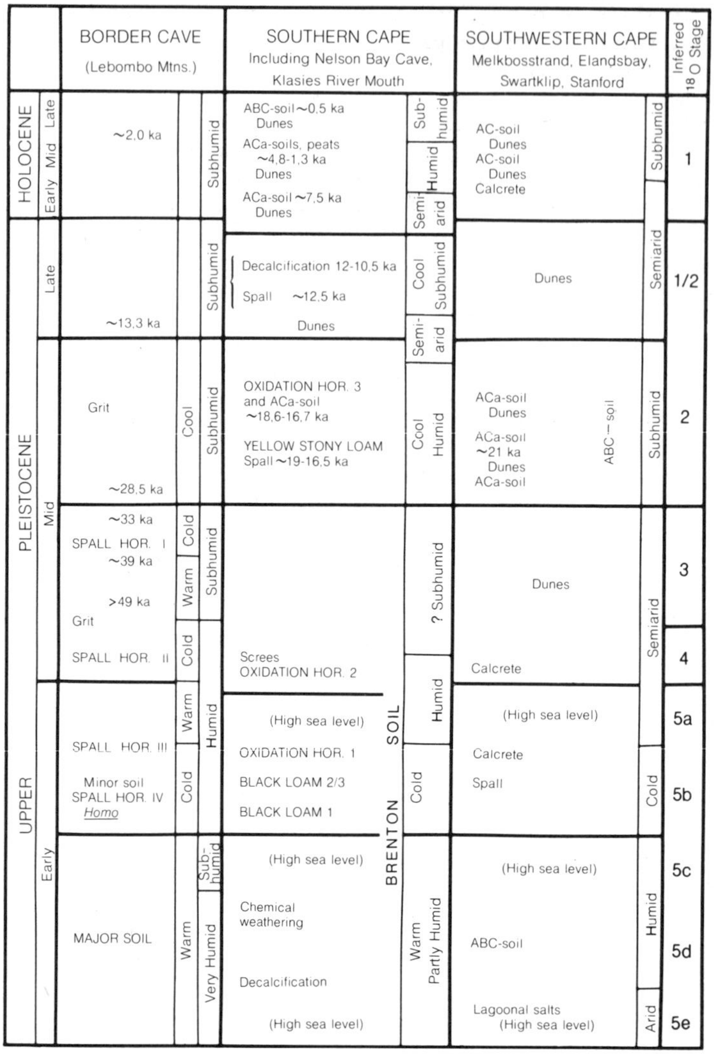

		BORDER CAVE (Lebombo Mtns.)		SOUTHERN CAPE Including Nelson Bay Cave, Klasies River Mouth		SOUTHWESTERN CAPE Melkbosstrand, Elandsbay, Swartklip, Stanford		Inferred 18 O Stage
HOLOCENE	Early Mid Late	~2.0 ka	Subhumid	ABC-soil ~0.5 ka Dunes ACa-soils, peats ~4,8-1,3 ka Dunes ACa-soil ~7,5 ka Dunes	Sub-humid / Humid / Semi-arid	AC-soil Dunes AC-soil Dunes Calcrete	Subhumid	1
PLEISTOCENE	Late	~13,3 ka	Subhumid	Decalcification 12-10,5 ka Spall ~12,5 ka Dunes	Cool Subhumid / Semi-arid	Dunes	Semiarid	1/2
	Mid	Grit ~28,5 ka	Cool Subhumid	OXIDATION HOR. 3 and ACa-soil ~18,6-16,7 ka YELLOW STONY LOAM Spall ~19-16,5 ka	Cool Humid	ACa-soil Dunes ACa-soil ~21 ka Dunes ACa-soil	ABC–soil / Subhumid	2
		~33 ka SPALL HOR. I ~39 ka >49 ka Grit SPALL HOR. II	Cold / Warm / Cold		? Subhumid	Dunes	Semiarid	3
UPPER				Screes OXIDATION HOR. 2	Humid	Calcrete		4
		SPALL HOR. III Minor soil SPALL HOR. IV Homo	Warm Humid / Cold	(High sea level) OXIDATION HOR. 1 BLACK LOAM 2/3 BLACK LOAM 1		(High sea level) Calcrete Spall	Cold	5a 5b
	Early	MAJOR SOIL	Warm Sub-humid / Very Humid	(High sea level) Chemical weathering Decalcification (High sea level)	BRENTON SOIL / Warm Partly Humid	(High sea level) ABC-soil Lagoonal salts (High sea level)	Humid / Arid	5c 5d 5e

Figure 2. Late Pleistocene sequences and paleoclimatic inference for the coastal sector of South Africa. Chronometric dates or date ranges are cited in ka or 1000 years.

The southern Cape record outlined in Fig. 2 indicates a relatively dry Early Holocene, except for modest pedogenesis c. 8000-7000 BP (ACa--profile). Moister conditions are indicated by cave flowstones, ACa soil profiles, and valley-floor peats c. 4850-1300 BP, with stability terminated by Afrikaner land use (grazing and burning after AD 1750). The broad lines of this pattern find confirmation in pollen sequences near Knysna (Martin 1968; Scott 1983; A Scholtz, in prep.) and in the microfauna of Byneskranskop Cave, which suggest slightly warmer and drier conditions about 6500-3500 BP compared with the preceding and subsequent period (Avery 1982a).

A moist spell during the terminal Pleistocene is suggested by extensive decalcification of shell in the Brown Stony Loam unit in NBC, dated c. 12 000-10 500 BP, although transgressional dunes continued to accumulate outside. A period of marked leaching and drip-line erosion is further indicated in the interior cave of Boomplaas, where it is dated 12 000-10 000 BP (HJ Deacon 1979).

From about 19 000-16 500 BP some frost-weathered roof spall accumulated in NBC (Yellow Stony Loam), with simultaneous or subsequent impregnation by hydrated iron compounds, suggesting a perennially moist cave microenvironment, although the microfauna indicates a harsh, presumably cold and dry microclimate (Avery 1982a). Since an organic soil was forming outside of the cave prior to 16 000 BP (on "The Island", Butzer and Helgren, 1972), the rodents probably responded more to the temperature environment than to the relative improvement of moisture conditions, at a time of low evaporation.

Much of the time spanning the three transgressive phases correlated with oxygen isotope stage 5 was marked by pedogenesis, including horizons of anhydrous iron in NBC. The key external paleosol of this approximate age, the Brenton Soil, probably reflects the cumulative impact of humification, decalcification, and iron mobilization during this complex period. It is an alfisol, with a 120 cm B2-horizon of strong brown to reddish yellow, partly ferruginized loamy sand on eolianite substrates.

A complex spell is indicated in NBC (Fig. 2) by Black Loams 1-3, which predate the last stage 5 transgression and are therefore assigned to substage 5b. Black Loam 1 has one major horizon of frost-weathered spall, separated from a second at the Black Loam 2/3 contact by a fine-grained, highly organic horizon. Spall concentrations in the upper part of Black Loam 3 are less angular and more spherical, suggesting a warming trend. Such a double cold interval within 5b is evident in some of the deep sea cores and shows up in European pollen and cave profiles (see Butzer 1981). Frost today is unknown in the area, with lowest annual temperatures averaging +1.4 to 5.3°C. To produce effective freeze-thaw alterations within a deep cavern would require a minimum 10°C reduction of mean winter temperatures, which is compatible with more intermediate, annual temperature depressions indicated by O18 deviations in a dated speleothem of Cango cave and in dated fossil groundwater at Amanzi, namely 5.5.°C judging by latitudinal shifts of precipitation belts (Vogel, in press). These frost interferences from NBC are based on morphometric statistics on quartzite cobbles derived from the cave roof conglomerate, and which are shattered from multiple, oblique angles independent of incipient weaknesses related to jointing properties. The fact that some levels of fallen roof cobbles are not fractured, and that similar cobbles on the present floors of the cave and nearby overhangs are never shattered, precludes other agents (Butzer 1973a).

Part of the slope denudation and frost-weathered screes on top of Brenton Soils in the Tsitsikama Forest may date from this interval, but at KRM major slope screes were activated after the last stage 5 transgression.

Border Cave. Border Cave lies just inside the KwaZulu frontier, overlooking the Swaziland Lowveld, near 600 m elevation. The climate is humid semitropical, with wet summers. The archeo-sedimentary sequence is dated by 28 C14 assays which place the top 30 % of the sediment column in the 0 to 50 000 BP time range. By adjusting sedimentation rates for fines and coarser deposits, allowing for sedimentary breaks, and considering differential compaction, the C14 dated part of the column suggests an extrapolated base for the MSA at 195 000 BP. The sedimentary column has been published in detail by Butzer, Beaumont and Vogel (1978).

Three horizons of the 'éboulis secs', lacking fine matrices, provide good indicators of frost-weathering in the southeastern interior of the cave where salt efflorescences are not present on the roof. Another spall horizon (IV) is limited to the lower deposits in the front of the cave; although this debris is angular, it is embedded in fine sediment, and salt efflorescences are noticeable nearby. Other climatic indicators are provided by soil horizons.

Publication of the micromammalian fauna (Avery 1982b) amplifies interpretation of Border Cave and further suggests an alternative climatostratigraphic interpretation for the lower part of the column. The microfauna of levels 1b to 4, dated from younger than 13 300 to well beyond 49 000 BP and including Spall Horizon 1, corresponded to an open grassland; this suggests oxygen isotope stages 2 and 3. For levels 5 to 10, including Spall Horizons II, III and IV, a dense and damp grass cover with more extensive forest is indicated, presumably spanning isotope stages 4 to 5b. Here there is little correspondence between the more tenuous rodent interpretations for "cold" versus "mild" on the one hand, and the frost spall horizons on the other. Spall Horizon II is assigned to stage 4, while III and IV are both attributed to substage 5b, in analogy to the spall complex in NBC; the minor chemical weathering event that led to gypsum mobilization is placed between spall horizons III and IV, or penecontemporaneous with the latter. The revision of the Border Cave column, incorporating the rodent inferences for wetter and drier vegetation, is given in Fig. 2.

The rodents of level 11b unequivocally indicate a dry savanna woodland as well as a warm, dry climate, i.e., characteristic non-glacial conditions. The major paleosol older than level 11 should therefore fall between substages 5c and 5e (Fig. 2). Alteration produced an acidic pH (5.0 compared with a modal value of 7.0 for levels 1-11) and a doubling of the clay fraction, despite relatively light occupation in levels 12-15. The requisite moisture was probably obtained from rain splash, capillary seepage, and condensation moisture, and is inconceivable without warm waters offshore.

Discussion. Comparing the three columns of Figure 1, it is apparent that the southern and southwestern Cape records are essentially similar, with a moist Mid-Holocene, a dry terminal Upper Pleistocene, and a mainly moist (and cold) glacial maximum c. 25 000-16 000 BP. The period equivalent to isotope stage 3 was dry, but much of the previous, warmer "interglacial complex" was moist. The cold phenomena related to substage 5b were remarkably intense. The Border Cave record lacks notable sedimentary or pedogenic events during the last 30 000 years, but includes four major frost spall horizons beyond 33 000 BP, the oldest of which is correlated with substage 5b. Evidently the freeze-thaw parameters and moisture conditions promoting major frost weathering were different in the Cape Province and Natal. But in both areas the major soils, as indicative of optimal vegetative cover and stable slopes, relate to the periods of non-glacial climate. Glacial-age environments were generally drier, geomorphologically active, and to a good extent characterized by open vegetation.

The patterns reported here are not restricted to a few sites. Several caves in the Cape Ranges exhibit similar sequences. In Boomplaas Cave (795 m elevation) several horizons of coarse, subangular to angular spall have a central date of about 22 000 BP; another spall horizon has a date of 42 000 BP, probably a minimum value; a major complex of (anhydrous) ferricrete spall horizons is well beyond the range of C14 but includes the same Howieson's Poort industry associated with the cold horizons at NBS (Black Loam), KRM (I-22), and Border Cave (Spall Horizon III) (J Deacon 1979). In Kangkara Cave (450 m) a much shorter sequence reveals an angular spall horizon dated 12 500 and 12 330 BP, with the subjacent deposits completely decalcified (HJ Deacon and Brooker 1976; Butzer, unpubl.). In Paardeberg Cave (130 m) there are two angular spall horizons older than a minimum date of 44 000 BP (Butzer, unpubl.). In Buffelskloof Rock Shelter (395 m) a major ferricrete horizon separates the Early LSA and MSA and is somewhat older than 23 000 BP (Opperman 1978). In Die Kelders sesqui-oxide staining is common in the sterile, terminal Pleistocene transgressive sands (Schweitzer and Tankard 1974).

The paleobiological evidence supports these interpretations from the sediments and soils. For the Cape region the mammalian faunas from Melkhoutboom, NBC, Boomplaas, Buffelskloof, Byneskranskop and Elands Bay caves all show a high proportion of large gregarious herbivores, particularly alcelaphine antelopes and equids, during the Later Pleistocene, indicating that grass was substantially more important in the vegetation than it was during the Mid- and Late Holocene (Klein 1980a}. The transition to a browsing, bush or woodland fauna began about 12 000 BP and was only completed during the Early Holocene. Simultaneously, the size of small browsing mammals increased, while that of small grazers decreased, again indicative of increasing moisture across the Pleistocene-Holocene transition (Klein 1983, and in prep.). This picture is supported by both the rodent fauna and pollen from Boomplaas, suggesting a cold and dry microclimate as well as an absence of trees c. 30 000-15 000 BP, whereas the subsequent trend to more mesic Holocene conditions was only completed 9000-5000 BP (Avery 1982a; HJ Deacon 1979). The pollen record of a Podocarpus woodland around the Cape Flats, once believed to be little more than 40 000 years old (on the basis of a "barely infinite" C14 age)(Schalke 1973), can be no younger than mid-stage 5 on the basis of the soils stratigraphy (see also discussion by Scott 1983). Periods of relatively low sea level during the early Upper Pleistocene at several Cape sites (Swartklip, Sea Harvest, KRM) coincided with higher proportions of larger grazing ungulates than did times of higher sea level (KRM) (Klein 1980a).

Border Cave lies in a more tropical climate where summers are wet and the faunal thresholds are correspondingly different. Interestingly stages 2-3 corresponded to an open grassland with little woodland whereas 4, 5a and 5b coincided with dense grass cover and bush or forest expansion; neither pattern resembles the modern vegetation (Avery 1982b).

The Late Quaternary geomorphologic evolution of the Natal Coastal Plain (Maud 1968) suggests notable similarities with that of the southern Cape, specifically a major ABC-soil postdating the major transgression of stage 5 and several subsequent eolianites or dunes. In Swaziland, Price-Williams, Watson and Goudie (1982) have identified extensive footslope colluvia below granitic outcrops along the upper margins of the Lowveld. Formation implies an ineffective groundcover and a prior mantle of weathered soil or sediment upslope -- this is conspicuously absent today. The mass of these deposits is atypical in southern Africa, where colluvial deposits are extensive but thin and invariably interdigitated with alluvial facies; these anomalous Swazi examples must be attributed to the ready supply of sandy to gritty debris provided by the regional bedrock. The younger, Mphunga colluvial complex rests on the remnants of a striking red paleosol recalling that of Natal, and includes MSA artifacts. C14 dates of less than 31 000 BP (Price-Williams et al. 1982) refer to post-depositional carbonates, in part related to polygonal fissures with calcium enrichment, in part to concretions and even typical watertable nodules, reflecting lateral seepage or throughflow. The main Mphunga colluvium is therefore substantially older. Its significance lies in its implications for a harsher environment with reduced vegetation cover, much the same as that verified in Border Cave. Holocene deposits in the Mkhondvo Valley include convex floodplain fills indicative of a distinctly wetter and better vegetated watershed.

In overview, the non-glacial climates of the coastal sector of South Africa generally tended to favour a more complete vegetation cover, geomorphic stability, and soil development. By comparison, the glacial-age climates were drier, and less luxuriant vegetation allowed considerable geomorphic dynamism. In detail, substage 5c or 5d coincided with the most mesic environment of the Late Quaternary, as recorded in the striking paleosol generated. Substage 5b was equally conspicuous, as the most dramatic "cold" horizon. The modest soil horizons developed during stage 2 indicate slightly more mesic conditions in the Cape coastal regions; comparable weak weathering features appear to be indicated in the upper part of the Mphunga colluvium at Mashila, Swaziland.

THE SUBHUMID INTERIOR

The eastern half of the South African interior, including most of the Free State and Transvaal, is subhumid by the Thornthwaite classification (Fig. 1). Annual rainfall averages between 500 and 1000 mm, with 75-85 % received during the summer half-year. This environment is primarily grassy, with increasing forest cover on the humid, montane fringes. Records of Pleistocene and Holocene climatic change are provided by valley floors, springs and caves, but they are less dramatic and often more equivocal than those of the semiarid interior or in the coastal sector. To date, fewer such environmental records have been studied or published, and the picture they project has only barely begun to come into focus.

Rose Cottage Cave. Rose Cottage (RCC) is a large sandstone overhang at 1660 m near Ladybrand, where sediment has accumulated behind a great, fallen roof block. Upland drainage waters intermittently emerged from a major joint intersection at the back of the cave, to drain out through a sediment-choked sluice along the eastern base of the block. The archeological sequence ranges from Howieson's Poort (MSA) to Pottery Wilton, and there are 11 C14 dates (Beaumont and Vogel 1972a, b; JC Vogel, unpubl.). The stratigraphy and sedimentology were studied by the writer in collaboration with L Scott (Butzer 1984).

The sediments of RCC consist of fine and coarse roof debris as well as intrusive wash.

(a) Material is detached from the sandstone roof, in part as a result of grain-by-grain weathering, in part through roof spalling. Gradual rock disintegration now produces sand-size particles with some subrounded, grit to gravel-sized spall. Angular roof rubble suggests more rapid, freeze-thaw weathering which did not, however, affect the sand fraction in a systematic way. Morphometric statistics allow a replicable distinction between angular spall, due in part to frost-weathering, and subangular to subrounded spall that relates to other weathering processes.

(b) Material washed in along the drip-line or from occasional spring activity at the back tends to be markedly laminated, with distinct dips down to the sluice. The sand fraction of the laminated spring deposits is relatively coarse and poorly sorted; the coarser sands are over 50 % polished quartz and may include carnelian, both derived from the upland above the site. Such laminated beds indicate heavy and protracted rains, and at least a seasonally high water table. Some of the strata are rich in clays and silts of non-archeological origin; these suggest a lower-energy but more persistent spring influx, presumably as a result of reduced precipitation seasonality.

(c) The bulk of the sand fraction was, however, derived from incremental, granular disintegration of the sandstone roof, and much of the statistical "noise" apparent in the sediment parameters is best attributed to cultural intervention in the cave.

The RCC record is summarized in Fig. 3. Three intervals with increased spring influx and two with sufficient moisture for decalcification within the cave can be identified during the last 30 000 years. At least two of these spells of wetter climate are recorded by an average of over a metre of sediment, building up across the alluvial fan below the sandstone escarpment, fed by drainage from the upland; the penultimate fan alluvium is C14 dated to before 2310 BP by its superposed organic paleosol.

Roof spall horizons with distinctive, frost-weathered components extend from before 30 000 to about 22 000 BP (late stage 3 to mid-stage 2), and are again identified beyond non-finite C14 dates of >50 000 BP (probably stage 4), in a late MSA level, and near the bottom of the sequence (possibly substage 5b), associated with the Howieson's Poort level.

The record of spring influx in the lower half of the sedimentary column is impressive, and generally includes considerable clay in levels assigned to stages 4 and 5; beds with prominent laminations here indicate inflow of variable intensity, presumably as a result of less effective groundcover above the cave and stronger seasonality of rain-fall. Eventually, during early stage 3, the clay component drops out and laminated sands argue for strong spring influx.

Bushman Rock and Heuningneskrans Shelters. Bushman Rock Shelter (BRS) and Heuningneskrans (HNK) are dolomite caves near Ohrigstad, near the eastern Transvaal escarpment, at 900-930 m elevation. There are 6 C14 dates for HNK and 15 for BRS (Beaumont and Vogel 1972a, b; Plug 1981). The two stratigraphic columns, studied by the writer, complement each other and record most of the Later Quaternary (Butzer 1984). Speleothem data from nearby Echo Cave (Brook 1982) supplement the information base.

Extrapolation of the dates from the HNK profile indicates sedimentation began by 30 000 BP. Disposition of the beds links them to a large, intermittent karst spring at the back of the cave. Eventually, as the bedding shows, the deposits coalesced with alluvium from the Ohrigstad River that aggraded a +5 m floodplain terrace c. 19 000-11 500 BP (interpolated from C14 dated cave profile). Moderate energy

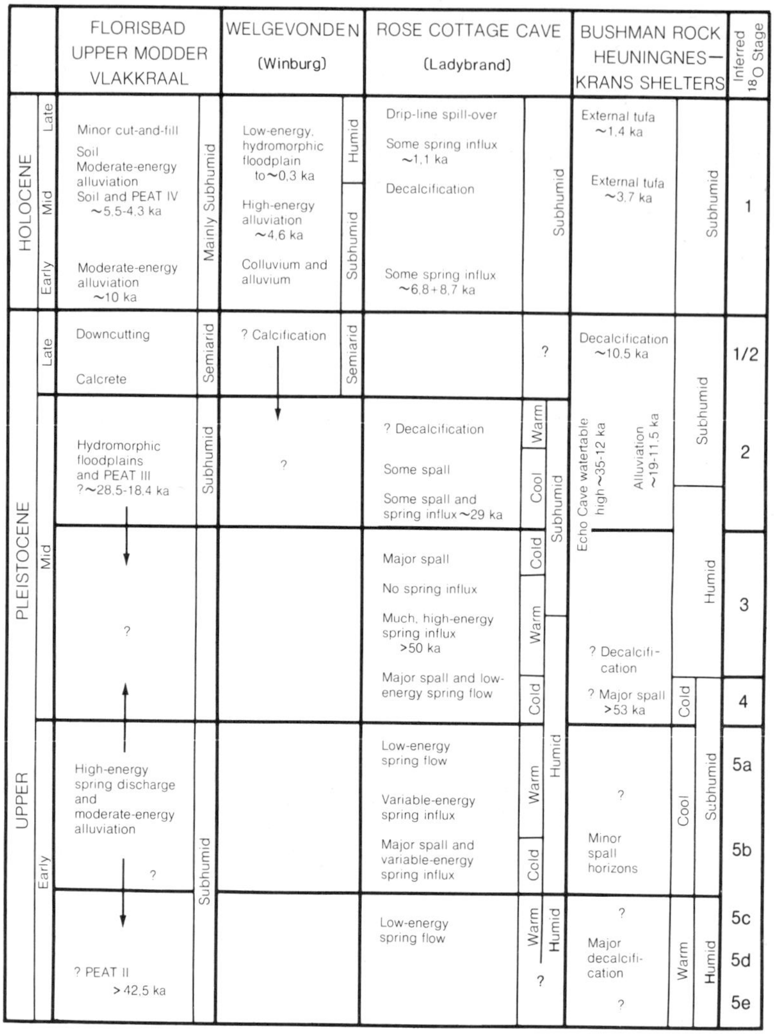

		FLORISBAD UPPER MODDER VLAKKRAAL		WELGEVONDEN (Winburg)		ROSE COTTAGE CAVE (Ladybrand)		BUSHMAN ROCK HEUNINGNES-KRANS SHELTERS		Inferred ^{18}O Stage
HOLOCENE	Late	Minor cut-and-fill Soil Moderate-energy alluviation Soil and PEAT IV ~5.5-4.3 ka	Mainly Subhumid	Low-energy, hydromorphic floodplain to ~0.3 ka	Humid	Drip-line spill-over Some spring influx ~1.1 ka Decalcification	Subhumid	External tufa ~1.4 ka External tufa ~3.7 ka	Subhumid	1
	Mid			High-energy alluviation ~4.6 ka	Subhumid					
	Early	Moderate-energy alluviation ~10 ka		Colluvium and alluvium		Some spring influx ~6.8+8.7 ka				
PLEISTOCENE	Late	Downcutting Calcrete	Semiarid	? Calcification	Semiarid		?	Decalcification ~10.5 ka	Subhumid	1/2
		Hydromorphic floodplains and PEAT III ?~28.5-18.4 ka	Subhumid	?		? Decalcification Some spall Some spall and spring influx ~29 ka	Warm / Cool / Subhumid	Echo Cave watertable high ~35-12 ka	Alluviation ~19-11.5 ka	2
	Mid	?				Major spall No spring influx Much, high-energy spring influx >50 ka Major spall and low-energy spring flow	Cold / Warm / Cold	? Decalcification ? Major spall >53 ka	Humid / Cold	3
										4
	UPPER Early	High-energy spring discharge and moderate-energy alluviation ?	Subhumid			Low-energy spring flow Variable-energy spring influx Major spall and variable-energy spring influx	Warm Humid / Cold	? Minor spall horizons	Cool Subhumid	5a
										5b
		? PEAT II >42.5 ka				Low-energy spring flow	Warm Humid	? Major decalcification ?	Warm Humid	5c
										5d
							?			5e

Figure 3. Late Pleistocene sequences and paleoclimatic inferences for the subhumid interior of South Africa.

alluviation implies strong precipitation seasonality. Ultimately the river began to downcut again, the water table dropped 2.5 m, and soil decalcification proceeded (C14 date of 11 220 BP on related pedocarbonates). Later HNK deposits form a mound just inside the drip-line. Roof spall is essentially absent throughout, and the sediment statistics suggest that their variability was primarily related to the intensity of LSA occupation.

The BRS profile represents the infilling of a very dry cave; external deposits are absent, and all non-cultural sediment is derived from breakdown of the shaley dolomite bedrock. Because of the small range of effective, non-cultural variables, the only significant variability is provided by interbedded roof spall and weathering horizons. There is a problem in dating the column between the uppermost spall horizon

(Level 19), and Level 14 with C14 dates of c. 12 000 BP. Charcoal assays suggest a limited time interval of about 12 000-13 000 years for this zone, although bone apatite dates extend this to beyond 31 900 BP. There is far too much variability among the sediment parameters to compress these 26 cm into one millennium, and dispersal of charcoal in rodent burrows may explain the discrepancy. The artifactual assemblage of this zone appears to be a mixture of MSA and LSA, an interpretation supported by the microstratigraphy in the Eloff excavations, which clearly shows that the earliest LSA occupations disturbed the underlying spall horizon (with MSA).

The roof spall horizons of the lower BRS column consist of classic 'plaquettes de gel' or frost-weathered, flattish, and (initially) angular rock. These beds thicken against the back wall of the cave and dip slightly down into the deposits, indicating mechanical weathering of both the ceiling and walls. The base of the upper spall horizon (Level 18-22) has a minimum C14 age of 53 000 BP, and the top 20 cm were partly decalcified by subsequent weathering. The water table of Echo Cave was 20-30 cm higher than today between roughly 35 000 and 12 000 BP (Brook 1982), but since the sediments accumulating in BRS after about 32 000 BP show no evidence of damp conditions within that cave, decalcification of the spall probably is somewhat older. Levels 22-30 are interdigitated with several, minor spall units and were preceded by a major weathering episode during which 1 m of earlier sediment were extensively decalcified. In Fig. 3 the major spall horizon has been correlated with stage 4, the minor spall interbeds with 5b, and the wet micro-environment implied by decalcification is assigned to the early part of stage 5. The preceding spall horizon (Levels 31-37) and the essentially rubble-free units (Levels 38-101) down to bedrock continue to record MSA occupations of presumed late Middle Pleistocene age.

Although BRS and HNK have not been sensitive to all but the most dramatic environmental changes, speleothem growth in Wolkberg and other east-central Transvaal caves was not prominent from before 50 000 to 30 000 BP, at around 21 000-17 000 BP, and again after 10 400 BP. O18 isotopic temperatures suggest it was 8-9°C colder c. 30 000 BP, 7.5°C colder c. 18 000 BP (Talma, Vogel and Partridge 1974; Vogel and Visser 1981). In combination with the Echo Cave record, these differences underscore the importance of geomorphic thresholds in determining sedimentation patterns.

Welgevonden (Winburg District). The Kalkbankspruit drainage of the central Free State is a tributary of the Vet River. Due to the absence of eolian veneers (Hutton Sands) the regional development of alluvial facies and soils resembles that of the eastern Free State in general, and of the Cornelia Formation holotype in particular. The stepped terraces of the Riverton Formation in the lower Vaal drainage are replaced by long, smooth concavo-convex slopes under which complex alluvial and colluvial deposits interfinger. Exposures on the farm Welgevonden are studied by L Scott and the writer; three C14 dates are available. The later units of the Cornelia Formation, which suggest many parallels, unfortunately are undated (see Butzer, Clark and Cooke 1974).

The facies represented at Welgevonden include (i) dark, organic flood silts/clays, (ii) channel beds, mainly interbedded loams and gravels, (iii) colluvial loams and clay loams with reworked calcareous concretions, and (iv) organic slope soils, mainly clay loams, at least in part of colluvial origin. These facies indicate a measure of slope soil mobility, linked with relatively low-energy stream transport and deposition in well-vegetated riparian meso-environments. This contrasts with recent donga cutting and transport of sandy to gravelly bedloads in clearly-defined channels, presumably as a consequence of over-intensive cattle grazing since the mid-19th century.

Mid-Holocene aggradation began c. 5000 BP, shifting from moderate to low-energy accretion by perhaps 4000 BP, at which time the floodplain essentially stabilized, with hydromorphic conditions prevailing until the 19th century (Fig. 3). This draws attention to the important fact that several large, undissected hydromorphic flood plains can still be observed in the easternmost Free State and the central Transvaal today.

The Early Holocene appears to be represented by reddish, colluvial loams or clay loams, with reworked concretions, but undated. This earlier infilling extends well up the slopes; it was separated from the Mid-Holocene alluvium by some 3 m of valley downcutting.

No identifiable Pleistocene deposits are exposed at Welgevonden, but the reworked concretions suggest that at least part of the Later Pleistocene was drier than Holocene conditions were. A more detailed Pleistocene alluvial record, with pollen, is currently being studied by Scott and the writer near Clarens, in the upper Caledon drainage.

Florisbad and Vlakkraal. The fossil and artifactual site of Florisbad is a 7 m mound of spring sands with intermediate or primary derivation from a Karoo System aquifer, moving up along the local dolerite intrusion. The latter is verified by the heavy mineral composition of the spring sands. Major phases of accelerated discharge reflect the regional aquifer, although episodes of flush flow and "eye" eruption may also relate to earthquakes. The spring record is substantially amplified by the sediments and soils in the small donga immediately downstream, as well as in the nearby Modder River and a tributary originating at the spring site of Vlakkraal. The close concordance of hydrological and geomorphic behaviour in all the local subsystems demonstrates that the Florisbad spring history is primarily related to regional environmental history. The present synopsis does not include the column below Peat II which appears to be late Middle Pleistocene; this segment includes six extinct mammalian species or genera (Klein 1983).

The sediment analyses (Butzer 1984) show that the basic Florisbad sequence consists of several cycles, each beginning with sands that rapidly become coarser and that, after a period of strong but fluctuating flow, become finer again until accretion is minimal, when organic matter begins to build up in the relatively stable spring eye. Laterally, such subaquatic or semiaquatic "peats" grade into humic, terrestrial soils. These cycles consequently reflect spring energy and, indirectly, provide a measure of discharge from the deep-seated Karoo aquifer. The peat horizons should consequently record the most luxuriant vegetation around the spring eyes, despite reduced water flow. It is also probable that sporadic, high-intensity rains of considerable duration would recharge the aquifer more effectively than more regular, lighter rainfalls that might favour a more complete vegetation mat in the surrounding landscape. Some of these interpretations are directly verified. So, for example, aquatic mammals such as water mongoose, clawless otter, and lechwe are found in Peat I (Klein 1980); similarly aquatic snails are concentrated in the clayey terminal beds of the donga sequence below Florisbad, while calcretes close off two of the donga aggradations, indicating gradual failure of the oligohaline spring discharge. Interpretation of the alluvial deposits suggests close similarities with the Riverton Formation in the Lower Vaal drainage (see below).

During the Holocene there were two cycles of greater spring discharge, paralleled by moderate-energy alluviation in the Modder River, and terminated by development of moderately prominent flood plain soils (Fig. 3). The first of these depositional phases lies between about 10 000 and 5500 BP, the second is not precisely dated.

The mid-Upper Pleistocene is recorded by some low-energy alluviation but more characteristically by hydromorphic flood plains and peat accumulation (Vlakkraal; Peat III at Florisbad), for which dates of 28 500 and 18 400 BP are probably applicable (Ox. Is. stage 2). The pollen associated with Peat III argues for a dense grass sward on the uplands (Van Zinderen Bakker 1956) but gives no indication of the luxuriant fringing vegetation implied by the soils. Eventually, calcification superseded these wet environments, indicating a drying trend.

The early Upper Pleistocene record may include the subaquatic Peat II, which is well beyond the range of C14, and it certainly encompasses the phenomenally high-energy spring discharge that followed Peat II and probably parallels the major, moderate-energy alluviation of the +12 m Modder terrace. Strong but periodic rains are suggested. The Florisbad pollen suggests gradual replacement of a semiarid, open vegetation by dense grass on the upland or sedge stands near water (Van Zinderen Bakker 1957), although the sediments and a fauna with lechwe (Klein 1980) would presume somewhat wetter micro-habitats.

Discussion. The Holocene record of the various flood plains and springs, and of Rose Cottage Cave, is sufficiently clear and speaks for several climatic shifts; but the modal environment was at least as moist as that of the 19th century, prior to Afrikaner settlement. The same settings suggest that the terminal Pleistocene was drier, a trend not paralleled in HNK. The full glacial climate of late stage 3 and early stage 2 was cool to cold, and relatively moist, possibly in response to reduced evaporation. The interglacial to glacial transition, probably equivalent to stages 5b through 4, was often cold and mainly wet, while parts of early stage 5 were strikingly moist.

This generalized summary obscures the fact that dating in the interior is far more insecure than in the coastal sector. There may, therefore, be significant inconsistencies between, for example, the eastern Transvaal and Free State that remain unidentified. The greatest confidence can be attached to the RCC record, and this most closely resembles that of Border Cave. Other potential support may come from excavated caves in the subalpine zone of southeastern Lesotho (Carter and Vogel 1974), once these have been more fully published. The impression, and it is no more, is that the major spall horizons in these caves as well as the cold-climate phenomena variously described for the high Drakensberg (Butzer 1973c; Fitzpatrick 1978; Marker and Dyer 1979) pertain to either stage 5b or 4, rather than 2.

Comparisons may be made with pollen sequences in the central Transvaal. At Wonderkrater, near Naboomspruit, Scott (1982a) identifies a mesic woodland with expanded montane forests beyond 34 000 BP (pollen zone W1), possibly in stage 4 or earlier. Open grassland with ericaceous elements indicates a relatively moist climate from before 25 000 to c. 11 000 BP, as much as 5-6 °C colder than at present. From 11 000-7300 BP climate was drier than today and becoming warmer, then warmer than today and a little wetter 7300-4000 BP. A cooling trend is apparent 4000-2000 BP, possibly corroborated in the Soutpansberg of the northern Transvaal (Scott 1982b, and this volume). In the southern Transvaal at Rietvlei, south of Pretoria, a grassland vegetation from before 10 300 BP was temporarily replaced by broadleaf tree-savanna (bushveld) c. 7000-3000 BP (Scott and Vogel 1983), while in the nearby Moreletta Valley, woodland pollen were more abundant from before 5300 to 400 BP than they were before or after (Scott, this volume). Based on extensive sampling of modern pollen rains, Scott's (1983) palynological inferences are far more sensitive to the complex interdependency of temperature, rainfall averages, and seasonal concentration or variability of precipitation than are the geomorphic and soils indicators. But neither record is sufficiently detailed and accurately dated to allow specific one-to-one comparisons between vegetational and geomorphic trends for the last 25 millennia or so.

The available information for the eastern Free State and for the Transvaal represents a first approximation only. It is sufficient to demonstrate continuing changes in the environment, between a cooler Late Pleistocene and a warmer Holocene, each in part moister, in part drier. This record was particularly complex during the last 12 000 years or so, with biological or geomorphic thresholds apparently crossed every few millennia.

THE SEMIARID INTERIOR

The semiarid interior of South Africa encompasses the western Free State, the northern Cape, and most of the Karoo, areas averaging 200-500 mm of precipitation. This zone lies athwart the transition from over 75 % winter rainfall in the southwest to over 80 % summer rainfall in the northeast. The writer has studied a series of sites or settings along a transect from RCC and Florisbad in the east, past Kimberley, to the Kuruman Hills and Kathu (Sishen) in the west. These sites are discussed here. Some of the evidence from Botswana and Namibia is reviewed by Butzer (1984), but no relevant sequences have yet been studied in the Karoo.

Alexandersfontein, Voigtspost, Deelpan, Klipfontein. The Alexandersfontein Basin, east of Kimberley, represents one of the largest erosional depressions of the western Free State. Three sequences have been intensively studied here in terms of geoarcheology; a perennial spring seep near the eastern end of plat Uitzigt 124 (US), a defunct spring near the wind-driven pumps of Benfontein 442 (BT), and a fossil shore zone of 4.3 ha area (28°52'21"S, 24°27'58"E) southeast of the modern pan, on Mauritsfontein 26 (MS). In combination with transects, detailed mapping of the basin, and 33 C14 dates, these key sites document spring and stream activity, lake fluctuations, eolian processes, and soil formation in the basin (Butzer 1984):

(a) The BT sequence represents a series of spring tufas, clays and soils resting against and over an old wave-cut nip and bench at + 19 m.

(b) The US sequence is developed in a valley head that allows linkage of spring and slope processes with alluvial terraces and, indirectly, lake levels.

(c) The MS sequence documents the range of vertical, longitudinal and transverse facies variation spanning the Later Pleistocene and Holocene. These facies include lake marls and shore sands, eolian and colluvial cover sediments, spring beds or calcification by seepage waters, multiple soil horizons, as well as transitional sediments of various types.

The Alexandersfontein record is complemented by those from Voigtspost, Deelpan, and Klipfontein. The Liebenbergspan, on the farm Voigtspost 373 (VP) near Petrusburg, provides a sequence of shallow but informatively interdigitated Holocene lacustrine, colluvial, and eolian deposits, with several soil horizons (Horowitz, Scott and Vogel 1978; Butzer 1984). The margins of the Deelpan (DP), near Brug, expose a longer and more massive suite of eolian, colluvial, lacustrine, and spring sediments, with several soils; actual lake levels are a little less clearly demarcated (Butzer 1984; L Scott, unpubl.).

Bushman's Fountain on Klipfontein, near the Vaal River, is an artificially enlarged pond, that is fed by two spring eyes, and is now drying out after a succession of dry years. The location is a major centre of prehistoric rock engravings. Two defunct spring heads are visible in the bedrock, presumably active at times of higher water table. The natural pond was initially dammed behind a Pleistocene alluvial fan, projecting from an adjacent stream. Two lacustrine horizons are present in the bottom of the pond; they have approximate C14 dates on carbonates, as well as pollen (Butzer, Fock, Scott and Stuckenrath 1979).

These several pan and spring sites of the western Free State and Griqualand West provide a detailed and reasonably coherent sequence of geomorphic and pedogenic indicators for the last 20 millennia or so (Fig. 4). Three principal moist intervals can be singled out:

(i) About 4500-1300 BP springs were more active, shallow but mainly perennial lakes occupied the pans (Alex to +2 m, VP +1-2 m, DP +2.5 m), and black organic soils formed in wet microenvironments. The pollen data are inconsistent; some spectra indicate more grassy upland vegetation, while others record a vegetation similar to that of today (Scott, unpubl.).

(ii) An earlier period of active spring flow began about 11 500 BP, with development of ABCa soil profiles on well-drained sites; by 7700 BP lakes were relatively deep (Alex as much as +7 m, VP +2 m, DP +3.5 m), until after 6300 BP. The pollen spectra suggest expansion of bush (Klipfontein) or karooid vegetation, and possibly thornveld (Voightspos) (Scott 1983), which matches the enigmatic record of active springs and periodically deep lakes, both rich in carbonates but lacking related organic beds. A somewhat different constellation of rainfall amount, seasonality, and evaporation is implied by this Early Holocene wet phase than for that after 4500 BP.

(iii) From before 16 000 (and probably before 21 000) to 13 600 BP the Alex lake formed shorelines at +12 to +16 m, while springs were active and dongas locally developed hydromorphic flood plains. Rare pollens in BT include Myrica (L Scott, unpubl.). Even with a 6°C reduction of mean temperatures, a paleohydrological budget based on sound analog data shows that such a lake would require 670-860 mm rainfall over the basin, compared with 400 mm today (Butzer, Fock, Stuckenrath and Zilch 1973).

These major moist intervals were briefly interrupted as well as separated by drier spells (Fig. 4). During one dry interlude c. 14 500 BP, spring activity temporarily halted and Lake Alexandersfontein

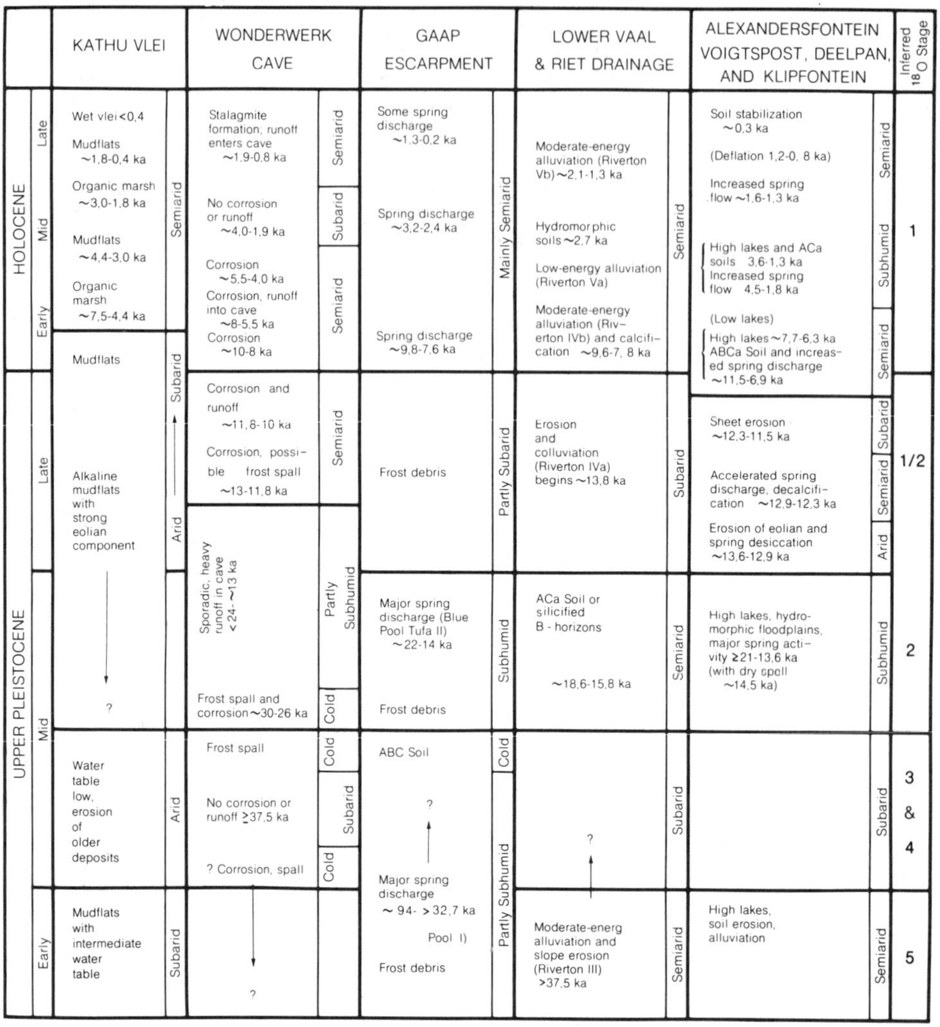

		KATHU VLEI		WONDERWERK CAVE		GAAP ESCARPMENT		LOWER VAAL & RIET DRAINAGE		ALEXANDERSFONTEIN VOIGTSPOST, DEELPAN, AND KLIPFONTEIN		Inferred ^{18}O Stage
HOLOCENE	Late	Wet vlei <0,4 Mudflats ~1,8-0,4 ka Organic marsh ~3,0-1,8 ka Mudflats ~4,4-3,0 ka Organic marsh ~7,5-4,4 ka	Semiarid	Stalagmite formation, runoff enters cave ~1,9-0,8 ka	Semiarid	Some spring discharge ~1,3-0,2 ka	Mainly Semiarid	Moderate-energy alluviation (Riverton Vb) ~2,1-1,3 ka	Semiarid	Soil stabilization ~0,3 ka (Deflation 1,2-0, 8 ka) Increased spring flow ~1,6-1,3 ka	Semiarid	1
	Mid			No corrosion or runoff ~4,0-1,9 ka	Subarid	Spring discharge ~3,2-2,4 ka		Hydromorphic soils ~2,7 ka		High lakes and ACa soils 3,6-1,3 ka Increased spring flow 4,5-1,8 ka	Subhumid	
	Early			Corrosion ~5,5-4,0 ka Corrosion, runoff into cave ~8-5,5 ka Corrosion ~10-8 ka	Semiarid	Spring discharge ~9,8-7,6 ka		Low-energy alluviation (Riverton Va) Moderate-energy alluviation (Riverton IVb) and calcification ~9,6-7, 8 ka		(Low lakes) High lakes ~7,7-6,3 ka ABCa Soil and increased spring discharge ~11,5-6,9 ka	Semiarid	
UPPER PLEISTOCENE	Late	Mudflats Alkaline mudflats with strong eolian component	Subarid	Corrosion and runoff ~11,8-10 ka Corrosion, possible frost spall ~13-11,8 ka	Semiarid	Frost debris	Partly Subarid	Erosion and colluviation (Riverton IVa) begins ~13,8 ka	Subarid	Sheet erosion ~12,3-11,5 ka Accelerated spring discharge, decalcification ~12,9-12,3 ka Erosion of eolian and spring desiccation ~13,6-12,9 ka	Subarid / Semiarid / Arid	1/2
	Mid	?		Sporadic, heavy runoff in cave ~24-~13 ka Frost spall and corrosion ~30-26 ka	Partly Subhumid	Major spring discharge (Blue Pool Tufa II) ~22-14 ka Frost debris	Subhumid	ACa Soil or silicified B-horizons ~18,6-15,8 ka	Semiarid	High lakes, hydromorphic floodplains, major spring activity ≥21-13,6 ka (with dry spell ~14,5 ka)	Subhumid	2
		Water table low, erosion of older deposits	Arid	Frost spall No corrosion or runoff ≥37,5 ka ? Corrosion, spall	Cold / Subarid / Cold	ABC Soil ? Major spring discharge ~94->32,7 ka Pool I)	Cold / Partly Subhumid	?	Subarid		Subarid	3 & 4
	Early	Mudflats with intermediate water table	Subarid	?		Frost debris		Moderate-energy alluviation and slope erosion (Riverton III) >37,5 ka	Semiarid	High lakes, soil erosion, alluviation	Semiarid	5

Figure 4. Late Pleistocene sequennces and paleoclimatic inferences for the semiarid interior of South Africa.

subsequently fluctuated beneath a level 4 m lower. About 13 600 BP the springs dried out completely and colluvially-reworked eolian sands penetrated deep into the dehydrated crack networks that developed in the old spring beds. Pollen are over 90 % Aizoaceae and compare with those of modern karoo-type vegetation (L Scott, unpubl.) Spring activity was temporarily renewed 12 900-12 300 BP, after which sheet erosion attacked the desiccated beds until about 11 500 BP. This same pattern of wet-dry oscillations is replicated in the pollen profile at Aliwal North (Coetzee 1967), except that the spring organic matter here yielded dates systematically 500 years younger than at Alex. The Aliwal North pollen indicate three alternations of grass and karoo vegetation 13 000-10 000 BP.

Later dry intervals at Alex and elsewhere are indicated after 6000 BP c. 3100, c. 1800, and about 1300-400 BP, when conditions were at least as dry as today. By contrast, the wettest periods, namely before 13 600 BP and c. 4500-1900 BP, were substantially moister, with a hydrology more comparable to that of the eastern Free State during the 19th century.

As yet undated within the early Upper Pleistocene, probably belonging within stage 5, was a +12 m lake at Alex and a +4 m lake in DP. The Alex lake was linked to a hydromorphic tributary flood plain that laterally includes colluvial sands of eolian derivation and a karoo-type pollen spectrum (L Scott, unpubl.). This enigmatic pattern of abundant water but sparse cover vegetation recalls that of the Early Holocene. The only MSA occupation of the pan region coincides with this period.

The Lower Vaal and Riet Drainage. The later Pleistocene of the Lower Vaal is recorded by the lower-energy flood plain silts and sands of the 5-unit Riverton Formation, best exposed at Windsorton, Riverton, the Harts-Vaal confluence, and along the lower Riet River (Butzer, Helgren, Fock and Stuckenrath 1973; Helgren 1977, 1978, 1979; Butzer, Fock, Scott and Stuckenrath 1979; Butzer 1984, and unpubl.). The Riverton Formation documents accumulation of increasingly fine, sandy and silty alluvia since the Mid-Pleistocene. Such overbank deposits, with repeated evidence for local slack water environments, imply frequent, moderate-magnitude events that generally entrained only limited sediment from relatively stable slopes. Intervening episodes of downcutting into older alluvium suggest infrequent, high-magnitude events at times of minimal sediment supply.

Member V consists of two parts, the older (Va) terminating before 2700 BP, the younger (Vb) before 1300 BP (Fig. 4). The older, in particular, documents widespread, low-energy aggradation of the Vaal and its several-order tributaries, with local development of wet flood plain conditions. The donga fills and their colluvial counterparts have a similar texture of semi-eolian upland sediments, and appear to represent reworked, humified materials of such origin. To explain both the erosion of these soils and the marshy flood plain environment, a modest increase in surface run-off and river discharge provides an economical hypothesis. One pollen spectrum from this unit indicates considerable grass (L Scott, unpubl.).

Member IV also represents two distinct fills, separated by downcutting, and reflecting different climatic impulses. Subunit IVa consists primarily of reworked, weathered eolian materials (Hutton Sands), and accumulation was concentrated downslope and downslope of such surface mantles. Judging by accumulation limited to valley floors, a lack of eolian bedding, and the presence of local pebble lenses, mobilization was primarily due to surface run-off. However, massive erosion is implicated, pointing to a marked deterioration of cover vegetation. Extensive lowland sedimentation and the cutting of a secondary channel, south of the Riet, also point to rivers choked with sediment. Subunit IVa sedimentation was under way by 14 000 BP and had terminated well before 9000 BP, when alluviation of IVb seems to have peaked. This younger phase was largely confined to flood silts and sands along the flood plains of trunk streams, and suggests a return to seasonal over-bank discharge at a time of essential slope soil stability; although valley water tables were higher there is no evidence of hydromorphic flood plain microenvironments. Once established, the new fluvial regime of IVb continued to maintain steady state conditions for several more millennia, with only modest downcutting prior to Member V alluviation.

Whereas subunit IVa records a markedly drier environment centered c. 13 500 BP, probably as a result of less reliable rains, subunit IVb suggests a return to a hydrology similar to that of recent centuries not long after 10 000 BP. But pollen from IVb indicates a relatively xeric vegetation, comparable to that of the present (L Scott, unpubl.).

Member III represents a complex aggradation, characteristically fluvial, but incorporating much tributary sediment, including crude slope rubble as well as reworked eolian components. There is no evidence of typical slack-water environments. Considerable surface run-off and moderate stream energy are indicated. Most of the sediment was derived from valley-margin denudation, by heavy and persistent rains; an incomplete vegetation mat can be explained by an accentuated dry season. Member III lies beyond the range of C14 dating, and probably pertains to some part(s) of isotope stage 5. In the Riet Valley it appears to be subdivided into two major alluvial bodies. It was coeval with the MSA settlement of the area.

The long interval between accumulation of III and IVa was marked by development of ACa soils (with dates of 18 600-15 800 BP) at Riverton and in the Riet Valley, or by a striking reddish B-horizon with subsoil silicification at Windsorton, as well as some 8-15 m of downcutting into alluvium.

The Gaap Escarpment. The Gaap Escarpment bounds the great dolomite plateau extending 275 km along the western margin of the Vaal and Hartz valleys. Springs that emerge along the 70-120 m high cliffs have periodically deposited sheets (aprons) or waterfalls (carapaces) of calcareous tufa, either below shallow drainage lines coming off the plateau or where the cliffs intersect subterranean aquifers. Major depositional sequences have been studied at Buxton-Norlim (Taung), Ulco, Grootkloof, Gorrokop and Mazelsfontein. C14 dating of the tufas provides an approximate chronology for the Late Quaternary sedimentary units. Apparent ages less than 2000 years are probably several centuries "too old", due to contamination by "dead" CO_3, while nominal ages greater than 3000 years are increasgly contaminated by younger bicarbonates and therefore represent minimum dates; apparent ages of 30 000 or more years most likely are "infinite".

The Gaap tufas include several facies that record a regular, six-part sequence of deposition and erosion (Butzer 1974; Butzer, Stuckenrath, Bruzewicz and Helgren 1978). (1) More frequent, high-intensity and protracted rains favour run-off that erodes soils and surface sands that then accumulate as colluvial sheets or as fillings in karst fissures. (2) Intensive mechanical weathering, due to severe frosts, produces crude, angular rock rubble below cliff faces (that today completely lack comparable footslope deposits). (3a) Repeated, high-energy run-off locally spilling over the escarpment reworks footslope rubbles into alluvial fans. (3b) Aquifers carry more of the precipitation received while stream flow becomes more persistent, with lower flood peaks; less debris is transported and sheets of calcium carbonate begin to accumulate along footslopes. (4) As discharge becomes more or less perennial, waterfall tufas build out as carapaces over earlier tufa sheets, except at times of maximum flow, when reduced bicarbonate saturation favours temporary corrosion. (5) As rainfall decreases, the drainage network disintegrates, tufas only accumulate in localized spring seeps, and rainwater corrosion enlarges joint systems and attacks rock cavities. This cycle is partly a function of precipitation amount, intensity and duration; however, lower temperatures would reduce evaporative losses during and after rain periods, probably facilitating tufa formation. With modern thermal conditions, a 50 % increase in precipitation (i.e. 600-650 mm per year) would allow adequate dolomite solution without excessive dilution of bicarbonates. A better vegetative ground cover would also increase percolation to the groundwater table, particularly if evaporative losses are reduced.

Six major geomorphic cycles of this type can be identified, spanning the Mio-Pliocene to the present. During the Holocene, three modest phases of accelerated spring activity are suggested with adjusted dates of c. 9800-7600, 3200-2400 and 1300-250 BP, in declining order of importance (Fig. 4). But pollen from archeological deposits at Taung with dates of 7480 and 2390 BP indicate little change in Holocene vegetation (L Scott, in prep.) Only minor erosional activity, with no sediment accretion, accounts for the period after 14 000 and before 10 000 BP.

The mid-Upper Pleistocene record, represented by Blue Pool Tufa II at Taung, indicates a protracted period of accelerated spring activity c. 33 000-14 000 BP, preceded as well as followed by significant mechanical weathering, with cliff-base accumulation of angular rubble at Grootkloof and Taung, and progradation of local, foreset-bedded rubbles at Gorrokop, all suggesting more severe frosts. Uncertain is whether additional tufa carapaces at Gorrokop (5A: 24 400 BP) and Mazelsfontein (2A: 26 840, 25 900 and 22 000 BP) represent an older substage of approximately 30 000-25 000 BP, or whether these deposits are of substantially greater age. The volume of these mid-Upper Pleistocene tufas decreases from north to south along the escarpment, as does modern precipitation. This seems to suggest that the balance of summer and winter rains was similar to that of today.

The early Upper Pleistocene is best represented at Taung by Blue Pool Tufa I, which has a 32 700 BP minimum C14 date at the top, and post-dates the Oxland Tufa complex with its Th/U approximation of 250 000 years (JC Vogel, pers. comm.). New uranium series dates from the top and base of Blue Pool I provisionally suggest bracketing ages of about 94 000 to 30 000 BP (Vogel and Partridge, this volume). Subsequently, but prior to Blue Pool II, a deep gorge was re-excavated, a deep reddish paleosol formed on the upland, and rockfalls accumulated below the resulting cliff. Furthermore, Blue Pool Tufa I is underlain by up to 30 m of block rubble, suggesting major dolomite backwearing and frost-weathering within isotope stage 5. In total, these events indicate substantial geomorphic change between 250 000 and 33 000 BP.

Wonderwerk Cave. Situated at the eastern margin of the Kuruman Hills near 1680 m elevation, Wonderwerk Cave is a large dolomite cavern dipping about 1° into a mountainside. The sedimentology and geochemistry of two composite sequences have been studied (Butzer 1984). The first is from excavated sections 18-32 m inside the cave entrance, the second from existing outcrops at 41-62 m. C14 dates for the latter sections were obtained by R Stuckenrath, for the former sections (Peabody, Beaumont and Thackeray excavations) by JC Vogel. The front profile (WB) spans the last 13 000 years or so and then picks up again with Acheulian strata. The interior section (WA) is 1.2 m higher and terminates about 12 000 BP and extends back with breaks, well into the Acheulian levels. The missing Late Pleistocene strata in the front of the cave have been eroded by water periodically moving in from the eastern side of the cave entrance, across the drip-line and then down towards a presumed lateral sluice. Re-examination of the bedding in the extensive excavation exposures shows variable inclinations of as much as 5° to the interior, in front of the stalagmitic boss – consistent with intermittent influx from the northeastern corner of the entrance; behind, the stalagmite dips are 1-2° to the interior and down to the eastern wall.

The properties of WA and WB columns show systematic differences, even for temporally equivalent strata. The interior sections have three times as much clay, twice as many organic colloids, a finer fraction in the greater than 4.75 ϕ category (3.45 instead of 3.20 ϕ), better sorting (σ 0.6 instead of 1.2), finer "tails" (Ski -0.35 instead of -0.30), and a different modal distribution (Kg +0.75 instead of +2.00).

These differences plus the extraneous mineralogy show that most of the sediment finer than 6 mm was washed in from the entrance, rather than weathered from the roof; however most of the organic component, responsible for much of the clay, was apparently derived from bat guano; a subordinate eolian component is probable but cannot be quantified. Gravel-sized components include some banded ironstone material (5-10 mm fraction) from the outside, but consist mainly of dolomite debris due to roof disintegration. Since the cave now is bone dry, edge corroded rubble indicates greater moisture, either penetrating the ceiling or affecting beds already in place; the unique, great stalagmitic column also indicates roof-water percolation. Angular roof rubble suggests pressure release as well as frost weathering. Frost is more likely to account for angular gravels mixed with occasional slabs, than for combinations of large roof blocks and interspersed gravel.

The Wonderwerk record (Fig. 4) indicates modestly or substantially wetter conditions in the cave 13 000-4000 BP and again 1900-800 BP. Much but not all of the period between 30 000 and 13 000 BP was also relatively moist, with major episodes of frost weathering c. 35 000-26 000 BP. Another wet and cool episode pertains to the early Upper Pleistocene. The Holocene segment of the record, which differs in terms of its emphasis and exact timing with that east of the Gaap Escarpment, is corroborated by the palynological evidence for more trees c. 10 000-5500 BP, with a complete dominance of grasses c. 5500-1900 BP, and more intermediate conditions thereafter (Van Zinderen Bakker 1982, dating revised).

Kathu Vlei. Kathu Vlei is an organic marshland amid the converging Molopo drainage next to Kathu, 15 km north of Sishen. A covered karst is developed in over 50 m of Tertiary calcrete that cements gravel, sands, and shale mantling a broad erosional surface. Recent groundwater pumping has lowered the water table 8 m; previously, the water table fluctuated within 2-3 m of the surface. There is no evidence of significant surface run-off, and the coalescent piedmont fans stretching to the Kuruman Hills and Koranaberg rarely carry surface water. This, in combination with spring-eye intrusions penetrating the Pleistocene fillings of the covered sinkholes from below, indicates underground waters rising under pressure, presumably in response to artesian seepage. A fossil sinkhole of this type (Kathu 1) was excavated to -11.7 m by PB Beaumont, and the sediments studied by the writer (Butzer

1984). Other exposures are provided by several sediment collapses in adjacent, buried sinkholes that reflect the recent water table drop and resulting fill compaction. Renewed excavations by Beaumont of several such sinkholes in 1983 provided new exposures that amplify the late Pleistocene part of the sequence described in Butzer (1984).

In terms of interpretation, periods of high water table with accumulation of fine, organic sediment, have alternated with (a) episodes of substantially lower water table, (b) episodes of intermediate water level that favoured sand accumulation by eolian processes or lateral flow through the interconnected karst passages, and (c) episodes of rapid, groundwater influx that periodically led to the eruption of spring vents under hydrostatic pressure. Since the bedrock constraints to the upper Molopo drainage preclude periodic incision of channels by 10-15 m or more, the Kathu sediment and water table record must be primarily attributed to long-term trends of groundwater recharge. A substantial proportion of the Kathu waters derives from artesian sources, further complicated by a subterranean threshold set by a major dike at nearby Sishen (BT Verhagen, pers. comm.); C14 determinations also indicate these waters are fossil. In general, large-scale aquifer changes tend to reflect long-term changes of absolute precipitation better than more complex, short-term environmental changes, and Kathu Vlei was apparently less sensitive to shifts in rainfall seasonality or evaporation.

The Holocene sediments of Kathu I consist of organic, sandy loam, clay loam, and loamy sand; organic carbon varies from 2.8 to 23.3 %, calcium carbonate equivalent 32 to 58 %. A near-basal C14 date of 7350 BP (Pta-3073) on organic matter indicates that the Holocene has generally been relatively moist. By interpolating ages and using three additional Pretoria dates (JC Vogel, pers. comm.), the higher organic carbon and clay values attained c. 7500-4400 BP and probably indicate maximum vegetation and perennial standing water (Fig. 4). Lower organic carbon and maximum calcium carbonate values, combined with high clay ratios, suggest alkaline and seasonal mud-flats with less vegetation cover c. 4400-3000 and 1800-400 BP.

Under the Holocene organic beds of Kathu 1, but on the top of gravelly deposits with reworked MSA artifacts, are non-organic sediments conformable with the upper but not the lower unit. These deposits begin with moderately calcareous fine sandy loam and terminate with 20 cm of highly calcareous loam. The dominant, fine sands are well sorted and eolian, either primary or derivative. They suggest a much lower but gradually rising water table, culminating in seasonal, alkaline mud flats with little vegetation. Extrapolating sedimentation rates, these beds suggest accumulation between very roughly 15 000 and 7500 BP, although a new C14 date at Kathu 5 indicates that related sediments were already accumulating about 27 000 BP (Beaumont, Van Zinderen Bakker and Vogel, this volume). The local environment was drier than it was during most of the Holocene, and the regional aquifer was eventually recharged towards the end of the major dry interval.

At Kathu 8, similar terminal Pleistocene deposits, but more organic towards the base, include early LSA artifacts (of Robberg and Albany type, according to Beaumont) above a diffuse Ca-horizon. At Kathu 6 these beds are absent and the Holocene organic deposits rest disconformably on a hard Ca-horizon or on lenticular, concretionary chalk suggesting small pans related to a low water table. Below this distinct horizon are slightly organic sandy sediments, interbedded with a prominent organic lens and including coarser, reddish sands near the base. These units include Howieson's Poort artifacts and, below that, early MSA, according to Beaumont. All these Upper Pleistocene deposits indicate low or intermediate water tables. The pan appears to have disintegrated completely subsequent to the chalk level (over the deposits with MSA in Kathu 6), at which time older beds were eroded and the artifactual residue concentrated in water reworked lags such as the MSA of Kathu 1.

Discussion. Since these sequences of the semiarid interior are arranged along an east-west transect, they allow comparison and contrast, both of the exact timing of environmental shifts and of longer-term trends.

In terms of details, the Early Holocene wet phase is shared at all sites, while the Mid-Holocene was dry west of the escarpment, moist to the east. Moister conditions c. 1000 BP, along the Gaap and in Wonderwerk coincided with drier conditions elsewhere. It is not clear whether these differences reflect spatially distinct trends or different thresholds, that respond to different combinations of rainfall amount, intensity, seasonality, or variability.

The similarities during the terminal Pleistocene, c. 14 000-10 000 BP are greater than the differences. Geomorphic processes were active, hydrological conditions fluctuated rapidly, and the vegetation mat was on the whole less effective than during the Holocene. At the time of the glacial maximum, c. 25 000-14 000 BP, geomorphic change was slow and relatively subdued, despite abundant surface water, suggesting effective ground cover. But the deeper Kathu aquifers were less productive, even though pans in the adjacent Kalahari were filled with water during much of this period (see reviews by Lancaster 1979 and Butzer 1984). This suggests that precipitation was actually less than during the Holocene, but somehow quite effective, possibly as a result of more, low-intensity winter rains and reduced evaporation.

The early Upper Pleistocene moist phase appears to have been general in the eastern half of the region: If the filling of Equus Cave at Taung is correctly correlated with Blue Pool Tufa I, then both the pollen and the dominant springbok fauna (L Scott, in prep.; Klein 1983) indicate a more grassy environment on the uplands. However, the pans and the Vaal-Riet alluvia suggest close analogs with the early Holocene, i.e. abundant water but an incomplete vegetation mat.

There are, then, many unresolved contradictions and ambiguities that caution against erecting a regional paleoclimatic framework, despite the tempting parallelisms. What can be generalized at this point is that most of the Holocene was moist, compared with a glacial Upper Pleistocene that was relatively moist at times, but more often dry. Cave and cliff environments sensitive to frost weathering indicate several Upper Pleistocene cold spells. An apparently non-glacial, early Upper Pleistocene moist period is indicated but its record is incomplete and inadequately dated.

GENERAL EVALUATION AND CONCLUSIONS

The paleoenvironmental data encapsulated in Fig. 2-4 provide a first approximation of the Late Quaternary data base for South Africa. It is essentially a record of geomorphic events and soils that reflect the broader environmental system and, ultimately, climate. Geomorphic dynamism versus soil stability more than anything reflect rainfall intensity or duration and the effectiveness of the vegetation mat. For caves, the observed phenomena are even more difficult to interpret because their peculiar microenvironments interject several filters that substantially modify the impact of the climatic forces. The geomorphologist considers vegetation in terms of raindrop interception, plant spacing, and the density of rooting, none of which can actually be reconstructed. The effectiveness of the groundcover is therefore inferred through a "black box" argument, despite the regular use of modern analog samples and due sensitivity to the range of local edaphic factors. This concept of vegetation differs from that of the palynologist or zoo-archeologist whose concern is with vegetation physiognomy or floristic composition or both. Interpretation of such paleobiological phenomena is complicated by other, equally distressing problems that also require a heavy dependence on analog studies. These sources of inference are therefore also not foolproof, and they are no less fragmentary nor less plagued by problems of dating. Ironic, too, is that the best study media for long records or pollen, faunas, and sediments rarely coincide.

There has evidently been a quantum jump in paleoenvironmental evidence for South Africa since 1970, and chronometric controls have kept apace. But the picture that emerges, even for the last 25 000 years, is indeed no more than a first approximation. There appear to be some general trends that are comparable over larger areas, but there is a great deal of variability in detail -- sometimes small shifts in temporal definition, sometimes differences in amplitude, and sometimes radical differences in the direction of change. It is not generally clear whether these divergences represent different threshold levels to a multidimensional environment or whether they reflect real climatic patterns in time and space.

Several broad observations can be offered. They serve more to draw attention to the difficulties than to contribute to a neat synthesis.

(a) The non-glacial intervals were generally wetter than the glacial time spans. This should not be surprising since the warm Agulhas Current, a major source of atmospheric moisture today, did not wash the southern and eastern coasts of South Africa during glacial times, when offshore waters averaged 5 °C cooler than after 10 000 BP (Vincent 1972; Huston 1980). However, the Benguela Current was weaker (Peterson et al. 1976) with reduced upwelling and atmospheric stability along the western littoral. In any event the glacial maximum c. 25 000 - 15 000 BP did experience several episodes of more effective vegetation and modest to substantial surface water in many parts of South Africa. The pattern appears to resemble that of the southern Mediterranean Basin and the northern Sahara. But it must be emphasized that the available empirical evidence remains inconclusive as to whether such moisture was due to winter or summer rains or a combination of both.

(b) There are some divergences between climatic trends in the coastal sector and in the interior, but these now appear to be less fundamental as the record is increasingly fleshed out and evaluated (contrary to earlier views expressed in Butzer, Stuckenrath, Bruzewicz and Helgren 1978). So, for example, optimal soil development or wetter conditions in both areas can be identified for the Mid-Holocene and for a substantial part of the early Upper Pleistocene (substage 5c or 5d to early stage 4). The modest soils of stage 2 can also be favourably compared in both areas. The Early Holocene in the Cape coastal sector was characterized by unstable dunes and slopes, or by calcification, whereas there was abundant water in the interior. However, slopes were also unstable in the interior and both the spring and lake beds of this time are strikingly calcified. These differences now appear to be more a matter of thresholds than they are of divergent climatic trends. The same applies to the terminal Pleistocene, when the coastal belt was overwhelmed by dunes c. 14 000-10 000 BP: More sensitive cave environments indicate spells of wet or cool climate or both, much like the generalized versus the specific records available in the interior. The temporal position of frost-spall horizons within the 50 000 year control of C14 dating and, as inferred, for earlier periods (stages 4 and 5b), also appears to be similar in coastal and interior caves; differences in development can be attributed to frost and moisture thresholds varying from cave to cave. This does not mean that all of South Africa responded in much the same way to an identical paleoclimatic trajectory. It does imply that, as data continue to be generated, we must continue to reassess the assumptions upon which interpretations are based.

(c) As the records are refined in resolution the rapidity of environmental change becomes increasingly dramatic. Variations in "effective" moisture had a wave length of only several hundred to several thousand years since about 30 000 BP. It is probable that earlier changes were equally rapid, but chronometric controls are too loose and many of the details in the preserved sediment packages have been lost or simply overlooked. Thick Pleistocene deposits such as a 20 m tufa body or 10 m of alluvium should be given much more detailed attention, and not considered as homogeneous blocks of time. The impression now obtains that the paleoclimatic continuum resembles that of tropical Africa, with its wildly fluctuating lake levels and alluvial traces. Unlike in Europe or North America, there is no representative mode with which to characterize glacial or non-glacial climates.

We should therefore not attempt to coerce the South African data into an over-simplified model of glacial/interglacial cycles (however they are labelled) such as have been applied for the British vegetation (Turner and West 1968) or the Czechoslovakian loess (Kukla 1975). At the other extreme, we should not attempt to identify parallels in the complex paleoenvironmental systems of continental South Africa for every oscillation in one or other deep sea O18 isotope curve. Reducing the paleoclimatic record of South Africa to a direct function of external thermal controls, as is implied in the recent analysis of HJ Deacon (1983), unfortunately obscures other equally important variables such as the frequency of the six distinct, large-scale circulation patterns that now determine weather patterns over southern Africa (P Tyson, unpubl.). The importance of Vogel's (1983) paper on C13 deviations in Pleistocene equid bone from Lesotho and Namibia lies in its implications for C3 versus C4 grasses as a function of rainfall seasonality. It is rainfall seasonality and intensity, as well as evaporation, that more than anything else control the density of groundcover, the ratio of grazing to browsing resources, and overall productivity. Such an edaphic perspective, at the local as much as at the regional level, is the most critical to an understanding of biotic resources for prehistoric populations in South Africa. Geomorphic processes and soil formation are intimately linked with such local, edaphic factors and consequently remain of central importance for understanding the paleoecology of southern Africa.

In conclusion, a note of cautious optimism. During the last ten years the body of empirical evidence for Late Quaternary paleoclimates has grown enormously in South Africa, in no small part due to the swelling ranks of researchers in many subfields. The Workshop "Evidence for Late Quaternary Climatic Change in Southern Africa" that followed this symposium has provided a wealth of complementary detail, but not necessarily a more coherent overview of the state of the art. It has suggested new hypotheses to test and will therefore help define more effective strategies to follow up in individual research and in matters of coordination. This will continue to require researchers based both inside and outside of South Africa. It will hopefully continue to encourage more collaboration and constructive discussions, but less polemic criticism and territoriality. Collectively, and only so, can we maintain the momentum to generate that quantity and quality of information over the next decade which will allow a second and then a third approximation of the temporal and spatial patterns of South African paleoclimates.

ACKNOWLEDGEMENTS

The writer appreciates information from and discussion with RG Klein, L Scott and JC Vogel during the preparation of this paper. Figures were drawn by Elizabeth Brooks.

REFERENCES

Avery DM 1982a. Micro mammals as paleoenvironmental indicators and an interpretation of the late Quaternary in the southern Cape Province, South Africa. Ann. S. Afr. Mus. 95:183-374.
Avery DM 1982b. The micromammalian fauna from Border Cave, KwaZulu, South Africa. J. archaeol. Sci. 9:187-204.
Beaumont PB and Vogel JC 1972a. On a new radiocarbon chronology for Africa south of the equator (1). Afr. Stud. 31:66-89.
Beaumont PB and Vogel JC 1972b. On a new radiocarbon chronology for Africa south of the equator (2). Afr. Stud. 31:155-182.
Brook GA 1982. Stratigraphic evidence for Quaternary climatic change at Echo Cave, Transvaal. Catena 9:343-351.
Butzer KW 1973a. Geology of Nelson Bay Cave, Robberg, South Africa. S. Afr. archaeol. Bull. 28:97-110.
Butzer KW 1973b. Re-evaluation of the geology of the Elandsfontein (Hopefield) site, South-western Cape, South Africa. S. Afr. J. Sci. 69:234-238.
Butzer KW 1973c. The question of Pleistocene 'periglacial' phenomena in South Africa. Boreas 2:1-11.
Butzer KW 1974. Paleoecology of South African australopithecines: Taung revisited. Curr. Anthrop. 15:367-382, 413-416.
Butzer KW 1978. Sediment stratigraphy of the Middle Stone Age sequence at Klasies River Mouth. S. Afr. archaeol. Bull. 33:141-151.
Butzer KW 1979. Geomorphology and geo-archaeology at Elandsbaai, western Cape, South Africa. Catena 6:157-166.
Butzer KW 1981. Cave sediments, Upper Pleistocene stratigraphy, and Mousterian facies in Cantabrian Spain. J. archaeol. Sci. 9:133-183.
Butzer KW 1982. Archaeology as Human Ecology: Method and Theory for a Contextual Approach. Cambridge Univ. Press, Cambridge.
Butzer KW 1984. Archaeogeology and Quaternary environment in the interior of Southern Africa. In: RG Klein (ed.), Southern African Prehistory and Paleoenvironments. Balkema, Rotterdam, in press.

Butzer KW, Beaumont PB and Vogel JC 1978. Lithostratigraphy of Border Cave, KwaZulu, South Africa. J. archaeol. Sci. 5:317-341.

Butzer KW, Clark JD and Cooke HBS 1973. The geology, archaeology, and fossil mammals of the Cornelia Beds, O.F.S. Mem. Nas. Mus. (Bloemfontein) 9:1-84.

Butzer KW, Fock GJ, Stuckenrath R and Zilch A 1973. Palaeohydrology of late Pleistocene Lake Alexandersfontein, Kimberley, South Africa. Nature 243:328-330.

Butzer KW, Fock GJ, Scott L and Stuckenrath R 1979. Dating and context of rock engravings in southern Africa. Science 203:1201-1214.

Butzer KW and Helgren DM 1972. Late Cenozoic evolution of the Cape Coast between Knysna and Cape St. Francis, South Africa. Quat. Res. 2: 143-169.

Butzer KW, Helgren DM, Fock GJ and Stuckenrath R 1973. Alluvial terraces of the Lower Vaal River, South Africa: a reappraisal and reinvestigation. J. Geol. 81:341-362.

Butzer KW, Stuckenrath R, Bruzewicz AJ and Helgren DM 1978. Late Cenozoic paleoclimates of the Gaap Escarpment, Kalahari margin, South Africa. Quat. Res. 10:310-339.

Carter PL and Vogel JC 1974. The dating of industrial assemblages from stratified sites in eastern Lesotho. Man (N.S.) 9:557-570.

Climate of South Africa 1965. Part 9: Average monthly rainfall up to the end of 1960. South African Weather Bureau: Pretoria.

Coetzee JA 1967. Pollen analytical studies in East and Southern Africa. Palaeoecol. Afr. 3:1-146.

Deacon HJ 1979. Excavations at Boomplaas Cave - a sequence through the Upper Pleistocene and Holocene in South Africa. Wld. Archaeol. 10:241-257.

Deacon HJ 1983. Another look at the Pleistocene climates of South Africa. S. Afr. J. Sci. 79:325-328.

Deacon HJ and Brooker M 1976. The Holocene and Upper Pleistocene sequence in the southern Cape. Ann. S. Afr. Mus. 71:203-214.

Deacon J 1979. Guide to archaeological sites in the southern Cape. Occ. Publ. Dept. Archaeol. Univ. Stellenbosch 1:1-149.

Fairhall AW, Young AW and Erickson JL 1976. University of Washington dates IV. Radiocarbon 18:221-239.

Fitzpatrick RW 1978. Periglacial soils with fossil permafrost horizons in southern Africa. Ann. Natal Mus. 23:475-584.

Folk RL 1966. A review of grain-size parameters. Sedimentology 6:73-93.

Gifford DP 1981. Taphonomy and paleoecology: a critical review of archaeology's sister discipline. Adv. archaeol. Method and Theory 4:365-438.

Helgren DM 1977. Geological context of the Vaal River faunas. S. Afr. J. Sci. 73:303-307

Helgren DM 1978. Acheulian settlement along the lower Vaal River, South Africa. J. archaeol. Sci. 5:39-60.

Helgren DM 1979. Rivers of diamonds: an alluvial history of the Lower Vaal Basin, South Africa. Res. Pap. Univ. Chicago, Dept. Geogr. 185:1-389.

Helgren DM and Butzer KW 1977. Paleosols of the southern Cape coast, South Africa. Geogr. Rev. 67:430-445.

Hutson WH 1980. The Agulhas Current during the late Pleistocene: analysis of modern faunal analogs. Science 207:64-66.

Klein RG 1976. A preliminary report on the "Middle Stone Age" open-air site of Duinefontein 2 (Melkbosstrand, southwestern Cape Province, South Africa). S. Afr. archaeol. Bull. 31:12-20.

Klein RG 1980a. Environmental and ecological implications of large mammals from Upper Pleistocene and Holocene sites in southern Africa. Ann. S. Afr. Mus. 81:223-283.

Klein RG 1980b. The interpretation of mammalian faunas from Stone Age archaeological sites. In: AK Behrensmeyer and AP Hill (eds.), Fossils in the Making. Univ. Chicago Press, Chicago. p. 223-246.

Klein RG 1983. The large mammals of southern Africa: late Pliocene to Recent. In: RG Klein (ed.), Southern African Prehistory and Paleoenvironments. Balkema, Rotterdam, in press.

Kukla G 1975. Loess stratigraphy of Central Europe. In: KW Butzer and GL Isaac (eds.), After the Australopithecines. Mouton, The Hague. p. 99-188.

Lancaster IN 1979. Quaternary environments in the arid zone of southern Africa. Occ. Pap. Dept. Geogr. Environ. Stud. Univ. Witwatersrand 22:1-77.

Martin ARH 1968. Pollen analysis of Groenvlei lake sediments, Knysna (South Africa). Rev. Palaeobot. Palynol. 7:107-144.

Marker ME and Dyer TG 1979. The origin of Lesotho hollows: an application of factor analysis. Z. Geomorph. 23:256-270.

Maud RR 1968. Quaternary geomorphology and soil formation in coastal Natal. Z. Geomorph. Suppl. 7:155-199.

Mc Intyre A et al. 1976. Glacial North Atlantic 18 000 years ago: a CLIMAP reconstruction. Geol. Soc. Amer. Mem. 145:43-76.

Opperman H 1978. Excavations in the Buffelskloof Rock Shelter near Calitzdorp, southern Cape. S. Afr. archaeol. Bull. 33:18-28.

Peterson GM et al. 1979. The continental record of environmental conditions at 18 000 BP: an initial evaluation. Quat. Res. 12:47-82.

Plug I 1981. Some research results on the late Pleistocene and early Holocene deposits of Bushman Rock Shelter, eastern Transvaal. S. Afr. archaeol. Bull. 36:14-21.

Price-Williams D, Watson A and Goudie AS 1982. Quaternary colluvial stratigraphy, archaeological sequences and palaeoenvironment in Swaziland, southern Africa. Geogr. J. 148:50-67.

Schalke HJWG 1973. The Upper Quaternary of the Cape Flats area (Cape Province, South Africa). Scripta Geol. 15:1-57.

Schulze BR 1958. The climate of South Africa according to Thornthwaite's rational classification. S. Afr. geogr. J. 40:31-53.

Scott L 1982a. A late Quaternary pollen record from the Transvaal bushveld, South Africa. Quat. Res. 17:339-370.

Scott L 1982b. A 5000-year old pollen sequence from spring deposits in the bushveld at the north of the Soutpansberg, South Africa. Palaeoecol. Afr. 14:45-56.

Scott L 1983. Palynological evidence for Quaternary environments in southern Africa. In: RG Klein (ed.), Southern African Prehistory and Paleoenvironments. Balkema, Rotterdam, in press.

Scott L and Vogel JC 1983. Late Quaternary pollen profile from the Transvaal Highveld, South Africa. S. Afr. J. Sci. 79:266-272.

Singer R and Wymer J 1982. The Middle Stone Age at Klasies River Mouth in South Africa. Univ. Chicago Press, Chicago.

Talma AS, Vogel JC and Partridge TC 1974. Isotopic contents of some Transvaal speleothems and their palaeoclimatic significance. S. Afr. J. Sci. 70:135-140.

Tankard AJ and Schweitzer F 1974. The geology of Die Kelders Cave and environs: a paleoenvironmental study. S. Afr. J. Sci. 70:365-369.

Turner C and West RG 1968. The subdivision and zonation of interglacial periods. Eiszeitalter und Gegenwart 19:93-101.

Van Zinderen Bakker EM 1957. A pollen analytical investigation of the Florisbad deposits (South Africa). In: JD Clark (ed.), Proceedings of the 3rd Panafrican Congress on Prehistory (Livingstone, 1955). Chatto and Windus, London. p. 56-67.

Van Zinderen Bakker EM 1982. Pollen analytical studies of the Wonderwerk Cave, South Africa. Pollen et Spores 24:235-250.

Vincent E 1972. Climatic change at the Pleistocene-Holocene boundary in the southwestern Indian Ocean. Palaeoecol. Afr. 6:46-54.

Vogel JC 1983. Isotopic evidence for the past climates and vegetation of South Africa. Bothalia 14: 391-394.

Vogel JC and Visser E 1981. Pretoria radiocarbon dates II. Radiocarbon 23:43-80.

SASQUA International Symposium / Swaziland / 29 August - 2 September 1983

The evidence from northern Botswana of Late Quaternary climatic change

H.J.COOKE
University of Botswana, Gaborone

ABSTRACT. The landscapes of N Botswana display a wide variety of polygenetic landforms which have resulted from the operation in the past of widely differing processes, mainly fluvial, lacustrine and aeolian. The findings of a decade of field research on the caves and associated phenomena in the Kwihabe hills, and the Makgadikgadi lake basin are summarized. A chronological scheme for the evolution of these features is presented, and climatic inferences are drawn.

INTRODUCTION

The landforms of northern Botswana have been well described in a series of publications over the past fifteen years (Grove 1969; Cooke and Baillieul 1974; Cooke 1975, 1976; Grey 1976; Grey and Cooke 1977; Cooke and Verhagen 1977; Heine 1978; Ebert and Hitchcock 1978; Wright 1978; Baillieul 1979; Coates et al. 1979; Cooke 1980; Breyer 1982, 1983; Cooke and Verstappen, in press). A brief outline summary should suffice here as a preliminary to a consideration of the palaeoclimatic significance of these landforms.

Relief and structure

The major features are summarized in Figures 1 and 2. It will be seen that a large shallow depression is defined roughly by the 950 m contour, and opens to the north towards the Zambezi valley. The lower parts of the depression are in the Makgadikgadi Pans where the lowest elevation is c. 890 m. Drainage is entirely endoreic, though in the north the Chobe-Zambezi system impinges upon the northward opening of the depression. This drainage derives from a very large area lying roughly between latitudes 13 °S and 24 °S, and longitudes 20 °E and 28 °E.

Two main structural trends are evident, one roughly WNW to ESE, the other NE to SW. The first has been shown to be associated with a massive and remarkable dyke swarm which traverses the whole region, and which Reeves (1978) has interpreted as a failed Gondwana spreading axis. The second is associated with two axes of seismicity, one described as the Kalahari axis by Reeves (1972), which cuts across the Makgadikgadi Pans, and the other described by Scholz et al. (1975) as

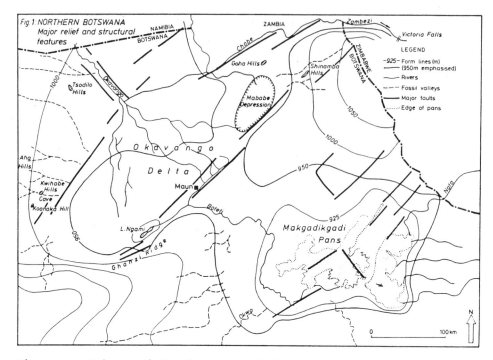

Figure 1. Major relief and structural features of N. Botswana.

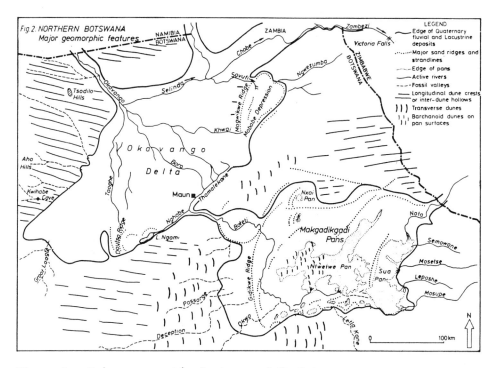

Figure 2. Major geomorphic features of N. Botswana.

marking a graben structure which is an incipient propagating rift. The Okavango Delta lies within this. Baillieul (1979) has also suggested that the pattern and alignment of the Makgadikgadi Pans is tectonically controlled with a series of fault-defined grabens likely.

Major landforms

Though Basement and Karoo age rocks outcrop in several parts of this large area, the greater part of the landforms with which this paper is concerned are developed in the mid-Tertiary to Recent Kalahari Beds, composed of marls, gravels, sands, calcretes, silcretes and alluvium. Localized cave deposits are of great interest. Aeolian, fluvial and lacustrine processes have been responsible for the deposition and subsequent extensive reworking of the superficial deposits, whilst the calcretes and silcretes whose vertical and horizontal disposition is of great complexity must reflect wide variations in environmental conditions. The broad areas of Quaternary fluvial and lacustrine deposition, and the major areas of aeolian landforms are shown in Fig. 2, and are briefly described below.

1. Aeolian landforms

These comprise very extensive systems of both longitudinal and transverse dunes. The most striking as landscape features are the former. They are prominent in the north-east where they are traversed by the Botswana-Zimbabwe border, but become less well-defined westwards where they appear to have been over-topped and inundated by northward extensions of a great lake which is referred to as lake Palaeo-Makgadikgadi. In the north-west they become clear again especially to the north and west of the Okavango Delta. Here they trend WNW to ESE. Elsewhere traces of linear features with the same alignment as the dunes are quite clear on air photos and LANDSAT images, due to vegetational differences, and presumably these mark the all but obliterated roots of ancient dunes. Major zones of transverse dunes lie to the west and south-west of the Makgadikgadi Pans and thus downwind of the prevailing north-easterly winds. They are confusing masses of sand hills up to 25 m high, and best developed to the west of the Gidikwe ridge. They fade away westwards and merge into the longitudinal dune traces already mentioned. Roughly barchanoid-shaped dunes occur on the surface of the present-day pans, especially in the western part of Ntwetwe pan where they form a remarkable conglomeration merging together westwards into an irregular mass of hummocky sand. These dunes show up spectacularly on air photos but on the ground are found to be very low features between 3.5 and 5 m high with very gentle slopes, slightly steeper to the west. Sets of dark parallel lines which show up well on the air photos are due to very slight vegetational changes associated with hardly discernible ridges which presumably mark strand lines of former flood levels. These dunes are uniformly grass covered.

2. Fluvial landforms

The river valleys of this region are of three kinds. The first are the perennial rivers such as the Chobe and the Okavango in the north, which have their main catchments in Angola. Second are the ephemeral rivers which flow seasonally, such as the Boteti which links the Okavango Delta to the Makgadikgadi Pans, and the rivers which enter Sua Pan from the east and north-east, of which the Nata is the chief. Finally, there are a number of totally dry valleys which orientate towards the Okavango-Makgadikgadi depression from the south and west. The former two types are fed by rainfall and run-off in areas well outside the region. The fossil dry valleys are floored by alluvium, with much calcrete along their flanks, and often strings of small pans which may seasonally hold water, along their beds. These fossil valleys are obviously relics from more humid climatic episodes in the past. The fact that they often cut through the dune fields is palaeoclimatically significant. The terminal points of some of these valleys are sometimes small relic fans or deltas which are useful in locating the former limits of lake transgressions.

3. Lacustrine landforms

LANDSAT imagery which has become available in the past decade, supported by good air photo cover has made it possible to delineate with reasonable confidence, the limits of very large areas of continuous alluvial deposition in two main zones. These are the Okavango Delta and the Makgadikgadi depression which are linked by a narrow belt about 10 km wide and 50 km long along the Boteti river valley (Fig. 2). A distinction may be made between the largely fluvial deltaic deposition in the Okavango Delta and the predominantly alkaline sandy clays and silts of the Pans. Much of the material is reworked Kalahari sand in one form or another. Very little detailed work has been done on these sediments, with the exception of Breyer's work in the lower Boteti area, and Staring's reports on the soils of the Delta (Breyer 1982, 1983; Staring 1978). It is clear that large areas outside the present-day active delta and beyond the limits of the bare pans have been subject in the past to fluvial and lacustrine processes on a large scale. The area of such deposition is in excess of 60 000 km^2. In the case of the Okavango, a large area to the west of the Taoghe, which is the westernmost distributary, and another to the north extending to the Chobe are a complex maze of old channels, sand mounds, and lake flats. In the former of these two areas the western edge is sharply defined by the Gomare fault-line which truncates the longitudinal dunes, and the fossil valleys descending from the west terminate at this line, often in small fans.

All around the Makgadikgadi Pans depression, traces are found of old lake shorelines in the form of massive sand ridges on the lee side of major sub-basins, which may represent old barrier beaches, and a variety of spits, bars, pebble beaches, and gravel ridges especially on the south and east sides. Minor strandline features are also found around all the present-day residual pans. These shorelines have been traced and mapped, and the major ones occur at c. 945 m, 920 m and 912 m. They clearly indicate the existence of a very large lake of c. 37 000 km^2 in extent at its highest level.

4. Karst landforms

These exist on a small outcrop of dolomite marble which forms the Kwihabe hills in the area between the Aha hills and the Okavango Delta. The fossil valley of the Kwihabedum cuts through these low hills. Minor karst features occur on the exposed rock, but the feature of principal interest is the large cave within the largest of the six hills (Fig. 3), while the calcretes along the valleys are also of great interest. The cave shows every sign of phreatic solution in a number of separate phases, much deposited sinter of different ages, and a partial infill of aeolian sand.

THE PALAEOCLIMATIC SIGNIFICANCE OF THESE LANDFORMS

That these diverse landforms are the result of the operation of widely varying geomorphic processes is obvious, as is the conclusion that these in their turn result from quite different climatic conditions. The matter is by no means simple however. Two other complicating factors need emphasis. In the first place the region is tectonically unstable. Reeves (1972), Greenwood and Carruthers (1973), and Scholz et al. (1975) have described the seismicity and tectonics of the area, and Cooke (1975, 1980) has assessed the role of neo-tectonics in landform evolution. The significant point to be made in the present context is that the dessication of the Makgadikgadi Pans cannot be attributed solely to climatic change. The formation of the Okavango Delta, and the reduction of inflow into the Makgadikgadi basin are almost certainly due to earth movements athwart the main direction of the regional drainage from the north-west towards the east.

The second complication arises from the immense size of this drainage basin occupying eleven degrees of latitude and eight degrees of longitude, and the fact that the major inflows of water into the central depression both at present and in the past derive from the Angolan highlands to the north. Inflow to the Okavango Delta at the present day has been calculated as 16×10^9 m^3 of which only 5×10^9 m^3 is of local origin. It is most unlikely that the direction and extent of climatic oscillation in the northern part of this large drainage basin has been similar or in phase with that in the centre and south. The elucidation of the sequence of climatic variations in relation to the pattern of evolution of the landforms becomes difficult. The wide fluctuations in the size of the palaeolakes in the region cannot have had a simple and direct relationship to the climate of the immediate environs. All that can be done at present is to describe the available evidence and assess its significance in very broad terms. Two major landform features have been examined in rather more detail, and these will now be referred to and their significance in palaeoclimatic terms assessed. They are the small karst assemblage at Kwihabe in the north-west, and the Makgadikgadi Pans in north-central Botswana. Most of what follows is derived from the literature already mentioned but particularly from Cooke (1975) for Kwihabe, and Cooke (1977, 1980), Grey and Cooke (1977), and Cooke and Verstappen (in press) for the Makgadikgadi.

The Kwihabe hills

The Kwihabedum fossil valley runs from the Aha hills in the west, cutting across the longitudinal dune field, to debouch across the faulted line marking an ancient extension of the Okavango wetland. It has a broad flat cross-section except where it cuts through the dolomite outcrop which forms the Kwihabe hills. Here it narrows into a short 800 m long gorge-like section with steep 10 to 15 m high walls of calcrete. Faulting has affected these hills at different times, and faults largely define the six hills, and also control a section of the valley here. Four distinct calcretes can be identified within the valley, and these have been dated by C14 assay from >45 000 BP (CI and CII) to c. 10 000 to 11 000 BP (CIII and CIV) (Table 1). The location of the calcretes may assist in the dating of the faulting episodes, thus CIII calcretes cement breccias in a fault which affects CI and CII calcretes. These calcretes are calcified sands and gravels, the sands having typical aeolian characteristics.

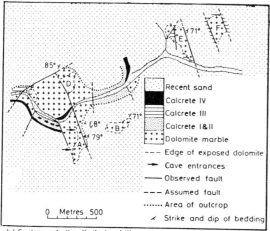
(a) Geology of the Kwihabe hills

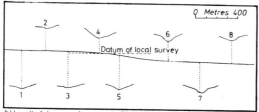
(b) Levelled long and cross profiles of the Kwihabe Valley near Drotsky's cave (V.E. on long profile × 10, on cross profile, × 5)

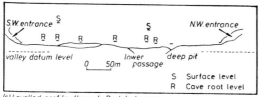
(c) Levelled profile through Drotsky's cave

Figure 3. Geology of the Kwihabe hills and profiles through Kwihabe valley and Drotsky's cave.

Within the largest of the six hills is located the large cave known as Drotsky's cave. It is controlled by major fault and joint weaknesses, but not by the general structure of the rock which locally dips

Table 1. C14 dates from Kwihabe material

Lab. No	Sample No	Sample description	Corrected age BP
Wits-164	C74/8(a)1	Slice of SIII/IV Stg.	15 200 ± 300
Wits-165	" 2	Same slice	15 100 ± 200
Wits-166	C74/8(b)	Lower slice of same Stg.	16 000 ± 200
Wits-171	C74/9(a)1	Slice of SIII/IV Stg.	13 200 ± 300
Wits-167	" 2	Same slice	13 600 ± 200
Wits-168	C74/9(b)2	Lower slice of same Stg.	14 000 ± 200
Wits-169	C74/10(a)1	Slice of SIII/IV Stg.	13 000 ± 200
Wits-170	C74/10(b)1	Lower slice of same Stg.	13 100 ± 200
Wits-195	C74/11(b)	Slice of SIV stg.	750 ± 100
Wits-163	C74/23(2)	Nodule on growing Stg.	20
Wits-122	C73 Stg(a)	Slice of SIV Stg.	2 200 ± 100
Wits-123	" (b)	Lower slice of same Stg.	2 550 ± 100
Wits-306	C75/1	Slice of SII Stg.	34 000 ± 1700
Wits-377	"	repeat	29 300 ± 1000
Wits-309	C75/5	Fragment SII sinter flowst.	>45 000
Wits-308	C75/16	Fragment SII sinter flowst.	41 900 ± 5500
Wits-386	"	repeat	37 000 ± 1700
Wits-310	C75/17	Fragment SII sinter	>45 000
Wits-313	C75/8	Hard Calcrete CII	>45 000
Wits-385	C75/9	Hard calcrete CI	34 700 ± 2000
Wits-315	"	repeat	22 700 ± 500
Wits-307	C75/13	Soft calcrete CIII/CIV	10 000 ± 200
Wits-375	"	repeat	9 800 ± 200
Wits-314	C75/22	Soft CIV calcrete	11 000 ± 100
Wits-376	"	repeat	10 800 ± 100

(Source: Cooke and Verhagen 1977)

steeply 80° to the west. The cave is horizontally developed and shows all the typical signs of solution under phreatic conditions. The lowest passages are approximately at present valley bottom level, while the upper passages have a mean elevation about 12 to 15 m above the same datum. Clearly the cave's development has been related to past water-table levels. Extensive sinter deposition is found in the cave, and a good deal of sediment, the latter divisible into older Phase I and younger Phase II material (following the nomenclature of Brain 1958). The sinters can be divided into four major age groups named SI to SIV, and samples of these have been C14 dated, the ages ranging from >45 000 BP to 20 BP (Table 1). In many cases the older sinters have been re-dissolved, and often clearly at the same time as the dolomite bedrock to which they are attached.

If certain assumptions are made regarding the solution and re-precipitation of carbonate rocks, and the formation of caves in them, some meaningful conclusions can be drawn from the evolutionary stages in the landforms of this small area which have considerable palaeoclimatic significance. These assumptions are as follows:

1. The solution of large cave passages under phreatic conditions, i.e. with water completely occupying the whole growing cavity, implies the presence of very large volumes of water, and where the cave is horizontal this development must have taken place below the local water-table. The latter will usually be determined by the degree of incision of local drainage lines.

2. The precipitation of cave sinter must take place within cavities drained of water but with a considerable vadose flow downwards through the cavity of solutions highly charged with bicarbonate derived from solution of the bedrock above. This movement depends on large supplies of water from the surface provided by a humid climate. Where the growth of sinter can be shown to have been very rapid, the logical conclusion is that the climate was particularly favourable in terms of a continuous water supply.

3. Where extensive re-solution of sinter takes place at the same time as further solution of the bedrock under phreatic conditions, then this must indicate a marked rise in the local water-table due either to a very large increase in groundwater flow, or a rise in the local valley bottom levels due presumably to valley infilling. Complete flooding of old cave passages must have occurred.

With these assumptions, and the observed relationships of the valley calcretes, the sinters and sediments in the cave, and the incidence of faulting, it is possible to derive a model relating climate and stages of landform evolution.

1. At the time of solution of the major cave passages whose mean elevation is 12 - 15 m above present valley-bottom datum, a water-table must be assumed which was well above this level. The Kwihabe river must have flowed in a valley about 25 m above the present level. A considerable precipitation to sustain high overland and groundwater flow must be assumed.

2. Draining of the caves must have followed stream incision which lowered the local water-table. Considerable flow of water continued, and vadose flow through the drained cave passages resulted in massive precipitation of sinter. Again a high precipitation must be assumed.

3. During sub-humid to semi-arid climatic episodes, the valley was infilled and calcretes formed on a number of occasions, cementing the sands and gravels. Some silcretization of calcretes also occurred.

4. In a more humid climate, cave passages were flooded again, and fresh solution of bedrock together with re-solution of earlier sinters occurred under phreatic conditions.

5. With a continuing humid climate, strong river flow incised through the valley infill lowering the local water-table again, thus causing draining of the cave. This in turn made possible the renewed growth of cave sinter in continuing wet conditions.

6. Faulting at a late date affected the cave, fracturing old sinters and bedrock together, and opening the caves by collapse to the subsequent entry of aeolian sand during dry climatic phases. Further calcrete growth took place along the valley in sub-humid to semi-arid conditions.

This sequence may have been repeated in whole or in part in a number of cycles. Major Late Quaternary climatic phases can be dated with some confidence. A consistent series of C14 dates was obtained from samples of SIII and SIV sinter stalagmites, and these lie between 16 000 and 13 000 BP. Stalagmite growth rates were measured as between 0.4 and 1.0 mm per annum, very high figures which indicate favourable conditions at the time. A sample of SII stalagmite gave dates of 29 300 ± 1000 and 34 000 ± 1700 less reliable than the group of SIII/SIV dates but nonetheless significant. Other SII material gave dates between 37 000 and 45 000 BP but this was from unshaped flowstone, a much less satisfactory material for dating than stalagmite. It is not possible to place the time at which the river initially incised its valley through the hills, but it must at least have predated the SII sinter age. It can however be confidently asserted that around 16 000 to 13 000 BP the climate in this locality must have been humid. Other stalagmite material gave dates around 2500 to 2250 BP and 750 BP, and wet conditions must be assumed for those times also.

Calcretes from the Kwihabe valley fall into the older CI and CII material which gave dates of 45 000 and 22 700 ± 500/34 700 ± 2000 BP, and more reliable younger CIII and CIV material with four consistent dates between 11 000 and c. 10 000 BP. The climatic significance of calcretes is perhaps debatable, but these fairly simple calcretized sands and gravels may indicate ephemeral river flow in a sub-humid to semi-arid climate.

In summary the following climatic inferences appear to be justified:

c. 30 000 BP	humid
16 000 - 13 000 BP	humid
11 000 - 10 000 BP	sub-humid/semi-arid
2500 - 2250 BP	humid
c. 750 BP	humid

The Makgadikgadi Pans

The major formative stages in the development of the Makgadikgadi Pans probably occurred in the Early and Middle Quaternary. The sequence may have been as follows (see Du Toit 1926; Bond 1963; Cooke 1975, 1977, 1980; Grey and Cooke 1977). A rise along the Zimbabwe-Kalahari axis of uplift in Late Pliocene times severed a proto-Limpopo whose headstreams included the Okavango, Chobe and upper Zambezi. The large endoreic basin thus formed became the site of a large lake which at its fullest extent reached a height of 945 - 950 m. This lake overflowed northwards and eastwards across the Victoria Falls escarpment, and the upper Zambezi was diverted from the central basin to link up with the middle and lower Zambezi to form the present large river. The major inflow to the lake was thus cut off and it began to shrink. Earth movements along the Thamalekane - Kunyere - Mababe line, which defines the distal end of the Okavango Delta, reduced flow from the Okavango basin to the Makgadikgadi basin, effectively creating two lakes. The Okavango Delta filled up the downfaulted trough in which it lies, and the Makgadikgadi lake waxed and waned with changes in inflow from the rivers to the south-east and west, and via the Okavango and the Boteti from the north. The later stages in the Makgadikgadi's development were probably of Late Quaternary age, and some C14 dates have made possible some tentative conclusions regarding climate. These dates were obtained mainly from calcretes, but also from shells and in one rare case from peat. At the time of writing (July 1983) they are in the process of being published (Cooke and Verstappen - in press), but are reproduced here for ease of reference.

Table 2. C14 dates from material from the Makgadikgadi basin.

Lab. No	Alt. of site (m)	Site and material	Age BP
GrN-11155	911	E. side Rysana pan	46 200 ± 1700 / 1400
GrN- 9018	912	N. of Toromoja, Calcrete	35 250 ± 550
GrN-10372	912	Sua spit. Calcrete	29 350 ± 650
GrN- 9015	912	N. of Toromoja, Calcrete	25 630 ± 180
GrN-10373	912	Sua Spit. Calcrete	21 920 ± 260
GrN- 9019	910	S. of Tsoi. Soft calcrete at base of dune.	19 680 ± 100
GrN- 9677	910	" "	19 169 ± 100
GrN- 9719	910	S. of Tsoi. Peat.	19 420 ± 150
GrN-11161	912	Xhumo. Soft calcrete in old channel bank.	3130 ± 50
GrN- 9676	912	Gwi Pan. Shells on beach terrace.	1710 ± 35
GrN- 9038	912	N. of Toromoga, Bones on terrace	1590 ± 70
GrN-10369	920	N. of Mmatsumo. Calcrete	34 600 +1300 / -1100
GrN-10371	920	E. side Sua Pan. Calcrete.	31 800 ± 850
GrN-10374	920	Sua Spit. Calcrete.	26 050 ± 550
GrN- 9017	920	Shoreline W. of Rakops.	10 070 ± 60
GrN-11159	920	SSE of Moremaoto in bank of Boteti. Calcrete	40 200 ± 900
GrN-11160	934	Terrace above 11 159 site Calcrete.	26 590 ± 200
GrN- 9016	945	E. side Sua Pan. Calcrete	38 000 ± 650
GrN-10370	945	" " "	29 300 ± 700
GrN-10375	945	Root of Sua spit.	21 930 ± 310

It should be emphasized that all the sample material was taken from precisely located sites at known altitudes in relation to proven former lake levels. The highest lake level was c. 945 m, and the other main levels were at 920 m and 912 m. These dates have been interpreted very simply on the basis of some obvious assumptions, as follows:
    1. Clearly when calcretes formed at very low elevations, the lake must have been dry above those levels at that time.
    2. Calcretes taken from clear beach terraces may be assumed to have formed when the lake was near that level.
    3. Calcretes taken from the base of beach ridges or at the foot of dunes and revealed by erosion of the overlying material must be presumed to pre-date those features.
    Some significant conclusions may be drawn from the dates on this basis. The group from sites at the 910-912 m level within the Makgadikgadi basin indicate times when the lake was dry, and calcrete actually forming at this low level. The climate was presumably sub-humid. Where a date from a 920 m site is close in time to others from the 912 m level, then clearly at that time the lake cannot have been close to the higher level. A group of dates from such sites lie between 35 250 and 25 350 BP, and it may be assumed that for much of this time the lake was low and inflow minimal. GrN 11 159 (40 200 BP) was from a sample at a height of 928 m close to the base of the Gidikwe ridge ex-

posed in the bank of the Boteti gorge where the river is beginning to debouch therefrom, and it may be assumed to pre-date the ridge. The Gidikwe ridge was formed when the lake was at a high 945 m level, so that clearly this major episode lay between 40 200 BP and c. 35 000 BP when the lake was dry again near 912 m. It presumably indicates a high water inflow via the Boteti and thus high rainfall in that river's ultimate source region in the Angolan highlands. Earlier, the lake was at a low level (GrN 11155, 46 200 BP, 911 m). GrN 11160 is from a sample taken on a terrace at 934 m cut into the Gidikwe ridge by the Boteti, so that by this time of 26 590 BP the river must have been well advanced in incising itself through the ridge. From this it must be assumed that the dry phase starting at c. 35 000 BP was ending, and humid conditions re-asserting themselves. GrN 9019, 9677 and 9719 are three consistent dates from the 910 m level, and indicate low lake levels at c. 19 420 to 19 169 BP, though the peat indicates cooler wetter conditions locally. GrN 10373 from material at 912 m also indicates dry conditions at a similar time of 21 920 BP.

There is much geomorphic evidence pointing to the 920 m lake level having been prolonged in time and thus a major phase. On the south and east flanks of the basin there are very extensive shoreline bars, beaches, spits and low cliffs at this level, while on Kubu island on the west side of Sua Pan a beach platform is covered with very well-rounded cobbles and gravel of the hard Basement metamorphic rock of which this isolated outcrop is composed, indicating that beach processes must have been operative for a long time. The height of this beach is 920 m. Kubu must have been an exposed rocky islet at this time.

It is tempting to ascribe the 920 m lake to the humid period between 16 000 and 13 000 BP indicated by the Kwihabe dates. The dates obtained from calcretes at this level, however, group mainly from 26 050 to 34 600 BP (GrN 10369, 10371 and 10374), though they conflict with the evidence already noted above for dry conditions at 912 m at this time. GrN 9017 is regarded as a more reliable date. The calcrete was obtained from a very good site, a low undercut cliff just above the very well-marked 920 m shoreline west of Rakops and yields a date of 10 070 BP. It has already been noted that c. 19 000 BP the lake was at a low level. One point which is worth emphasizing is that in very flat terrain of minimal relief relatively small changes in the volume of water inflow may cause shoreline migrations of considerable amplitude, a fact well borne out by contemporary evidence from the lake Ngami basin. It may thus be suggested, and all geomorphic evidence supports this, that the lake was at the 920 m level on a number of occasions during the period c. 25 000 to c. 10 000 BP, but with marked regressions to much lower levels within this period. The climate must have varied between humid and semi-arid locally, but with cooler episodes as indicated by the peat sample. Major rises must have been caused by increased Boteti inflow and this, as has been noted, is due to climatic conditions elsewhere.

The much more recent dates at 3130, 1710 and 1590 BP on material from sites at 912 m show that the Boteti was flowing strongly at 3130 BP (the sample was from the edge of an old channel), while the other dates on shell and bone material at 912 m fix the date of at least one lake level at this height.

Periods of aridity or semi-aridity, with dried out lake beds and aeolian processes dominant, cannot be accurately dated but only inferred. It would seem from the evidence that such periods occurred as follows: At c. 46 000 BP before the lake rose to a 945 m level; at the end of the period of the 945 m lake when aeolian processes affected the Gidikwe ridge; post-19 000 BP when dune formation took place at the 912 m level, burying earlier calcretes; and at a number of times between advances of the lake to the 912 m level, when extensive dune formation and reworking of dried-out lake floor sediments took place on the pan surfaces.

A broad chronology linking lake levels to climatic conditions may be suggested thus:

Date BP	Lake level	Climate
>46 000	low	sub-humid/semi-arid
40 000 - 35 000	high at 945 m	humid
35 000 - 26 000	low	sub-humid/semi-arid
26 000 - 10 000	high at 920 m on a number of occasions, with intervening lows at 912 m on a number of occasions	alternating humid to sub-humid with drier interludes
10 000 - present		sub-humid to semi-arid, with more humid interludes.

The problem remains of the sources of water required to fill this large lake basin to its various levels. Grove (1969) suggested that under present temperature conditions an inflow of c. 50 km^3 would be required to fill the lake to the 945 m level, and that only the Zambezi with a mean annual discharge of 40 km^3 could come anywhere near supplying this. The drainage lines to the west and south, from the nature of their valleys and the size of their catchments could not have supplied such flows, and the same applies to the rivers draining into Sua Pan from the east. The Boteti however is at present a misfit stream whose valley indicates a capacity for much larger volumes of flow, and as already suggested above, this is the most likely source of major inflows from the Okavango headstreams far to the north.

It is useful to compare the evidence described above with other work, especially by Lancaster (1979) and Butzer et al. (1973, 1978), and the conclusions of Heine (1983). Lancaster obtained dates from Urwi Pan (22°50'S) and Butzer from Alexandersfontein (28°50'S) and the Gaap escarpment (27°07'S to 29°10'S), all places in the Kalahari or its margins. Dates from these areas indicate humid and possibly cooler conditions in the same range 21 000 BP to 14 000 BP, which correlate well with the Kwihabe dates of this age. The Gaap sequence suggests sub-humid conditions again at c. 9700 to 6500 BP, which roughly overlap with the Kwihabe dates from calcretes CIII and CIV at 11 000 to 9800 BP. Another recent date not listed above, from the lake Ngami basin - GrN-11157 6445 $\pm$ 35 BP - from a calcrete at the base of the Dautsa ridge strandline marking a high level of Lake Ngami, also correlates with the Gaap dates, suggesting that the period from c. 11 000 BP to c. 6500 BP probably had a semi-arid climate initially, changing to subhumid.

Heine has suggested firstly that humid conditions prevailed in the period from 30 000 BP to 19 000 BP, and secondly that this was followed by drier conditions from 19 000 BP to 13 000 BP. In the Kwihabe situation described above the wetter period 30 000 to 19 000 might possibly be accepted to explain a period of river activity and valley incision preceding the SIII sinter, but the necessity for continuing high precipitation to account for this massive sinter makes Heine's second suggestion suspect. Furthermore the Makgadikgadi evidence for low lake levels on one or more occasions within his 30 000 to 19 000 BP period casts some doubt on his first proposition, at least without some modification. Without going into any detailed argument, it is clear that the evidence at present available is inconclusive, and detailed schema are premature. However, a tentative model for the central Kalahari region in the Late Quaternary might be suggested on the following lines.

Locally humid conditions sufficient to generate modest local river flows created low swampy lakes from time to time. At other times large inflows of water adequate to fill the Makgadikgadi to high levels were brought down by the northern rivers from the Angolan highlands where humid climates prevailed. At certain periods of overlap, humid conditions in both areas sustained high levels of water in the lake for lengthy periods such as for the 920 m lake. Similarly, very dry conditions may have resulted from low precipitation in both areas, leading to the dominance of aeolian processes. Since there appears to be a degree of correlation between the dates of Butzer, Cooke and Lancaster from the region roughly between $20\,^\circ$S and $29\,^\circ$S, it may be that the climatic shifts in the central and southern Kalahari were broadly in phase.

ACKNOWLEDGEMENTS

Grateful acknowledgement is made for research funding from the University of Botswana, De Beers Botswana Pty Ltd, and NUFFIC, and for the C14 dating to Dr BT Verhagen of Witwatersrand University, and Prof. WG Mook of Groningen University.

REFERENCES

Baillieul TA 1979. The Makgadikgadi Pans complex of central Botswana. Bull. geol. Soc. Am. 90:133-36.
Bond G 1963. Pleistocene environments of southern Africa. In: African ecology and human evolution. Chicago. p. 308-335.
Brain CK 1958. The Transvaal ape-man bearing cave deposits. Trans. Mus., Pretoria. Mem. No. 11.
Breyer JIE 1982. Reconnaissance geomorphological terrain classification of the lower Boteti region, Botswana. ITC J. (Verstappen Jubilee ed.)
Breyer JIE 1983. Soils in the lower Boteti region, Central District, Botswana. Working Pap. No. 47, Natn. Inst. Dev. Res., Univ. Botswana, Gaborone.
Butzer KW, Fock GJ, Stuckenrath R and Zilch A 1973. Palaeohydrology of a late Pleistocene lake, Alexandersfontein, Kimberley. Nature 243:328-30.
Butzer KW, Stuckenrath R, Bruzewicz AJ and Helgren DM 1978. Late Cenozoic palaeoclimates of the Gaap escarpment, Kalahari margin, South Africa. Quat. Res. 10:310-39.
Coates JNM et al. 1979. The Kalatraverse One Report. Director, Geol. Surv. Lobatse, Botswana.

Cooke HJ 1975. The palaeoclimatic significance of caves and adjacent landforms in the Kalahari of western Ngamiland, Botswana. Geogr. J. 143:431-44.

Cooke HJ 1978. The palaeogeography of the middle Kalahari of northern Botswana and adjacent areas. In: Proc. Okavango Delta Symp. Botswana Soc. Gaborone, Botswana.

Cooke HJ and Verhagen BTh 1977. The dating of cave development - an example from Botswana. Proc. 7th Int. Spel. Congress. Sheffield, UK. p. 122-24.

Cooke HJ 1980. Landform evolution in the context of climatic change and neo-tectonism in the middle Kalahari of north central Botswana. Trans. Inst. Brit. Geogr. NS 5:80-99.

Cooke HJ and Verstappen HTh (in press). The landforms of the western Makgadikgadi basin in northern Botswana, with a consideration of the chronology of the evolution of lake Palaeo-Makgadikgadi. Z. Geomorph.

Ebert J and Hitchcock RK 1978. Ancient lake Makgadikgadi, Botswana: mapping, measurement and palaeoclimatic significance. Paleoecol. Africa. 10/11:47-57.

Greenwood PG and Carruthers RM 1973. Geophysical Surveys in the Okavango Delta. IGS Geophys. Div. London UK. Rep. No. 15.

Grey DRC and Cooke HJ 1977. Some problems in the Quaternary evolution of the landforms of northern Botswana. Catena 4:123-133.

Grove AT 1969. Landforms and climatic change in the Kalahari and Ngamiland. Geogr. J. 135:191-212.

Heine K 1978. Radiocarbon chronology of late Quaternary lakes in the Kalahari, southern Africa. Catena 5:145-49.

Heine K 1983. The main stages of the late Quaternary evolution of the Kalahari region, southern Afrca. Palaeoecol. Africa. 14:53-75.

Lancaster IN 1979. Evidence for a widespread late Pleistocene humid period in the Kalahari. Nature 279:145-6.

Lancaster IN 1979. Quaternary environments in the arid zone of southern Africa. Dept. Geogr. Witwatersrand Univ. occ. Pap. No. 22.

Reeves CV 1972. Evidence of rifting in the Kalahari. Nature 237:96.

Reeves CV (ed.) 1978. Reconnaissance aeromagnetic survey of Botswana 1975-77. Terra Surveys, Canada, and Director, Geol. Surv. Lobatse, Botswana.

Scholz CH et al. 1975. Seismicity, tectonics, and seismic hazard of the Okavango Delta. UNDP/FAO Bot./71/506 proj. rep. UNDP, Gaborone, Botswana.

Wright EB 1978. Geological studies in the northern Kalahari. Geogr. J. 144:235-50.

SASQUA International Symposium / Swaziland / 29 August - 2 September 1983

Geomorphic evidence of Quaternary environmental changes in Etosha, South West Africa/Namibia

UWE RUST
Universität München, Germany

ABSTRACT. The fundamentals of the landscape evolution of Etosha are outlined, and models of the soil sequences and landform sequences are presented. An ancient fully developed paleosol was modified during the Late Quaternary and the present super pan was formed by the fusion of several pans. C14 dates on calcretes, chalks, etc. document stages of landscape evolution induced by environmental changes. The time ranges 33 000 - 28 000 BP, 22 000 - 18 000 BP, and 10 500 - 9000 BP were periods of geomorphic stability, while from 18 000 - 10 500 BP and 9000 - 3500 BP denudation of slopes and accumulation of basin sediments occurred. The latter period was also a time of effective eolian morphodynamics.

BACKGROUND: A MODEL OF MORPHOSEQUENCES

Between September 1979 and January 1980 the author carried out fieldwork on landscape evolution of the Etosha National Park and the neighbouring Kaokoveld. As a result a model of morphosequences can be offered (Fig. 1). Details of this model, and especially the field and laboratory data on which the model is based, will be discussed elsewhere.

The main aspects of this model are as follows: The geomorphic history started with a surface of unknown age. This surface is represented by a fully developed soil (MBv-ca-C). The ca-horizon, dipping gently (0.35 - 0.08 %), covers the Precambrian rocks ('Otavi', 'Nosib') as well as the 'Kalahari' strata of the Ovamboland basin (Hedberg 1979), and thus clearly represents a continental unconformity. Red sands (iron-stained quartz grains) represent the upper horizon of the soil (= Sandveld). These are soil sediments (= MBv), because they are not autochthonous layers but sediments reworked by eolian, mixed eolian-pluvial, and pluvial processes. Calcretes represent the lower horizon of the soil (= Kalkveld). As will be shown later, several distinct stages of calcrete formation produced thick calcrete horizons (several metres). It is of great geomorphic importance that the two pedogenetic horizons are equivalent with distinct lithologic and geomorphologic differences. The MBv-horizon is a morphologically soft, very mobile layer, whereas the ca-horizon is a morphologically hard rock.

On this old surface, land forms of predominantly fluvial origin developed below the Great Escarpment (= exorheic relief with valleys,

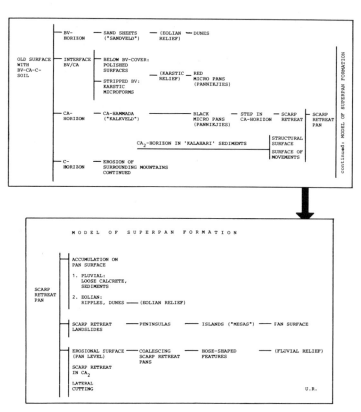

Figure 1. Endorheic relief of Etosha. A model of morphosequences.

canyons, sedimentary basins). Above the Great Escarpment, complex land forms developed (Fig. 1), but the more advanced ones were of predominantly pluvial origin (endorheic relief).

The morphogenesis of the endorheic relief was strongly influenced by the above-mentioned pedogenetically inherited lithologic varieties above, at, and below the MBv-ca-interface. In W. Etosha the first stages of the morphosequences prevail, whereas in E. Etosha advanced stages predominate. Compared with the exorheic relief below the Great Escarpment, the geomorphic effects of the denudational processes within the endorheic relief were not spectacular. Height differences of only 20 metres (e.g. Eto 61 in Fig. 2) are maximum values, whereas the Kaokoveld valleys were eroded to a depth of hundreds of metres.

Concerning the model of morphosequences, the most advanced denudational land forms in the area of the endorheic relief are the scarp retreat pans. As can be deduced by land form analysis of the eastern Etosha pan area, these scarp retreat pans tend to coalesce by means of broad valleys and hose-shaped features, whereas the surrounding old surface (and succeeding surfaces) correspondingly tend to dissolve into peninsulas and islands. The Etosha pan itself is a super pan formed by the fusion of several scarp retreat pans.

C14-DATED EVENTS

Some 35 C14 dates were obtained from different materials (calcrete, sinter, chalk, stromatolites) at selected sites within the research

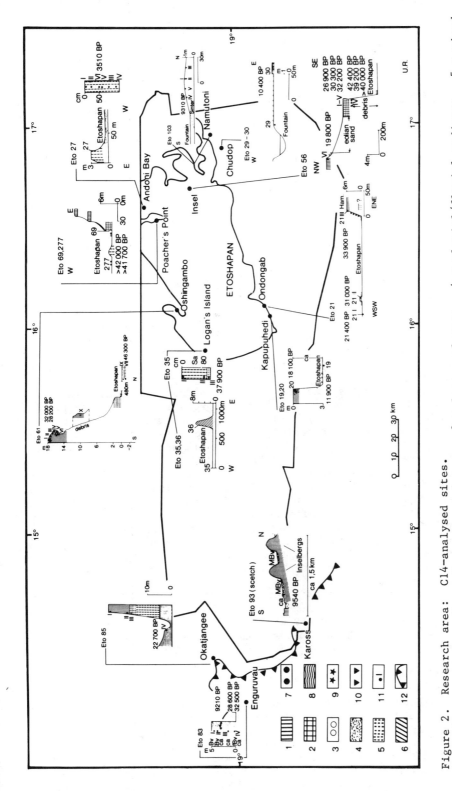

Figure 2. Research area: C14-analysed sites.
1 = calcareous crust, 2 = calcareous sandstone, 3 = loose calcrete, 4 = pluvial/fluvial sediment, 5 = mixed eolian-pluvial/fluvial sediment, 6 = eolian sediment, 7 = conglomeratic facies resp. fluviatile gravels, 8 = Precambrian rocks of different ages, 9 = stromatolites, 10 = snails, 11 = sample, 12 = Great Escarpment.

area (Table 1, Fig. 2). The sites are as follows: Exorheic relief: Enguruvau, Kaross, Okatjangee; endorheic relief: Logan's Island, Kapupuhedi, Ondongab, Oshingambo, Poacher's Point, Andoni Bay, Insel, Chudop, Namutoni. From the dates it can be seen that within the reliability of the method, the morphosequence introduced above was marked by several distinct events. From the geomorphic and stratigraphic positions of the dated samples it can be concluded that the morphosequences developed discontinuously because of environmental changes.

INTERPRETATION

A palaeoenvironmental interpretation of geomorphic-sedimentological findings can best be presented by means of Rohdenburg's (1970) approach. He developed a model, which may be described as a 'model of ecomorphodynamic alternatives'. He differentiated between a 'period of morphodynamic stability' (with soil formation, dense vegetation, low run-off/infiltration ratio) and a 'period of morphodynamic activity' (with slope erosion, open vegetation, high run-off/infiltration ratio). Perhaps the correlation between the two alternatives and their transitional conditions can be derived from the model presented by Rohdenburg and Sabelberg (1980, Fig. 4). It can be seen that the total amount of rainfall is of minor importance (with the exception of very arid areas). The authors assume that the precipitation regime can lead to either morphodynamically active conditions (irregularly distributed rainfall) or morphodynamically stable conditions (evenly distributed rainfall).

Pedogenous calcretes represent events of morphodynamic stability. The dates of pedogenous calcretes can be grouped as follows (Fig. 3) : c. 33 000 to 27 000 BP, c. 22 000 to 18 000 BP, and c. 10 000 to 9000 BP, thus clustering three periods of morphodynamic stability. The two first periods fit roughly into the more humid periods in the Namib desert as reported by Vogel (1982). The field evidence at Enguruvau reveals a distinct morphodynamically active period (accumulation) which lies between 32 000 and 28 000 BP. Perhaps the pluvial denudation post-dating 31 000 BP at Ondongab and the accumulative sequence 30 000 to 27 000 BP at Insel are corresponding events. But on the other hand, alternative explanations should also be considered. Either the data gap in the endorheic relief at 29 000 BP came about by chance, or the morphodynamic activity at Enguruvau is a response restricted to the area below the Great Escarpment, i.e. the 'drier' area of greater ecomorphodynamic sensibility.

The events of the time span between 22 000 and 18 000 BP were reconstructed from three pedogenous calcretes and one chalk sample. The events from 10 000 to 9000 BP were reconstructed from three pedogenous calcretes and the Namutoni sinter. As a result it can be seen that the calcretes in Etosha are of different ages. Thus the thick calcrete horizons in W. Etosha (e.g. Eto 85 in Fig. 2), where the above-mentioned old surface is best preserved, are presumably the effect of several calcrete forming periods.

The chalks taken from the surface of the Etosha pan (21 400 BP Ondongab, 11 900 BP Kapupuhedi) and the near surface evaporitic calcareous sandstones (37 900 BP Logan's Island, 46 300 BP Oshingambo) document the existence of the Etosha pan at those sites. The date for Eto 61 VII (Fig. 2) documents additionally that the Etosha pan of 46 300 BP at Oshingambo consisted of a surface eroded into the old surface. Furthermore these dates prove a polycyclic evolution of the Etosha pan.

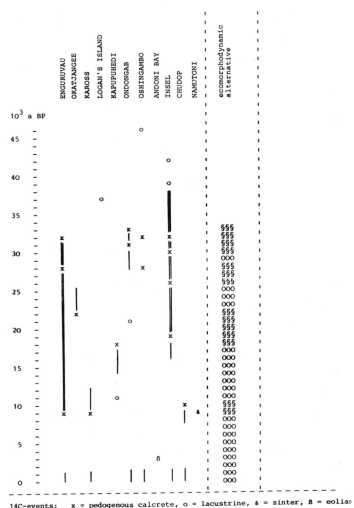

Figure 3. Palaeo-ecomorphodynamic synopsis of C14-dated events. Sites grouped roughly W-E.

Comparison of the two Kapupuhedi dates (calcrete Eto 20 = 18 100 BP, chalk Eto 19 = 11 900 BP) provides evidence for scarp retreat post-dating 18 100 BP. The chalks of the neighbouring sites Kapupuhedi and Ondongab were taken from the surface of the Etosha pan. Because the event of Ondongab, on the one hand, is contemporaneous with a morphodynamically stable period, whereas the event of Kapupuhedi happened at the end of a morphodynamically active period, one can conclude that palaeoenvironmental deductions cannot be drawn from chalk events.

The stromatolites (Eto 56 IV in Fig. 2) at Insel document a lacustrine phase c. 8 m above the level of the present Etosha pan. Wilczewski and Martin (1972) assumed these stromatolites to be of Pliocene age! From stromatolites at Poacher's Point (Eto 277), only maximum ages were obtained (Table 1). Although these stromatolites were found at the surface of the Etosha pan (Fig. 2) the dating unfortunately does not give any hint concerning the age of the pan.

Table 1. C14 dates for samples from Etosha.

Site/Sample	Lab No (Pta)	Corr. Age BP	δ C13 (‰)	CaCO$_3$ (%)	Gypsum (%)	Material
ENGURUVAU (14.15° E/18.98° S)						
Eto 83 IV	3074	32 500 ± 800	- 2.4	35.9	0.21	pedogenous calcrete
Eto 83 III	3058	28 600 ± 480	- 4.4	72.9	0.28	pedogenous calcrete
Eto 83 II	3103	9 210 ± 90	- 3.5	23.2	0.0	pedogenous calcrete
OKATJANGEE (14.38° E/18.86° S)						
Eto 85 IV	3039	22 700 ± 240	- 2.8	89.3	0.26	pedogenous calcrete
KAROSS (14.53° E/19.39° S)						
Eto 93	3040	9 540 ± 100	- 1.7	27.2	0.56	pedogenous calcrete
LOGAN'S ISLAND (15.88° E/18.78° S)						
Eto 35 III	3066	37 900 ± 1550	+ 0.8	21.5	1.77	evaporitic calcrete
KAPUPUHEDI (16.02° E/19.13° S)						
Eto 20	3046	18 100 ± 190	- 3.6	71.6	1.59	pedogenous calcrete
Eto 19	3041	11 900 ± 120	+ 0.3	69.8	0.40	chalk
ONDONGAB (16.15° E/19.07° S)						
Eto 21 III	3052	33 900 ± 730	- 0.7	91.6	0.32	pedogenous calcrete
Eto 21 II	3065	31 000 ± 710	+ 1.2	95.2	0.35	pedogenous calcrete
Eto 21 I	3042	21 400 ± 230	+ 0.5	95.2	1.26	chalk
OSHINGAMBO (16.11° E/18.65° S)						
Eto 61 VII	3047	46 300 + 4060 − 2680	+0.4	78.3	0.76	evaporitic calcrete
Eto 61 II fr 1	3140	28 700 ± 550	+ 0.4	68.5	0.49	pedogenous calcrete
Eto 61 II fr 2	3141	28 200 ± 890	+ 0.2			
Eto 61 I	3134	32 000 ± 780	- 0.6	72.5	0.38	pedogenous calcrete
ADONI BAY (16.65° E/18.52° S)						
Eto 27 VI	3044	3 510 ± 120	+ 1.3	17.1	-	evaporitic calcrete
INSEL (16.69° E/18.74° S)						
Eto 56 IV nod.	3038	42 400 ± 1950	+ 3.4	10.0	7.96	Stromatolith
Eto 56 IV fr 1	3035	39 300 ± 1470	+ 2.8			Stromatolith
Eto 56 IV fr 2	3036	>40 000	+ 3.7			Stromatolith
Eto 56 III	3100	32 200 ± 770	- 0.3	52.9	1.26	pedogenous calcrete
Eto 56 V	3048	30 300 ± 720	- 0.1	43.0	0.04	pedogenous calcrete
Eto 56 VI	3053	19 800 ± 180	- 2.2	31.1	0.0	pedogenous calcrete
POACHER'S POINT (16.52° E/10.58° S)						
Eto 277 (i)		>42 000	+ 4.0			Stromatolith
Eto 277 (iii)		>41 700	+ 4.4			Stromatolith
CHUDOP (14.53° E/18.87° S)						
Eto 30	3050	10 400 ± 90	- 3.1	96.5	0.12	pedogenous calcrete
NAMUTONI (16.94° E/18.86° S)						
Eto 103 IV	3043	9 310 ± 90	- 7.0	98.5	0.68	sinter

+ Radiometric data by Dr JC Vogel, CSIR, Pretoria.   CaCO$_3$, gypsum by Inst.Geogr. Univ. Munich

The date 3510 BP was obtained from a powder calcrete within a barkhan of the Andoni barkhan field. The barkhan nowadays has a grass cover and is being eroded by pluvial processes. The event documents a Late Holocene stage of intense eolian morphodynamics. Because of the orientation of the barkhan it can be concluded that at that time strong winds blew from E to SE.

The time span between c. 27 000 and 22 000 BP, and 18 000 and 10 000 BP (Fig. 3) were periods of morphodynamic activity. At Enguruvau, 32 000 to 28 000 BP was also a period of morphodynamic activity. The following processes, forms, and sediments are characteristic for morphodynamically active conditions at the various sites.

Pluvial:
scarp retreat and correlative sedimentation of loose calcrete, development of valley slopes and correlative accumulation of basin sediments, transformation of calcareous crusts into hammadas ("Hammadaisierung", boulder calcrete formation in the sense of Netterberg 1978).

Eolian:
barkhan and derivates,
longitudinal dunes (W. Etosha),
accumulative dunes at the border of scarp retreat pans ("Pfannenranddünen"),
ripple fields

Because land forms evolve under morphodynamically active conditions, and because of the events discussed above, which document stable conditions, it can be concluded that in the study area the morphosequences developed discontinuously (Fig. 1).

Nowadays scarp retreat, formation of eolian ripples, and denudation of loose soil sediments (Le Roux 1977) can be observed in the endorheic relief. Below the Great Escarpment the loose basin sediments are affected by strong badland formation; locally coppice dunes are accumulated. During field work the author observed dust storms and ripple transports as well as illuviation of sediments into the Etosha pan from the surrounding rims, i.e. depending on weather conditions, both eolian and pluvial processes can act at the same site (azonality of geomorphic processes, see Rust 1980). With the exception of the barren surfaces of the scarp retreat pans the pluvial morphodynamics seem to be predominant.

Below the Great Escarpment the above-mentioned morphodynamics is at least partially induced by man's interference (man-made desertification, overgrazing by cattle, goats). Overgrazing by wild animals (Le Roux 1977) in the Etosha National Park may have had the same effect. Nevertheless, the present ecomorphodynamic conditions are to be classified as morphodynamically active (Fig. 3). Damage to the vegetation, especially the destruction of the grass cover seems to be an important boundary condition.

Judging from the C14 dates (Fig. 3), these conditions have obtained since c. 9000 BP. Concerning the discussion of the high ages of calcretes in southern Africa (Netterberg 1969), the author's results could offer an explanation. In the study area, at least, calcretes did not develop after 9000 BP, since calcrete formation (pedogenesis) is linked to morphodynamically stable conditions!

Taking into account the assumptions modelled by Rohdenburg and Sabelberg (1980), and assuming the author's inferences to be admissible, it must be concluded that:

1. The present climatic conditions stand for unevenly distributed precipitation. Strong seasonal precipitation (summer rains) (Schulze and McGee 1978) and the spatial and temporal variability are characteristic features of the present regime.

2. The climatic conditions during morphodynamically stable periods must be characterized by fundamentally different precipitation regimes, either more precipitation and/or more regularly distributed precipitation.

The sinter event of Namutoni (9310 BP) during a morphodynamically stable period can be interpreted by both alternatives (more, or more regularly distributed rains). According to Netterberg (1969) calcrete formation needs rain of ≥ 500 mm per year. Today the mean values for Etosha do not reach this limit.

Apart from this, during morphodynamic stability the precipitation regime and the relationship overland flow/infiltration must have been different, and consequently non-actualistic atmospheric circulation patterns have to be postulated for northern South West Africa. But because the intention of this paper was to offer a brief review of own field results, the author does not want to speculate about such circulation patterns here (compare Rust et al. 1983).

ACKNOWLEDGEMENTS

The author wants to thank The Director, Nature Conservation and Tourism, Windhoek, the South African Defence Force, Ohopoho and Oshakati, the Deutsche Forschungsgemeinschaft, Bonn, Dr JC Vogel, Pretoria, and F Kestler, Cham, for logistical, financial, technical and personal support.

REFERENCES

Hedberg RM 1979. Stratigraphy of the Ovamboland Basin South West Africa. Precambr. Res. Unit Cape Town, Bull. 24:1-325.

Le Roux CJG 1977. 1976/77 Ann. Rep. Plant Ecology of the Etosha National Park, grazing pressure experiments, veld burning experiment. Div. Nature Conservation and Tourism, SWA, Windhoek. 19 pp.

Netterberg F 1969. Ages of calcrete in Southern Africa. S. Afr. Archaeol. Bull. 24:88-92.

Netterberg F 1978. Dating and Correlation of Calcretes and other Pedocretes. Trans. geol Soc. S. Afr. 81:379-391.

Rohdenburg H 1970. Morphodynamische Aktivitäts- und Stabilitätszeiten statt Pluvial- und Interpluvialzeiten. Eiszeitalter und Gegenwart 21:81-96.

Rohdenburg H and Sabelberg U 1980. Northwestern Sahara margin: Terrestrial stratigraphy of the Upper Quaternary and some paleoclimatic implications. Palaeoecol. Africa 12:267-275.

Rust U 1980. Models in Geomorphology - Quaternary Evolution of the Actual Relief Pattern of Coastal Central and Northern Namib Desert. Palaeontol. Africana 23:173-184.

Rust U, Schmidt HH and Dietz KR 1983. Palaeoenvironments of the Present Day Arid South Western Africa 30 000 - 5 000 BP: Results and Problems. Palaeoecol. Africa 16, in press.

Vogel JC 1982. The age of the Kuiseb river silt terrace at Homeb. Palaeoecol. Africa 15:201-209.

Wilczewski N and Martin H 1972. Algen-Stromatolithen aus der Etoscha-Pfanne Südwestafrikas. Neues Jb. geol. Paläont., Mh. 12:720-726.

The occurrence of ferricrete at Witsand in the south-eastern Kalahari

T.H.VAN ROOYEN & E.VERSTER
University of South Africa, Pretoria

ABSTRACT. An investigation of the sands and other surface phenomena in the Witsand area led to the discovery of ferricrete outcrops. In this part of the Kalahari the formation of ferricrete is clearly out of phase with the present climate. The ferricrete occurs approximately 15 m above the present water table and overlies several metres of light yellowish sand. Witsand is known not only for its roaring dunes, but also for an ample water supply and a high water table originating as runoff from the Langeberg range. The ferricrete, 15-25 cm thick, has a predominantly vesicular structure. Goethite is the dominant mineral present with a low degree of aluminium substitution. Based on these data, supported by geomorphic evidence, there is little doubt that the ferricrete is the result of the accumulation of iron on a wet bottomland situation formed during wet periods of the Upper Pleistocene. Subsequent to wind erosion and/or lowering of the water table during dry periods of the Late Upper Pleistocene, the ferricrete hardened upon exposure. The evidence supplies significant information about climatic changes during the Quaternary.

INTRODUCTION

A ferricrete (or hard plinthite) is a massive, horizon-type hardpan enriched and strongly cemented by sesquioxides, mainly iron oxides. Soft and hard plinthite are diagnostic subsoil horizons defined in the South African binomial soil classification system (MacVicar et al. 1977). Plinthite forms by segregation of iron in hydromorphic environments or more specifically within the zone of seasonal fluctuation of the water table. Subsequent to exposure and/or the lowering of the water table, plinthite would change irreversibly to hard plinthite or ferricrete.

In the course of a soil survey ferricrete outcrops were observed in the Witsand area - an accumulation of whitish and yellowish sands with a dune-like appearance situated in the south-eastern Kalahari Region (28° 34'S, 22° 28'E), west of the Langeberg range and south-west of Olifantshoek (Van Rooyen and Verster 1983). This concentration of light-coloured sands (Witsand proper) covers approximately 3200 ha and is elevated 20 - 30 m above a vast plain of yellowish and reddish aeolian sandy soils. Another characteristic of some of the sand dunes is their tendency to 'roar' when disturbed.

From a genetic viewpoint, the presence of ferricrete in an arid region is a definite indication of climatic change. In addition, the occurrence of ferricrete well above the present water table could be indicative of past climatic conditions considerably wetter than those of today. The aim of this paper therefore is to present evidence of Quaternary climatic change as well as an attempt to formulate a theory regarding the origin of this relict ferricrete.

PHYSICAL ENVIRONMENTAL FEATURES OF THE WITSAND AREA

Terrain morphology

Broadly speaking, the surface morphology is dominated by three types of terrain, viz. a featureless sandy plain to the west of Witsand; a chain of open, low ridges associated with the characteristic compound dunes of Witsand proper, parallel to the Langeberg range; and the hills of the Langeberg itself striking NNE-SSW (Fig. 1).

Geology and parent materials

The solid geology of the Langeberg and low ridges amid the Witsand dunes comprises the Matsap Formation (Fig. 1), a sequence of white, grey and pink quartzites, grey and brown coarse-grained subgraywacke, and conglomerate (Geol. Surv. 1977). Upon weathering these rocks yield medium- to coarse-grained sands as parent material for the sandy soils.

The aggradational landsurface of the plains is composed of yellowish and reddish sands of Quarternary age. These sands are of aeolian origin and were blown in from the north-west, probably during recurrent arid periods of the Middle Pleistocene (Van Rooyen and Verster 1983). The sands in and around Witsand, being somewhat coarser, differ from those of the plains with respect to grain size. They are probably an admixture of the aeolian sand and sandy parent materials derived from the Matsap Formation.

Soils and other surface phenomena

Various types of sandy soils have developed from the sandy parent materials. Generally speaking the soils can be subdivided, according to the South African soil classification system (MacVicar et al. 1977), into three broad groups, viz. (i) reddish (2.5YR 4/8-5YR5/8 - Munsell notation) fine and medium sands (clay content 5-11 %) of the Hutton form on the footslopes of the Langeberg; (ii) stabilized yellowish (5YR 6/8-7.5YR 7/8) medium sands (clay content 3-5 %) of the Clovelly form on the sandy plains surrounding Witsand; and (iii) yellowish (7.5YR 6/6-7/8) and whitish (10YR 7/2-7/4) medium sands (clay content 0 %) of the Clovelly and Fernwood forms respectively, occurring as active dunes at Witsand proper. (See Fig. 1 for a schematic representation of the soil associations in the Witsand landscape.)

Layers of friable, very dark grey (10YR 3/1) peat were observed in bottomland sites between dunes where they are inevitably associated with a high water table. C14 dating suggests an age of 880 ± 70 BP (Pta - 3100. Van Rooyen and Verster 1973).

Assemblages of stone implements wer observed at several sites during

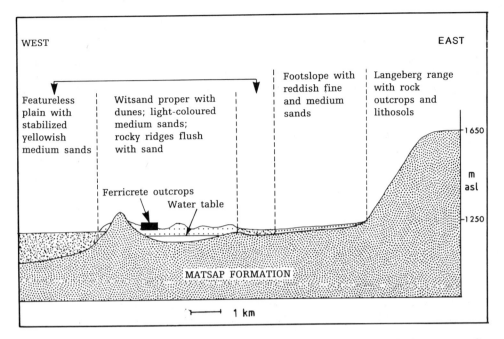

Figure 1. Schematic representation of some of the physical features of the Witsand landscape.

the survey phase. Stratigraphically these sites appear to be at a lower level with respect to the ferricrete outcrops. According to Fock (1961) the stone implements belong to the Middle Stone Age and Wilton cultures. The implication is that Witsand was inhabited during the Upper Pleistocene and the Holocene, and a constant water supply must have been accessible during at least some of this time to enable Stone Age people to live here on the fringe of the Kalahari.

Climate

The climate may be described in the terms of Poynton (1971) as "semi-arid" and "cooler-temperate". It is estimated from weather stations in the vicinity that the average annual rainfall is approximately 340 mm, and the mean maximum and minimum air temperatures are $31.8\,^\circ C$ and $16.5\,^\circ C$ for January and $17.6\,^\circ C$ and $-0.6\,^\circ C$ for July respectively (Weather Bureau 1954). It is also estimated from data by Kriel (1967) that the mean class A pan evaporation is in excess of 3300 mm per annum. Comparing rainfall and evaporation, it is obvious that a very large deficit of water is characteristic of the annual water budget of the Witsand area.

Drainage and water supply

Although several ephemeral watercourses are visible on the footslopes of the Langeberg range, they disappear within a short distance in the porous sand mantle. Drainage is therefore effected by means of seepage to the groundwater zone.

Witsand is something of an enigma: whereas surface water is extremely scarce in this arid region, an ample water supply can be observed in the high water table at several localities between the dunes. The source of this water is concentrated in two basins. These basins filled with windblown sand, overlying hard bedrock and bordered by quartzite ridges, conform to the theory of a natural dam (Circle Engineer: Lower Orange River Circle 1954). The authors are of the opinion that runoff from the Langeberg range was mainly responsible for the source of water, particularly during wet periods of the Upper Pleistocene.

The important point is that the conditions have been ideal for water to accumulate in the basins. This water was, among other factors, responsible for the discoloration of the sands by means of the process of hydromorphism, the formation of ferricrete and peat, and the creation of an environment with ideal living conditions for Stone Age man (Van Rooyen and Verster 1983).

Vegetation

The Witsand area generally supports a plant community that can be described as an acacia grass savannah. The dunes sustain very little vegetation, Stipagrostis amabilis being the most conspicuous plant species.

CHARACTERISTICS AND GENESIS OF THE FERRICRETE

Only one exposure of the ferricrete was located on the western margin of the northern basin, covering an area of about 0.5 ha. This more or less continuous layer lies approximately 15 m above the level of the present water table (Fig. 1). In turn, the ferricrete overlies several metres of light yellowish (10YR 6/4) sand. The ferricrete has an extremely hard consistency, displays a dark brown (7.5YR 4/4) colour and is non-magnetic (Fitzpatrick, pers. comm. 1980). The structure is predominantly vesicular with some pipestems and cellular zones. A fracture pattern of polygonal appearance is also evident, probably due to desiccation. Quartz particles can be seen randomly distributed throughout. The ferricrete is generally covered by an Fe-rich, slightly varnished patina. It varies in thickness from 15 - 25 cm.

Total element analysis by means of X-ray fluorescence shows a composition of c. 76 % Fe_2O_3, c. 20 % SiO_2, less than 2 % Al_2O_3, less than 2 % TiO_2, c. 80 % mg/kg Zr and traces of Ce. On the other hand, an estimate by means of XRD indicates that goethite is the dominant mineral present, with quartz subdominant and traces of anatase and rutile. A detailed analysis of the 111 reflection suggests the presence of a relatively crystalline goethite with a low degree of substitution by aluminium. No weatherable, primary minerals or clay minerals were detected.

Polished thin sections (Fitzpatrick and Milnes, pers. comm. 1983) show that it is a fairly homogeneous material composed of well-sorted quartz grains, sub-rounded to sub-angular in shape and up to medium sand size (0.5 mm) in a matrix of goethite. The goethite appears to be comparatively uniformly structured with a very low degree of orientation. Only traces of red, remnant haematite are visible in the matrix, but it lacks conspicuous structure. Although pipestems are uncommon, they are lined with goethite structures.

Based on the above-mentioned data a few conclusions with reference to the genesis of the ferricrete can be drawn. The low degree of aluminium substitution in the goethite structure and its high degree of crystallinity suggest that the goethite originally formed in a wet bottomland situation, i.e. where Fe accumulated in a reduced state without interference by Al according to the model by Fitzpatrick and Schwertmann (1982). The presence of fossil pipestems also confirms the possible formation under wet bottomland conditions. The absence of Mn, however, indicates that the ferricrete is presently not subjected to waterlogged conditions, confirming its relict nature.

There is little doubt that the ferricrete is the result of the accumulation of Fe in a wet bottomland situation. It is considered likely that the Fe was derived from coatings of sand grains during the discoloration of the sand (Van Rooyen and Verster 1983). Otherwise it would be difficult to explain accumulation of Fe in view of the extreme paucity of weatherable iron-bearing minerals. An influx of Fe from adjacent higher-lying deposits could also be visualized. Subsequent to erosion and/or lowering of the water table the ferricrete became exposed, hardened and attained a patina-like appearance.

LANDSCAPE EVOLUTION AND PALEOCLIMATIC CORRELATIONS

Interpretation of the surface phenomena and geomorphic setting records

Table 1. Tentative stratigraphic framework and paleoclimatic correlations of the Witsand features.

Events	Chronostratigraphy (Dates in yr BP)	Possible or inferred climate
Peat formation	Holocene (0 - 10 000) (880 ± 70)	Altern. warm & dry and warm & wet[1]
Active dune formation		(Warm & dry)
Recharge of groundwater		(Warm & wet)
Habitation by Wilton people	Holocene[2] (8000 - 10 000)	
Exposure of plinthite and hardening into ferricrete; erosion; lowering of water table	Late Upper Pleistocene (10 000 - 14 000)	Warm & dry[1]
Formation of plinthite; accumulation of groundwater in the two basins	Upper Pleistocene (20 000 - 45 000)	Cold & wet to very wet[1]
Habitation by Middle Stone Age people	Upper to Middle Pleistocene[2] (>40 000)	
Deposition of aeolian sand and admixture with locally derived sandy parent materials	Middle Pleistocene (>50 000)	Warm & dry

1. Brook (1982)   2. Klein (1977)

a distinct period of deposition of aeolian sand, accumulation of groundwater, pedogenesis (ferrugination), habitation by Stone Age people and erosion (mainly wind). A tentative stratigraphic framework and paleoclimatic correlations are presented in Table 1.

It seems likely that the sand of the Witsand area originated from a source to the north-west and was blown in to form seif dunes with a south-easterly to easterly trend during arid periods of the Middle Pleistocene (Van Rooyen and Verster 1983). It is assumed that the deposition of sand was completed during the Early Upper Pleistocene. At that stage the open low ridges to the west of the Langeberg range lay flush with the sand, forming two natural dams at Witsand in which runoff from the Langeberg accumulated and created a habitat for Middle Stone Age people.

The ferricrete developed in these sandy materials under the influence of a fluctuating water table. It is postulated that this high water table (about 15 m above the present one) resulted from periods of higher rainfall, although alternating wet and dry seasons would have been a necessary prerequisite. Seasonal distribution of rainfall is typical of the South African summer weather pattern even today, so it is likely that these conditions also existed during the Upper Pleistocene. Because of the formation of the ferricrete over a layer of porous sand, it can be deduced that a long period of a very stable water table existed. From all the available evidence, it is possible that these events may be correlated with the wet to very wet periods of approximately 45 000 - 20 000 BP inferred by Brook (1982). The hardening of the ferricrete may be ascribed to subsequent drier phases towards the transition of the Upper Pleistocene to the Holocene causing, among others, wind erosion and the lowering of the water table.

It is well-known that whitish sands and to a lesser extent yellowish sands are more subjected to wind erosion than, for example, reddish sands of a similar texture. Thus, after discoloration by hydromorphism, the light-coloured sandy soils were subjected to wind erosion and Witsand attained its characteristic dune-like appearance. These events probably took place during recurrent dry periods of the Holocene (Table 1).

CONCLUSIONS

The evidence presented above seems to indicate that the Witsand area was subjected to distinct events including deposition and admixture of sandy parent materials; accumulation of groundwater in two natural dams; pedogenesis (discoloration of sandy deposits by means of hydromorphism and plinthite formation); erosion and subsequent hardening of plinthite into ferricrete due to exposure; at the same time the general reduction of the water table to the present level about 15 m lower; habitation by two Stone Age cultures; and recent active dune formation, recharge of groundwater and peat formation. These events provide information about climatic changes during the Quaternary and can tentatively be correlated with the record of climatic evolution of this region as described by Brook (1982).

In itself, the ferricrete represents an irreversible, self-terminating feature (Yaalon 1971) and is an indicator of paleoclimatic conditions much wetter than those of today. It is therefore out of phase with the present climate and thus a paleosol with relict features. The stratigraphy of the ferricrete presents evidence of the extent to which water accumulated during wet periods of the Upper Pleistocene and the long period in which the water table was stationary with seasonal fluctuations.

ACKNOWLEDGEMENTS

The authors acknowledge financial support of the University of South Africa and Mrs MC Holsten is thanked for technical assistance.

REFERENCES

Brook GA 1982. Stratigraphic evidence of Quaternary climatic change at Echo Cave, Transvaal, and a paleoclimatic record for Botswana and northeastern South Africa. Catena 9:343-351.

Circle Engineer: Lower Orange River Circle 1954. Witsands. Unpubl. Rep. No. 432/H.15. Dept. Water Affairs, Pretoria.

Fitzpatrick RE and Schwertmann U 1982. Al-substituted goethite - an indicator of pedogenic and other weathering environments in South Africa. Geoderma 27:335-347.

Fock GJ 1961. A preliminary archaeological survey of the Witsand. Ann. Cape Prov. Museums 1:81-85.

Geol. Surv. 1977. 1:250 000 Geol. ser. 2822 Postmasburg. Dept. Mines. Govt. Printer, Pretoria.

Klein RG 1977. The ecology of Early Man in Southern Africa. Science 197:115-126.

Kriel JP 1967. Monthly rainfall and evaporation records of evaporation stations up to September, 1967. Div. Hydrol. Dept. Water Affairs, Pretoria.

MacVicar CN, De Villiers JM, Loxton RF, Verster E, Lambrechts JJN, Merryweather FR, Le Roux J, Van Rooyen TH and Harmse HJ Von M 1977. Soil classification. A binomial system for South Africa. Sci. Bull. Dep. Agric. Rep. S. Afr. No. 390.

Poynton RJ 1971. A silvicultural map of southern Africa. S. Afr. J. Sci. 67:58-60.

Van Rooyen TH and Verster E 1983. Characteristics and origin of the sands and other surface phenomena of the Witsand area. Tech. Comm. Dept. Agric. Rep. S. Afr. No. 180:107-114.

Weather Bureau 1954. Climate of South Africa. Part 1. Climate statistics, W.B.19. Govt. Printer, Pretoria.

Yaalon, DH 1971. Soil-forming processes in time and space. In: DH Yaalon (ed.), Paleopedology: origin, nature and dating of paleosols. Israel Universities Press, Jerusalem. p. 29-38.

Cainozoic palaeosols in the Naboomspruit area, Transvaal

E.VERSTER & T.H.VAN ROOYEN
University of South Africa, Pretoria

ABSTRACT. The study of soil layers with relict properties and associated features near Naboomspruit provides data for the stratigraphic reconstruction and climatic evolution of the area. Geomorphic evidence would seem to indicate that the landform on which these features occur qualifies as a high level terrace of Late Tertiary age. Relict properties are indicated by the base status of the soils that is considerably lower than may be expected in the present climate, and by the presence of ferricrete and an underlying mottled layer in a landscape that is at present well drained. The palaeosols and surrounding soils, for comparison, have been characterized by morphological, chemical, mineralogical and granulometrical analyses. The genesis of the palaeosols is ascribed to distinct events including cyclic periodicities involving deposition, erosion and various intensities of pedogenesis.

INTRODUCTION

A palaeosol is a soil which has formed on a landscape of the past (Yaalon 1971). Basically two types of palaeosols can be distinguished:
  1. buried soils which occur where the land surface has been covered by younger sediments, and
  2. relict soils which have one or more soil features reflecting environmental conditions different from those of the present day.
  Two kinds of palaeosols in the Naboomspruit area belonging to the latter type have been described previously (Verster 1975). The first consists of red, porous, apedal loams and clays of the Hutton soil form (MacVicar et al. 1977), displaying mesotrophic base status that is considerably lower than may be expected in the present-day climate. The second is a ferricrete or hard plinthite in a landscape that is at present well drained. Subsequently a detailed study was undertaken uncovering two additional layers, also with palaeo-features. Some of these features indicate past climatic conditions considerably wetter than those of today. This paper presents stratigraphic evidence of Quaternary climatic change as well as a record of cyclic periodicities involving deposition, erosion and pedogenesis since the Late Tertiary.

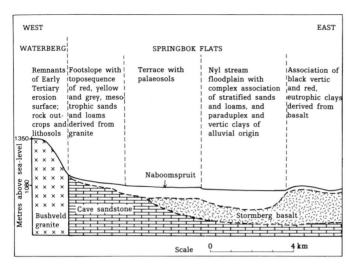

Figure 1. Position of terrace with palaeosols in the Naboomspruit landscape with associated environmental features.

PHYSICAL ENVIRONMENTAL FEATURES

The palaeosols occur over an area of 3400 ha in and around Naboomspruit (24° 31'S, 28° 42'E). It has been postulated that they occupy a high level terrace of the Nyl stream which is probably of Mio-Pliocene age (Verster 1975). In Fig. 1 the relative position of this palaeo-terrace in the Naboomspruit landscape as well as underlying rock formations and associated soils are shown. This terrace has a very slight gradient (slope 1-2 %) eastward and cuts across geological divergent rocks such as basalt and sandstone. In this region the adjoining Waterberg mountains are composed of granite of the Bushveld igneous complex. It is likely that the flat-topped crests of the Waterberg comprise remnants of the Early Tertiary erosion surface and that various other features which can be ascribed to Late Tertiary erosion and depositional phases (the high level terrace being an example of the latter) are discernible on the Springbok flats (Verster 1975).

Except for the palaeosols, all the surrounding soil associations are apparently in phase with present-day environmental conditions. These include, among others, the mesotrophic sands and light loams of the Hutton, Avalon, Longlands and Westleigh soil forms on the footslope of the Waterberg derived from granite colluvium, as well as the black, vertic clays of the Arcadia soil form and the red, eutrophic clays of the Shortlands and Hutton soil forms derived mainly in situ from basalt. The soils of the Nyl stream floodplain are typically soils developed in recent alluvial sediments and include soil forms such as Dundee, Oakleaf, Valsriver and Rensburg.

The climate, characteristic of the interior summer rainfall zone of South Africa, may be described in the terms of Poynton (1971) as "dry subhumid, warm temperate". The annual rainfall is about 600 mm, whereas the estimated average maximum and minimum air temperatures are

30 °C and 16.5 °C for January and 21.5 °C and 2.5 °C for June respectively (Weather Bureau 1954). According to Schulze (1965) the mean class-A pan evaporation for Potgietersrus is about 2100 mm per year. Comparing rainfall and evaporation, it is obvious that a large deficit of water is typical of the annual water budget of the Naboomspruit area. From the climatic data it is assessed that the soils have an ustic moisture regime (Soil Survey Staff 1975).

FEATURES AND PROPERTIES OF THE STRATIGRAPHIC LAYERS

A detailed study revealed that the high level terrace is composed of four stratigraphic layers (Fig. 2). Analytical data of each layer are presented in Table 1.

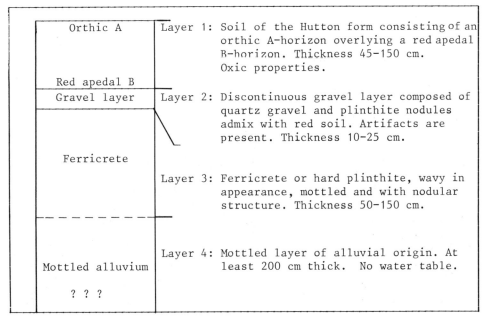

Figure 2. Vertical sequence of the four stratigraphic layers on the Naboomspruit paleoterrace.

Layer 1.

Soil of the Hutton form consisting of an orthic A horizon overlaying a red apedal B horizon. The orthic A is a thin (less than 20 cm), dark red to dark reddish brown (10R-2.5YR), apedal, slightly hard to hard sandy clay loam to sandy clay (clay content 25-40 %); no samples were analysed. The transition to the B horizon is diffuse and smooth.
 The B horizon is a red (10R3/4-2.5YR3/6), apedal, slightly hard to hard sandy clay loam to clay (clay content 30-50 %). Clay eluviation from the lower part of the layer is shown by a decrease in clay content. As already stated above, the palaeo-character is manifested inter alia by the mesotrophic base status (average S-value me/100 g clay = 10.6, see Table 1). In total, the thickness of the solum varies

Table 1. Mean analytical data of the four stratigraphic layers.

	Hutton B horizon (Layer 1) $\bar{x}$ (n = 14)	Gravel layer (Layer 2) $\bar{x}$ (n = 4)	Ferricrete (Layer 3) $\bar{x}$ (n = 2)	Mottled alluvium (Layer 4) $\bar{x}$ (n = 6)
Total fraction coarser than 2mm (% of whole soil)	7	65	69	34
Quartz gravel (2-16mm)(%)	6.8	26	10	23
Relict rounded Fe-rich nodules (%)	0.2	6	1	2
Angular Fe-rich nodules (%)	-	33	58	9
Statistical grain-size parameters				
Graphic mean (mm)	0.41	0.99	0.55	0.96
Graphic standard deviation ($\emptyset$)	1.72	1.94	1.78	1.97
Graphic skewness	-0.05	0.17	0.16	0.23
Graphic kurtosis	0.74	0.89	0.74	0.85
Fraction finer than 2mm				
Particle size distribution				
Clay (<2µm) (%)	39	35	26	37
Silt (2-20µm) (%)	8	9	17	7
Sand (20µm-2mm) (%)	53	56	57	56
Exchangeable bases (me/100g soil)				
Na	0.07	0.01	0.10	0.12
K	0.18	0.77	0.10	0.41
Ca	1.93	2.23	2.10	2.10
Mg	1.96	1.83	1.47	2.14
S-value (me/100g soil)	4.14	4.84	3.77	4.77
S-value (me/100g clay)	10.6	13.8	14.5	12.9
CEC (me/100g soil)	4.44	5.11	4.00	5.45
CEC (me/100g clay)	11.4	14.6	15.4	14.7
Base saturation (%)	93	95	94	88
pH.H_2O	5.7	5.7	5.8	5.8
Electrical conductivity (mS/m)	8	7	23	11
Free Fe-oxides (%)	5.28	-	-	-

from 45-100 cm with an average thickness of 90 cm. A conspicuous feature is the coarse fraction which contains about 7 % (of whole soil) of angular to sub-rounded quartz gravel (2-5 mm particle diameter) and, on average, less than 1 % of rounded, very hard, dark grey (10YR3/1) Fe-rich nodules (2-7 mm particle diameter) of relict origin. XRD analysis shows that the clay fraction is dominated by kaolinite, quartz and goethite with minor amounts of haematite and mica traces.

All of the diagnostic criteria, i.e. consistence, massive structure, diffuse horizon boundaries, presence of pores, chemical analysis, especially the CEC (less than 16 me/100 g clay, see Table 1), and the minerals in the clay fraction suggest that this subsoil represents an oxic horizon. The Hutton profile therefore belongs to some type of Eutrustox great group (Soil Survey Staff 1975).

Interpretation of the grain-size parameters (Table 1) indicates that the analysed sediment comprising the B horizon is bimodal, poorly sorted, near-symmetrical and platykurtic (Folk and Ward 1957). A complicated history of two or more sources (probably granite, sandstone and felsite), alluvial deposition and relatively short distance of transport, to account for the lack of roundness of particles, may be postulated.

The transition to Layer 2 is smooth and clear, a lithological discontinuity.

Layer 2

Gravel layer. It contains approximately 65 % coarser fragments (2-15 mm particle diameter) subdivided into 26 % sub-angular to sub-rounded quartz gravel; 6 % rounded, relict Fe-rich nodules; and 33 % angular, very hard, brown to dark brown (10YR4/3) Fe-rich nodules; and red loamy to clayey soil similar to Layer 1 (see Table 1). It occurs discontinuously in the sequence and varies from 10-25 cm in thickness.

In comparison with Layer 1, the coarser nature of the sediment is shown by an average mean grain size of 0.99 mm (Table 1). The sediment is also bimodal, poorly sorted, fine-skewed and platykurtic.

An important feature of the gravel layer is the presence of stone implements, thinly scattered in the form of an indistinct stoncline on the contact with the ferricrete. These implements which are made mainly of felsite and quartzite consist of various scrapers and grindstones. They probably belong to a younger phase of the Middle Stone Age, or more specifically, some of the implements correspond to assemblages typical of the so-called Bambata complex (JA Aukema, pers. comm. 1983). It is estimated that this complex is older than 30 000 BP (Sampson 1974).

The transition to Layer 3 is wavy and abrupt, a lithological discontinuity.

Layer 3

Ferricrete. From 50-150 cm thick and wavy in appearance, the ferricrete is characteristically variegated with many, distinct mottles of various sizes. Colour variegations such as very dark grey (10YR3/1 and 5YR3/1), dark reddish brown (5YR3/2), yellow (10YR6/8 and 2.5YR7/8), brown (10YR5/3) and light grey (10YR7/1) are common. Furthermore, red (2.5YR3/6) clayey soil, obviously derived from Layers 1 and 2 by illuviation, has filled some channels, voids and cracks and subsequently imparted an overall reddish colour to certain sections. The structure is predominantly nodular with some pipestems, although some exposures also display vesicular structure locally. Commonly it has a very hard consistence although, in a few localities, it was possible to auger through the ferricrete. In these cases it seems as if the ferricrete has undergone weathering into a loose, Fe-oxide nodular horizon.

The components of the coarse fraction are shown in Table 1. The grain-size parameters indicate that the ferricrete is bimodal in origin, poorly sorted, fine-skewed and platykurtic. Chemically the ferricrete is also similar to Layer 1 (Table 1).

Polished thin-sections of ferricrete (RW Fitzpatrick and AR Milnes, pers. comm., 1983) show that it is a heterogeneous material composed of poorly-sorted quartz grains, sub-angular to sub-rounded in shape and up to 4 mm in size, in a matrix of goethite and clay. Some structures represent fragments of relict ferricrete. Within the matrix, there are conspicuous differences in the distribution of goethite and clay minerals, with some areas being composed exclusively of clay. Red haematite occurs mainly as ferrans around some voids or channels. Many of these voids or channels are also lined with black manganese oxide representing the final depositional phase.

Total element analysis by means of X-ray fluorescence show a composition of 60.9 % SiO_2, 26.5 % Fe_2O_3, 7.4 % Al_2O_3 and all other elements less than 1 %, indicating that quartz is the dominant mineral in the clay fraction. Kaolinite is the dominant layer silicate with minor amounts of mica. Trace amounts of feldspar, rutile and anatase were also identified. A detailed analysis of the 111 reflection suggests the presence of a relatively crystalline goethite with a small amount of aluminium (5-8 mole % AlOOH) incorporated in the goethite structure.

The transition to Layer 4 is smooth and gradual.

Layer 4.

Mottled alluvium. At least 200 cm thick, this layer is typically variegated with many, faint mottles of various sizes. Colour variegations such as reddish (2.5YR5/8 and 5YR4/6), yellowish (10YR6/6 and 7.5YR5/6), greyish (10YR7/4, 7/3 and 6/4) and some blackish (7.5YR2/0) occur commonly. In addition, dark red brown (10YR3/4 and 2.5YR3/4) cutans are encountered. These argillans, filling voids, channels and cracks, are also proof of the argillic nature of especially the upper part of the layer.

According to Table 1, this layer has similar chemical properties to Layer 1. It therefore displays some features characteristic of an oxic horizon which obviously is relict. Furthermore, the presence of plinthite in this situation, where no water table is operative at present, is also proof of palaeo-conditions.

The heterogeneous character and the alluvial origin of the material are shown by the granulometric properties: apart from 23 % sub-angular to sub-rounded quartz gravel (2-8 mm particle diameter); 2 % relict, rounded Fe-rich nodules; and 9 % angular, dark yellow brown (10YR4/4) to brownish yellow (10YR6/6), hard Fe-rich nodules, it also contains sub-rounded to rounded gravel and stones (size 30-160 mm) composed of quartzite (origin unknown), sandstone and quartz. Although the interpretation of the average grain-size parameters (Table 1) indicates that the sediment is bimodal, poorly sorted, fine-skewed and platykurtic, widely divergent individual figures were obtained. For example, the graphic mean value varies from 0.37 mm to 1.85 mm with an average of 0.96 mm (Table 1).

Though the contact to the bedrock was nowhere exposed, borehole information suggests a total thickness of 18-20 m for the four stratigraphic layers.

DISCUSSION OF GENESIS AND PALAEOCLIMATIC CORRELATIONS

Interpretation of the layers suggests three distinct periods of deposition (Layers 1, 2 and 4) and one of erosion (between Layers 2 and 3), habitation by Middle Stone Age people (on Layer 3) and pedogenesis. The following is an attempt to reconstruct the genesis of the high level terrace and its associated palaeosols, with a tentative strati-

Table 2. Tentative stratigraphic framework and palaeoclimatic correlations of the Naboomspruit palaeosols.

Events	Chronostratigraphy	Possible or inferred climate
Present-day soil-forming processes: eluviation from Layers 1 and 2 and illuviation into Layers 3 and 4; ferrugination of Layers 1 and 2; homogenization of stratifications.	Holocene to Late Upper Pleistocene	Alternating warm and dry and warm and wet; also with cooler and moister intervals
Deposition of Layer 1 derived from preweathered, oxic materials.	Upper Pleistocene	Cold and wet
Deposition of Layer 2.	Upper Pleistocene	Cold and wet to very wet
Habitation by Middle Stone Age people (Bambata complex).	Upper Pleistocene (>30 000 BP)	
Erosion of topsoil and hardening of upper part of plinthite into ferricrete; lowering of water table.	Early Upper Pleistocene	Hot and dry
Development of plinthite (Layers 3 and 4); high fluctuating water table.	Middle to Lower Pleistocene	Wet, strongly seasonal
Pedogenesis and oxic weathering of deposits on high level terrace (Layers 3 and 4) and materials on stable land surfaces on top of Waterberg mountains.		Warm and wet
Deposition of alluvial materials (Layers 3 and 4).	Lower Pleistocene to Late Tertiary	Wet
Formation of terraces along the Nyl stream.	Late Tertiary	

graphic framework and palaeoclimatic correlations presented in Table 2.

Since the advent of the high level terrace, thick deposits of alluvium (especially Layer 4) accumulated, probably during Late Tertiary times. It seems likely that they were laid down by the lateral migration of the Nyl stream channel, with overbank deposition, thus accounting for the fine sediments, in a manner similar to that of the classic floodplain model proposed by Mackin (1937). Whether or not the formation of this terrace is to be attributed to fluctuations in climate, tectonic activity or to critical thresholds intrinsic to an alluvial system being exceeded (Young and Nanson 1982), is unclear.

The plinthite developed in these deposits under the influence of a fluctuating water table obviously connected to the water level in the Nyl stream. It is postulated that this high water table was associated with periods of higher rainfall, although alternating wet and dry seasons would have been a necessary prerequisite. In fact, it seems almost certain that the plinthite formed mainly by absolute Fe-accumulation in a hydromorphic environment according to the model by Fitzpatrick and Schwertmann (1982). Irreversible hardening of the upper part of the plinthite into a ferricrete resulted from the stripping of the topsoil by erosion and/or the lowering of the water table. Both these processes could presumably be ascribed to drier periods.

Subsequent to the erosional phase which caused the exposure of the ferricrete, and the habitation by Middle Stone Age people, the gravel layer (Layer 2) was deposited in a higher energy environment indicated by the coarser nature of the deposits, and followed by the deposition of finer material of Layer 1.

Of special interest to this study is the deposition characteristics of Layer 1, which is interpreted as being mainly an overbank deposit. Apparently aggradation took place during wet periods while the Nyl stream channel maintained a constant position. Also of importance is whether or not the mesotrophic or oxic properties were inherited or developed in situ. Previous research has noted that the mesotrophic nature was inherited from preweathered materials which were deposited on the high level terrace (Verster 1975). Present research which includes, among others, additional granulometric and stratigraphic analyses is in accordance with this hypothesis. The authors are still of the opinion that the deposits of Layer 1 were derived primarily from very old and stable landscapes, probably from the Early Tertiary surface on top of the Waterberg mountains (see Fig. 1). Furthermore it is argued that the inferred climate, according to Brook (1982), since the habitation by Middle Stone Age people, was not conducive to oxic weathering, apart from the fact that very long periods of intensive weathering are apparently required to produce oxic properties (Soil Survey Staff 1975). It is postulated that this type of weathering as well as pedogenesis occurred during long periods of the Middle to Lower Pleistocene (Table 2).

Since deposition, Layer 1 has been modified by pedogenesis. Some of the more important processes are listed in Table 2. Illuviation of clayey materials into the ferricrete and upper part of Layer 4 forms part of these recent processes.

CONCLUSIONS

The evidence given above seems to suggest that the Naboomspruit area is a veritable palaeosol museum, displaying palaeo-terraces along the Nyl stream floodplain composed of stratigraphic layers, each with relict features. These features record distinct events, including cyclic periodicities involving deposition, erosion and various intensities of pedogenesis since the Late Tertiary. Some of the soil-forming processes give rise to oxic horizons and plinthite formation which are irreversible, self-terminating features and according to Yaalon (1971) the best indicators of palaeopedogenetic conditions.

Alghough these features provide significant indications of Cainozoic climates, the construction of a chronostratigraphic framework is somewhat speculative. However, the presence of MSA stone implements on the contact between the ferricrete and the overlying soil is to assign an Upper Pleistocene age to the deposits of Layers 1 and 2.

In addition, it is not self-evident that the relict features and associated events could be used to make meaningful statements about the climatic evolution of the area. The problem seems to be one of interpretation, i.e. whether or not terrace formation is attributable to fluctuations in climate, tectonic activity or any other factor; or whether episodes of deposition and erosion can be related to specific climatic conditions. On the other hand the period referring to the last 40 000 years has a relatively well-documented record of climatic evolution for this region due to research on cave deposits (Brook 1982) and pollen spectra (Scott 1982). The deposition of Layers 1 and 2 can therefore be explained in terms of the wet to very wet conditions that apparently prevailed during the Upper Pleistocene. Other palaeoclimatic correlations to periods older than the Upper Pleistocene are likely, however, to be very speculative. Nevertheless some of the palaeo-features indicate specific climatic conditions. Hence this study presents stratigraphic evidence of Quaternary climatic changes and also contributes to a better understanding of the history of the area.

ACKNOWLEDGEMENTS

The authors acknowledge financial support of the University of South Africa and Mrs MC Holsten is thanked for technical assistance.

REFERENCES

Brook GA 1982. Stratigraphic evidence of Quaternary climatic change at Echo Cave, Transvaal, and a palaeoclimatic record for Botswana and northeastern South Africa. Catena 9:343-351.
Folk RL and Ward WC 1957. Brajos River Bar. A study in the significance of grain size parameters. J. sedim. Petrol. 27:3-27.
Fitzpatrick RW and Schwertmann U 1982. Al-substituted goethite - an indicator of pedogenic and other weathering environments in South Africa. Geoderma 27:335-347.
Mackin JH 1937. Erosional history of the Bighorn Basin, Wyoming. Bull. geol. Soc. Am. 48:813-894.
MacVicar CN, De Villiers MN, Loxton RF, Verster E, Lambrechts JJN, Merryweather FR, Le Roux J, Van Rooyen TH and Harmse HJ Von M 1977. Soil classification. A binomial system for South Africa. Sci. Bull. Dep. Agric. Rep. S. Afr. No. 390.
Poynton RJ 1971. A silvicultural map of southern Africa. S. Afr. J. Sci. 67:58-60.
Sampson CG 1974. The Stone Age Archaeology of Southern Africa. Academic Press, New York.
Schulze BR 1965. Climate of South Africa. Part 8. General Survey, WB28. Govt. Printer, Pretoria.
Scott L 1982. A late Quaternary pollen record from the Transvaal Bushveld, South Africa. Quat. Res. 17:339-370.
Soil Survey Staff 1975. Soil Taxonomy. US Dep. Agric. Handbk 436. US Govt. Printing Office, Washington D.C.
Verster E 1975. Palaeosols in die Naboomspruitomgewing. Proc. 6th Congress, Soil. Sci. Soc. S. Afr.: 317-323.
Weather Bureau 1954. Climate of South Africa. Part 1. Climate statistics, WB19. Govt. Printer, Pretoria.
Yaalon DH 1971. Soil-forming processes in time and space. In: DH Yaalon (ed.), Palaeopedology: origin, nature and dating of palaeosols. Israel Universities Press, Jerusalem. p. 29-38.
Young RW and Nanson GC 1982. Terrace formation in the Illawarra Region of New South Wales. Aust. Geographer 15:212-219.

Lake level fluctuations during the last 2000 years in Malawi

R.CROSSLEY, S.DAVISON-HIRSCHMANN & R.B.OWEN
University of Malawi, Zomba

P.SHAW
University of Botswana, Gaborone

ABSTRACT. Current progress in identifying fluctuations, over the last 2000 years, in the levels of Lake Chilwa and Lake Malawi is described. High levels are identified by surveying and excavating of beach ridges, low levels by piston-coring and grab-sampling of the floors of the lakes and of hydrologically connected lagoons. Dating of fluctuations is based on types of Iron Age pottery in stratified sequences, radio-carbon dating, baobab sizes, and sedimentation rates.

Evidence for water level fluctuations of 14 m in Lake Malawi, and of 12 m in Lake Chilwa, is presented. The last major rise of Lake Chilwa, which occurred about 160 BP and reached about 9 m above modern mean water levels, does not correlate with the last major rise of Lake Malawi. Some of the palaeoclimatic implications of these findings are discussed.

INTRODUCTION

Our interest in fluctuations over the last 2000 years in the levels of Malawi's lakes (Fig. 1) was prompted by the need to obtain reliable estimates of extreme water levels for development planning. A palaeo-hydrological approach was adopted because of the repeated failure of standard hydrological methods of extreme level prediction (Crossley, in press). Lake level variations naturally have a bearing on climatic fluctuations and although the period examined is short in palaeoclima-tological terms, it is apparent that Malawi's climate was not static over the 2000 year interval.

LAKE LEVEL RECORDS

Records have been kept since 1896 for Lake Malawi and since 1950 for Lake Chilwa. A barrage has been operated on the Shire River since 1966 with the result that Lake Malawi levels were 15 cm higher in 1979 than they would otherwise have been (Drayton 1979). In order to show 'natural' levels, the data plotted in Fig. 2 have had the effects of barrage operation removed, following Drayton (1979).

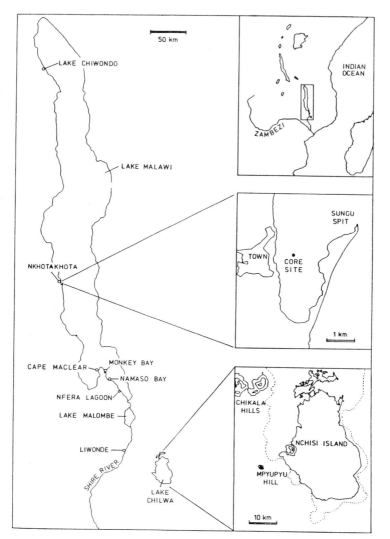

Figure 1. Map showing localities around Lake Malawi and Lake Chilwa.

HISTORICAL EVIDENCE

Lake Chilwa

The lake level fluctuations shown in Fig. 2 are based on data in Lancaster (1979) and from three additional sources. Firstly, it is certain, from an article in the missionary newspaper "Life and Work" (Anon. 1903), that Lake Chilwa was dry in the early 1900's. Secondly, whilst it can be confirmed that Lake Chilwa was high at the time of Livingstone's visit in 1859, the suggestion that the lake was significantly above the 1978 peak (Lancaster 1979) is based largely on a widely-held, but mistaken, opinion regarding the site from which Kirk's painting of the lake was made. This painting, reproduced in Wallis (1956), is often reported as having been painted from Mpyupyu Hill.

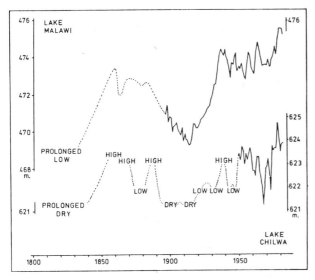

Figure 2. Variations in the maximum levels of Lakes Malawi and Chilwa based on hydrological records and historical data.

Comparison of Kirk's picture with various views from the west margin of the lake suggests that it was painted from the vicinity of the Chikala Hills – and certainly not from Mpyupyu Hill. The broad expanse of water represented in the painting can then be interpreted as being due to limited reed mat growth rather than extremely high lake level. An extremely high lake level in 1859 is also ruled out by the size of baobabs only a metre or so above the 1978 peak lake level.

Thirdly, there is some evidence for a prolonged low level in the early 1800's. A hard layer at least 1.2 m thick underlies about 1 m of softer, mainly muddy, sediments at the 18 sites cored so far in the lake. The hard layer is considered to be the result of an extremely prolonged dry phase (see below). It is inferred that in 1903 only a small amount of sediment overlay the hard layer. This inference is based on an account of a walk across the dried lake floor to Nchisi Island (Anon. 1903), which mentions that the crust occasionally broke, causing walkers to sink into the mud beneath which was only ankle deep; presumably the hard layer prevented them from sinking farther.

The thinness of sediment above the hard layer carries implications for the date of the end of the prolonged dry phase. At a rough guess, the 1903 mud crust would have been between 100 and 200 mm thick (if it were to support men's weight yet rupture periodically). At modern sedimentation rates, the sediment in such a crust, plus the underlying 100 to 200 mm of mud, would have taken between 50 and 100 years to accumulate. One can therefore, somewhat cautiously, infer that the major pre-1903 dry phase ended some time between 1800 and 1850.

Lake Malawi

The Lake Malawi levels between 1867 and 1896 shown in Fig. 2 are based on Drayton (1979) after Pike (1965). The 1863 level is based on soundings by Kirk in Lake Malombe because that lake level is, at moderate to high water levels, virtually identical to Lake Malawi values. The peak level around 1860 is based on coincidence between the lake level deduced from Livingstone's soundings in Lake Malombe in 1861 and Dixey's (1924) report of marks (presumably aquatic algal lines) on lakeshore rocks up to that level.

The extremely low lake levels around the early 1800's are based on two sources: Dixey's (1924) report of very low levels in the 1830's (the precise source of his data is not known), and a report by Swann (1894).

Swann states: "The gigantic trees still standing erect at the North End (of Lake Malawi), submerged in three foot of water, remain mute witnesses to a past life on dry land". Swann's observations were made in 1892-93, when the lake stood at a level of about 470 m - its lowest since Livingstone's 1861 visit. The implication is that the waters of Lake Malawi were, at some time prior to 1860, continuously at a level of less than 469 m for a period long enough to allow the growth of "gigantic" trees. Even allowing for the relatively rapid growth of certain trees in the tropical climate a low level period of at least 50-100 years must be envisaged, presumably spanning the late 1700's and early 1800's.

EVIDENCE FROM THE LAKE MARGINS

Lake Chilwa

Initial findings on both low and high levels of Lake Chilwa during the Iron Age are described by Shaw et al. (in press). The most dramatic discovery reported is a beach gravel representing a lake level of 631 m, some 9 m above modern mean level. This gravel immediately overlies a sand layer, which has yielded charcoal giving the following radiocarbon results: 160 ± 50 BP (PTA-3316) and 160 ± 90 BP (HAR-4701). Vogel (pers. comm., 1982) comments that, allowing for uncertainties in calibration, the isotopic ratios could represent dates anywhere in the range AD 1750 ± 100. Lancaster (1981) calculated that a 631 m Lake Chilwa level represents, assuming modern evaporation rates, a 35-40 % increase in rainfall. It might seem surprising that a climatic fluctuation of that order could have occurred within the last few centuries, so we have examined other dating evidence.

The archaeological material obtained in the course of excavation of the 631 m beach provides an independent age check. The beach gravel overlies a colluvial unit containing unrolled potsherds mainly of the "Longwe" type. This ceramic tradition is known to span the 10th to 16th centuries, though a later persistence of the tradition in the Chilwa area is possible. Rolled "Longwe-type" potsherds occur throughout the beach gravel. Unrolled pottery occurs in the upper part of the gravel and consists predominantly of traditional "Lomwe-type" potsherds, of the 19th or 20th centuries. The beach gravel has also yielded "Indian red" trade beads, likely to be post-16th century, and a "slave trade" bead, common in 19th century deposits in Malawi. The archaeological material therefore gives no reason to doubt an AD 1750 ± 100 date for the 631 m lake level.

If Lake Chilwa had indeed reached that level around the mid-18th century, then its effect might be expected to show up in the age distributions of high longevity trees occurring below the 631 m level.

Observations on the Lake Malawi shore in recent years have shown that baobabs are extremely intolerant of waterlogging and collapse readily when encroached upon by rising water. This factor, together with their longevity, clear growth-ring patterns and relative abundance on Nchisi Island, made baobabs the natural choice for a search for evidence of disturbed age distributions.

Southern Malawi annually has a single clearly defined wet season, so tree ages can be estimated by counting growth rings. On the basis of 131 ring thickness measurements, the average width of baobab growth rings on Nchisi Island is 83 mm. This figure has been used to calculate, from their girths, the ages of 74 baobabs on the island. Their total age distribution is shown in Fig. 3, below a graph of data from 173 baobabs measured by Caughley (1976) in the Luangwa Valley, Zambia. The average ring width of the Luangwa baobabs was 75 mm but the maximum age of trees in that population is similar to that found for baobabs on Nchisi Island, thereby supporting the ages calculated for the island trees.

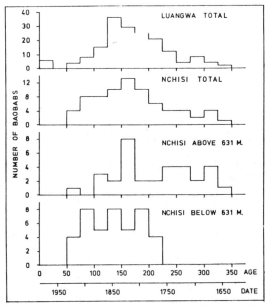

Figure 3. Age distributions of baobab trees from Nchisi Island in Lake Chilwa and from the Luangwa Valley, Zambia (after Caughley 1976).

There is a marked difference between the age distributions of baobabs occurring below and above the 631 m level, which is illustrated in Fig. 3. Using the Chi squared test, the difference between the two populations is significant at the $P<0.001$ level. The contrast is because the oldest measured baobab below the 631 m level is only 223 years whereas above that level ages ranging up to 326 years have been found.

Flooding by a 631 m Lake Chilwa level would seem a likely cause of the age distribution difference. The radiocarbon results and the baobab ages therefore bracket the date of the 631 m Chilwa level at somewhere between AD 1650 and 1750.

Reworking of previous beach deposits during the last 631 m lake level has posed problems for obtaining unequivocal evidence for the number and size of earlier high Chilwa levels. Nkope-type (Early Iron Age) potsherds, of probably pre-10th century age, in the 631 m beach gravels consistently show greater abrasion than Longwe-type sherds in the same gravels, suggesting that the Early Iron Age sherds have suffered more than one phase of rolling.

A cobble beach, at altitudes of at least 631 m, which pre-dates the Iron Age sequences at Sonko Bay (Shaw et al., in press) and the beach sands, occurring up to a maximum of 633 m, which form the foundations of the Chilwa sand bar (Lancaster 1981) testify to prolonged or repeated high stands of the lake in pre-Iron Age times. The precise age of these early high levels is at present a matter of conjecture.

Not surprisingly, there is little evidence onshore regarding low levels of Lake Chilwa. Shaw et al. (in press) do however report evidence for the lake level being around or below 624 m during part of the Early Iron Age.

Lake Malawi

Three prehistoric high lake levels within approximately the last 2000 years have already been identified and assigned approximate ages on the basis of associated artefacts (Crossley and Davison-Hirschmann 1981, 1982):
 1. A lake level of 476.1 m was reached late in the period of the Mawudzu pottery tradition, which is currently thought to span the mid-12th to mid-18th centuries.
 2. The lake peaked at a level of 475.6 m shortly before, or very early in, the Mawudzu period.
 3. A level of 476.8 m was reached shortly before the period covered by the Nkope pottery tradition, which is presently believed to span the 3rd to 10th centuries.

The late Mawudzu high lake phase is now considered to have been a double peak event but it is not known whether the interval between the twin events was merely a few years, or many tens of years. Tree-ring counting and radiocarbon dating have now been used in an attempt to refine dating of these peak level events.

The oldest measured tree standing on the late Mawudzu beach ridge, an Acacia albida, has an age of about 360 years. We have not obtained a statistically significant number of baobab ages from the late Mawudzu ridge but the oldest specimen examined is about 225 years. A radiocarbon date of "modern" (HAR-4705) has been obtained from charcoal which post-dates the late Mawudzu beach sand. Charcoal from a horizon which, since it is within the beach sand unit, may represent the interval between the twin peak levels, has given an age of 480 ± 80 BP (HAR-4703).

Charcoal from a stone structure, first reported by Crossley and Davison-Hirschmann (1981), beside the Shire river at Liwonde has given a radiocarbon age of 480 ± 80 BP (HAR-4702). The structure is interpreted as a boat "wharf" and/or "pavement" for washing and water collection. In view of its location, on river silts about 40 cm above the 1980 river level, the interpretation implies a high river level. The date correlates with the late Mawudzu high level of Lake Malawi and the inferred river level implies that the flow rate of the Shire at that time was closely comparable to that of the modern river. This is important since it shows that the late Mawudzu high lake level was likely to be due to climate rather than to blockage of the overflow.

A charcoal sample from above the pre-, or early Mawudzu beach sand has yielded a date of 990 ± 90 BP (HAR-4704). Two radiocarbon dates on charcoal have been obtained from samples which post-date the pre-Nkope beach. One associated with a cluster of potsherds considered to be typologically "late" in the Early Iron Age, has given a date of 1340 ± 50 BP (PTA-3315). The other, associated with pottery believed on typological grounds to be "early" in the Early Iron Age, has yielded a date of 1720 ± 90 BP (HAR-4707).

EVIDENCE FROM THE LAKE FLOORS

Lake Chilwa - coring

In an attempt to investigate past dry phases of Lake Chilwa, piston-coring has been attempted at 18 sites in the modern lake floor. A consistent picture emerges: the upper 1 m is relatively easily penetrated and shows distinct stratification; below this depth penetration is extremely difficult and no clear stratification is seen to the limit of coring (2.2 m).

The stratified sequences include homogeneous muds, water-rolled mud conglomerates, and quartz sands deposited by water, together with quartz sands, dried mud clasts and ostracod valves which were reworked and deposited by wind. Salts are common and mud cracks are encountered occasionally. Extrapolation of sedimentation rates following the latest low phases (1968 and 1973) suggests that the stratified sequences probably represent fluctuations in lake level since the late 19th century.

The underlying unit comprises structureless sandy clays and its homogeneity is interpreted as the result of mixing during an extremely prolonged dry phase. There are a variety of reasons for this conclusion. Firstly, we know of no way in which homogeneous sandy clays can be deposited in the centre of Lake Chilwa - either clays and silts are deposited when the lake is full, or sands can be deposited when the lake is dry. It would therefore appear that mixing has occurred - yet the lake floor benthos is remarkably sparse and the stratified unit testifies to the ineffectiveness of bioturbation by aquatic organisms in this lake. Secondly, there is a marked drop in uncombined water from around 60 % to around 40 % (hence the penetration problems) when the homogeneous unit is reached, suggesting that the mixing is associated with drying.

Inspection of the fringes of the modern lake reveals that three mixing processes can operate. Firstly, Chilwa clays are expanding lattice type, so "heaving" of exposed clays occurs with repeated seasonal wetting and drying. Secondly, ants and termites excavate substantial quantities of sediment, up to coarse sand size, in the grasslands surrounding the present lake. Thirdly, plant-root growth can, at least on a minor scale, disrupt sediments.

None of these mixing processes operates rapidly, so the fact that the homogeneous unit is at least 1.2 m thick indicates an extremely prolonged period of relatively dry climate. The historical information suggests that the low levels responsible for the hard layer ended around the 1850's and the sedimentological data indicate that this dry phase was far more prolonged than any which has occurred within the last 100 years.

Lake Malawi - grab sampling

Grab sampling at three locations with very gentle bottom gradients in the SW part of the Lake Malawi: Namaso Bay, Monkey Bay and Cape Maclear, consistently shows beach sands and gravels down to levels of 461 m (approximately 15 m depth). The negligible pelagic cover on these sediments suggests that the low levels implied by the submerged beaches may well have occurred within the last few thousand years.

Submerged beach sands and gravels are also found on relatively steep bottom slopes to levels of 428 m (approximately 48 m depth) to the NE of Monkey Bay. However, it is not yet certain that these particular sediments are 'in situ' since slumping could conceivably occur on these bottom gradients.

Lake Malawi - coring

Lake Malombe, Nfera Lagoon and Lake Chiwondo, which are all hydrologically continuous with Lake Malawi, were subjected to an exploratory coring programme. Each water body is comparatively shallow, with floor levels above 469 m, and thus dried out during the brief period of low water levels around 1915 (minimum 468.5 m).

Cores in each case showed that a few hundred mm of soft mud overlay a horizon which contained clay breccia and/or shell concentrations and which rested in turn on very hard clay containing carbonate concretions and soluble salts. Diatoms are present in the soft muds, but absent from the underlying hard clay, perhaps because of dissolution in the high pH pore waters in the hard clay. In view of the absence of diatoms and problems of penetrating the hard clays, these sites were abandoned.

A new site was chosen off Nkhotakhota which is within the main lake but protected from strong wave action by Sungu sand spit. The lake floor level at the site is 468.3 m so it was not exposed in 1915. The sediments cored are dominantly sandy micaceous silts and, unlike the clays in the smaller water bodies, do not appear to form a concrete-like substance upon desiccation during some very low levels pre-1915 (see below). Consequently three metres of sediment were penetrated before a sand-gravel unit prevented further progress.

Information from the Nkhotakhota cores is summarized in Figs. 4 and 5. Samples of 0.8-1.0 g (air-dry weight) were taken for sediment analysis at 40 mm intervals throughout the upper 750 mm of the sequence. Below that level a 40 or 80 mm sampling interval was used, depending upon the degree of stratigraphic variation noted during examination of the split cores under the binocular microscope. For diatom analysis, samples were taken at 50 mm intervals, with additional materials from particularly interesting horizons. The diatom percentages are based on counts of 200 per slide.

The terrigenous fraction is divided for purposes of this analysis into "equant" grains, which consist mainly of quartz with minor amounts of feldspar, "pyribole" and garnet, and the "platey" grains which consist largely of bleached biotite with minor amounts of muscovite and graphite.

The "equant" grains show an overall fining-upward trend in the maximum clast size, from gravel at 2600 mm depth to fine sand near the sediment-water interface. It seems likely that this trend is due primarily to increased protection from wave action afforded by growth of the Sungu sand spit complex.

Sandy horizons at 270-300 mm depth and at 750 to at least 850 mm depth, interrupt the fining-upward trend. A third sandy horizon occurs at 1720 mm depth but is not unusually coarse grained. Discontinuous sand patches occur at 150 mm and 1600 mm depth. The sandy units could be caused by wave action, which in this protected situation implies very shallow water depths.

Alternatively, exposure could result in similar sediments through the action of overland flow. Either interpretation implies that the sand horizons represent low lake levels.

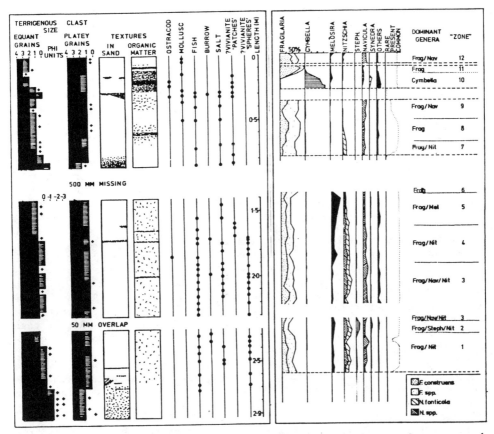

Figure 4. The Nkhotakhota cores, Lake Malawi: sedimentary features and diatom distributions. Grain size analysis keys: black - 50 or more grains; hatched - 49 to 10 grains; crosses - 9 to 1 grains. Filled circles indicate presence of feature indicated.

Variations in organic matter textures are most prominent in the upper 400 mm of the sequence. There is a thin plant-rich layer between 85 and 95 mm depth, and a major plant-rich layer between 105 and 180 mm depth which contains abundant remains of Ceratophylum spp. (identification by Prof G Berry).

The restriction of salt to depths greater than 180 mm is significant. Since the sediments in this freshwater lake are unlikely to be strongly saline at the time of accumulation, it is considered probable that the salts in the cores were precipitated in response to evaporative concentration of interstitial liquors. In other words, the salt indicates that the site was above water level at the time represented by the 280 mm level in the core.

Seven diatom assemblages can be recognized, some of which occur at several levels. These are: 1 Fragilaria assemblage; 2 Fragilaria/Navicula assemblage; 3 Fragilaria/Navicula/Nitzschia assemblage; 4 Fragilaria/Nitzschia assemblage; 5 Fragilaria/Melosira assemblage; 6 Fragilaria/Stephanodiscus/Nitzschia assemblage; 7 Cymbella assemblage. The Fragilaria of assemblage 1 are commonly cited as benthic forms; however, they are often swept up into the plankton of shallow lakes. Benthic/epiphytic floras such as those represented in assemblages 2, 3 and 7 reflect relatively low levels or perhaps less turbid conditions and consequent increased light penetration. In contrast, assemblages 4, 5 and 6 include planktonic diatoms that suggest greater water depth. Other factors probably contributing to floral changes through the core include competition from Cyanophyceae and variations in water chemistry. More detailed comments may be possible after study of additional cores which have now been obtained from Nkhotakhota.

Although this work is preliminary in nature, certain interpretations may be made:

a) The thin organic-rich layer around 90 mm may represent the brief period of low lake level around 1915. The absence of diatoms in this horizon could reflect competition from blue-green algae since, during that period, the core site was in a lagoon seasonally separated from the main lake.

b) The Ceratophylum-rich layer between 105 and 280 mm depth, which is dominated by an epiphytic Cymbella in its lower half, clearly represents an important event. The layer rests on cross-bedded sands which in turn overlie burrowed salty sediments. The burrows are near-vertical, about 2 mm across, and lined with clean sand; they resemble structures made by ants around the present lake shore. There can be little doubt that this sequence represents a period when the site was above lake level, probably for a substantial period. It is likely that this low lake level phase is the one inferred on historical grounds which spanned the late 1700's and early 1800's.

c) It is suspected that the sand horizons around 800, 1720 and 2600 mm depth also represent periods of emergence or near-emergence. The absence of a major organic-rich layer may be because the configuration of the sand spits did not allow the formation of a lagoon or swamp during the earlier low lake level events.

d) If our interpretation (a) is correct, then the upper 85 mm of the sequence accumulated in the last 68 years. If this sedimentation rate is extrapolated to the base of the sequence, then the cores reach back about 2300 years. However, the assumption of a constant sedimentation rate is unlikely to be strictly valid.

CONCLUDING REMARKS

Our present understanding of Lake Malawi and Lake Chilwa levels may be summarized as follows:

1. A variety of evidence indicates that both lakes were low for an extended period some time between AD 1750 and 1850. This suggests a relatively dry and widespread climatic event.

2. Lake Chilwa was extremely high some time between AD 1650 and 1760. According to Lancaster's (1981) calculations such a level represents a 35-40 % increase in rainfall, assuming modern evaporation rates. We do no have unequivocal evidence that Lake Malawi was particularly high at that time.

3. Lake Malawi reached its highest level within the last 1700 years some time around the 15th century. This appears to have been a double peak event and the time spanned by the twin rises is not known, but Acacia tree-rings suggest that the high lake phase had ended by AD 1620. The high Shire river level indicated around the 15th century shows that overflow blockage was not likely to have been the cause of the high lake level. A relatively wet climatic phase is therefore inferred. The climate need not have been substantially wetter than that of the late 1970's.

4. High Lake Malawi levels are indicated around the 10th century and around the beginning of the 1st millennium AD but in the absence of data on the character of the overflow at these times it is not possible to be sure that relatively wet climatic periods are the cause.

5. Lake Chilwa was high on at least one occasion during the Iron Age prior to the 18th century. Dating is presently insufficient to determine possible correlations with Lake Malawi levels.

6. Lake Chilwa was around or below modern mean levels during part of the Early Iron Age.

7. The Nkhotakhota cores indicate that Lake Malawi was very low (at levels of less than 467.5 m) on at least three occasions over a period of perhaps 2000 years prior to the 18th century.

8. Grab sampling indicates that Lake Malawi has been at levels as low as 461 m within the last few millennia.

So, although the last few thousand years are generally considered 'unremarkable' by palaeoclimatological standards, climatic variations over this period have had a substantial impact on the levels of Malawi's lakes.

ACKNOWLEDGEMENTS

This work was funded by the University of Malawi and by the International Development Research Centre of Canada. Prof D Livingstone kindly donated the coring equipment. P Mosley, R Noble, D and S Tweddle, A Seymour, M Swords, L Paine, J Wilson, T Jones and M Vaughan all helped in various ways.

REFERENCES

Anon. 1903. A mission-week on Mchisi Island. Life and Work in British Central Africa. Mission Press, Blantyre. 13 pp.
Caughley G 1976. The elephant problem - an alternative hypothesis. E. Afr. Wildl. J. 14:265-283.
Crossley R, in press. The implications for development of variations in the levels of Lakes Malawi and the Shire River. In: G Williams (ed.), Proc. South. Afr. Conf. Commonw. Geograph. Bur., Lusaka, 1982.
Crossley R and Davison-Hirschmann S 1981. Hydrology and archaeology of Lake Malawi and its outlet during the Iron Age. Palaeoecol. Africa 13:123-126.

Crossley R and Davison-Hirschmann S 1982. High levels of Lake Malawi during the late Quaternary. Palaeoecol. Africa 15:109-115.

Dixey F 1924. Lake Nyassa level in relation to rainfall and sunspots. Nature 1-3.

Drayton R 1979. Study of the causes of the abnormally high levels of Lake Malawi in 1979. Malawi Govt, Water Resour. Div., Tech. Rep. No. 5. 47 pp.

Lancaster N 1979. The Changes in the lake level. In: M Kalk, AJ McLachlan and C Howard-Williams (eds.), Lake Chilwa: Studies of Change in a tropical Ecosystem. Monographae Biologicae 35. Junk, The Hague. p. 27-34.

Lancaster N 1981. Formation of the Holocene Lake Chilwa sand bar, southern Malawi. Catena 8:369-382.

Pike JG 1965. The sunspot-lake level relationship and the control of Lake Nyasa. J. Inst. Wat. Engrs 19:221-226.

Richardson JL 1968. Diatoms and lake typology in East and Central Africa. Int. Rev. Ges. Hydrobiol. 53:299-338.

Richardson JL 1969. Characteristic planktonic diatoms of the lakes of tropical Africa. Int. Rev. Ges. Hydrobiol. 54:175-176.

Shaw P, Crossley R and Davison-Hirschmann S, in press. A major fluctuation in the level of Lake Chilwa, Malawi, during the Iron Age. Palaeoecol. Africa.

Swann A 1894. Two African Lakes. British Central Africa Gazette, 1 October:2.

Wallis JCP 1956. The Zambesi Expedition of David Livingstone. II. Journals, Letters and Despatches. Chatto and Windus, London.

Late Quaternary palaeoenvironments in the Transvaal on the basis of palynological evidence

L.SCOTT
University of the OFS, Bloemfontein, South Africa

ABSTRACT. New palynological results from the Moreletta Stream near Pretoria are presented with a brief review of those previously reported for Late Pleistocene and Holocene spring sites in the Transvaal. The evidence from different areas is used to reconstruct the palaeoenvironments of the region. The phase of amelioration in temperatures at Wonderkrater in the central Bushveld during the Late Glacial (zone W5, Scott 1982a) is probably partly coeval with pollen zones M1 (Moreletta Stream) and R1 (Rietvlei, Scott and Vogel 1983) from the Pretoria area. M1 indicates vegetation with reduced tree cover and slightly cooler and possibly drier conditions. The climatic significance of increased numbers of Stoebe-type pollen in the Glacial, Late Glacial and Early Holocene from sites in Transvaal and elsewhere in South Africa is tentatively evaluated. The Early Holocene pollen spectra from Wonderkrater indicate an increase in temperature, followed by wetter conditions, which are reinterpreted as having developed gradually as early as c. 7300 BP. This is comparable with similar changes as early as 6580 BP at Rietvlei and earlier than 5220 BP in zone M2 at the Moreletta Stream. The latter zone indicates relatively moist warm conditions but it is uncertain how these differed from the present climate. The absence of warm, dry indicators below zone M2 in the Moreletta profile possibly suggests an unconformity. No pollen was recorded for the interval between c. 5000 to 1000 BP in the Moreletta profile but a detailed pollen record of roughly the last 1000 years was preserved (zone M3). Except for showing a possible reduction in tree cover which tentatively implies a less favourable climate around 400 to 500 BP, this unit indicates conditions similar to the present.

INTRODUCTION

The purpose of this report is to present some new data from the Moreletta Stream site near Pretoria (Fig. 1), and to incorporate the information in a brief review of the existing palynological evidence for the Late Pleistocene and Holocene of the Transvaal. Palynological research in southern Africa is still in an early stage and there are some important problems hampering the detailed reconstruction of palaeoenvironments. In the Transvaal region only a few isolated pollen profiles have been studied. One of the important requirements is to find more polleniferous deposits. Other objectives are to obtain a more complete coverage of modern surface pollen spectra and to develop

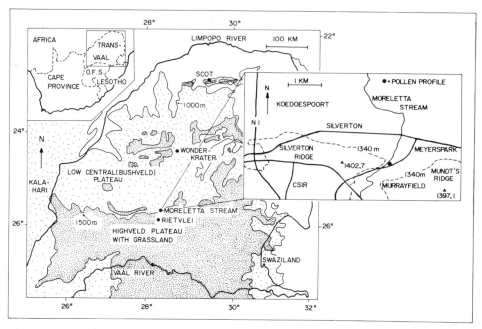

Figure 1. Locality map of the Transvaal showing Late Quaternary pollen sites and the position of the pollen profile at the Moreletta Stream.

reliable techniques of using the complex data to interpret fossil spectra and reconstruct past vegetation types and climates. This reconstruction will remain problematical because past environments often appear to have contained plant combinations for which there are no modern equivalents (Scott 1983) and also because fossil spectra are usually derived from isolated swampy areas which are difficult to relate to the surrounding environment. Another problem which has been identified is that of anomalous radiocarbon dates which appear regularly in the polleniferous deposits (e.g. Scott 1982a). It has been suggested that this is due to excessive rootlets in the swamp deposits (Scott and Vogel 1983). A further problem is that of understanding sedimentation processes and detecting possible unconformities or the reworking of pollen-rich deposits which can be expected in spring sequences and along drainage lines like the Moreletta Stream.

LATE PLEISTOCENE POLLEN ZONES (W1 - W4) OF WONDERKRATER

The pollen zones W1 - W4 at Wonderkrater in the central Transvaal Bushveld at 1100 m altitude provide the first direct evidence of the vegetation in the region during the Late Pleistocene (Scott 1982a). The earliest zone, W1, represents woodland, with expanded montane forests containing Podocarpus on the nearby Waterberg slopes and bushveld with relatively high proportions of Combretaceae and Olea. Small numbers of macchia elements are also present. This points to a past vegetation type under slightly cooler but significantly wetter conditions. The next zone, W2, shows a decrease in the tree pollen percentages and a different composition consisting of more elements characteristic of relatively drier savanna, such as Tarchontheae and

Capparaceae. The narrow zone W3 represents an interval when Podocarpus pollen dominated but the climatic significance of this event is not clear. In zone W4, the arboreal pollen content is even lower than in W2, consisting mainly of Podocarpus and Myrica. The non-arboreal pollen becomes prominent, consisting especially of Stoebe-type, Tulbaghia-type and other Compositae and also more macchia elements like Ericaceae, Passerina and Cliffortia. The reconstructed vegetation can be compared with types occurring in the Drakensberg escarpment about 1000 m higher than Wonderkrater, and therefore suggesting a decrease in temperature of 5 - 6°C.

Inconsistencies in the radiocarbon age determinations impose some uncertainty on the precise age of the above deposits (Scott 1982a) but it is not unlikely that pollen zones W1 and W2 formed during the Middle Pleniglacial interstadial phase, while W4 represents the full Glacial Maximum phase. Preliminary sediment analyses and description of the Wonderkrater sequence suggest fairly uniform characteristics for the section covering W1 - 4 in Borehole 4 (K W Butzer, unpubl.). It is implied that the sediment body represents an intact homogeneous peat with the only possibility of unconformity or reworking of peat at the narrow gritty level of zone W3. It is therefore fairly unlikely that the inconsistent radiocarbon dates in this section were the result of disruption or redeposition of peat. Another contamination mechanism, such as that suggested for the Rietvlei deposit (Scott and Vogel 1983), is probably more likely.

TERMINAL PLEISTOCENE AND HOLOCENE POLLEN SEQUENCES

Moreletta Stream

The latest palynological results from the Holocene in the Transvaal are obtained from deposits along the Moreletta Stream. The site is situated in Meyers Park, an eastern suburb of Pretoria, at 25° 44' S, 28° 18' E where the Moreletta Stream flows along the Silverton Ridge (Fig. 1). The area lies at c. 1310 m alt. and 15 km to the north of the Rietvlei Dam site which is situated at c. 1480 m alt. (Scott and Vogel 1983). While the latter is near the northern limits of the high-veld grassland, the Moreletta Stream site lies within the southern extremities of the Transvaal bushveld vegetation, c. 5 km south of the Magaliesberg, and reflects slightly warmer conditions than those occurring at Rietvlei. The natural vegetation of the area around the site has largely been destroyed by urban expansion. However, a steep rocky slope to the west still contains remnants of the natural woodland, while some indigenous trees also occur along the stream. Some of the typical species are Acacia caffra, Rhus spp., Combretum spp., Ilex mitis, Dombeya rotundifolia, Olea africana, Burkea africana and many others. The swampy basin (vlei) which contains the pollen profile is covered mainly by Cyperaceae, Gramineae including some Phragmites, and a variety of herbs such as Compositae, Umbelliferae, Liliaceae, Leguminosae, Cuscuta, Rumex, Polygonum, Typha and others.

Organic deposits and alluvium in the basin accumulated behind a quartzite barrier where strong springs emerge. During 1969/70 the deposits were exposed by excavation trenches for sewerage pipes on the eastern side of the stream. Material for dating was collected by P Verhoef and JC Vogel, and EM van Zinderen Bakker collected the samples on which the present palynological study is based. The

section rests on Daspoort quartzite and consists of the following stratigraphical units (from top to bottom):
1. 0 - 45 cm . Brownish grey clay with Phragmites roots.
2. 45 - 150 cm . Black peaty clay.
3. 150 - 245 cm . Grey sand with dark bands becoming coarser towards the bottom.
4. 245 - 310 cm . Coarse sand and pebbles.
5. 310 - 330 cm . Black peaty clay.
6. 330 - 380 cm . Grey sand with pebbles.

This description does not match that of Verhoef (1972) exactly but it is essentially similar. The difference may partly be due to a more generalized description by Verhoef while the present sequence represents a specific vertical section. Verhoef reported the following radiocarbon dates for the two peaty layers:

The bottom of the upper peaty horizon, -140 cm : 440 ± 40 BP (Pta-129).

The top of the lower peaty horizon, -280 cm : 5220 ± 55 BP (Pta-128).

According to Verhoef both these horizons wedge out in the upstream direction so that the sequence is reduced to a twofold division of mineral-rich sediment consisting only of a combination of layers 3 and 4, which overlie unit 6.

The methods and procedures for interpreting the present pollen profile on the basis of surface pollen spectra is similar to those employed previously (Scott 1982a,b,c; Scott and Vogel 1983). The arboreal and non-arboreal pollen percentages belonging to the so-called pollen sum are presented in Figs. 2 and 3 respectively while those of the swamp vegetation and grasses are presented in Fig. 4. Figure 5 provides a summary of the pollen data of the sequence.

No determinations of pollen concentration per sample were made but the lowest unit (6) is relatively poor in palynomorphs while units 1, 2, 3, and 5 are relatively rich. Unit 4 was not sampled. The lower four samples in unit 6 produced only small numbers of palynomorphs viz. 225, 108, 160 and 92 per sample (from the bottom upwards). The rest contain rich assemblages and in most cases counts of at least 200 palynomorphs for the total spectrum and of at least 200 for the scarcer and regional elements were made per sample. The only exceptions were samples at 215 cm (unit 3) and 55 cm (unit 2) where 24 and 250 palynomorphs in total were counted respectively.

The pollen in the Moreletta deposits can be expected to represent mainly the local swampy area and the immediate surroundings such as the steep rocky slope. However, a certain proportion was probably also transported to the site by the stream which has its origins along the slopes of the highveld to the south. Some pollen probably also reached the swamp by long-distance wind transport.

The sequence can be subdivided into three pollen zones M1, M2 and M3 (Figs. 2-5). The deepest one, M1, which is probably older than 6000 years, corresponds to the sand unit 6. Although low pollen counts rule out reliable percentages, the zone is clearly different from the younger ones. It is characterized by relatively low, inconsistent numbers of tree pollen including Oleaceae, Rhus and Acacia. A single Pinus pollen at 370 cm is probably a contaminant. The non-arboreal pollen comprises mainly Stoebe-type (up to 60%), other Compositae (up to 56%) and some Thymelaeaceae including Passerina. The local environment produced low numbers of aquatic pollen (Typha, Polygonum and Gunnera) but more pollen of Cyperaceae (up to 15%), and monolete fern spores (up to 62%). Up to 60% grass pollen was recorded. The scarcity of typical bushveld tree-pollen suggests cooler conditions

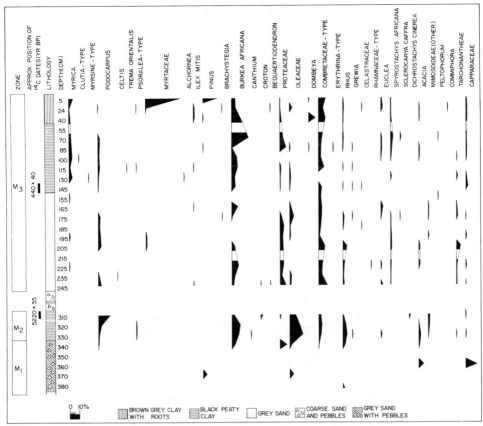

Figure 2. Pollen diagram of the arboreal pollen in the Moreletta Stream sequence. The relative abundance of taxa is presented as percentages of the pollen sum which includes some non-arboreal form shown in Fig. 3.

than at present. The spectra seem to suggest a relatively open drier, possibly karoid, vegetation.

Zone M2 corresponds to the peaty clay unit (5) and probably accumulated during a period starting around roughly 6000 BP and ending c. 5220 BP. This zone shows the appearance of more arboreal pollen like Podocarpus (up to 12%), Oleaceae (up to 14%), Rhus (up to 5%), Acacia and other Mimosoideae, and also some typical bushveld types like Combretaceae (up to 4%) and Burkea africana (up to 9%). The most important non-arboreal elements are Artemisia (up to 17%) and other Compositae (up to 43%) and smaller numbers of Acanthaceae, Acalypha, Chenopodiaceae-type, Aizoaceae-type and Anthospermum. The local palynomorph production differed from that of zone M1 in containing relatively abundant monolete fern spores (up to 86%) and Cyperaceae (up to 25%). Relatively smaller numbers of grass pollen (up to 25%) were deposited. The increase in the tree pollen probably represents warmer, wetter conditions than those implied by M1 which were probably relatively close to present conditions.

Zone M3 represents the upper sandy and peaty units (1-3) and is separated from M2 by the coarse sandy unit (4). It is estimated that

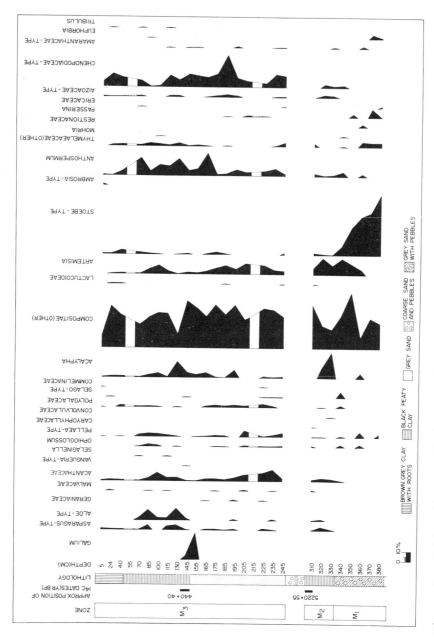

Figure 3. Pollen diagram of the non-arboreal pollen in the Moreletta Stream sequence. The relative abundance of taxa is presented as percentages of the pollen sum which includes the arboreal pollen shown in Fig. 2.

it accumulated during approximately the last 1000 years. The zone contains bushveld tree pollen approximately equal in numbers to those of M2 but declining to lower percentages in the central section. The arboreal pollen peak in the uppermost sample (Figs. 2 and 5) represents the recent introduction of exotic trees like Pinus and Eucalyptus (Myrtaceae). Important woody elements are Burkea africana (up to 18%), Combretaceae (up to 9%), smaller numbers of Proteaceae, Podocarpus, Spirostachys and (in the upper section) Myrica. The zone differs from M2 in that Proteaceae pollen is more important while that of Oleaceae is reduced. Other Compositae (up to 48%) are especially prominent in the non-arboreal pollen but Artemisia (up to 11%), Chenopodiaceae (up to 32%), Anthospermum (up to 23%), Acalypha and Acanthaceae are also relatively important. Cyperaceae pollen (up to 81%) is abundant in the local environment and shows a gradual increase towards the top of the zone where the aquatic Typha (up to 30%) and also monolete fern spores (up to 96%) show peaks. Grass pollen (up to 60%) is generally prominent in the lower section but shows a gradual decrease towards the top of the zone. Some Zea pollen in the uppermost brown clay unit (1) indicates that this deposit formed after the advent of modern agriculture in the area. The pollen assemblage of zone M3 suggests generally similar environmental conditions to the present, except perhaps in the central part of the zone, at 400 - 500 BP, when moisture and/or temperature conditions were slightly less favourable. The younger spectra suggest that the local swampy environment, at least, was slightly wetter during roughly the last 400 years.

General interpretation

The results from the Moreletta Stream, together with those from other sites can be used to reconstruct the terminal Pleistocene and Holocene environments in the Transvaal.

Zone W5 at Wonderkrater probably accumulated during the Late Glacial and the beginning of the Holocene, suggesting a relatively cool, dry and low savanna vegetation (Scott 1982a) which developed under warmer conditions than the vegetation of the previous zone (W4). It is not known what the age of M1 in the Moreletta Stream is but considering the suggested relatively cool, dry conditions, it could tentatively be assigned to roughly the same age as W5 and probably also zone R1 at Rietvlei which presumably accumulated between 10 300 and c. 8500 BP (Scott and Vogel 1983) and which shows similarities to M1. During the next stage at Wonderkrater (W6a, 8390 BP), it became slightly warmer and the bushveld became denser with Capparaceae and Tarchonantheae, characteristic of the dry Kalahari Thornveld. It is likely that this zone corresponds to R2 at Rietvlei which shows the same elements in smaller numbers and which accumulated roughly between c. 8500 and c. 7500 BP. There seems to be no clear equivalent for this zone at the Moreletta Stream, possibly as a result of a sedimentation break between zones M1 and M2. The next phase at Wonderkrater (W6b, 6330 BP) shows an increase in arboreal Combretaceae pollen, while the numbers of Kalahari Thornveld elements remained fairly constant. Around this time (R4, 6580 BP) woodland expanded over the northern parts of the highveld at Rietvlei, probably as a result of warmer and wetter conditions. The increase in Combretaceae and Gramineae pollen, starting at approximately 7300 BP during zone W6b at Wonderkrater, was therefore possibly not only the result of increased temperatures but also of more moisture, perhaps slightly wetter than the "semi-arid" conditions pro-

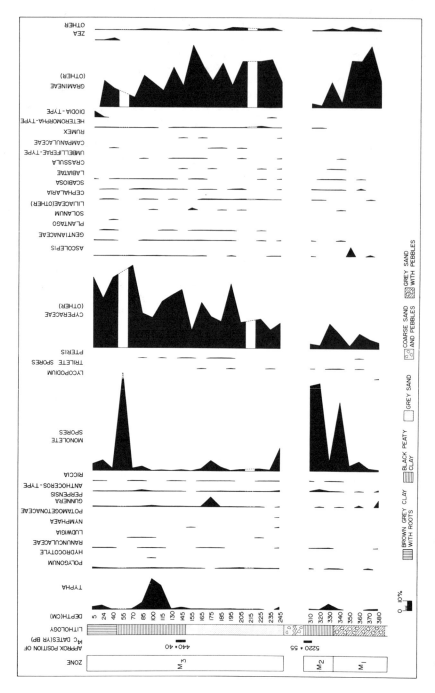

Figure 4. Pollen diagram of the local, grass and other pollen in the Moreletta Stream sequence. The relative abundance of taxa is presented as percentages of the total composition.

posed previously by Scott (1982a). The lack of traces of forest pollen (other than that of Podocarpus) which occur higher up in the sequence, however, excludes the possibility of conditions any wetter than "dry sub-humid". Pollen zone M2 which formed between c. 6000 and 5220 BP at the Moreletta Stream compares reasonably well with zones W7 at Wonderkrater and the upper part of zone R4 at Rietvlei which are both thought to be of the same age.

The absence of detailed pollen and radiocarbon data for the periods between c. 5000 and 1000 BP at both Rietvlei and the Moreletta Stream does not allow any reliable environmental reconstruction for this period in the southern Transvaal. However, at other bushveld pollen sites (Fig. 1) at Wonderkrater (zones W8 and W9) and Scot (S2 to S5) there are indications of substantial increases in Compositae pollen relative to arboreal pollen (Scott 1982a,c). Based on surface pollen data, slightly cooler, subhumid conditions for the period covering c. 4000 to c. 2000 BP have been postulated. Pollen in zone W9 at Wonderkrater suggests that conditions remained relatively mesic between 2000 and 1000 BP while temperatures changed to normal.

The relatively high numbers of Proteaceae pollen in M3 during the last 1000 years at the Moreletta Stream may be an indication that especially Protea caffra became more prominent while Oleaceae trees were reduced in the area probably before the start of the zone. Since proteas are not numerous on the rocky slopes adjacent to the Moreletta site (Fig. 2), and since Faurea is absent, it can be assumed that some of the Protea pollen was transported to the site along the stream from the northern highveld slopes, where proteas are abundant. As at Wonderkrater (W10), the suggested vegetation during the last 1000 years at the Moreletta Stream (M3) resembled the modern type with the exception of the period between 400-500 BP when possibly reduced tree cover occurred at Moreletta Stream.

DISCUSSION

One of the salient aspects emerging from the pollen sequences from the Transvaal is that Stoebe-type pollen is more abundant in three sections which are older than 6000 BP. This also seems to be the case in the supposedly cooler, Pleistocene assemblages at Aliwal North, OFS (zone U, Coetzee 1967) and Taung, NW Cape (Scott, unpubl.). In Rietvlei and Moreletta it is prominent in the terminal Pleistocene and early Holocene, while at Wonderkrater it is especially abundant in the full glacial period (Table 1). In the former two younger sites these pollen grains are more likely to represent Stoebe vulgaris, at present in the Transvaal on the escarpment and higher-lying areas, rather than Elytropappus which belongs to the southern Cape Province. Their climatic implications in terms of temperature or seasonality and intensity of precipitation are not exactly clear but they do seem to represent relatively open vegetation with shrubby (karoid) elements and relatively cool climates.

In connection with the determination of the warmest phase during the Holocene at Wonderkrater (Scott 1982a), criteria such as scarcities of Stoebe-type and other Compositae (including Artemisia) pollen as well as upland woodland elements such as Proteaceae and Oleaceae, can be considered. These forms all diminish markedly in zone W6b at Wonderkrater, indicating a warm period between roughly 7000 and 6000 BP. A swamp-forest situation similar to that at the relatively warm site Scot (Fig. 1; Scott 1982c) never developed at the Wonderkrater

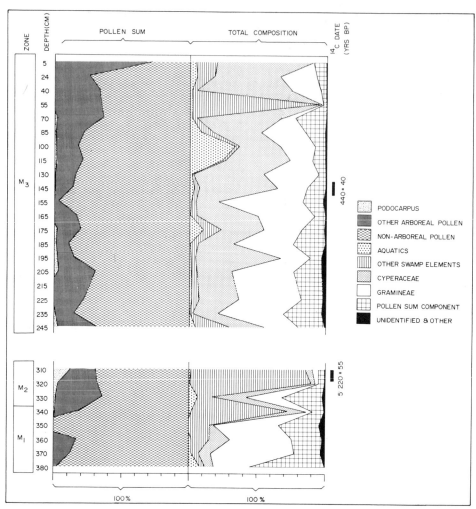

Figure 5. Condensed pollen diagram of the Moreletta Stream sequence.

spring. The warmest phase in the Holocene was probably not warm enough (therefore only very slightly, if at all, warmer than today) for forest vegetation to develop. The high percentages of supposedly cool Oleaceae pollen at the Rietvlei site during the suggested warmest phase at 6580 BP (Scott and Vogel 1983) are understandable, since this high-lying site is cooler than Wonderkrater where Oleaceae pollen disappeared around that time. The replacement of Oleaceae in M2 by more Proteaceae in M3 at Moreletta River is difficult to explain but a subtle change in temperature and moisture patterns may have played a role. M2 possibly represents relatively warm conditions in the process of cooling after optimum temperatures were reached earlier around 7000 to 6000 BP.

The pollen results from the Moreletta Stream do not accord with Verhoef's (1972) tentative palaeoclimatic interpretation of the sequence. The proposal that the peaty sections represent relatively drier situations while the fluvial units represent phases with more flood energy, is not supported by the pollen data for the lower sandy

Table 1. Average percentages of Stoebe-type pollen at some sites in the Transvaal (number of samples in parenthesis).

Scot		Wonderkrater		Moreletta		Rietvlei	
Age (yr BP)	%	Age (yr BP)	%	Age (yr BP)	%	Age (yr BP)	%
2000-5000	0.07(16)	0-7500	0.1(35)	0-6000	1.1(21)	0-6000	1.7(7)
		7500-11 000	1.4(18)	older	37(5)	older	17(23)
		11 000-25 000	11(26)				
		older	2.4(10)				
			1.0(22)				

zone M1 and the overlying peat, M2. There is also no indication of increasing dryness corresponding to the change from mineral-rich to peaty sediment in zone M3 of the upper cycle. However, because of the absence of pollen data for the intermediate coarse sands (unit 4) the sequence of events in the upper cycle cannot be reliably evaluated.

ACKNOWLEDGEMENTS

I am grateful to Prof EM van Zinderen Bakker for sampling the Moreletta Stream sediment profile and for allowing me to study the material.

REFERENCES

Coetzee JA 1967. Pollen analytical studies in East and Southern Africa. Palaeoecol. Africa 3: 1-146.
Scott L 1982a. A Late Quaternary pollen record from the Transvaal Bushveld, South Africa. Quat. Res. 17: 339-370.
Scott L 1982b. Late Quaternary fossil pollen grains from the Transvaal, South Africa. Rev. Palaeobot. Palynol. 36: 241-278.
Scott L 1982c. A 5000-year old pollen sequence from spring deposits in the bushveld north of the Soutpansberg, South Africa. Palaeoecol. Africa 14: 45-55.
Scott L 1983. Palynological evidence for vegetation patterns in the Transvaal (South Africa) during the Late Pleistocene and Holocene. Bothalia 14: 445-449.
Scott L and Vogel JC 1983. A Late Quaternary pollen profile from the Transvaal Highveld, South Africa. S. Afr. J. Sci. 79: 266-272.
Verhoef P 1972. A stream deposit with interstratified peat, at Pretoria. Palaeoecol. Africa 6: 147-148.

Environmental changes since 32 000 BP at Kathu Pan, northern Cape

PETER B.BEAUMONT
McGregor Museum, Kimberley, South Africa

E.M.VAN ZINDEREN BAKKER, Sr.
University of the OFS, Bloemfontein, South Africa

JOHN C.VOGEL
CSIR, Pretoria, South Africa

ABSTRACT. Excavations at Kathu Pan near Sishen by PBB have focussed on the contents of two dolines in the partly peat-mantled karstic surface of the pan and on the stratigraphy within what is probably an infilled palaeostream channel that drained into it from the south. The up to 12 m deep sequences in these depositories document sediment and faunal shifts since the Middle Pleistocene onset and also contain a composite cultural succession that presently comprises 2 phases of the ESA and 3 phases each of the MSA and LSA. We here confine ourselves to a consideration of the strata that provide information concerning climatic changes since 32 000 BP in this transitional region on the southern margin of the Kalahari.

INTRODUCTION

An interdisciplinary research endeavour directed at producing environmental reconstructions with particular emphasis on multiple data sets from contexts that are more susceptible to reliable isotopic dating (see Butzer 1978) has been underway since 1978 within a still relatively confined portion of the northern Cape (Beaumont 1979, 1982). This article deals largely with our preliminary findings from various superficial sediment sequences at Kathu Pan and how these correlate with the compatible palaeoclimatic evidence from three other sites to produce a reasonably detailed Holocene succession that includes good evidence of the age of the climatic optimum (Van Zinderen Bakker 1980).

Kathu Pan is situated on a flat north-dipping plain drained by the Ga-Mogara, some 5.5 km NW of the town Kathu in the Postmasburg district of the northern Cape, at about 27° 39'50"S, 23° 0'30"E (Fig. 1). ISCOR boreholes there reveal that the superficial pan sediments are underlain by over 40 m of calcretes followed by some 30 m of sands, clays and basal gravels that collectively belong to the Tertiary-aged Kalahari Group. The natural mid-winter water-table within the southern part of this c. 30 ha pan is maintained at about -1 m by a local rainfall of about 350 mm. The pan is largely covered by dwarf grass species while fringing it are thickets of Acacia karoo that lead southwards to an extensive Acacia erioloba open woodland, but to the west the vegetation is more open and typified by Acacia mellifers, Grewia flava and Tarchonanthus camphoratus (A Gubb, pers. comm.). The location of the various sites is shown in Fig. 2. and are here discussed in terms of three geo-

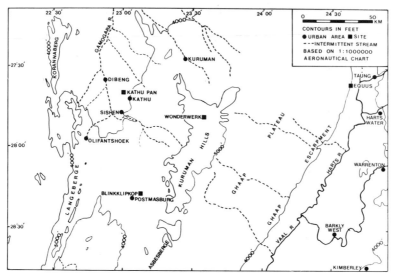

Figure 1. Map showing the location of Kathu Pan in the northern Cape.

morphic groupings, viz. pan-surface sites, doline infill sites and palaeochannel sites.

PAN-SURFACE SITES

Large portions of the calcrete pan substrate are mantled by essentially unstratified peats (Fig. 2) that have been sampled from pits at the following four localities

K P 2: About 1.8 m of peat with a more sandy lower 30 cm and a basal scatter of sporadic small subrounded pebbles on flat calcrete bedrock. It is located in a low-lying area of limited extent that is usually under water and was noted to contain rare bone fragments and stone flakes (LSA ?). Pollen was found by vZB to be present in most of the 10 cm interval samples and his analysis of their arboreal and non-arboreal components has resulted in the identification of three distinct zones, namely:
   (a) Pollen zone III. 4 spectra. 0-50 cm. (2700 BP to present).
NAP: Grasses average 25 %; major grassveld types are Compositae, Acanthaceae and Liliaceae; typical wet habitat sporomorpha are Typha and Cyperaceae (up to 39 %).
AP : The very few found include rare palm and Acacia.
   (b) Pollen zone II. 3 spectra. 50-120 cm. (4500-2700 BP).
NAP: Grasses only 4-12 %; Compositae 21-54 %; some Aizoaceae, Crassulaceae and Chenopodiaceae.
Nearly no swamp taxa and practically no AP.
   (c) Pollen zone I. 4 spectra. 120-180 cm. (7400-4500 BP).
NAP: Grasses 14-24 %; Compositae very variable (up to 39 %); other grassveld types as in Zone III; also Typha and Cyperaceae (up to 56 %).
AP : Well represented by Olea, Combretaceae, Anacardiaceae, Heteromorpha, cf. Euclea and 1 % Acacia.

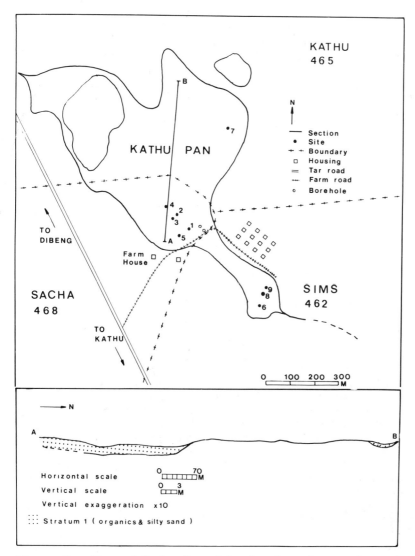

Figure 2. Plan of Kathu Pan with site locations and section along line A - B.

Radiocarbon dates for this section are:
```
Pta-3504    K P 2,  17- 20 cm    Zone 3, Upper    1830 ±  50 BP
Pta-3510    K P 2,  57- 60 cm    Zone 2, Upper    2980 ±  60 BP
Pta-3518    K P 2, 117-120 cm    Zone 2, Base     4420 ±  60 BP
Pta-3073    K P 2, 157-160 cm    Zone 1, Lower    7350 ±  90 BP
```

K P 3: A 1.85 m thick unsampled peat comparable to K P 2 except that no basal scatter of small pebbles was noted on level calcrete bedrock.

K P 4: A 2.25 m thick peat with occasional small calcareous nodules and often compact lower levels that directly overlies uneven calcrete bedrock. Preliminary processing indicates that good pollen preservation is confined to levels equivalent to Zone I of K P 2. The base of this section at 210-225 cm has yielded dates of 6130 ± 70 BP (Pta-3545) and 6330 ± 110 BP (I-13055).

K P 7: A 2.8 m thick peat sequence situated on the sloping rim of a doline, with a lower metre that has a markedly higher incidence of sandy silt and calcareous nodules. An incomplete systematic excavation of this lower zone revealed a fair number of lightly to heavily abraded rock fragments that include artefacts ascribable to the MSA and ESA. With these were found segments in mint condition which are here taken to indicate that this entire deposit, except perhaps for its basal 0.5 m, postdates the regional Wilton onset at c. 8500 BP.

DOLINE INFILL SITES

Two such localities on the southern karstic margin of the pan proper owe their identification to the post-1974 (?) collapse of a portion of their old sediment content into phreatic-zone cavities in the underlying calcretes. This was possibly the result of large-scale pumping from nearby boreholes during the past decade and more, particularly during peak demand in summer when the pan water-table was noted to sink to at least -12 m.

K P 1: 1978-82 excavations here showed the doline base to be at -11.1 m and exposed a sequence of five units extending back to the lower Middle Pleistocene or beyond, on the basis of late Elephas recki remains in Stratum 4b and 4c (R Klein, pers. comm.), but only the following two levels are relevant here (Fig. 3):

(a) Stratum 1 : 145-160 cm of interdigitating peats and grey calcified silty sands dipping at 1-2° to the north, with a fair amount of subunit variability but best typified by the square Y18 face sequence shown in Fig. 3, containing a localized concentration of Iron Age or Pottery LSA ceramic fragments at c. 50 cm.

(b) Stratum 2 : 170-190 cm of grey calcareous silty sands dipping at 1-2° to the north which show only slight vertical changes in terms of organic content or calcification and with an extremely low incidence of small rounded pebbles, bone fragments, ostrich eggshell pieces and artefacts probably ascribable to the LSA.

K P 5: A pit dug here during 1982-3 was discontinued at 6.45 m without having reached doline base in a fill that shows a sequence of four units which patently match Stratum 1-4 of K P 1 as can be seen from the Main Face Section depicted in Fig. 3 and with the following pertinent details:

(a) Stratum 1: 200-210 cm of interdigitating peats and grey calcified silty sands showing almost no variation within a 5 m wide exposure and seemingly bereft of any cultural or faunal remains.

(b) Stratum 2: 90 cm of deep grey and only slightly calcified silty sands with a uniformly higher organic content than Stratum 2 at K P 1, except for paler subunit 2a which contained a marked number of small (<3 cm dia.) subangular-subrounded clasts, mainly calcrete, while bone fragments and artefacts were rare throughout but did include a probably 'Early LSA' scraper at c. 290 cm/

Radiocarbon dates for these two levels are:

Pta-3582	K P 5, 60- 65 cm	Peat 2, Top	2690 ± 50 BP
I-13036	K P 5, 105-110 cm	Peat 3, Top	3620 ± 80 BP
I-13039	K P 5, 210-215 cm	Stratum 2, Top	5980 ± 120 BP
Pta-3586	K P 5, 230-232 cm	Stratum 2, Upper	19800 ± 280 BP

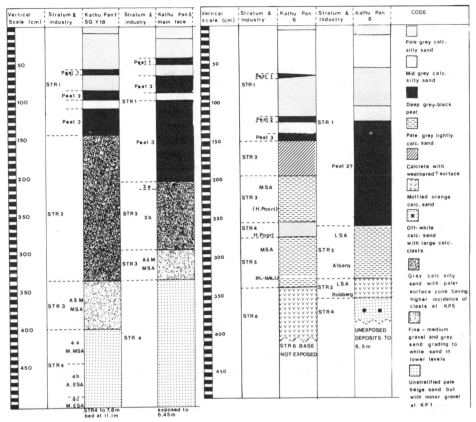

Figure 3. Lithostratigraphy at Kathu Pan 1 and 5 (left) and Kathu Pan 6 and 8 (right). In column 2, A = abraded, M = mint.

Pta-3566	K P 5, 265-270 cm	Stratum 2, Lower	27500 ± 530 BP
Pta-3591	K P 5, 285-290 cm	Stratum 2, Lower	32100 ± 780 BP
I-13040	K P 5, 290-300 cm	Stratum 2, Base	26930 ± 750 BP

PALAEOCHANNEL FILL SITES

In 1980-1 (?) three sinkholes formed some 400 m to the SE and 7.5 m above K P 1 by the slumping of up to 6.5 m of superficial deposits into surface cavities in the underlying calcrete within a linear depression that seems to represent the lower reaches of a palaeostream channel which can be traced southwards for some kilometres. K P 6, 8 and 9 are aligned at a very approximate right-angle to drainage direction and thereby provide a 'ready-made' fill cross-section which is still in the process of being investigated by way of excavations adjacent to the original exposures (Fig. 2). It would seem that the exotic subrounded pebbles which constitute a major component of levels like Stratum 3 at K P 1 and 5 were fluvially transported along this stream from now subsurface Dwyka occurrences near Kathu (M Bissett, pers. comm.)

K P 6: 1983 excavations here were temporarily discontinued at 6.40 m without having reached bedrock in a succession which presently comprises six units that range back to before Is. stage 5 on the basis of a 'Stillbay'-type aggregate in Stratum 5 but only the following two levels are relevant here (Fig. 3):

(a) Stratum 1: 150 cm of grey silty sand with three intercalated peatzones that appear to correspond to those at K P 1 and 5 and which include sporadic bone fragments, ostrich eggshell pieces and amorphous artefacts that probably relate to the LSA.

(b) Stratum 2: 45 cm of calcrete with a compact and perhaps slightly weathered surface which postdates Is. stage 5b in terms of Howiesons Poort aggregates in Stratums 3 and 4.

K P 8: A preliminary pit on the west side of this sinkhole was taken down to 3.8 m and has exposed the following four units within that depth, but deposits can be seen to extend down to the calcrete surface at 6.5 m (Fig. 3).

(a) Stratum 1: 260 cm of grey silty sand with a more calcified midzone and a thick basal peat that equates largely or entirely with Peat 3 of K P 1, 5 and 6. No artefacts were noted but bone occurs sporadically and includes the perhaps complete skeleton of an as yet unidentified ungulate.

(b) Stratum 2: 70 cm of pale grey lightly calcified sand with some bone fragments and ostrich eggshell pieces as also an abundant 'classic' Albany dated to c. 11 500-10 000 BP on the basis of the temporal data at Wonderwerk Cave (Fig. 4).

(c) Stratum 3: 25 cm of mottled orange calcified sand with a fairly high incidence of artefacts tentatively ascribed to the Robberg.

(d) Stratum 4: 30 cm+ of off-white calcified sand with large calcrete clasts that would seem to be bereft of bone and artefacts.

INTERPRETATION

The internally consistent but preliminary data cited above are here taken to indicate the following sequence of environmental shifts reflecting essentially on long-term ground-water trends (Butzer 1983):

(a) >32 000-11 000 BP. The low organic content of Stratum 2b when compared to that of Stratum 1 peats overlying it at K P 5 is taken to indicate that the water-table when it formed was lower than at present (by c. 2 m+) and the case for a fairly stable period of relatively dry conditions with a slow rate of sediment build-up (c. 20 % of that for Stratum 1) is further strengthened by the dominance within it of well-sorted fine to medium grade sands suggestive of a strong primary or derivative aeolian component (Butzer 1983).

(b) c. 11 000-7400 BP. The marked increase in calcrete clasts and reduced organic content of Stratum 2a at K P 5 is here taken to indicate accelerated sheet erosion from the bordering plain surface as a result of a disrupted vegetation cover, possibly resulting from a shift in rainfall regime patterning during a markedly drier than present environment, with a terminal date of c. 7400 BP based on Pta-3073 which is preferred to the stratigraphically comparable date of 6130 ± 70 BP (Pta-3545) at K P 4.

(c) c. 7400-4500 BP. Pollen spectra of Zone I at K P 2 clearly indicate an open grassland with wet habitats and a more varied tree growth, which suggests a climate that was warmer and wetter than at present, to an extent which seemingly raised the wet season water-table to the pan surface, thereby permitting peat to commence accumulating on the calcrete substrate within its southern sector (Fig. 2).

(d) c.4500-2700 BP. Pollen spectra of Zone II at K P 2 are taken to reveal a climate that was perceptibly drier than at present as is shown by the virtual absence of trees and of species indicating wet habitats in Zone III. Even greater aridity may be implied by the absence of pollen from 20 cm zones at its base and surface, while locally brackish conditions are suggested by the presence of Aizoaceae, Crassulaceae and Chenopodiaceae.

(e) c.2700 BP-Present: The Zone III pollen spectra at K P 2 represent a grassveld with locally wet conditions as existed before the pan was persistently drained by mainly summer pumping during the past two decades.

REGIONAL COMPARISONS

Equus Cave:

1976 and 1982 excavations at this locality near Taung (Fig. 1) exposed four strata within an up to 2.5 m deep accumulation that yielded one of the world's largest Pleistocene macrofaunal samples, representing some 50 large mammal species that reflect on its protracted use as a maternity den by Hyaena brunnea (R Klein, pers. comm.). Stratum 1a has mid-level and basal dates of 2390 ± 55 BP (Pta-2452) and 7480 ± 80 BP (Pta-2495) respectively while the still undated underlying units are considered to span much or all of Is. stages 2-5 (late) on the basis of sporadic MSA-type artefacts in Stratum 2b. The sediment and coprolite derived pollen spectra indicate Kalahari Thornveld conditions like those present in Stratum 1a, whereas the Stratum 1b and 2b values show less tree cover with more small shrubs, that are taken to indicate a climate which was cooler than that prevailing since 7500 BP (L Scott, pers. comm.). This deduction is further supported by the greater than present mean size of Canis mesomelas and Vulpes chama in Stratums 1b-2b (R Klein, pers. comm.) while the presence of the aquatic Kobus leche in these units may suggest that reduced evaporation significantly offset less rainfall then in terms of the evidence from Stratum 2b at K P 5.

Wonderwerk Cave

Renewed excavations since 1978 at this massive 140 m deep cave situated south of Kuruman (Fig. 1) have revealed a complex sequence of strata with a cumulative thickness of over 5 m (Fig. 4) that ranges back to well within the Middle Pleistocene on the basis of refined handaxe aggregates associated with traces of fire in the lower units (Beaumont 1979; Thackeray 1981; Thackeray et al. 1981). The following summary of the environmental deductions for the LSA-linked levels there is largely based on the concordant small mammal (Avery 1981) and pollen (Van Zinderen Bakker 1982) evidence, set within a somewhat revised and refined temporal framework currently constituted of some 30 C14 dates.

(a) Stratum 5 of Strips 6-34/Stratums 1 and 2 of Strips 49-63. c. 13 000-11 500 BP. Robberg industry : The lower age limit of this grouping remains uncertain but may refer to basal Is. stage 2 on the evidence of a seemingly related reading of $28 200 \pm 1000$ BP (SI-2021) derived from an exposure in Strip 62 (Butzer et al. 1978; Butzer 1983). Multiple éboulis zones suggesting conditions markedly cooler than those at present occur, with the most recent of these being firmly dated to c. 12 000 BP. (Butzer et al. 1978), but related decisive microfaunal and pollen bearing samples remain to be studied.

(b) Stratum 4d of Strip 19-34. c. 10 500-9000 BP. Late Albany industry: An open scrub savanna with only occasional trees probably restricted to drainage lines as indicated by the pollen and microfaunal data is taken to reflect dry conditions while more regular and/or severe frosts are inferred from an increase in roof spall relative to the overlying units.

(c) Stratum 4c of Strip 6-34. c. 8000-6000 BP. Wilton industry: A higher or more effective rainfall with short grass and more shrubs/trees is indicated and supported by evidence for enhanced cave humidity deduced from a higher fern count and the maximum expansion then of the stalagmite boss centered on Strip 19.

(d) Stratum 4b of Strips 6-34. c. 6000-5000 BP. Wilton industry: Still warmer and wetter conditions at this time seemingly sustained an outside vegetation cover that included a significant savanna woodland component.

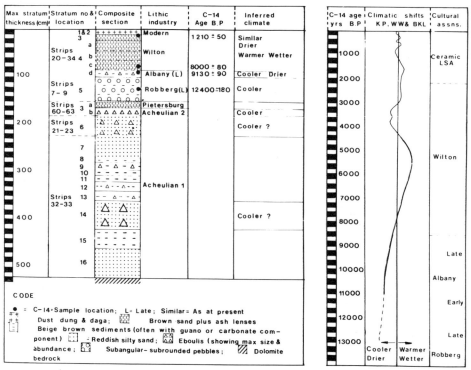

Figure 4. Composite stratigraphy at Wonderwerk Cave.

Figure 5. Inferred climatic shifts since 13 000 BP in the northern Cape.

(e) Stratum 4a of Strips 6-34. c. 5000-4000 BP. Wilton industry: A drier and probably cooler arid to semi-arid climate supported an almost treeless grassland.

(f) Stratum 3b of Strips 6-34. c. 4000-2500 BP. Wilton industry: No dating is available for the surface of this subunit within which the microfaunal evidence indicates little change except for a possible minor increase in bushes/trees.

(g) Stratum 3a and 2 of Strips 6-34. c. 2500-1000 BP. Wilton industry: The pollen evidence is taken to imply that rainfall was at times slightly higher than before. This interpretation is supported by the presence of post-depositional travertine sheets radiating from the stalagmite boss in Strip 19 which indicate humid conditions at c. 1950-1450 BP and c. 850-750 BP, if a correction value of -1500 yrs. is applied to these dates.

(h) Stratum 1 of Strips 6-34. After AD 1900. Modern: Distinctly drier and/or more overgrazed as indicated by short grass and some scrub but with few trees.

Blinkklipkop Cavern

1980 excavations at this large specularite working near Postmasburg (Fig. 1) revealed 2.7 m of mainly mining rubble that, in the investigated area, accumulated rapidly between 1150 ± 40 BP (Pta-2835) and the present (Thackeray et al. 1983). The microfauna indicates a brief drier oscillation at 830 ± 45 BP (Pta-3419) and a wetter interval at 260 ± 45 BP (Pta-2833).

CONCLUSIONS

The combined evidence from the northern Cape sites suggests the following local sequence: The cold and dry environment of the Late Pleistocene improved only gradually until optimal warm and wet conditions were reached between 8000 and 5000 BP. After 4500 BP it again became drier and slightly cooler to be followed by some summer oscillations up to the present. The Mid-Holocene warm phase seems also to be documented at Wonderkrater and Rietvlei in the central Transvaal (Scott 1982a and b; Scott and Vogel 1983). Figure 5 provides a 'most probable' composite reconstruction of the largely compatible environmental data provided by Kathu Pan 1-8, Equus Cave, Wonderwerk Cave and Blinkklipkop Cavern for the past 13 000 yrs. within a limited portion of the northern Cape and is clearly concordant with other somewhat less certainly dated Holocene evidence from Namibia (Brain and Brain 1977; Vogel and Visser 1981) and the Transvaal (Scott 1979, 1982a and b; Scott and Vogel 1978, 1983). We anticipate that further pollen and microfaunal studies at some of the above sites will, with time, provide unique additional data bearing on climatic change back to the Middle Pleistocene, particularly if complemented by O18 readings on 'bracketed' stalagmite fragments at Wonderwerk Cave (Beaumont 1982) and C13 measurements on the bones of 'sensitive' species like zebra and springbok (Vogel 1978, 1983).

ACKNOWLEDGEMENTS

Financial assistance from the HSRC towards the cost of the research is hereby acknowledged, but opinions expressed or conclusions arrived at are those of the authors only.

REFERENCES

Avery DM 1981. Holocene micromammalian faunas from the northern Cape Province, South Africa. S. Afr. J. Sci. 77:265-273.

Beaumont PB 1979. A first account of recent excavations at Wonderwerk Cave. Paper, S. Afr. Ass. Archaeol. Conf., Cape Town, June 1979.

Beaumont PB 1982. Aspects of the northern Cape Pleistocene Project. Palaeoecol. Africa 15:41-44.

Butzer KW 1978. Climate patterns in an un-glaciated continent. Geog. Mag. 51 (9):201-208.

Butzer KW 1983. In: RG Klein (ed.), Southern African palaeoenvironments and prehistory. Balkema, Rotterdam.

Butzer KW, Stuckenrath R, Bruzewicz AJ and Helgren DM 1978. Late Cenozoic palaeoclimates of the Gaap Escarpment, Kalahari margin, South Africa. Quat. Res. 10:310-339.

Hall M 1976. Dendrochronology, rainfall and human adaptation in the Later Iron Age of Natal and Zululand. Ann. Natal Mus. 22:693-703.

Scott L and Vogel JC 1978. Pollen analyses of the thermal spring deposits at Wonderkrater (Transvaal, South Africa). Palaeoecol. Africa 10:155-162.

Scott L 1979. Late Quaternary pollen analytical studies in the Transvaal (South Africa). Unpubl. Ph.D. thesis, Univ. O.F.S.

Thackeray AI 1981. The Holocene cultural sequence in the northern Cape Province, South Africa. Unpubl. Ph.D. thesis, Yale Univ.

Thackeray AI, Thackeray JF, Beaumont PB and Vogel JC 1981. Dated rock engravings from Wonderwerk Cave, South Africa. Science 214:64-67.

Thackeray AI, Thackeray JF and Beaumont PB 1983. Excavations at the Blinkklipkop specularite mine near Postmasburg, northern Cape. S. Afr. archaeol. Bull. 38:17-25.

Van Zinderen Bakker (sr.) EM 1980. Some problems of Quaternary studies in Southern Africa. S. Afr. Soc. Quat. Res. Newsletter No. 11:1-2.

Van Zinderen Bakker (sr.) EM 1982. Pollen analytical studies of the Wonderwerk Cave, South Africa. Pollen et Spores 24 (2):235-250.

Vogel JC 1978. Isotopic assessment of the dietary habits of ungulates. S. Afr. J. Sci. 74:298-301.

Vogel JC 1983. Isotopic evidence for the past climates and vegetation of Southern Africa. Bothalia 14 (3 & 4):391-394.

Correlation of palaeoenvironmental data from the Late Pleistocene and Holocene deposits at Boomplaas cave, southern Cape

H.J.DEACON, J.DEACON, A.SCHOLTZ, J.F.THACKERAY & J.S.BRINK
University of Stellenbosch, South Africa

JOHN C.VOGEL
CSIR, Pretoria, South Africa

ABSTRACT. The excavated sequence at Boomplaas Cave dates back about 80 000 years and includes biological and sedimentological materials which reflect environmental changes during the Last Glacial and Present Interglacial cycle and enable comparisons to be drawn between different lines of evidence. The harshest conditions pertained between about 25 000 and 16 000 BP when there was a low diversity of plant and small mammal taxa and oxygen isotope ratios from the nearby Cango Caves indicate a temperature reduction of about 5 degrees C. Amelioration of temperatures had begun by 14 000 BP when charcoals and small mammal remains suggest higher precipitation than at present. Holocene climate fluctuated around the present-day mean with the warmest temperatures in the early-mid Holocene and somewhat cooler conditions within the last 2 000 years.

INTRODUCTION

Excavations at Boomplaas Cave have provided a unique record from a number of different indicators of climatic and environmental changes in the Cango Valley of the southern Cape over the last 80 000 years. Spanning much of the Last Glacial and the Present Interglacial cycle, these data offer a land-based set of parameters with which we can compare the sequence of changes from deep-sea and Antarctic ice cores. The sequence includes materials that have accumulated both with and without the bias of human selection, and they reflect both temperature and rainfall fluctuations indirectly through the vegetation cover and directly through weathering and sedimentary processes. The samples recovered comprise remains of larger mammals hunted and charcoals of wood selected for camp fires by the people living in the cave, and small mammals hunted by owls, bones collected by carnivores other than man, pollens blown into the cave by the wind, sediments accumulated through natural weathering and materials contributed by people occupying the cave, changes in oxygen isotope ratios in water precipitated as speleothems in the nearby Cango Caves, and natural weathering processes outside the cave. The research at Boomplaas Cave, continued over 10 years, was designed to provide a number of complementary but independent measures of the scale of changes in the local habitat and regional climatic conditions in the past. The merit lies in the better resolution of change afforded by the conjunction of these different lines of evidence and the completeness of the sequence for the Late Pleistocene and Holocene means it can serve as a benchmark for wider correlation.

BOOMPLAAS CAVE

Boomplaas Cave is situated about 800 m a.s.l. and 80 km inland in limestone country rock in the foothills of the Swartberg range about 5 km from the Cango Caves, a local tourist attraction. Boomplaas Cave itself is an enlarged opening of a fissure that forms a large domed rock shelter which offers very adequate protection from wind and rain over a floor area of about 150 sq m. The deposits in the cave floor have built up behind a rock tumble at the mouth to a depth of between 5 and 6 m and comprise both naturally weathered sediments and materials accumulated during times when the cave was inhabited.

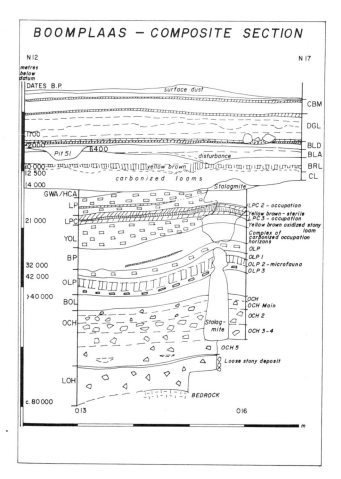

Figure 1. Generalized composite section through the deposits at Boomplaas Cave.

The stratigraphy (Fig. 1) has been described in detail (Deacon 1979) in terms of major lithostratigraphic subdivisions called, in order of importance, Members, Units and Sub-units. They were recognized during excavation by changes in colour and texture. The deposit is well stratified with several clear marker horizons that are masked only where leached and cemented. Red-brown loams have typically accumulated under natural conditions and low human occupation intensity, and may include clasts of roof rock and abundant bones of small mammals. The periods during which there has been more intense use of the cave by people are marked by thicknesses of ash and carbonized organic materials that are variously leached, as well as concentrations of artefacts and food waste.

The culture stratigraphy of the deposit has been described elsewhere (Deacon 1979; J Deacon 1982). The cave has functioned at various times in the prehistoric past as a base for herders and for Middle and Later Stone Age foragers. Particular interest lies in any correlation between the human occupation of the cave and changes in the local habitat conditions. It would seem however that potential food resources for foragers have never been superabundant in the valley and occupation has been episodic both in the short and long term. The Howiesons Poort substage of the Middle Stone Age is represented in the base of the sequence (Member OCH) and although episodic use of the cave is evidenced in subsequent times, the next significant occupation represents a late substage of the Middle Stone Age dated to about 32 000 BP (Member BP). The major technological change between the Later and Middle Stone Ages is in the YOL Member, bracketed between dates of 32 000 and 22 000 BP. This major change which is a marker for the appearance of essentially modern norms of behaviour correlates broadly with the onset of the harshest environmental conditions of the Late Pleistocene (Deacon 1983). Later Stone Age peoples occupying the cave since some time prior to 22 000 BP show a general shift in lifeways from mobile hunters of plains game to more settled-in patch foraging, as plant foods and territorial bovids became more important in their economy with the progressive amelioration of climates in the end Late Pleistocene and Holocene.

SOURCES OF PALAEOENVIRONMENTAL DATA

Larger mammal remains

Analysis of the remains of animals with a mean adult bodymass exceeding 2 kg was undertaken by RG Klein (1978) and JS Brink (unpublished). The principal agents of accumulation of the larger mammal remains, distributed throughout the deposit, have been the human inhabitants and large carnivores. Although selection undoubtedly played a role in the species represented, the remains do reflect some general changes in environmental conditions through time.

Table 1 lists the relative abundance of ungulate species in four sets of excavated members which correspond to the following intervals: the Holocene (DGL-BLA), the end-Pleistocene period of climatic amelioration immediately post-dating the Last Glacial Maximum (BRL-CL), the period including the Last Glacial Maximum (GWA-LPC), and the preceding period extending back in time to the base of the sequence (YOL-OCH) estimated to date to 80 000 B.P. Small browsing non-gregarious antelope such as klipspringer (Oreotragus oreotragus) and steenbok/grysbok (Raphicerus spp.) occur in the youngest period. These taxa indicate a well vegetated shrubland habitat analogous to the present. The Holocene fauna contrasts with the equid and alcelaphine fauna from the end-Pleistocene (BRL-CL) in which eland (Taurotragus oryx) is also prominent, and that of the coldest interval dating between about 18 000 and 22 000 BP (GWA-LPC) in which alcelaphines are the major component. These latter faunas are indicative of more grassy habitats and the difference between them is notably the high alcelaphine frequency in the earlier levels which would be consistent with other lines of evidence of drier climates at the Last Glacial Maximum. Little fauna was obtained from YOL and thus the earliest period dates to older than 30 000 BP and the rhebuck (Pelea capreolus) shows a notably increased representation in these lower stratigraphic layers (Fig. 2c). The occurrence of the rhebuck, a small browser found in montane habitats such as the Drakensberg and the most common antelope in the Bontebok Park (Bigalke pers. comm.) as well as Oreotragus and Raphicerus may reflect carnivore selection as the intensity of occupation is lower in these units, but they do also indicate a significant browse component in the vegetation. The presence of Pelea would not be inconsistent with cool and moister conditions relative to the Last Glacial Maximum as indicated by other lines of evidence. Late Pleistocene carnivore lairs from the same valley include equids, alcelaphines and buffalo as well as Raphicerus and Pelea and are probably of similar age to the lower part of the Boomplaas sequence.

Table 1. Relative abundance of ungulate taxa (excluding sheep/goat, elephant and rhinoceros) represented at Boomplaas Cave within four general periods, expressed as a percentage of the total number of ungulates under consideration.

	6400 – 1500 BP	14 000 – 10 000 BP	22 000 – 18 000 BP	Older than 25 000
Equidae	11.22	20.00	12.07	4.55
Alcelaphini	16.10	31.76	51.72	27.27
Pelorovis antiquus	0.00	2.35	3.45	0.00
Syncerus caffer	0.98	4.71	3.45	0.91
Antidorcas sp.	0.00	2.35	3.45	1.82
Hippotragus sp.	4.39	8.24	6.90	0.00
Redunca fulvorufula	8.78	3.53	0.00	8.18
Redunca arundinum	0.49	0.00	0.00	2.73
Taurotragus oryx	5.37	15.29	5.17	1.82
Pelea capreolus	6.83	3.53	3.45	23.64
Tragelaphus strepsiceros	0.98	1.18	0.00	0.00
Suidae	2.93	2.35	0.00	0.00
Oreotragus oreotragus	20.49	0.00	8.62	11.82
Raphicerus spp.	21.46	4.71	1.72	17.27
Total number of ungulates	205	85	58	110

Small mammal remains

The analysis of small mammal remains (rodents, insectivores and bats) excavated from the Boomplaas deposits was undertaken by DM Avery (1979, 1982). Avery (1982:207-9) argues that owls (notably Tyto alba) are likely to have been the principal agents responsible for the accumulation of small mammals represented in the Boomplaas cave deposits and, if it is assumed that the same agents were responsible for the accumulation of small mammal bones in all levels of the site, one may infer that major changes in the relative abundance of taxa represented in the cave deposits are a reflection of changes in the availability of small mammals in the local environment. Assuming further that any change in the availability of prey was itself a function of the combined influence of changes in physical factors (temperature, rainfall and seasonality in these parameters), as well as changes in the structure of vegetation which affected the distribution and abundance of prey, change in the relative abundance of small mammal taxa may help to identify the environmental factors concerned.

The vlei rat (Otomys irroratus) and common mole-rat (Cryptomys hottentotus) are the most common rodent taxa and the red musk shrew (Crocidura flavescens) is the most common insectivore in the assemblages dating from 14 000 BP to the uppermost units (CL-DGL). Their abundance contrasts sharply with their relative scarcity in the Glacial Maximum units of GWATBF, GWA and LPM in which Saunders' vlei rat (Otomys saundersae) is the dominant rodent and the forest-shrew (Myosorex varius) is the most common insectivore (Fig. 2a).

At present Saunders' vlei-rat appears to have an isolated distribution, occurring in regions of the eastern and south-western Cape as well as in Lesotho, inhabiting "heath country on high mountain slopes" (Shortridge 1934; De Graaf 1981:152). Its modern distribution at high altitudes and its abundance in assemblages dating to the Last Glacial Maximum at both Boomplaas and Nelson Bay Cave on the southern Cape coast (Avery 1979, 1982) provide a basis for suggesting that this species is able to withstand relatively cold temperatures. The forest-shrew and Saunders´ vlei-rat continue to be the dominant taxa in levels predating the Last Glacial Maximum, but percentages for these two species reach maxima or secondary maxima at different times during the Late Pleistocene. The forest-shrew is particularly abundant in assemblages from OLP 1-3, BP 1 and LPC 2B, while Saunders' vlei-rat peaks in BP 3, BP 4 and BOL 4. This inverse relationship is of interest since modern studies indicate that the forest-shrew is dependent on moist habitats (Walker 1964; Kingdon 1981; Rautenbach 1982) and we may infer that levels in which it is more abundant are likely to correspond to periods that were relatively moist, while those with higher percentages of Saunders' vlei-rat were relatively dry.

Summary statistics based on factor analysis scores obtained for each species have been calculated from the relative abundance of microfaunal taxa represented at Boomplaas. Two factors (F1 and F2) together accounted for 84% of the total variance in a three-factor solution. F1 may be identified with temperature by recognizing that all those taxa with high loadings on this factor occur most frequently in Holocene levels and are found in relatively warm habitats today, whereas those with low loadings on F1 occur most commonly in assemblages dating to the Last Glacial Maximum and earlier units. Summary statistics based on F1 in Fig. 2b correspond to "composite statistics" described by Nie et al. (1975), although changes in the value of the summary statistic (SS) based on F1 have been normalized so that the lowest value is 0 and the highest value in 100.

Changes in the value of the summary statistic based on F2 (accounting for 17.9% of the variance in a three-factor solution) are also shown in Fig. 2b. Microhabitats preferred by species with high loadings on this factor, for example the dwarf mouse (Mus minutoides) and the dwarf shrew (Suncus varilla), are associated with relatively moist conditions. The taxa with low loadings on the same factor, Saunders' vlei-rat and the bush Karoo rat (Otomys unisulcatus), appear to be able to tolerate relatively harsh environmental conditions, including cold and dry extremes. In a general way then, it is inferred that this factor is identifiable with a xeric-mesic gradient.

Variations in the values of the summary statistics based on F1 and F2 (Fig. 2b) indicate that assemblages dating to the Last Glacial Maximum (notably GWA) are characterized by relatively low values of both summary statistics, from which it may be concluded that cold and dry conditions prevailed at that time. By contrast, Holocene assemblages suggest warmer and relatively moister conditions. Within the Late Pleistocene cool and moist conditions are indicated for OLP 1-3 and BP 1.

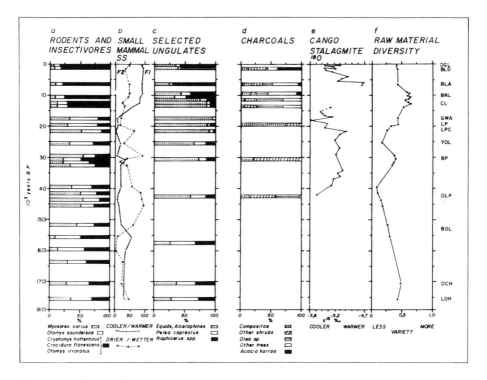

Figure 2. The scale and timing of changes in different palaeoenvironmental indicators at Boomplaas Cave (a, b, c, d, f) and oxygen isotope ratio changes in the Cango Cave speleothem (e). Taxa typical of the Holocene are marked by solid bars and those typical of the Last Glacial Maximum are dotted. The summary statistics (SS) for the small mammals are explained in the text.

Charcoals

Charcoals have been studied in samples systematically collected from members OCH, dating to more than 40 000 BP, to BLD, dating to about 2000 BP, and the results are presented in Table 2 and summarized in Fig. 2d. Marked changes in the composition of the woody vegetation over time are evidenced in the sequence. The present-day woodland of the valley with Acacia karroo as a dominant proves to be a Late Holocene phenomenon because A. karroo is present in low frequencies earlier in the Holocene. Supplementary samples analysed from the

Buffelskloof rock shelter 20 km to the west show that the increased frequency of A. karroo in the later Holocene is not simply linked to the appearance of herding but precedes this event by several thousand years. Thicket taxa like Maytenus/Pterocelastrus and Euclea/Diospyros are relatively prominent in the Early Holocene in the Boomplaas samples and there is a notable increase in Olea in the end Pleistocene samples, particularly in the CL Member. These data evidence changes in the association of woodland taxa with the amelioration of climates after the Last Glacial Maximum. Samples from the latter time range drawn from the GWA, LP and LPC members dating between about 17 000 and 22 000 BP show a virtual absence of woodland from the Cango Valley and charcoal samples are dominated by woody composites. These data are interpreted as being a response to the harsh climatic conditions of the Last Glacial Maximum.

Samples from BP dated to 32 000 years ago indicate that the vegetation in the surrounds of the cave was shrubland with a taxon with attributes similar to the Ericaceae family prominent and woody composites very much less common. The oldest samples analysed suggest that in prior times woodland with Olea prominent, but somewhat different in composition from that of the end Pleistocene, existed in the valley. It is inferred that climatic conditions were neither as dry nor as cold as at the Last Glacial Maximum. The apparent shifts from woodland to shrublands and again to woodland of somewhat different character suggests dynamic changes in the woody vegetation in the Cango Valley, as reflected by the charcoal analyses. These changes are best explained as the result of climatic forcing. The association of taxa in the vegetation mosaic represents a temporary alliance that has changed markedly through time.

Table 2. Percentage frequencies of identified charcoals from Boomplaas Cave.

	BLD	BLA	BRL	CL	LP/LPC	BP	OLP
Salix cf. mucronata	-	-	-	9.09	4.90	0.40	1.82
Protea arborea	12.09	1.56	4.65	4.13	2.58	5.13	8.45
Other Proteaceae	-	-	-	4.13	-	6.33	3.65
Acacia karroo	38.60	4.69	6.98	-	-	-	-
Nymania capensis	4.65	1.56	-	-	-	-	-
Maytenus/Pterocelastrus	2.79	14.09	4.65	-	-	-	0.50
Rhus cf. undulata	4.65	3.13	-	1.65	-	-	-
Other Rhus spp.	6.05	9.38	23.26	23.14	-	-	0.63
cf. Rhamnus	-	-	2.33	2.48	5.09	0.53	-
Passerina spp.	2.79	-	-	-	-	-	3.53
cf. Erica spp.	-	1.56	-	5.78	15.72	34.52	6.30
Euclea/Diospyros	1.40	12.50	16.28	1.65	-	-	-
Olea group	3.26	15.63	18.60	38.86	-	-	33.69
cf. Buddleia glomerata	4.19	9.38	2.33	0.83	-	8.20	9.27
cf. Buddleia salvifolia	4.19	-	-	-	-	-	-
cf. Lycium	-	-	-	-	-	2.93	-
cf. Nemesia fruticans	-	-	-	-	0.72	3.26	2.52
Compositae	11.17	17.19	13.96	4.96	67.16	18.52	12.02
Type A	-	1.56	-	-	0.72	9.40	0.31
Other	4.20	7.81	6.98	3.31	3.47	11.25	16.46
Total number	215	64	43	121	129	217	185

Pollens

Cave sites are generally not ideal for the study of vegetation changes through pollen analysis because the pollen rain comes from a restricted area in the immediate vicinity of the site and because use of the cave by people, animals and insects can cause anomalies in the range and redistribution of the pollen. Nevertheless, a pilot study of pollens at Boomplaas shows some interesting patterns that can be used together with other lines of evidence.

In comparing pollen spectra from the Holocene (BLD and BLA) and Last Glacial Maximum (LPC) members there is a clear contrast in the species diversity with between 30 and 40 different pollen types in the Holocene samples, but only nine in LPC. Of these nine types, five were Compositae and a single type, viz. Elytropappus (renosterbos), was the most common. This suggests that species diversity was low in the vicinity of the site during the Last Glacial Maximum when renosterbos, prominent also in the charcoals, was possibly the dominant component of the local vegetation.

Pollen was preserved down to about 3 m below the present surface, i.e. back to c. 32 000 BP in the BP Member. In older deposits (between OLP and LOH) conditions were evidently not conducive to pollen preservation. The specimens attributed to Elytropappus are relatively common in the lowest BP sample, but are not as common as in LPC and the diversity of pollen types is relatively higher in BP indicating that the environment at this time was not as harsh as it was during the Last Glacial Maximum. Within the last 11 000 years changes were of a lower order than those observed between the Late Pleistocene and Holocene.

Human settlement pattern

The very fact that people occupied Boomplaas Cave at various times in the past is taken to be indicative of favourable conditions in the vicinity of the site, while the density of the occupation debris can be taken as a rough guide to the intensity of occupation and, possibly, of the size and composition of human groups. By far the greatest density of artefactual material comes from the CL Member and, thereafter, from the early Holocene BRL Member and from the mid- and late Holocene BLA and DGL members. The stratigraphically lower Late Pleistocene members, in contrast, show many more non-occupation deposits where sediments have built up through natural weathering processes and through the deposition of owl pellets. Where people have occupied the site in the earlier part of the Late Pleistocene, the density of stone artefacts is generally low.

Although quartz is the most commonly used raw material throughout, the diversity of raw materials changes through time. The highest diversity is in the CL 3 to BRL 1 units dating broadly between 14 000 and 8000 BP. Diversity is relatively high in the Holocene members and in OCH at the base of the sequence, and is moderately high in BP dated to c. 32 000 BP, but is lowest in LP-LPC dated to the Last Glacial Maximum and in the other remaining Late Pleistocene units. A raw material diversity index has been calculated and changes through time are shown in Fig. 2f.

The raw materials used for artefact making are not all equally available and can be graded on a scale from most readily available (quartz), through local materials (quartzite, hornfels and chalcedony) to the least available (silcrete), the known sources of which are outside the valley. The high incidence of silcrete in OCH and GWA-CL 3 may indicate relatively mobile groups occupying the cave. A reduction in home range is suggested by the replacement of silcrete by chalcedony for bladelet manufacture in the CL unit and the high diversity of raw material usage in the BRL to DGL members can be viewed as a reflection of the raw material usage pattern of settled-in populations living in the more productive and diverse habitats of the end-Pleistocene and Holocene.

The low raw material diversity in the Middle Stone Age levels from BOL to OLP correlates with a virtual absence of chalcedony and silcrete. The explanation may lie in the low intensity use of the site in this time range and generally low productivity of the effective environment for hunter-gatherers. The briefly increased diversity in the BP Member correlates with interstadial environmental conditions noted also in the small mammals (Avery 1979, 1982) and a well-developed occupation deposit including carbonized organic layers suggesting a relatively productive habitat.

Sediments

Soil samples from the excavated sequence at Boomplaas were analysed by L Webley (1978). Except where masked by secondary sedimentation, the deposits are well stratified and consist of sandy loams with inclusions of coarser clasts of roof rock. Grading analysis shows little difference through the sequence in the finer component (-2 mm fraction) in the mean grain size, sorting, skewness and kurtosis. This is as would be expected in a situation such as this cave on a slope above a valley where the sources contributing to the build-up of the deposit are limited and have remained constant. The major source of coarser material has been the cave interior and some horizons are marked by a higher component of roof debris than others. Angular roof clasts in levels OCH, OLP, YOL and LP may reflect cold periods but there is no evidence for intense frost shattering phenomena producing clast supported "eboulis sec" as found in European or Tasmanian sequences (Goede, pers. comm.).

Cold conditions are also indicated in the LP sample by cleavage planes and conchoidal fractures on quartz grain surfaces studied with the aid of a scanning electron microscope, while the low carbonate, phosphate and pH measurements in GWA are indicative of leaching, possibly during the accumulation of CL.

Stalagmite formation

Calcium carbonate has cemented the Boomplaas deposits locally below drip points fed by roof fissures. Radiocarbon dates on cemented materials are in agreement with age determinations on charcoal from the uncemented facies and this shows that cementation has kept pace with the build-up of the deposit through much of the Late Pleistocene. Cementation of a column in the main excavation (the stalagmite in Fig. 1) ceased before the deposition of the CL Member at c. 14 200 BP, when there is evidence from the site that the amelioration of climates had set in. It is of interest to note that a speleothem collected from about 1 km from the entrance to the nearby Cango Caves cavern system for palaeotemperature studies (Vogel 1983) shows a hiatus in growth between 14 000 and 5000 BP. Cementation of deposits dating to the last 5000 years has not been recorded in Boomplaas Cave, but the aeration of the cave, one of the factors on which speleothem formation depends, is very different from that in the deep recesses of Cango.

As the availability of drip water and the bicarbonate content of the ground waters can change with changing climate, the growth and hiatuses in the growth of speleothems or cementation of a deposit may have palaeoclimatic significance. No cavestone would have been formed under conditions of low precipitation when water did not percolate down into the cave. The factor governing the bicarbonate content of the soil water is the partial pressure of the carbon dioxide in the soil which decreases when the vegetation cover becomes sparse. It would be expected on these grounds that the cessation of stalagmite formation could be coupled to low rainfall and decreased plant cover. This expectation however is not met because the biological evidence from Boomplaas shows conditions during the deposition of the CL Member to have been wetter than the Last Glacial Maximum and the vegetation of the Holocene, although different in structure from the Late Pleistocene, was not reduced. This suggests that the controls of speleothem formation are relatively subtle, and are possibly linked to changes in the structure and composition of the vegetation as well as to the source of the groundwater feeding the fissures of the cavern system. Although the controls are not fully understood at present, a pattern of more continuous speleothem formation under cooler glacial conditions seems indicated.

The Cango stalagmite (Vogel 1983) has provided an indication of the scale of mean annual temperature depression at the Last Glacial Maximum and this is of the order of 5 degrees C lower than the present. Although only the last 5 000 years of the Holocene record is represented, the oxygen isotope curve shows a high amplitude of change that may reflect changes in the sources of precipitation rather than temperature fluctuations. Precipitation is enriched in the lighter oxygen isotope relative to sea water the further the rain-bearing storms have to travel inland and the more mountain barriers that have to be crossed. Any change in the present mix of elements that bring non-seasonal rains to the inland Cango Valley would be expected to be reflected in the isotope curve giving this additional significance.

Local geomorphology
Most of the data assembled from Boomplaas Cave and the Cango Caves have confirmed the existence of significantly lower temperatures in the Cango Valley at the Last Glacial Maximum. Rainfall data are less easy to come by, but indications from the pollens and charcoals as well as from the larger and smaller mammals are that the Last Glacial Maximum was also somewhat drier than the present. The observation that true ´eboulis sec´ are not present in the Boomplaas cave deposit has suggested that there was insufficient moisture during the coldest season for large scale frost fracturing, and this is supported by the observation that the slopes of the surrounding Swartberg mountains do not have extensive scree mantles. Winter rainfall does not seem to have been significantly increased at the Last Glacial Maximum.

SUMMARY AND CONCLUSIONS

An essential part of the concept of the research project at Boomplaas Cave and the Cango Valley was to provide a number of lines of complementary evidence measuring the scale of environmental change in the Late Pleistocene and Holocene and the human response to it. The evidence has been drawn not only from the analysis of biological remains and sediments, but also from a speleothem in the nearby Cango Caves. Some measures have proved more sensitive to local habitat changes, for example the analyses of charcoals and small mammals, and others are more influenced by agents of accumulation such as the large mammals. However, the merit and the strength of the study lies in the combined weight of the evidence for environmental and inferred climatic change over the last 80 000 years.

The sequence covers isotope stages 1 to 4 and possibly 5a. The best evidence for an amelioration at the base of the sequence that could represent 5a comes from the small mammal data, and the occurrence of ferricrete nodules in OCH may be indicative of an episode of higher precipitation in the stage 4-5 range. The BOL unit may correlate with stage 4, but again for this time range only small mammal data are significant. Several lines of evidence indicate that OLP dates to more than 40 000 BP and correlates with stage 3. It was cooler than the present and moister than the period coeval with the Last Glacial Maximum. The forest-shrew is prominent in the small mammal samples in this time range, rodents are found in relatively low proportions and the charcoal samples include a relatively high proportion of Olea. It is probable that the kind of environment in this time range has no modern analogue. It is also interesting that environmental conditions do not appear to have been favourable for human settlement in the Cango Valley and indeed this period is poorly represented in the archaeological record generally in the southern Cape and elsewhere in southern Africa (Deacon 1983; Deacon and Thackeray, this volume). There is some evidence for interstadial conditions in the time range of 32 000 in the small mammal data, the pollen diversity and the intensity of occupation in the BP Member. However woodland had already been replaced by shrubland on the charcoal data.

Perhaps the most significant trend indicated in the Boomplaas sequence is the abrupt decline in diversity of all biological indicators in the members YOL to GWA dating between about 25-30 000 BP and about 17 000 BP. The lowest diversity of pollens, charcoals and small mammals is correlated with the harshest climatic conditions in GWA/LP at about 18 000 BP. It is inferred that climates were both cold and dry under the harshest climates which is essentially the concept of a Last Glacial Maximum. An indicator of dry conditions amongst the small mammals is the prominence of the bush Karoo rat and the dominance of composites and absence of woodland taxa in the charcoal samples. Also significant is the evidence for the amelioration of climates prior to 14 000 BP with the re-establishment of woodland in the valley.

The charcoals in the CL Member show a marked increase in taller woody taxa such as Olea indicative of more effective precipitation after about 14 000 BP. The diversity and species composition of both plant and small mammal samples increased in CL relative to GWA/LP and this is also interpreted as the result of a warming of temperatures and an increase in precipitation. The reasons for the absence of calcium carbonate cementing of deposits at Boomplaas and the cessation of growth of the Cango speleothem after 14 000 BP are not clear, but the biological evidence does not accord with this being due to reduced precipitation relative to the Last Glacial Maximum.

From about 11 000 BP and coincident with the accumulation of the BRL Member, charcoals show that thicket taxa became common in the valley. These were in turn replaced in the Late Holocene by Acacia karroo. After about 10 500 BP the larger mammals reflect the change in the hunting pattern common to all Holocene sites excavated in the southern and eastern Cape, namely a concentration on small, non-gregarious browsing antelope, in particular grysbok/steenbok and klipspringer (Deacon 1972, 1976; Klein 1974, 1980), that contrasts with the large grazers (equids and alcelaphines) hunted in the late and terminal Pleistocene. The smaller mammals of the mid-Holocene reflect the warmest conditions of the sequence with possibly a cooler interval in the Late Holocene that may be matched in the oxygen isotope record from the Cango Caves. The scale of the temperature shift at the end of the Holocene in the Cango oxygen isotope curve is anomalously high, however, and the isotope record in this time range may be more an indication of the source of precipitation and thus of synoptic climates than simply a measure of past mean temperatures.

Although evidence from other sites in the southern Cape is patchy because sequences are generally not as complete as the one at Boomplaas, several of the climatic "events" noted at Boomplaas can be traced further afield than the Cango Valley. The larger mammal fauna, for example, reflects human hunting strategies and these are in turn dependent on the organization of people in a wide social and demographic network that includes a large part of southern Africa. The shift from hunting large grazers to the trapping of smaller browsers, and the accompanying increase in the importance of gathered plant foods at the end-Pleistocene/Early Holocene, is a response to the development of more heterogeneous or patchy habitats generally in southern Africa that was stimulated by the amelioration of climates after the Last Glacial Maximum (Deacon 1972, 1976, 1983).

Another feature which can be found at several sites in the southern Cape is the occurrence of ferricrete nodules or ferruginous crusts. That at Boomplaas in the lower OCH Member can be linked to similar deposits at Nelson Bay Cave (Butzer 1973), Kangkara (Deacon 1982) and Buffelskloof (Opperman 1978) because they all indicate a period of ferruginous crust development as the result of watertable hardpans forming under wet conditions in the earlier part of the Late Pleistocene. The crusts at all the above sites cap Middle Stone Age deposits, but the more complete sequence at Boomplaas shows that Middle Stone Age occupations post-dated their formation as well.

One of the most rewarding results of the analysis of the Boomplaas sequence has been the confirmation of the timing of gross palaeoenvironmental changes between 25 000 and 16 000 years ago that clearly relate to the Last Glacial Maximum and the periods immediately preceding and succeeding it. The dating of these changes at both Boomplaas and the Cango Caves agrees so closely with changes of similar scale measured in the oxygen isotope record from Dome C in Antarctica (Lorius et al. 1979) that there can be no doubt that they relate to the same southern hemisphere and worldwide phenomena.

ACKNOWLEDGEMENTS

We are grateful to all those who have been involved in the Boomplaas project from its inception. In particular we would like to thank DM Avery, LD Daitz, VB Geleijnse, AC Goede, RG Klein, and L Webley for their contributions. The research has been supported by the University of Stellenbosch, the CSIR through the Cooperative Scientific Programmes National Programme for Weather, Climate and Atmosphere Research (J Deacon) and the Fynbos Biome Project (A Scholtz), and the HSRC (HJ Deacon and JF Thackeray).

REFERENCES

Avery DM 1979. Upper Pleistocene and Holocene palaeoenvironments in the southern Cape: the micromammalian evidence from archaeological sites. D Phil thesis, Univ. Stellenbosch.

Avery DM 1982. Micromammals as palaeoenvironmental indicators and an interpretation of the late Quaternary in the southern Cape Province, South Africa. Ann. S. Afr. Mus. 85:183-374.
Butzer KW 1973. Geology of Nelson Bay Cave, Robberg, South Africa. S. Afr. archaeol. Bull. 28:97-110.
Deacon HJ 1972. A review of the post-Pleistocene in South Africa. S. Afr. archaeol. Soc. Goodwin Ser. 1:26-45.
Deacon HJ 1976. Where hunters gathered: a study of Holocene Stone Age people in the eastern Cape. Monogr. Ser. S. Afr. archaeol. Soc., Claremont.
Deacon HJ 1979. Excavations at Boomplaas Cave: a sequence through the Upper Pleistocene and Holocene in South Africa. World Archaeol. 10:241-257.
Deacon HJ 1983. The peopling of the fynbos region. In: HJ Deacon, QB Hendey and JJN Lambrechts (eds), Fynbos palaeoecology: a preliminary synthesis. S. Afr. Natn. Scient. Progr. Rep. 75:183-204.
Deacon HJ and Thackeray JF 1984. Late Pleistogene environmental changes and implications for the archaeological record in southern Africa, (this volume).
Deacon J 1982. The Later Stone Age in the southern Cape, South Africa. PhD thesis, Univ. Cape Town.
De Graaf G 1981. The rodents of southern Africa. Butterworths, Durban.
Kingdon J 1981. East African mammals. Academic Press, London.
Klein RG 1974. Environment and subsistence of prehistoric man in the southern Cape Province, South Africa. World Archaeol. 5:249-284.
Klein RG 1978. A preliminary report on the larger mammals from the Boomplaas Stone Age cave site, Cango Valley, Oudtshoorn District, South Africa. S. Afr. archaeol. Bull. 33:66-75.
Klein RG 1980. Environmental and ecological implications of large mammals from Upper Pleistocene and Holocene sites in southern Africa. Ann. S. Afr. Mus. 81:223-283.
Lorius C, Merlivat L, Jouzel J and Pourchet M 1979. A 30 000-yr isotope climate record from Antarctic ice. Nature 180:644-648.
Nie HH, Hull CH, Jenkins JG, Steinbrenner K and Bent DH 1975. Statistical package for the social sciences, 2nd ed. McGraw-Hill, New York.
Opperman H 1978. Excavations in the Buffelskloof rock shelter near Calitzdorp, southern Cape. S. Afr. archaeol. Bull. 33:18-28.
Rautenbach IL 1982. Mammals of the Transvaal. Ecoplan, Pretoria.
Shortridge GC 1934. The mammals of South West Africa. 2 vols. Heinemann, London.
Vogel JC 1983. Isotopic evidence for the past climates and vegetation of South Africa. Bothalia 14:391-394.
Walker EP 1964. Mammals of the world. John Hopkins Press, Baltimore.
Webley L 1978. Sediment analysis of Boomplaas Cave. BA (Hons.) thesis, Univ. Stellenbosch.

Investigations on archaeological charcoals from Swaziland, using SEM techniques

JULIET PRIOR
Imperial College of Science and Technology, London, UK

ABSTRACT. Charcoal remains from the mainly Holocene site of Siphiso have been identified using SEM techniques and compared with a collection of charred modern woods from the area. The results yield both ethnological and palaeoclimatic information and demonstrate the value of this type of investigation. It is seen that the occupants of the site selected comparatively few woods for highly specific purposes. The sequence suggests alternation of dry, moist, dry phases since an estimated 12 000 BP.

INTRODUCTION

The rock shelter at Siphiso has provided a rich source of archaeological charcoal fragments, from Recent to Late Pleistocene in age. Details of the site, which lies at an altitude of 320 m, are shown in Fig. 1. The contours relate to a datum point outside the shelter and the shaded areas represent the extent of excavations during the 1981 and 1982 field seasons of the Swaziland Archaeological Research Association (Price Williams 1980).

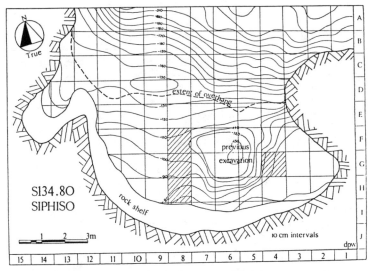

Figure 1. Plan of rock shelter at Siphiso.

Siphiso lies on a steep north facing slope and a detailed transect and frequency distribution survey in the immediate vicinity showed that plants in the area today are predominantly Bushveld species, as defined by Acocks (1975). These include both deciduous and drought tolerant trees, interspersed with sweet grassland, capable of supporting a high biomass. Present-day rainfall, approximately 500 mm per annum, falls in the mid or late summer. Annual evapo-transpiration rates exceed precipitation, forming salt-rich upper soil layers in what is effectively a semi-arid area. The greatest density of vegetation is to be found on the alluvium adjacent to the headwaters of the Nkumbane. This river, which is today swollen by summer rains, must have provided an important water source, both for browsing and grazing animals and for past hunter-gatherer populations of Siphiso.

The charcoal fragments examined since January 1982 have been individually excavated from squares F8 and H8 (Fig. 1). The various strata, rich in archaeological artefacts, bone and charcoal, span the Late Stone Age, which corresponds to the period of time which has elapsed since the Last Glacial Maximum. There is a considerable amount of spall between strata, due partly to pressure unloading but also to efflorescent salts contained within the rhyolites of the roof. The pH of the deposit is greater than 8, because of high concentrations of both sodium and calcium. Details of the strata are shown in Table 1. Radiocarbon dates, obtained from CSIR Pretoria, are 7600 ±80 for the early Wilton period and 8700 ±120 BP for the Wilton-pre-Wilton interface (Vogel, pers. comm.).

Table 1. Chronology of the strata in Siphiso rock shelter.

STRATUM NO.	ARCHAELOGICAL PERIOD	DATE
I and II	Recent, + lead shot, snare pins, birds nests, etc.	50-100 BP
III	Late Wilton/Smithfield	Late Holocene
IV	Climax Wilton	Mid-Holocene
V and VI(top)	Early Wilton	Early Holocene
VI(base)	Pre-Wilton	Late Pleistocene

Of recent years, charcoals have been used extensively as a source of material for radiometric dating but have been greatly underutilized as a source of palaeoclimatic information, though their importance has been shown at sites such as Boomplaas (Deacon 1979). It is generally agreed that the flora of Southern Africa has remained essentially unchanged throughout archaeological time. Comparatively minor shifts in vegetational zones have however occurred throughout the Holocene (Is. Stage 1), whilst greater changes may be evident during the isotopic stages of the Pleistocene (Van Zinderen Bakker 1978). Such shifts in climate are likely to be reflected in the assemblages of trees represented by charcoals within the succession of strata. The archaeological charcoals in any level represent the remains of comparatively few local trees and shrubs selected by each successive group of hunter-gatherers, either for fires or for other utilitarian purposes. With such small assemblages, the importance of ecological studies, resulting in an increased knowledge of the micro-distribution of woody species, can hardly be overemphasized. Palaeoclimatic interpretations also depend upon accuracy in the identification of taxa. It is for this reason that a comprehensive reference collection of the extant woody species of Swaziland has been made. The anatomical structure of some of these has been described (Kromhout 1975; Miles 1978) but for the majority of the 600 species few details exist, as the woods concerned are of little economic importance. Twig and trunk wood samples were collected for each species, since both may be represented amongst the charcoals and they may exhibit differences in structure. Further, in some instances, the same wood was sampled from more than one of the four vegetational zones of Swaziland, since environmental factors may sometimes affect structure (Robbertse et al. 1980).

METHODS

Figure 2 illustrates the way in which ancient material is prepared. Charcoal fragments, which vary in size from 1 cm to just a few mm, are manually fractured in three planes: transverse, tangential longitudinal and radial longitudinal. Specimens are then mounted on a single aluminium stub to display these freshly cleaved planes, before being coated with 100 Å of gold in an atmosphere of argon and examined by scanning electron microscopy (SEM), using magnifications from 50 to 6400 X. This technique gives good results with opaque charcoals, whereas the alternative technique of incident light microscopy requires a flat surface, which is usually difficult to obtain. The scanning electron micrographs shown in Figs 3-10 illustrate a wide range of structural characters by which comparison can be made with charred modern woods. Each fragment can then be identified to a particular genus if not to a species. Such comparisons with charred material are necessary, since experimental work has recently shown that certain

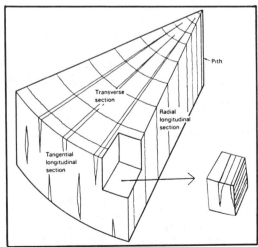

Figure 2. Wood segment illustrating planes of fracture.

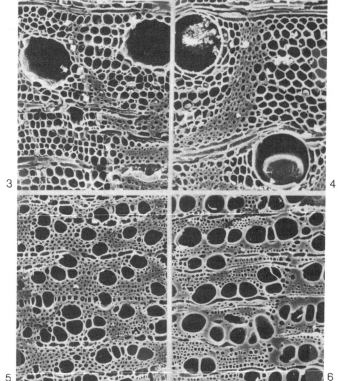

Mag. X135

Figs. 3 and 4. Acacia nigrescens. 3. Transverse fracture of modern wood showing vessels, living thin walled cells, small thick walled fibres and ray cells running in a transverse plane. 4. Transverse fracture of Early Wilton charcoal fragment.
Figs. 5 and 6. Androstachys johnsonii. 5. Transverse fracture of modern wood. 6. Transverse fracture of Early Wilton charcoal fragment showing some fusion of intercellular material.

structural features occur simply as a result of the charring process (Prior and Alvin 1983). These must be clearly distinguished from characters used in the identification of taxa. Particularly fragile charcoal fragments may be hardened using artificial resins before the fracturing treatment described above. This does not occlude any of the structural details subsequently observed under SEM.

RESULTS

Figures 3-10 are scanning electron micrographs of ancient and modern charcoals. With each pair of photographs, the ancient fragment is on the right-hand side and the modern comparative material is on the left. The transverse sections of Acacia nigrescens, the knobthorn, illustrate the structure of a typical wood. The largest cells are the

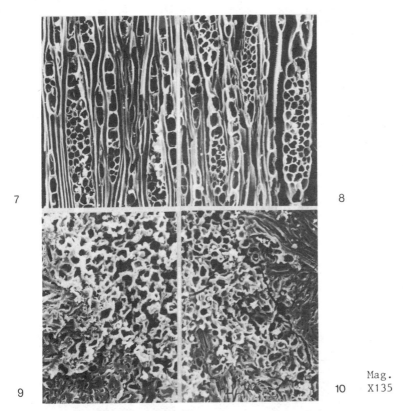

Mag. X135

Figs. 7 and 8. Bridelia cathartica. 7. Tangential longitudinal fracture of modern wood showing detailed structure of rays. 8. Bridelia sp. Tangential longitudinal fracture of Late Wilton charcoal fragment. Fig. 9 and 10. Morula seeds. 9. Fracture of charred modern seed. 10. Fracture of charred Early Wilton seed. The cracks indicate the fragile nature of this material.

vessels, responsible for the conduction of water, the small thin-walled cells represent a living tissue capable of supplying nourishment to the tree and the small thick-walled cells form the supporting fibrous tissue. The rays, rows of living cells providing a means of horizontal conduction within the wood, can be seen running in a transverse plane. It is the size, arrangement and structural details of these and sometimes other cell types that provide the characters necessary for secure identification. The transverse sections of Androstachys johnsonii, the Lubombo ironwood, illustrate a common effect of charring. The ancient fragment shows a considerable amount of fusion of intercellular materials when compared with the modern. This is probably caused by a high charring temperature.

The woods represented in the strata examined are shown in Table 2. These results must still be regarded as preliminary, though during the next 18 months corroborative evidence will become available from square G9. Woods represented by +++ indicate the most important taxa, those

Table 2. Frequency of woods represented in charcoal spectra from squares F8 and H8.

Stratum no.	Dichrostachys cinerea	Combretum apiculatum	Lonchocarpus or Acacia caffra or A. nigrescens	Sclerocarya birrea (seeds only)	Androstachys johnsonii	Diospyros sp.	Grewia sp.	Maytenus sp.	Bridelia sp.	Ozoroa engleri	Gardenia sp.	Rubiaceae
I and II RECENT	+++	+++	++	++	++	++	++	++	+	+		
III LATE HOLOCENE	+++	+++	++	+	+	+		+	+			+
IV MID HOLOCENE	+++	+++	+	+++								+
V and VI(top) EARLY HOLOCENE	++		++	+++	+					+	+	+
VI(base)+ LATE PLEISTOCENE	+	+	+	+++								

Average sample size per stratum = 46 fragments

represented by ++ were also present in considerable amounts, whereas those designated by + were represented by only one or two fragments. Within each period, assemblages permit both ethnological and palaeoclimatic interpretations.

ETHNOLOGICAL INTERPRETATIONS

The comparatively small number of taxa represented is probably not explained by selective charcoal preservation, since almost all the fragments examined were well preserved. Rather, it suggests a high degree of selectivity on the part of successive occupants of the shelter. Dichrostachys cinerea, the sickle bush, Combretum apiculatum, the leadwood, and Acacia nigrescens, the knobthorn, are all preferentially selected as firewoods by the modern Swazis. The relative frequency of these species in the Recent and Late Holocene strata suggests a similar usage in the past. Latex-rich species such as Ozoroa engleri, the white resin tree, might well have provided a source of mastic for the hafting of microlithic Stone Age artefacts. Sclerocarya birrea, the morula, produces a large fruit, the soft outer part of which is fermented by the Swazis to produce beer whereas the seeds, which are embedded in a hard shell, are prized as a rich source of both oil and vitamin C (Palmer and Pitman 1972; Fox and Norwood Young 1982). The presence of large numbers of charred seeds found in both Early and Mid-Holocene levels suggests their past popularity. Androstachys johnsonii, the Lubombo ironwood, is a dense, termite resistant hardwood which in recent times has been used extensively for fencing posts. The presence of this wood in the Recent levels is interesting, since it is currently a relict species occurring in small, pure stands some distance from the site. A saw mill was present in the area, however, until about 40 years ago and this could have provided a source of posts which were later used as firewood by the Siphiso inhabitants.

PALAEOCLIMATIC INTERPRETATIONS

The assemblages represented in Strata I to III (Is. Stage 1) include trees, all of which are common around Siphiso today. The inference must therefore be that little climatic change has occurred in the area of the shelter since about 4000 BP. In the Mid-Holocene, the lack of riverine species such as Bridelia together with the disappearance of typical Bushveld shrubs such as Grewia and Diospyros may be coeval with a drier period around 6000 BP. This was due in part to a northward displacement of the westerlies, mentioned by climatologists in this volume.

In Stratum V the top of Stratum VI (Is. Stage 1), Combretum apiculatum, a tree characteristic of dry open woodland, has disappeared and this, together with the assemblages represented, would be consistent with the warmer, moister phase known to have occurred about 9000 BP in the Southern Hemisphere (Salinger 1981).

During the Late Pleistocene represented by the Strata below VI (Is. Stage 2-1), the assemblage is dominated by Androstachys johnsonii. This tree occurs today in both tropical and sub-tropical areas and in Swaziland it is at its southern limit of distribution (Coates Palgrave 1977). In both Swaziland and northern Natal it occurs in relict forest patches, predominantly in areas of rain shadow, which receive as little as 300 to 400 mm per annum. Its occurrence in considerable quantities at Siphiso in Late Pleistocene levels may therefore indicate a warm, dry period around 11 000 to 12 000 BP. This result would corroborate the findings of Heusser in Chile (this volume).

CONCLUSIONS

The current investigation illustrates the importance of charcoals as indicators of past vegetational communities in the area immediately surrounding Siphiso. Such studies will complement those based on palynological and semidentological data. Within Swaziland, many such charcoal-rich archaeological sites exist, spanning a large part of the Quaternary period. One such site is Sibebe shelter in the Highveld (Price Williams 1981). Here, information from charcoals will relate to more of the Pleistocene, since the earliest archaeological levels include at least two phases of the Middle Stone Age.

Swaziland is ideally situated for such studies, as it contains four highly distinctive vegetational zones, the Highveld, the Middleveld, the Bushveld and the Lubombo. It is in areas of such altitudinal and hence environmental sensitivity that past changes in climate, even comparatively minor ones, will be most clearly seen. Further, this diversity of climatic patterns and hence of vegetation in Swaziland is representative of a large area of the sub-continent.

ACKNOWLEDGEMENTS

Funding for the Siphiso charcoal investigations and for the associated field work is gratefully acknowledged from the Anglo American Corporation and from the National Trust Commission of Swaziland. Dr D Price Williams was responsible for the excavation of the shelter and for the initiation of the botanical work, which has been supervised by Dr KL Alvin of Imperial College, London.

REFERENCES

Acocks JPH 1975. Veld Types of South Africa, 2nd edn. Mem. bot. Surv. S.Afr. 40. Govt. Printer, Pretoria. 50 pp.
Coates Palgrave K 1977. Trees of Southern Africa. C Struik, Cape Town, Johannesburg.
Deacon HJ 1979. Excavations at Boomplaas Cave - a sequence through the Upper Pleistocene and Holocene in South Africa. World Archaeol. 10:241-257.
Fox FW and Norwood Young ME 1982. Food from the Veld. Delta Books, Johannesburg.
Kromhout CP 1975. 'n Sleutel vir die mikroskopiese uitkenning van die vernaamste inheemse houtsoorte van Suid-Afrika. Bull. 50. Dept. Bosbou, Pretoria.
Miles A 1978. Photomicrographs of world woods. Dept. of the Environment, Building Research Establishment. HMSO, London.
Palmer E and Pitman N 1972. The Trees of Southern Africa. Balkema, Cape Town.
Salinger MJ 1981. Palaeoclimates north and south. Nature 291:106-107.
Price Williams D 1980. Archaeology in Swaziland. S. Afr. archaeol. Bull. 35:13-18.
Price Williams D 1981. A preliminary report on recent excavations of Middle and Late Stone Age levels at Sibebe Shelter, north west Swaziland. S. Afr. archaeol. Bull. 36:22-28.
Prior J and Alvin KL 1983. Structural changes on charring woods of Dichrostachys and Salix from Southern Africa. IAWA Bull. n.s.4.:197-206.
Robbertse PJ, Venter G and Janse van Rensburg H 1980. The wood anatomy of the South African acacias. IAWA Bull. n.s.1.:93-103.
Van Zinderen Bakker EM 1978. In: W Junk (ed.), Biogeography and Ecology of Southern Africa. Vol. 1. Dr W Junk, The Hague. p. 133-143.

Micromammalian population dynamics and environmental change: The last 18 000 years in the southern Cape

D.M.AVERY
South African Museum, Cape Town

ABSTRACT. The population dynamics of two coexisting species of Otomys (vlei rats) was examined to assess its usefulness in palaeoenvironmental interpretation. The age profiles and mortality patterns of Otomys irroratus and O. saundersiae in micromammalian samples from the Cango valley dating from 18 000 BP to the present show significant variations, suggesting that the population structure contains palaeoclimatic information. The results may indicate cold dry conditions during the Last Glacial Maximum with general improvement thereafter.

INTRODUCTION

"Environment does not act directly on numbers as such but indirectly through its influence on fecundity at each age and survival over each interval of age" (Caughley 1977:1). Conversely changes in fecundity and/or survival over each age interval in the population should reflect change in the environment. This point has also been made by Dodd and Stanton (1981:337) who give a good review of the application of population studies in palaeoecology. Such studies have, however, generally been directed towards marine invertebrate populations. Exceptions are the pioneering work of Kurtén (1953) and Voorhies (1969) where the use of life tables in palaeontology is examined, using the basic work on modern populations by Deevey (1947) as a starting point. In both cases, however, the principal concern has been with the palaeobiology of the species involved rather than with the use of the information to interpret environmental conditions or change. The latter will in any case only be possible where a sequence of samples is available for internal comparison.

In the present instance it is proposed to examine the potential of Otomys irroratus (Mammalia: Rodential; vlei rat) as an indicator of environmental change, using samples from the last 18 000 years in the Cango valley near Oudtshoorn in the southern Cape Province. Four samples are available from Boomplaas Cave (BPA-LP,-BRL7,-BL and -BLD3), together with three modern comparative samples, one from below Boomplaas Cave (BPB/C) and the other two from near the south Cape coast, at Glentyre (GT) and Klein Hagelkraal (KH). The modern samples are from pellets cast by Tyto alba (barn owl), since this bird is thought to have been responsible for the accumulation of the remains in the archaeological sites (Avery 1982). The palaeontological samples are dated to approximately 18 000 BP (LP), 12 000 BP (BRL7), 5000 BP (BL) and 3000 BP (BLD3) (Deacom 1979).

There is a considerable body of data available on the biology of living O. irroratus (Curtis and Perrin 1979; Davis and Meester 1981; Perrin 1979, 1980a, b), as well as a very useful study of the prey of T. alba, including O. irroratus (Perrin 1982). On the other hand virtually nothing is known about the biology of living O. saundersiae, which appears to be uncommon and to survive in refugia. It will be necessary for the time being to make tentative extrapolations from O. irroratus in order to assess possible interpretations, with the proviso that these will probably require modification once more is known about the species itself.

Population structure, or the relative abundance of differently aged individuals in the population, is generally summarized in life tables and survivorship curves (Dodd and Stanton 1981:345; Deevey 1947). Dodd and Stanton (1981:344) suggest that the structure of a particular population can provide information about the environment of the species in terms of its relative favourability to that species. Thus a high proportion of dead young could indicate marginal or unfavourable conditions. This, however, depends on whether the sample constitutes a random sample of a living population or an accumulation of dead individuals. Hutchinson (1978:46) points out that increases in population due to increased birth rate will result in higher proportions of juveniles in a time-specific estimate; if increase is due to reduced mortality there will be a greater proportion of older individuals. This information could be useful for assessing the fitness of the population to its environment.

The point has also been made elsewhere that there are critical phases in the life history of the species. Clearly any factor operating on those phases will have a proportionately greater effect upon the population as a whole than will a factor affecting a less critical phase (Slobodkin 1964:159). The q_x series of a life table which shows specific mortality at different ages is advocated by Caughley (1966:909) as a useful tool, being independent of any bias or under-representation there may be in the youngest age class, and it is possible that this will indicate the cause of increase or decrease in a population. If, for instance, there is high mortality in the mature, reproductive stages, one may assume that the population is not fit, but is most certainly in decline.

The majority of palaeontological samples must be considered as representing stable populations because they are long-term averages. By definition such an average can only represent a stable population. Only under circumstances where it can be shown that the sample is the result of a discrete catastrophic event, will this not necessarily be the case. Whether or not the samples represent increasing or declining populations, would be much more difficult or perhaps impossible to assess from this type of evidence.

From the present data it would appear that in fact the Otomys irroratus population was increasing since the samples were progressively larger with the passage of time. This, however, is based more on the evidence that the barn owl tends to take prey in general proportion to its frequency on the ground (Craighead and Craighead 1956:138) rather than on any internal evidence from the population structure. It is seductively simple to suppose that a high proportion of young may indicate an increasing population and Caughley (1977:121) has warned that this can be misleading. He does, however, state (Caughley 1977:121) that when the stable age distribution changes, changes in survival and/ or fecundity will have changed over at least one age interval. Comparison of the age distribution at different times in the past should then point up these facts.

If one accepts that the palaeontological samples are most likely to have been accumulated by owls one will need to establish whether in fact the owl is accumulating an unbiased sample of the standing population. In the case of Otomys irroratus predation by the barn owl appears to aproach the random condition but is skewed towards greater emphasis on older-aged individuals (Perrin 1982:23). This conclusion was reached after comparison with the results of snap-trapping (Perrin 1982) and it would appear that in fact one could equally well argue that young O. irroratus, being inexperienced, are relatively more trap-prone than are the older individuals. However, for the purpose of the present study the results suggest that samples accumulated by barn owls may reasonably be considered close approximations of live populations.

METHODS

The basic data comprise measurements of the occlusal length and maximum length (at base of crown) of all right lower first molars ascribed to Otomys irroratus and O. saundersiae. The lengths were measured to the nearest tenth of a millimetre with needlepoint dial calipers under a microscope. The first stage in interpretation entails translating these measurements into estimated ages of the individuals represented by the teeth. Perrin (1979) established that the variation in the occlusal length of the first lower molar gives a good approximation to eye-lens weight which is commonly used to estimate age.

In the present study, however, this was amended for two main reasons. The coefficient of variation ($V = 100s/\bar{x}$) is about average (Simpson et al. 1960:91) for maximum length of tooth within samples (V is between 5.52 for BPA-LP and 4.00 in BPB/C) but there is in some cases, a significant difference between the mean maximum lenght in different samples (for BPA-LP and BPA-BRL7 't' is 4.96 with P less than .001 at d.f. 339). This would suggest that it is preferable to take the proportion of the occlusal length to the maximum length of the tooth in order to remove any differences that may be due to variation in the latter rather than in the former. In this case, the occlusal length increases with age relative to the maximum length until, potentially, both coincide. The square-root function suggested by Spinage (1971) was, however, applied to an index that was decreasing. For this reason the index ,y, employed here is the difference between the occlusal length and the maximum length, taken as a proportion of the maximum length; this difference, being the complement of the occlusal length, decreases with age, potentially reaching zero at maximum ecological longevity.

Spinage's (1971) square-root function was employed to transform the indices because the rate at which the occlusal surface lengthens with age is not constant through life. In order to transform y to age, using the formula $y = y_0 (1 - \sqrt{t/n})$ (where t is the age of the individual, n is the maximum age and y_0 is the index value at or soon after birth), it is necessary to make two assumptions. The first is that y_0 is 0.50, this being the maximum value that was observed. The second assumption is that this index will have reduced to zero at the age (n) of 24 months, this being the maximum observed longevity computed by

Davis (1973) for Otomys irroratus. In the case of O. saundersiae n was taken arbitrarily to be 21 months, on the assumption that smaller animals tend to have a shorter life span, and y_0 remained the same.

The maximum longevity was divided into age classes of three months each. This length of time was chosen partly because the size of the samples generally seemed to warrant this division and partly because it has been determined that Otomys irroratus becomes sexually mature at about three months of age (Davis and Meester 1981). This means that age class I comprises the pre-reproductives, thereby providing a potentially useful division. The values of the index were correlated with the age classes and the readings for the teeth in the samples allowed assignment of these to age classes.

Because the samples appear to represent approximately random samples of living populations it is appropriate to calculate time-specific life tables (Voorhies 1969). Indeed, Dodd and Standon (1981:349) are of the opinion that it is hardly even possible to calculate age-specific life tables for fossil material.

RESULTS

Among the present-day control samples of Otomys irroratus there is great similarity in l_x (survivorship curves: Fig. 3) and q_x (mor-

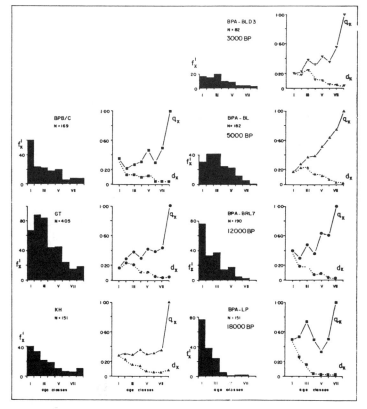

Figure 1. Numbers of specimens (n), probability of dying (d_x) and mortality rate (q_x) per age class for Otomys irroratus in modern samples (Boomplaas BPB/C, Glentyre GT and Klein Hagelkraal KH) and palaeontological samples (Boomplaas BPA-BRL7, BPA-BL and BPA-BLD3).

tality rate: Fig. 1). On the other hand, a contingency test of the age distribution of the samples shows significant difference (chi-squared: 36.37, d.f. : 14, P : less than .001) (Fig. 1), there being a relatively low proportion of pre-reproductives at Glentyre.

The four palaeontological samples of O. irroratus (Fig. 1), on the other hand, show a much greater degree of difference. The l_x and d_x series indicate that there is a progression through time from the early BPA-LP sample with a very high juvenile mortality to the most recent sample, BPA-BLD3, which exhibits a present-day pattern. The mortality rate (q_x) also varies quite considerably, again becoming noticeably lower for the two youngest age classes in the more recent samples (Fig. 1). There is also an increase in the actual longevity; the oldest age class is not represented in the two earlier samples but is represented in the two later samples. The mortality rate of the Holocene samples is essentially identical to that of the present-day samples. There is a significant difference in the age structure of the samples from BPA-LP and BPA-BRL7 (chi-squared : 20.53, d.f. : 6, P: .01>.001), as well as between BPA-BRL7 and BPA-BL (chi-squared: 28.87, d.f.: 7, P : less than .001) Between the latter sample and that from the youngest level, BPA-BLD3, there is no significant difference nor is there a difference between BPA-BLD3 and the modern BPB/C sample.

In Otomys saundersiae there is a significant difference in the age structure of the three samples (chi-squared: 88.31, d.f. 12, P : less than .001) (Fig. 2), there being a far higher proportion of young individuals in the BPA-LP sample than in the more recent samples. This trend is comparable to that shown by O. irroratus, as is illustrated by Figs 1 and 2. The tendency towards more animals living longer is also apparent in O. saundersiae, as it is in O. irroratus. In general, the mortality rate becomes progressively lower for any given age class as time passes

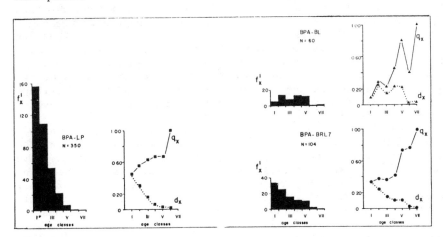

Figure 2. Numbers of specimens (n), probability of dying (d_x) and mortality rate (q_x) per age class for Otomys saundersiae in Boomplaas samples BPA-LP, BPA-BRL7 and BPA-BL.

DISCUSSION

It has been mentioned above that the mortality rate in samples of three modern populations of Otomys irroratus is very similar (Fig. 1). It seems reasonable to suppose, therefore, that this pattern can be held to be generally valid, at least for populations in the southern Cape Province today. The basic pattern here is that of a mortality rate of between approximately 0.20 and 0.50 for all except the last age class where it must be 1.00. Mortality is therefore relatively constant, especially in the KH sample. In the earlier palaeontological samples there is a great deal of fluctuation which may be due to small sample size. It is, however, noticeable that the most recent sample is not only the smallest but also exhibits a pattern similar to that of the modern samples. It could be postulated that such fluctuations, if real, are indicative of a population that is ill-adapted to its environment. In the present context this would seem to indicate that conditions were generally more favourable to the survival of the species during the later part of the Holocene when numbers were also higher than previously (Avery 1982). With O. saundersiae the situation is apparently reversed, with the amplitude of fluctuation increasing towards the present. Here, though, decrease in sample size is correlated with increased fluctuations and the evidence is less convincing. On the other hand, the fact that O. saundersiae declined in numbers during the Holocene (Avery 1982) suggests that such fluctuations may, in fact, be a feature of an ill-adapted species. It may not be possible to prove this statistically at present but it is proposed as a working hypothesis.

The next point that required consideration is the interpretation of the possible causes of changes in the mortality rate of Otomys irroratus. Davis and Meester (1981) refer to the fact that there is higher mortality in winter at higher latitudes and also note that reproduction virtually stops during May, June and July. Perrin (1980b) reported that there was a significantly greater litter size in spring compared with the rest of the year. There was probably greater seasonal variation in weather, with harsher winters in the southern Cape Province during the Last Glacial Maximum than is presently the case. One might therefore expect there to have been greater seasonality in births and deaths in Otomys irroratus, comparable perhaps to the current situation further north. The high mortality rate of age class III animals, approximately six to nine months old, and the apparent failure of the animals to live beyond the age of approximately 21 months could perhaps be interpreted as evidence of high winter mortality in first- and second-year animals.

The data supplied by Lynch (1975) suggest that O. irroratus can survive low temperatures; generally depressed temperatures during the Glacial Maximum would therefore perhaps be unlikely alone to have affected O. irroratus. It is, however, much more likely that long periods of drought, probably coinciding with winter, and a generally more marked seasonality of weather conditions could have had a highly detrimental effect upon the O. irroratus population. Periods of drought are stated to be unfavourable to reproduction in O. irroratus (Perrin 1980b). Although this species is adapted to eating low-quality food, it is clear that it feeds predominantly upon green plant matter (Perrin 1980a) upon which a period of drought, especially when coupled with depressed temperature, would have a deleterious effect. It is impossible to say at present what may have been the mechanism whereby Otomys saundersiae could have become increasingly ill-adapted during the Holocene. Information on the biology of O. saundersiae is sorely needed.

Apparently contradictory evidence is provided by the survivorship curves which show the same trends in both Otomys irroratus and O. saundersiae (Fig. 3). This coincidence may, in fact, be due to bias in some of the samples. The samples from BPA-BL and especially from BPA-BLD3 apparently did not include material from the finest sieves. This material, which would be mainly loose teeth, should increase representation in the younger classes disproportionately so that the overall pattern would be altered (Avery, in press), This, however, could hold true for O. irroratus as well as O. saundersiae, although possibly to a lesser degree because the former species is the larger. On the other hand mortality rates are different in the two species, which is more significant, since these will not be affected by under-representation of the younger age classes (Caughley 1966).

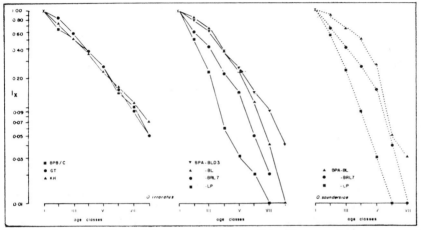

Figure 3. Survivorship curves (l_x) for Otomys irroratus and Otomys saundersiae (sites as listed in Fig. 1).

An alternative possibility is that the shape of the survivorship curve is relatively general, at least in closely related species and in samples averaged over a long period of time. If this were the case, one would expect the same trend to be revealed in different species in certain conditions. Thus, progressively greater expectations of survival in all age groups from the Last Glacial through to the present may be a pattern that has to do more with general conditions than with the fitness of a particular species. This might be testable by looking at the survivorship curves of different populations of the same species under different conditions.

It is clear from this brief examination that there are trends and patterns in micromammalian population dynamics during the past. These can be interpreted in terms of what is already known about palaeoenvironments in the southern Cape, specifically through small mammal community analysis (Avery 1982). It was suggested that during the Last Glacial Maximum vegetation in the Cango valley was relatively open with more than usual semi-arid scrub and that the climate was cold and dry (Avery 1982:316). Thereafter temperature and rainfall increased and dense mesic vegetation become more extensive on the floor of the valley. Otomys irroratus does not do well under drought conditions and xeric vegetation acts as a barrier to its movement (Perrin 1980a). Therefore it is reasonable to suggest that high juvenile mortality could be caused by drought and possibly by competition for space and food in a reduced habitat. As these constraints fell away or relaxed, mortality rate would be reduced.

It is possible that rainfall and temperature were also less variable at different times of the year, that the climate became more temperate, which could be indicated by reduced fluctuations in the mortality rate. K-selected species such as O. irroratus are essentially adapted to predictable climates (Perrin 1980a) and so a generally higher survival rate should indicate increased predictability. In this general way the previously established palaeoenvironmental pattern tends to be confirmed by the present data. Further data, both modern ecological and palaeontological samples, should serve to confirm and expand the basic picture.

ACKNOWLEDGEMENTS

Prof HJ Deacon kindly provided the samples from Boomplaas Cave. Dr TP Volman helped with the index for age estimation; Mr G Avery and Mr ML Wilson made useful comments on the manuscript.

REFERENCES

Avery DM 1982. Micromammals as palaeoenvironmental indicators and an interpretation of the late Quaternary in the southern Cape Province, South Africa. Ann. S. Afr. Mus. 85:183-374.
Avery DM, in press. Sampling procedures and cautionary tales. Proc. Conf. S. Afr. Assoc. Archaeol. Gaborone 1983.
Caughley G 1966. Mortality patterns in mammals. Ecology 47:906-918.
Caughley G 1977. Analysis of vertebrate populations. John Wiley, London.
Craighead JJ and Craighead FC 1956. Hawks, owls and wildlife. Stackpole, Harrisburg.
Curtis BA and Perrin MR 1979. Food preferences of the vlei rat (Otomys irroratus) and the four-striped mouse (Rhabdomys pumilio). S. Afr. J. Sci. 14:224-229.
Davis RM 1973. The ecology and life history of the vlei rat, Otomys irroratus (Brants, 1827), on the Van Riebeeck Nature Reserve, Pretoria. Unpubl. DSc thesis, Univ. Pretoria.
Davis RM and Meester J 1981. Reproduction and postnatal development in the vlei rat, Otomys irroratus, on the Van Riebeeck Nature Reserve, Pretoria. Mammalia 45:99-116.
Deacon HJ 1979. Excavations at Boomplaas Cave - a sequence through the Upper Pleistocene and Holocene in South Africa. World Archaeol. 10:241-257.
Deevey ES 1947. Life tables for natural populations of animals. Quart. Rev. Biol. 22:283-314.
Dodd JR and Stanton RJ 1981. Paleoecology, concepts and applications. Wiley-Interscience, New York.
Hutchinson GE 1978. An introduction to population ecology. Yale University Press, New Haven and London.
Kurten B 1953. On the variation and population dynamics of fossil and recent mammal populations. Acta zool. fenn. 76:1-122.
Lynch CD 1975. The distribution of mammals in the Orange Free State, South Africa. Navors. nas. Mus. Bloemfontein 3:109-139.

Perrin MR 1979. Ageing criteria and population age structure of co-existing populations of Rhabdomys pumilio and Otomys irroratus. S. Afr. J. Wildl. Res. 9:84-95.

Perrin MR 1980a. Ecological strategies of two co-existing rodents. S. Afr. J. Sci. 76:487-491.

Perrin MR 1980b. The breeding strategies of two co-existing rodents, Rhabdomys pumilio and Otomys irroratus. Acta oecol./Oecol. gener. 1:383-410.

Perrin MR 1982. Prey specificity of the barn owl, Tyto alba, in the Great Fish River valley of the eastern Cape Province. S. Afr. J. Wildl. Res. 12:14-25.

Simpson GG, Roe A and Lewontin RC 1960. Quantitative zoology. Harcourt Brace, New York.

Slobodkin LB 1964. Growth and regulation of animal populations. Holt, Rinehart and Winston, New York.

Spinage CA 1971. Geratodontology and horn growth of the impala (Aepyceros melampus). J. Zool. 164:209-225.

Stearns SC 1976. Life-history tactics: A review of the ideas. Quart. Rev. Biol. 51:3-47.

Voorhies MR 1969. Taphonomy and population dynamics of an early Pliocene fauna, Knox County, Nebraska. Univ. Wyoming Contrib. Geol. spec. Pap. 1:1-69.

Climatic change and mammalian fauna from Holocene deposits in Wonderwerk cave, northern Cape

J.F. THACKERAY
University of Stellenbosch, South Africa

ABSTRACT. Quantitative analyses of the macrofaunal assemblages from Wonderwerk cave suggest that environmental conditions in the area varied markedly during the Holocene.

INTRODUCTION

Wonderwerk cave (22°50'45"S; 23°33'29"E) is a large solution cavity in dolomitic limestone situated 43 km south of Kuruman near the highest point of the Ghaap Plateau in the northern Cape Province, South Africa. Well-preserved samples of large mammalian fauna were recovered from the cave more than forty years ago (Malan and Cooke 1941; Malan and Wells 1943; Malan 1944). This paper presents some of the results obtained from an analysis of faunal material recovered from excavations of Holocene deposits at the site undertaken by PB Beaumont, AI Thackeray and myself during 1979. The stratigraphy of these Holocene deposits and the results of the analyses of the microfauna, pollen and cultural material have been presented elsewhere (Avery 1981; Thackeray 1981; Thackeray et al. 1981; Van Zinderen Bakker 1982).

HOLOCENE MACROFAUNA

The ungulate faunal list for the Holocene deposits (Table 1) includes no species which had not previously been recorded from excavations at Wonderwerk cave (Malan and Cooke 1941; Malan and Wells 1943; Cooke 1963). However, two extinct species, Equus capensis and Megalotragus priscus, previously reported only from Late Pleistocene contexts at Wonderwerk cave, have been identified in the Holocene units. E. capensis is represented in layer 4d, dated to c. 10 000 BP and M. priscus is last recorded in layer 4c, dated around 7500 BP. The dominant taxa include equids (E. burchelli/quagga), alcelaphines (Alcelaphus/Connochaetes) and springbok (Antidorcas marsupialis), indicative of open grassland conditions.

The high ungulate:carnivore ratio, the low degree of bone damage caused by porcupines or carnivores, and the association with cultural material suggest that human population groups were the principal agents responsible for the accumulation of the excavated samples. The possibility that the cave was not occupied intensively by agents of accumulation during the Early Holocene is suggested by estimates of the accumulation rate of macrofaunal material. An index (AMA) reflecting this accumulation rate has been calculated from the mass of bone per unit volume of excavated deposit, together with estimates of sedimentation rate. Changes in AMA shown in Fig. 1 suggest that there were two major periods of accumulation in the Holocene, separated by an interval (represented within layer 4a) in which accumulation rates of macrofauna were relatively low. A radiocarbon date of 4240 $\pm$ 60 BP (Pta-2541) has been obtained for the top of layer 4a, while a date of 5180 $\pm$ 70 BP (Pta-2544) was determined for the underlying layer 4b. The low accumulation rate of macrofauna as well as microfauna within layer 4a (Fig. 1) falls within the interval of time for which no radiocarbon dates for archaeological deposits are known in southern Namibia (Vogel and Visser 1981) and in the Karoo biome (Deacon 1974). This period may in turn correspond to a time when population densities were relatively low in the dry interior regions of the country and when productivity was similarly low.

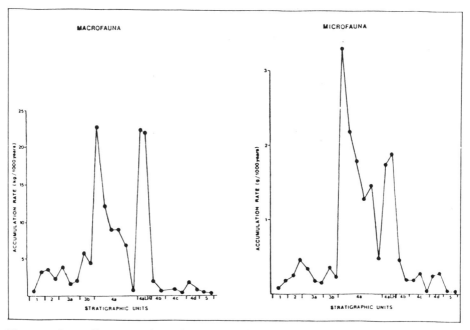

Figure 1. Changes in the accumulation rate of macrofaunal and microfaunal bone, Wonderwerk cave Holocene sequence.

DISCUSSION

One cannot assume that faunal assemblages characterized by low AMA values should necessarily correspond to periods of low primary productivity and low human population densities. Indeed, a major difficulty confronting faunal analysts concerns the problem of distinguishing the extent to which the observed patterns are a reflection of environmental changes (including changes in primary production and productivity), and the extent to which the data reflect processes of preservation or selection. However, a summary statistic based on mean ungulate bodymass (MUB), calculated by a method described elsewhere (Thackeray 1980), suggests that the "observed" MUB values for the Wonderwerk Holocene sequence correspond to those which might be "expected" from the number of ungulate species represented at the site (Thackeray 1984). This observation suggests that processes of selective predation need not necessarily have been a significant bias affecting the representation of ungulates at Wonderwerk, although selective transportation of body parts of variously sized ungulates has evidently occurred.

Table 1. Minimum numbers of individuals of ungulates and other fauna represented in Holocene deposits, Wonderwerk cave.

	Stratigraphic units									
	1	2a	2b	3	4a	4aLH	4b	4c	4d	5
Papio ursinus, baboon	1	-	-	-	5	1	-	1	-	-
Panthera pardus, leopard	-	-	-	-	1	-	-	-	-	-
Hyaena sp., hyaena	-	-	-	-	1	-	1	-	-	-
Canis cf. mesomelas, jackal	-	-	1	1	2	1	-	1	1	-
Felis libyca, wildcat	-	-	-	-	-	1	-	-	-	-
Viverridae, small carnivore	-	-	-	1	2	-	1	-	-	-
UNGULATE CLASS I	1	1	1	-	-	-	-	-	-	-
Oreotragus oreotragus, klipspringer	-	-	-	-	-	-	-	1	-	-
Raphicerus campestris, steenbok	2	-	-	3	2	3	2	3	2	-
Sylvicapra grimmia, duiker	-	-	-	-	1	-	-	-	-	-
Ovis/Capra, sheep/goat	2	1	-	-	-	-	-	-	-	-
Redunca fulvorufula, reedbuck	1	-	-	1	1	-	-	-	-	-
Antidorcas marsupialis, springbok	1	1	2	1	-	3	-	-	-	-
UNGULATE CLASS II	-	-	-	1	1	-	-	-	-	1
Equus quagga/burchelli, zebra	3	1	2	4	3	2	2	1	1	1
Equus capensis, extinct horse	-	-	-	-	-	-	-	-	1	-
Alcelaphus/Connochaetes, hartebeest/wildebeest	2	-	2	3	2	2	-	1	1	-
UNGULATE CLASS III	-	-	-	-	1	-	1	-	-	1
Phacochoerus aethiopicus, warthog	-	-	1	-	3	-	1	-	-	-
Taurotragus oryx, eland	-	-	-	2	-	-	-	1	-	-
Syncerus caffer, buffalo	-	-	-	3	1	3	-	1	1	-
Hippotragus equinus, roan	-	-	-	-	-	-	1	-	-	-
Kobus ellipsiprymnus, waterbuck	-	-	-	-	1	1	-	-	-	-
Megalotragus priscus, extinct alcelaphine	-	-	-	-	-	-	-	1	-	-
OTHER										
Lepus sp., hare	2	-	-	2	3	2	2	2	-	1
Procavia capensis, rock rabbit	1	1	1	5	2	2	-	2	-	1
Hystrix africaeaustralis, porcupine	2	-	1	3	2	2	1	2	-	1
Oryterops afer, aardvark	-	-	-	1	-	1	-	-	-	-
Testudo sp., tortoise	3	-	2	10	4	2	1	3	4	1
Pyxicephalus sp., bullfrog	-	-	2	5	4	2	2	2	1	-

The samples of macrofauna do not provide a direct means of identifying fluctuations in environmental conditions during the

Holocene, but the fact that changes in primary production and productivity did occur is suggested from the microfaunal samples (Avery 1981; Thackeray 1984). An increase in primary production between 10 000 and 5000 BP, fluctuating thereafter, may have occurred partly as a result of an increase in rainfall, although a decline in temperature after about 9000 BP (contributing to a decline in evaporation and to an increase in evapotranspiration) could also have been responsible for this increase during the Early Holocene. Certainly the distinction between temperature, rainfall and other factors contributing to environmental change may be resolved not so much by detailed examination of the macrofauna, as by statistical analysis of large samples of microfauna which are more sensitive to environmental fluctuations.

REFERENCES

Avery DM 1981. Holocene micromammalian faunas from the northern Cape Province, South Africa. S. Afr. J. Sci. 77: 265-273.

Cooke HBS 1963. Pleistocene mammal faunas of Africa, with particular reference to southern Africa. In: FC Howell and F Bourliere (eds.), African ecology and human evolution. Viking Fund Publications in Anthropology No. 36, New York. p. 65-116.

Deacon J 1974. Patterning in the radiocarbon dates for the Wilton/Smithfield complex in southern Africa. S. Afr. archaeol. Bull. 29: 3-18.

Malan BD 1944. Fieldnotes and report on excavations at Wonderwerk cave, Archaeol. Survey File B20/1/1. Govt Archs, Pretoria, S. Afr.

Malan BD and Cooke HBS 1941. A preliminary account of the Wonderwerk cave, Kuruman district. S. Afr. J. Sci. 37: 300-312.

Malan BD and Wells LH 1943. A further report on the Wonderwerk cave, Kuruman. S. Afr. J. Sci. 40: 258-270.

Thackeray AI 1981. The Holocene cultural sequence in the northern Cape Province, South Africa. PhD thesis, Yale Univ.

Thackeray AI, Thackeray JF, Beaumont PB and Vogel JC 1981. Dated rock engravings from Wonderwerk cave, South Africa. Science 214: 64-67.

Thackeray JF 1980. New approaches in interpreting archaeological faunal assemblages with examples from southern Africa. S. Afr. J. Sci. 76: 216-223.

Thackeray JF 1984. Man, animals and extinctions: the analysis of Holocene faunal remains from Wonderwerk cave, South Africa. PhD thesis, Yale Univ.

Vogel JC and Visser E 1981. Pretoria radiocarbon dates II. Radiocarbon 23: 43-80.

Van Zinderen Bakker EM 1982. Pollen analytical studies of the Wonderwerk cave, South Africa. Pollen et Spores 24: 235-250.

Late Pleistocene environmental changes and implications for the archaeological record in southern Africa

H.J.DEACON & J.F.THACKERAY
University of Stellenbosch, South Africa

ABSTRACT. As the archaeological record for the later Pleistogene in southern Africa has become better documented, it is apparent that there have been marked changes in the distribution and density of human populations through this time. The amelioration of climates after the Last Glacial Maximum from 16 000 BP saw the spread of populations over the greater area of southern Africa with a wide geographical dispersion of pre-agricultural peoples being achieved in the earlier Holocene. There was contraction in human range in the mid-Holocene in some biomes. Few sites can be dated to the coldest interval of the Late Pleistocene between 16 000 and 30 000 BP and in this period populations may have been confined to refugia. Although there is some evidence for wider spread populations between 30 000 and 40 000 BP it is probably only during the Last Interglacial that population densities were remotely comparable to those of the Holocene. This suggests that a broad correlation between environmental change, the productivity of environments, and human demography is reflected in the archaeological record in the latitudes of southern Africa.

INTRODUCTION

It is two decades since Lee (1963) reviewed the effects of "shifting Pleistocene climates" on human population distributions and densities in southern Africa in the Late Pleistocene. In view of advances in the understanding of climates of the Holocene and Pleistocene, i.e. Pleistogene (Harland et al. 1982), and in the light of progress in the archaeology of southern Africa since Lee published his review, it is worthwhile to reassess his conclusion and to consider the present state of knowledge of man-climate relationships. In this paper southern Africa is given a more restricted definition than was used by Lee as that part of the continent which lies south of the Zambezi. The reasons are twofold. Firstly, the archaeology of central Africa, including Angola, Zambia and Mozambique, is relatively poorly known and secondly, the plateau region dominated by deciduous miombo woodland is ecologically very different from the bushveld, thorn tree savannas, grasslands and shrublands found to the south of the Zambezi. It was Lee's conclusion that southern Africa exhibited extreme ecological stability during the later Pleistocene and that throughout this period there existed a large core area which was permanently favourable for the maintenance of a large standing human population. The implication was that although some geographical displacements may have occurred, the effect of Pleistocene climate changes on human demography in southern Africa was low.

A number of assumptions in Lee's study can be questioned and make his conclusion untenable. In the 1960s there was little direct evidence of the scale and timing of Late Pleistocene climatic changes in southern Africa and reconstructions of past habitat conditions were based on the now discredited pluvial hypothesis (Cooke 1957, 1958; Flint 1959; Bishop 1971), assuming zonal shifts of isohyets as well as vegetation types. Such reconstructions have little basis because changes in precipitation were emphasized without consideration of the interacting elements of synoptic climate, and vegetation changes involving changes in community composition were not taken into account. Little was then known of the subsistence ecology and determinants of human settlement in the Late Pleistocene and the distribution of what are still the poorest defined and dated archaeological entities, the Fauresmith and Sangoan industries, was offered as a test of the reconstructions. These culture-stratigraphic units, however defined, would certainly be older than the 50 000 years which Lee cited and could be of Middle Pleistocene age. If Lee's reconstructions of the human habitats in the Late Pleistocene in southern Africa and his conclusions on the effects of Pleistocene climate changes can be proved unacceptable, this should be seen not so much as an adverse criticism of his study, but as a sign of progress. The question of identifying the effect of Pleistogene climatic change on human populations living in southern Africa remains a valid one.

There are two sources of information for assessing the influence of long-term climatic changes on human populations, and these are palaeoenvironmental studies and the archaeological record. The time depth considered in the paper is the Late Pleistocene and Holocene, from 125 000 years to some 2000 years ago when food production was introduced into southern Africa. In terms of palaeoenvironmental studies this includes the Last Interglacial-Glacial-present Interglacial cycle and in terms of culture stratigraphic technological stage units, the Middle and Later Stone Ages. Interest lies in the extent to which palaeoenvironmental data can be used to retrodict periods of lowered productivity and stress, conditions which may have limited the distribution and density of human population groups in prehistory. Clarke (1972:14) has argued that population distribution is ever-changing and that the causes and the effects are not necessarily constant in time or in space. Although climate is an important limiting factor, deterministic correlations between climate, migrations and cultural evolution beg counter-arguments of the human ability to adapt to different climatic conditions with the aid of technology. In the case under consideration here, of Stone Age populations with a food gathering economy, simple (low energy) technology and fire as the only tools for resource management, a relatively direct relationship with the environment can be posited (Netting 1971). Given the scale of change in the diversity of habitat productivity due to climatic forcing that can be assumed in the later Pleistogene, it would indeed be surprising if relatively major changes in population densities and distributions had not occurred. The palaeoenvironmental evidence needs to be measured against that of the archaeological record.

A suggested relationship between climate, productivity and usable resources on the one hand and foraging strategies, community and group organization and demography on the other are given in a flow diagram

(Fig. 1) together with the relevant sources of information. Thus the input is knowledge of palaeoclimates derived from field observations and models, and the output is the archaeological record composed of food waste, site distribution, density and size data, coupled with information on occupation intensity, raw material usage and such like. Migration is used in the widest sense of the term to include any scale of individual or group movement in time. The link or interaction is how climates affect the production of usable resources which in turn constrains the maintenance of human populations.

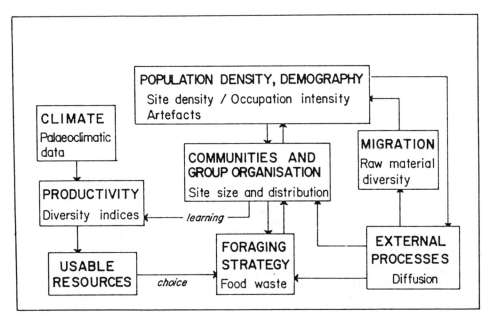

Figure 1. Flow chart showing the relationship between climate, productivity and human populations.

Past productivity of habitats cannot be measured directly but measures of species diversity of flora and fauna can be used as proxy indicators. It can be suggested that changes in productivity in the past, as indicated by changes in diversity, would have necessitated continuous evaluation of usable resources and learning about environmental potentials. The foraging strategy is a product of this evaluation process. In southern Africa, two foraging strategies which might be reflected in the archaeological record can be contrasted as either patch foraging strategies in which there is a high reliance on underground plant foods and territorial and ground game, or as mobile mixed strategies in which more search-time is invested in obtaining larger migratory herd animals. As a working hypothesis it is suggested that patch foraging strategies were more appropriate in relatively heterogeneous habitats (such as those which are indicated by more diverse plant and animal communites) and that these in turn allowed closer nesting of cells of local populations, defined as dialectic units by Birdsell (1973). It is proposed further that interglacial and interstadial climates in southern Africa favoured patch foraging and relative population increase. A consequence of higher populations is not only increased archaeological visibility but also increased effectiveness of cultural processes such as migration and diffusion. Conversely, mobile mixed strategies were appropriate under conditions characterized by relatively low habitat diversity, with corresponding greater home range of foraging groups, more dispersed populations and lower densities. The consequence is relatively lower archaeological visibility (Figs 2 and 3).

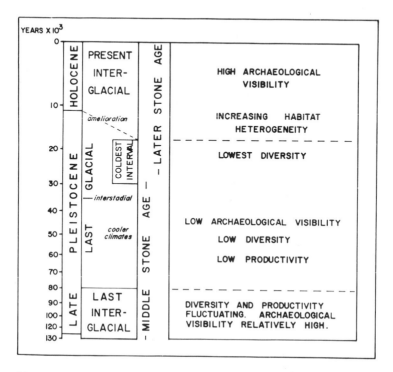

Figure 2. Characteristics and chronology of the later Pleistogene environments in southern Africa in relation to the archaeological record.

PALAEOENVIRONMENTAL CHANGE

A major advance in understanding Pleistogene climates has come from deciphering the record of the Ice Age, largely through the study of deep sea cores. The Late Pleistocene corresponds to the last pulse of the Ice Age (Imbrie and Imbrie 1979). The generalized oxygen isotope palaeotemperature curve for the Late Pleistocene (Shackleton 1969) shows that the Last Interglacial, isotope stages 5a-e dating to between 130 000 and 75 000 BP, was as warm or warmer than the present, with cooler intervals, and that the following period until some 12 000 years ago, corresponding to isotope stages 2-4, was cooler than the present. Information on global climates during the cold maximum of the Late Pleistocene has come from studies undertaken within the CLIMAP Project (1976) with the goal of mapping the sea-surface temperatures of the oceans at 18 000 BP. These data and others on ice distribution, land elevation, taking into account the regression of sea level, and the earth s albedo have been used to specify boundary conditions in several versions of mathematical global circulation models developed to hindcast climatic conditions of the Last Glacial Maximum (Gates 1976; Manabe and Hahn 1977; Heath 1979). All show global climates markedly colder and generally drier at the Glacial Maximum than they are at present. From the study of Antarctic ice cores (Lorius et al. 1979) and other evidence (Salinger 1981), it would appear that amelioration of climates after the Glacial Maximum in the Southern Hemisphere commenced about 16 000 BP, several thousand years in advance of the amelioration in the Northern Hemisphere with its more extensive continental ice sheets. The scale of palaeotemperature changes in the last 12 000 years has been relatively low.

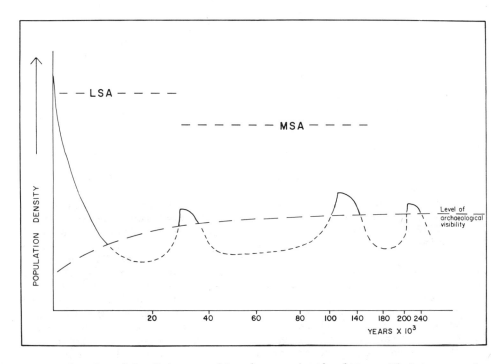

Figure 3. A model of demographic changes in the later Pleistogene in southern Africa.

The oceanic record is very much more complete than the record available for any climatic parameters from land-based observations, yet it is apparent that the scale of Pleistogene climatic change has had extremely marked effects on terrestrial ecosystems, even including the equatorial forest biome (Hamilton 1976, 1982; Simpson and Haffer 1978). Once the pluvial hypothesis had been discarded, Van Zinderen Bakker pioneered new thinking about recent climatic history in southern Africa. In papers written in the 1960s and 1970s he drew on data then available for the southern oceans, palynology and geomorphology to model the main elements of changes in atmospheric circulation, ocean currents and vegetation between glacial and interglacial climates (Van Zinderen Bakker 1967, 1976). His initial model was based on assumed significant zonal shifts of the westerlies and anticyclone cells, elements of circulation that have a profound influence on the climates of southern Africa. Shifts of climatic zones however are not supported in the evidence for oceanographic conditions in the South Atlantic relevant to the position of the westerlies at 18 000 BP (Morley and Hays 1979), nor in the imprint of anticyclone cells on dune alignments in the interior of southern Africa (Grove 1969; Lancaster 1979). In a more recent discussion of Pleistogene climates Van Zinderen Bakker (1982a) has stressed the increase in intensity of atmospheric circulation under glacial climates. The modelling of the changes in past climates in southern Africa remains an important area of endeavour and needs to move beyond the discussion of the penetration of frontal systems into the interior of the subcontinent, shifts in the positioning of the intertropical convergence zone, and north versus south viewpoints on the relative importance of winter versus summer precipitation in the past. Any model suggesting changes in the season of precipitation is difficult to test on the basis of evidence preserved in the ground.

Lacking in the 1960s and previous decades was a body of well-dated observations of palaeoclimatic significance for the Pleistogene in South Africa. The demise of the pluvial hypothesis had a very negative effect on palaeoclimatic studies from which recovery has been slow. A contributory reason is the dearth of obvious kinds of deposits that would yield well-dated information on past climates. This must always be a problem in a subcontinent lacking significant basins of deposition, and where the record is one of shreds and tatters.

The evidence for conditions warmer than the present in the Last Interglacial comes from the occurrence of thermophile mollusca, the Swartkops fauna in association with Last Interglacial beaches (Davies 1971, 1972), and from the isotopic analysis of shell from the base of the Klasies River Mouth cave sequence (Shackleton 1982). Reliable dating of deposits in the earlier part of the Late Pleistogene is problematical but ionium dates for the Klasies and Herolds Bay sequences suggest a Last Interglacial age (Vogel, pers. comm.). Herolds Bay is of interest in the occurrence of Antidorcas and Connochaetes in a deposit dated to more than 80 000 BP (Brink and Deacon 1982) because these are not typical taxa of the recent Interglacial fauna. This suggests conditions more comparable to the end-Pleistocene, and evidence for large mammal communities responding to climatic perturbations during the Last Interglacial is apparent in the Klasies River Mouth sequence (Klein 1976).

There is a problem in dating the earlier deposits coeval with the Last Interglacial and Glacial of the Late Pleistocene. On the Boomplaas evidence from the Cango Valley (Deacon 1979; Deacon et al., this volume), it can be suggested that the period corresponding to oxygen isotope stages 3 and 4 was marked by cool and at times relatively moist climates. Plant and animal communities however were different from those either at the cold maximum or during the present Interglacial. There is still the question of what periods of interstadial climates have been recognized between 75 000 and 30 000 BP. There is some evidence for an interstadial that may have been marked in the southern Cape by higher precipitation, judging from the reactivation of the Amanzi springs (Deacon 1970) and milder temperatures at Boomplaas (Avery 1982a; Deacon et al., this volume) at about 32 000 BP. One or more earlier periods of interstadial climate may have occurred in isotope stage 3 but the number and duration have yet to be well documented.

The most significant climatic event of the Late Pleistocene was its coldest interval which can be dated to between 25 000 and 16 000 BP and which includes the cold maximum at 18 000 BP. Relative changes in habitat conditions that are proxy evidence for the scale of climatic change prior to, during and following the coldest interval can be traced in the Boomplaas and Wonderkrater sequences (Scott 1979, 1982). In both these sequences, which are geographically far removed, there are relatively dramatic changes in vegetation with woodland of one kind being replaced by shrubland during the coldest interval and this in turn being replaced by woodland of a different and more modern kind after the end of the interval. Isotopic studies (Talma et al. 1974; Vogel 1983) show that mean annual temperatures were depressed by 5-9 degrees C at the cold maximum, contributing to the most extreme climatic conditions which would have affected the distribution and density of animal as well as plant populations. It is in this time range that extensive colluvial mantles were apparently deposited in the eastern half of the country (Price-Williams and Watson 1982).

The amelioration of climates after the coldest interval could have begun as early as 16 000 BP (Deacon et al., this volume) and the end Pleistocene and Holocene was the period when modern-type climates came into being and present relatively heterogeneous or patchy and more diverse interglacial habitats were created. The scale of change between 16 000 BP and the present is the maximum range of variation in the Late Pleistogene climates and could be expected to be strongly mirrored in the archaeological record.

ARCHAEOLOGICAL SOURCES OF DATA

Stone artefacts, trace fossils of human presence (Behrensmeyer 1982), are permanent markers of where people were distributed in the past and they can be dated relatively, within the known culture-stratigraphic framework, and chronometrically. Although artefacts are made by individuals, the individuals were members of communities, essentially local populations, and dated find sites thus carry palaeodemographic information. The chances of sampling the residues left behind by an itinerant group are low and artefact occurrences tend to be grouped in time and space, supporting the contention that they represent local populations. The archaeological survey of the large geographical area of southern Africa is by no means complete even at a reconnaissance level; some regions are virtually unknown archaeologically. However it is also apparent that in some of the better surveyed areas there are notable temporal gaps in the record when people were apparently absent. An example is the mid-Holocene hiatus in the occupation of large parts of the Karoo and grassland biomes of South Africa (Deacon 1974) and it can be asked whether such evidence for population flux requires explanations by climate-related factors. The response of local populations to climatic changes is not simple but this does not preclude climatic changes from being an important element in considering the maintenance of hunter-gatherer populations in the later Pleistogene.

THE CULTURE-STRATIGRAPHIC RECORD IN SOUTHERN AFRICA

The recognition of high level integrative technological stage units was the main contribution of the initial phase of systematic archaeological research in South Africa (Goodwin and Van Riet Lowe 1929). The temporal range of the units relevant to this paper, the Middle and Later Stone Age, have only become better defined through recent research. The association of the Middle Stone Age with the minor emergence recognized by Krige (1932) has long been known (Rogers 1905) and valuable confirmation of the Last Interglacial age of this 7 m beach and the Middle Stone Age was provided by Shackleton (1982). The dating of a number of Middle Stone Age occurrences to beyond the conventional range of C14 dating was pointed out by Beaumont and Vogel (1972a, b) and there is evidence now that this stage spans the earlier part of the Late Pleistocene.

The upper limit of the temporal range of the Middle Stone Age would seem to be at some 25 000 to 30 000 BP. It is unlikely that this major culture-stratigraphic boundary is notably time transgressive in southern Africa and there is a possible coincidence of the interface with the beginning of the coldest maximum. The best evidence comes from Boomplaas Cave (Deacon 1979; Deacon et al., this volume) where the stratigraphy is relatively continuous through this period of the Pleistogene. A similar conclusion is indicated by observations of Middle Stone Age occurrences associated with colluvial deposits in Swaziland (Price-Williams and Watson 1982). The Border Cave and Heuningneskrans evidence (Beaumont 1978, 1981) apparently does not fit such a chronology but there is the likelihood that the age of the latest Middle Stone Age at Border Cave is not significantly different from that at Boomplaas (Avery 1982b). Later Stone Age occurrences date from more than 20 000 BP to historic times and thus span the coldest period of the Late Pleistocene and the succeeding period of amelioration of climates.

There is a correspondence in the apparent dating of technological substages like the Howiesons Poort/Epi-Pietersburg and, more securely, the Robberg, Albany/Pomongwan/Oakhurst and Wilton substages that shows these to represent widely diffused norms of technology. Although the equivalent stage units can be recognized in central Africa, not all the substage divisions are represented and those that are broadly equivalent are markedly different in formal tool content. This suggests that through the Late Pleistocene and into the Holocene, southern Africa functioned as a unitary "cultural" (technological) region, implying the existence of an effective open social network linking technology of local populations and the absence of significant barriers to diffusion within the subcontinent. For this reason, cultural factors are an important constraint on the response shown by local populations to changes in their habitat conditions and indirectly to climatic changes.

Differences in the archaeological visibility of past populations in time and space in southern Africa is a dimension central to this paper. While an open network linked local populations in southern Africa, these populations were not uniformly distributed in time or space. The most persuasive pattern that emerges from consideration of time-space relationships is the overall low visibility of local populations anywhere in southern Africa during much of the period coeval with the Last Glacial. Higher archaeological visibility would seem to correspond to interglacial or interstadial times.

There are major disconformities in a number of Late Pleistocene sequences with Middle and Later Stone Age occupations that suggest a long time interval separating the relevant occupations. These temporal disconformities are marked by rock tumbles, elutriation of fines and lag concentrations of artefacts, or by the accumulation of culturally sterile deposits. This is particularly evident in the general southern

Cape region, certainly the best known archaeologically, where seven out of eight excavated sequences with both stages represented show prominent disconformities. While this does not prove that the older Middle Stone Age horizons are all of Last Interglacial age, it is a probability. Although more chronometric dating evidence is needed, it is an acceptable hypothesis that a large portion of the known Middle Stone Age occurrences are indeed of Last Interglacial age.

There are finite C14 dates for some Middle Stone Age occurrences that can be accepted as falling in the range between 25 000 and 40 000 BP. These include the Apollo 11 site (Wendt 1976), sites in the ecotone between the Karoo and grassland biomes such as Highlands (Deacon HJ 1976) and Grassridge (Opperman, pers. comm.), some sites along the Drakensberg escarpment (Carter 1978), and possibly colluvium-associated donga sites along the eastern margin of the subcontinent (Price-Williams and Watson 1982). At Boomplaas the relevant final Middle Stone Age occupation dated to some 32 000 BP is the richly organic BP stratigraphic unit (Deacon 1979) which is overlain by deposits of the cold maximum. The apparent age, dating to the end of isotope stage 3, can be supported on biostratigraphic grounds (Avery 1982a) and occupation appears to be coincident with interstadial conditions. The oldest Later Stone Age occurrences fall in the following coldest interval.

A plot of the number of C14 dated assemblages against time (Fig. 4) gives a crude but dramatic picture of the increase in archaeological visibility after 15 000 BP. Although the data are influenced by the vagaries of archaeological sampling, biased by preservation and over-representation of cave sequences, there is a marked exponential increase in the dated sites through time. This pattern suggets an increase in populations in southern Africa following the amelioration of climates and increased productivity of human habitats after the coldest interval of the Late Pleistocene.

On the basis of available data, regional patterns cannot be documented fully because of the smaller sample numbers, although they are of considerable interest in the study of the relationship between human populations and climates. The increase in the number of radiocarbon dates obtained from archaeological sites in various regions after 21 000 BP is indicated in Fig. 5. The southern Cape region shows a consistent increase in the number of sites as well as in the number of radiocarbon dates obtained for assemblages from these sites. This region has provided the most continuous culture stratigraphic record, contrasting with other areas where there are notable temporal discontinuities that do not appear to be a function of research intensity. Prior to 9000 BP populations appear widespread, including the Matopos area of Zimbabwe (Walker 1980), the Transvaal (Mason 1962, 1974, 1981; Plug 1981), the northern Cape (Thackeray 1981) and the southern Cape (Deacon 1982). Undated surface sites of the Lockshoek industry (Sampson 1974:271) in the Orange River basin are probably of similar age and indicate settlement in the interior plateau. This may represent the progressive stage-by-stage expansion of local populations into different habitats in southern Africa at the close of the Pleistocene.

There is an inflection in the plot of the number of radiocarbon dated assemblages evident in Fig. 4 that appears to be a product of the temporal discontinuities noted above. The figure suggests there were two phases of exponential population growth after 15 000 BP, the earlier one, terminating at some 10 - 9000 BP and the second initiated about 4000 BP. The inflection would seem to mark a contraction in the geographical spread of populations in southern Africa and the absence of established local populations in the Karoo and grassland biomes between 9000 and 4500 BP has been noted (Deacon 1974). The eastern Cape forelands (Deacon 1976) and Orange River basin (Sampson 1974; Humphreys 1979) were re-populated after 4500 BP but in the Transvaal highveld there is low visibility of Stone Age populations even in the Late Holocene prior to settlement by Iron Age people. The occurrence of stone tools in an Iron Age context as at Broederstroom (Mason 1981) raises the possibility that agriculturalists provided resources previously lacking to maintain hunter-gatherer groups there.

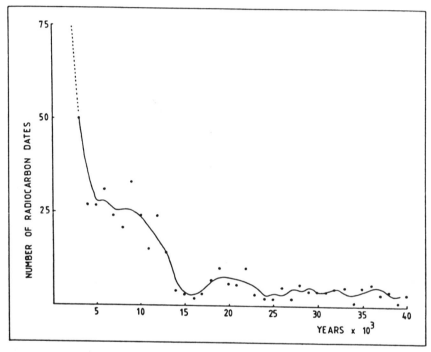

Figure 4. Number of radiocarbon dates obtained for archaeological assemblages in southern Africa between 40 000 and 3000 BP. Solid line is based on running means, calculated at intervals of 3000 years. Dashed line reflects the increase in radiocarbon dates for the period postdating the advent of agriculture and pastoralism.

Evidence is accumulating that the earlier half of the Holocene was drier and warmer relative to the later Holocene in parts of the interior of southern Africa (Avery 1981; Van Zinderen Bakker 1982b; Tusenius in prep.) and this in itself could have been a factor in determining that particular areas were marginal for habitation. There is also evidence of a shift towards increasing use of plant foods, non-migratory bovids and ground game documented in the southern Cape (Deacon 1976) in the mid-Holocene and such a change in subsistence in turn suggests a change in the territorial organization of the residential groups making up local populations. A re-structuring of community organization and adoption of a foraging strategy of this type, essentially foraging in resource-rich patches, would have altered the perception of the potentials of habitats and population densities. Areas lacking the diversity and concentration of reliable resources to maintain communities within defined home ranges through the seasonal round would have become marginal. This would seem to have been the case in the central interior region of South Africa, characterized by what Humphreys (1979) has termed unstructured environments. Essentially this region lacked the resource-rich patches to allow the maintenance of significant populations in the Holocene prior to 4500 BP and thereafter the potential for population expansion and growth attendant on the re-structuring of community organization seems to have been achieved under more favourable climatic conditions and productive habitats. A complex interplay of cultural and ecological factors, a topic worthy of study on its own, is involved in explaining the Holocene demographic patterns.

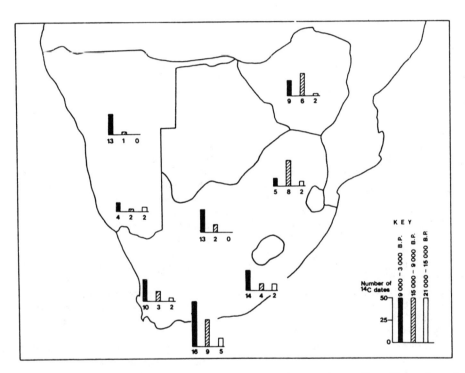

Figure 5. The general increase in the number of radiocarbon dates obtained from archaeological sites in regions of southern Africa since 21 000 BP. Numbers below each bar refer to the number of archaeological sites from which dates were obtained.

It is of interest to note that a very comparable Late Holocene increase in archaeological visibility has been recorded in Australia, another southern continent where hunter-gatherer populations survived into this time range. The Australian evidence has been interpreted in conflicting ways: on the one hand as the product of intensification of social and economic relations (Lourandos 1983) and on the other hand as a consequence of predictable natural population increment (Beaton 1983). It might be expected that explanation in this example would again involve an interplay of cultural as well as ecological factors.

A model of demographic changes in the later Pleistogene is shown in Fig. 3 which suggests the culture-stratigraphic record of southern Africa is made up of sets of observations that relate to periods when population densities in parts of the region were sufficiently high to lie above the limits of archaeological visibility. This suggests the archaeological record is inherently discontinuous. The model is derived from the kind of data outlined in this section and incorporates the idea that abrupt declines in population have occurred following exponential growth to some upper limit. The declines are viewed as a consequence of lowered usable productivity due to climatic change and carry the implication that the extinction of local populations on some scale was important in the past. It follows that the growth of Holocene hunter-gatherer populations would also reach some limiting value with a predictable decline as environmental diversity is lowered with the progression from the present interglacial conditions to those of the next glacial. Food production introduced in the last few thousand years in southern Africa has however caused a massive population overshoot. It remains to be seen whether advances in food production technology are able to maintain population levels in future times.

CONCLUSION

This paper argues against assumptions of stability of populations on time scales such as the duration of the Late Pleistocene, as suggested by Lee (1963). It is considered that changes in both population distribution and density have been important and that climatic change in the later Pleistogene through effecting changes in productivity of resources has been a major determinant of population fluxes.

Climatic changes on the scale of those of the later Pleistogene must be taken into account in assessing the archaeological record. The latter is the product of adjustments in the dynamics of small or local populations in response to usable resources with stochastic fluctuations in numbers, migrations and extinctions playing an important role (Ammerman 1975; Weiss 1975). There is even the possibility that fertility in some populations may be homeostatically controlled by productivity (Howell 1979). A limiting factor in the success of the foraging life-style and the reason why foragers remained closely tuned to natural productivity while being vulnerable to environmental perturbations is that the highest work loads are imposed when resources are seasonally scarce, contributing to higher mortality rates. The introduction of food production had very marked demographic consequences in southern Africa in the last 2000 years because it solved the seasonal shortfall in production.

The fluxes in populations that have occurred in southern Africa are important with respect to human genetics. The fragmentation and isolation of populations resulted in the evolution of the distinctive Khoisan genotype (Tobias 1978) in the later Pleistogene, although the relevant skeletal remains which may fully document this event are largely lacking. The relatively abrupt reduction in diversity of habitats at the beginning of the coldest interval of the Late Pleistocene may have been a factor in promoting such fragmentation and isolation.

Archaeology in southern Africa has been successively concerned with technology, subsistence ecology and settlement patterns on a regional scale. The larger scale patterning in archaeological data reflecting the establishment, maintenance and fluxes of local populations over larger areas represents another level of enquiry and an important one for the understanding of the meaning of the archaeological record. Although populations in the latitudes of southern Africa appear to have achieved their highest archaeological visibility, correlated here with highest density, in interglacial or interstadial times of increased environmental diversity, this timing clearly does not hold for higher latitudes if the example of Upper Palaeolithic population visibility in Europe during the colder phases of the Late Pleistocene is taken into account. The maintenance of human populations during the later Pleistogene merits investigation on a wider geographical framework than simply southern Africa.

ACKNOWLEDGEMENTS

HJ Deacon wishes to acknowledge the benefit of discussions with PB Beaumont, JC Vogel and TP Volman on this topic at a previous SASQUA meeting. The copy editing was done by J Deacon and the diagrams were drawn by VB Geleijnse. Research on the maintenance of human populations in the Late Pleistocene is a current project in the Department of Archaeology at the University of Stellenbosch and financial support from the University and the Human Sciences Research Council is gratefully acknowledged.

REFERENCES

Ammerman AJ 1975. Late Pleistocene population dynamics: an alternative view. Human Ecology 3:219-233.
Avery DM 1981. Holocene micromammalian faunas from the northern Cape Province, South Africa. S. Afr. J. Sci. 77:265-273.
Avery DM 1982a. Micromammals as palaeoenvironmental indicators and an interpretation of the late Quaternary in the southern Cape Province, South Africa. Ann. S. Afr. Mus. 85:183-374.
Avery DM 1982b. The micromammalian fauna from Border Cave, KwaZulu, South Africa. J. archaeol. Sci. 9:187-204.
Beaton JM 1983. Does intensification account for changes in the Australian Holocene archaeological record? Archaeol. Oceania 18:94-97.
Beaumont PB 1978. Border Cave. MA thesis, Univ. Cape Town.
Beaumont PB 1981. The Heuningneskrans shelter. In: EA Voigt (ed.), A guide to archaeological sites in the northern and eastern Transvaal. Pretoria, Transvaal Museum.
Beaumont PB and Vogel JC 1972a. On a new radiocarbon chronology for Africa south of the equator. Part I. Afr. Stud. 31:66-89.
Beaumont PB & Vogel JC 1972b. On a new radiocarbon chronology for Africa south of the equator. Part II. Afr. Stud. 31:155-182.
Behrensmeyer AK 1982. The geological context of human evolution. Ann. Rev. Earth Planet. Sci. 10:39-60.
Birdsell JB 1973. A basic demographic unit. Curr. Anthrop. 14:337-356.

Bishop WW 1971. The Late Cenozoic history of East Africa in relation to hominoid evolution. In: KK Turekian (ed.), The Late Cenozoic glacial ages. Yale University Press, New Haven.

Brink JS and Deacon HJ 1982. A study of a Last Interglacial shell midden and bone accumulation at Herolds Bay, Cape Province, South Africa. Palaeoecol. Afr. 15:31-39.

Carter PL 1978. The prehistory of eastern Lesotho. PhD thesis, Cambridge.

Clarke JI 1972. Population geography. Pergamon, Oxford.

Climap Project Members 1976. The surface of the Ice-Age earth. Science 191:1131-1137.

Cooke HBS 1957. The problem of Quaternary glacio-pluvial correlation in east and southern Africa. In: JD Clark and S Cole (eds), Proceedings of the Third Pan-African Congress on Prehistory, Livingstone 1955. Chatto and Windus, London. p. 51-55.

Cooke HBS 1958. Observations relating to Quaternary environments in east and southern Africa. Trans. geol. Soc. S. Afr. Annexure 60:1-73.

Davies O 1971. Pleistocene shorelines in the southern and southeastern Cape Province. Part 1. Ann. Natal Mus. 21:183-223.

Davies O 1972. Pleistocene shorelines in the southern and southeastern Cape Province. Part 2. Ann. Natal Mus. 22:255-279.

Deacon HJ 1970. The Acheulian occupation at Amanzi springs, Uitenhage District, Cape Province. Ann. Cape Prov. Mus. 8:89-189.

Deacon HJ 1976. Where hunters gathered: a study of Holocene Stone Age people in the eastern Cape. S. Afr. Archaeol. Soc. Monogr. Ser. 1, Claremont.

Deacon HJ 1979. Excavations at Boomplaas Cave: a sequence through the Upper Pleistocene and Holocene in South Africa. World Archaeol. 10:241-259.

Deacon J 1974. Patterning in the radiocarbon dates for the Wilton/Smithfield complex in southern Africa. S. Afr. archaeol Bull. 29:3-18.

Deacon J 1982. The Later Stone Age in the southern Cape, South Africa. PhD thesis, Univ. Cape Town.

Deacon HJ, Deacon J, Scholtz A, Thackeray JF and Vogel JC 1984. Correlation of palaeoenvironmental data from the Late Pleistocene and Holocene deposits at Boomplaas Cave, southern Cape (this volume).

Flint RF 1959. On the basis of Pleistocene correlation in East Africa. Geol. Mag. 96:265-284.

Gates WL 1976. The numerical simulation of ice-age climate with a global general circulation model. J. atmosph. Sci. 33:1844-1873.

Goodwin AJH and Van Riet Lowe C 1929. The Stone Age cultures of South Africa. Ann. S. Afr. Mus. 27:1-289.

Grove AT 1969. Landforms and climatic changes in the Kalahari and Ngamiland. Geogr. J. 135:191-212.

Hamilton AC 1976. The significance of patterns of distribution shown by forest plants and animals in tropical Africa for the reconstruction of Upper Pleistocene palaeoenvironments: a review. Palaeoecol. Afr. 9:63-97.

Hamilton AC 1982. Environmental history of East Africa. Academic Press, New York.

Harland WB, Cox AV, Llewellyn PG, Pickton CAG, Smith AG and Walters R 1982. A geologic time scale. Cambridge University Press, Cambridge.

Heath GR 1979. Simulations of a glacial palaeoclimate by three different atmospheric general circulation models. Palaeogeogr.,

Palaeoclimatol., Palaeoecol. 26:291-303.
Howell N 1979. Demography of the Dobe !Kung. Academic Press, New York.
Humphreys AJB 1979. The Holocene sequence in the northern Cape and its position in the prehistory of South Africa. PhD thesis, Univ. Cape Town.
Imbrie J and Imbrie KP 1979. Ice Ages: solving the mystery. Macmillan, London.
Klein RG 1976. The mammalian fauna of the Klasies River Mouth sites, southern Cape Province, South Africa. S. Afr. archaeol. Bull. 31:74-98.
Krige LJ 1932. The geology of Durban. Trans. geol. Soc. S. Afr. 35:37-67.
Lancaster IN 1979. Evidence for a widespread late Pleistocene humid period in the Kalahari. Nature 279:145-146.
Lee RB 1963. The population ecology of man in the early Upper Pleistocene of southern Africa. Proc. prehist. Soc. 29:235-257.
Lorius C, Merlivat L, Jouzel J and Pourchet M 1979. A 30 000-yr isotope climate record from Antarctic ice. Nature 180:644-648.
Lourandos H 1983. Intensification: a late Pleistocene-Holocene archaeological sequence from southwestern Victoria. Archaeol. Oceania 18:81-94.
Manabe S and Hahn DG 1977. Simulation of the tropical climate of an ice age. J. geophys. Res. 82:3889-3911.
Mason RJ 1962. Prehistory of the Transvaal. Witwatersrand University Press, Johannesburg.
Mason RJ 1974. The last Stone Age San (Bushman) of the Vaal-Limpopo basin. S. Afr. J. Sci. 70:375.
Mason RJ 1981. Early Iron Age settlement of southern Africa. S. Afr. J. Sci. 69:324-326.
Morley JJ and Hays JD 1979. Comparison of glacial and interglacial oceanographic conditions in the South Atlantic from variations in calcium carbonate and radiolarian distributions. Quatern. Res. 12:396-408.
Netting RM 1971. Cultural ecology. Cummings, Menlo Park Ca.
Plug I 1981. Some research results on the Late Pleistocene and early Holocene deposits of Bushman Rock Shelter, eastern Transvaal. S. Afr. archaeol. Bull. 36:14-21.
Price-Williams D and Watson A 1982. New observations on the prehistory and palaeoclimate of the Late Pleistocene in southern Africa. World Archaeol. 13:373-381.
Rogers AW 1905. A raised beach deposit near Klein Brak River. Ann. Rep. geol. Comm. Cape of Good Hope 10:293-296.
Salinger MJ 1981. Palaeoclimates north and south. Nature 291:106-107.
Sampson CG 1974. The Stone Age archaeology of southern Africa. Academic Press, New York.
Scott L 1979. Late Quaternary pollen analytical studies in the Transvaal (South Africa). PhD thesis, Univ. Orange Free State.
Scott L 1982. A late Quaternary pollen record from the Transvaal Bushveld, South Africa. Quatern. Res. 17:339-370.
Shackleton NJ 1969. The Last Interglacial in the marine and terrestrial records. Proc. R. Soc. Lond. B174:135-154.
Shackleton NJ 1982. Stratigraphy and chronology of the Klasies River Mouth deposits: oxygen isotope evidence. In: R Singer and J Wymer, The Middle Stone Age at Klasies River Mouth in South Africa. University of Chicago Press, Chicago. p.195-199.

Simpson BB and Haffer J 1978. Speciation patterns in the Amazon forest. Ann. Rev. Ecol. Systematics 9:497-518.

Talma AS, Vogel JC and Partridge TC 1974. Isotopic contents of some Transvaal speleothems and their palaeoclimatic significance. S. Afr. J. Sci. 70:135-140.

Thackeray AI 1981. The Holocene cultural sequence in the northern Cape Province, South Africa. PhD thesis, Yale Univ.

Tobias PV 1978. The San: an evolutionary perspective. In: PV Tobias (ed.), The Bushman: San hunters and herders of southern Africa. Human and Roussouw, Cape Town. p. 16-32.

Van Zinderen Bakker EM 1967. Upper Pleistocene and Holocene stratigraphy and ecology on the basis of vegetation. In: WW Bishop and JD Clark (eds), Background to evolution in Africa. University of Chicago Press, Chicago. p. 125-147.

Van Zinderen Bakker EM 1976. The evolution of late Quaternary palaeoclimates of southern Africa. Palaeoecol. Afr. 9:160-202.

Van Zinderen Bakker EM 1982a. African palaeoenvironments 18 000 years BP. Palaeoecol. Afr. 15:77:99.

Van Zinderen Bakker EM 1982b. Pollen analytical studies of the Wonderwerk Cave, South Africa. Pollen et Spores 25:235-250.

Vogel JC 1983. Isotopic evidence for past climates and vegetation of South Africa. Bothalia 14:391-394.

Walker NJ 1980. Later Stone Age research in the Matopos. S. Afr. archaeol. Bull. 35:19-24.

Weiss KM 1975. The application of demographic models to anthropological data. Human Ecol. 3:87-103.

Wendt WE 1976. Art mobilier from the Apollo 11 cave, South West Africa: Africa s oldest dated works of art. S. Afr. archaeol. Bull. 31:5-11.

Evidence for Late Quaternary climatic change in southern Africa
Summary of the proceedings of the SASQUA Workshop held in Johannesburg, September 1983

JANETTE DEACON
University of Stellenbosch, South Africa

N. LANCASTER
University of Cape Town, South Africa

L. SCOTT
University of the OFS, Bloemfontein, South Africa

A Workshop on "Evidence for Late Quaternary climatic change in southern Africa" was held at the University of the Witwatersrand in Johannesburg on September 3 and 5, 1983, under the auspices of the CSIR Cooperative Scientific Programmes National Programme for Weather, Climate and Atmosphere Research and SASQUA. The session was organized by JC Vogel who arranged for the preparation and circulation of data sheets summarizing the nature of the evidence, dating and interpretation of information on Late Pleistocene and Holocene palaeoclimate. At the meeting regional rapporteurs summarized the data for each of six time zones, and after each summary the evidence was open for general discussion amongst the 80 or more delegates. A brief resume of the results of these discussions is given here, J Deacon being responsible for the Cape ecozone, N Lancaster for the Kalahari and the Namib, and L Scott for the central and eastern interior.

TIME ZONES 1 AND 2
The Last Interglacial, Early Glacial and Lower Pleniglacial. Oxygen isotope stages 5e-4, c. 130 000 - 40 000 BP

There is little reliable palaeoclimatic information available for this time period and for the most part the dating of what is available is uncertain or is often assumed on the basis of an association between the palaeoclimatic data and Middle Stone Age (MSA) artefacts. MSA material is clearly dated to the Last Interglacial and Early Glacial at Klasies River Mouth on the southern Cape coast and widespread technological changes within the MSA are used as a means of relative dating elsewhere in the subcontinent.
 In the Kalahari, lake levels were apparently low at the end of this period in the Makgadikgadi Depression (see Fig. 1 for location of all sites mentioned in the text). At Gi in the north-west Kalahari sediments containing MSA material suggest a semi-arid regional climate with a productive savanna ecosystem indicated by the mammalian fauna (Helgren and Brooks 1983). On the Gaap escarpment Butzer (this volume) suggests that the Blue Pool tufa formed during oxygen isotope stage 5e indicating greater spring discharge and regional rainfall during this period. Erosion of tufas, in sub-humid climates, occurred during stage 4 and and possibly stage 3.
 In the Namib MSA artefacts are widely distributed in areas which are

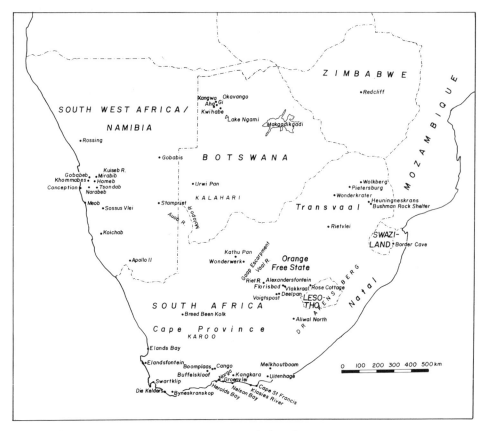

Figure 1. Location of sites mentioned in the text.

today hyperarid (Korn and Martin 1957) suggesting that more favourable conditions prevailed, at least seasonally, at intervals during this time zone.

Sedimentary sequences from the central and eastern interior (the lower Vaal River basin, Alexandersfontein pan, Florisbad, Rose Cottage Cave, Heuningneskrans, Bushman Rock Shelter and Border Cave) have been analysed and summarized by Butzer (this volume). The palaeoclimatic interpretation suggests a complex series of fluctuations in temperature and moisture conditions that cannot be correlated in detail due to a lack of precise absolute dating, but that nevertheless show a gradual reduction of temperatures that is assumed to date within the time span equivalent to oxygen isotope stages 5e through 4. For the most part this time period also seems to have been moister than the succeeding period. Roof spall horizons at Rose Cottage, Border Cave, Heuningneskrans and Bushman Rock Shelter, interpreted (Beaumont et al. 1978; Butzer, Beaumont & Vogel 1978) as indicative of frost action and cold temperatures, are assigned to stage 4. There was considerable discussion at the Workshop on the reliability of roof spall or clast-rich horizons as cold temperature indicators and a number of researchers felt that non-climatic factors such as the release of clasts due to pressure unloading could be involved.

Biological data in the form of pollen samples from Florisbad (Van

Zinderen Bakker 1967), micromammals from Border Cave (Avery 1982a) and larger mammals from Border Cave (Klein 1978) and Redcliff in Zimbabwe (Cruz-Uribe 1983) confirm cooler conditions in deposits associated with MSA artefacts and are assumed to date to stage 4. The effect of cooler temperatures on larger mammal populations was not as marked as it was farther south, but it was sufficient to enable some species (blesbok/bontebok, the common springbok and the mountain reedbuck) to extend their ranges northwards into Zimbabwe. At Border Cave on the border between Swaziland and KwaZulu, cooler temperatures favoured species which prefer bushier habitats where today the fauna is dominated by grazers (Klein 1977).

In the Cape ecozone generally it has been noted that cooler temperatures in the past have encouraged the spread of grassland at the expense of shrubland and this is evident particularly in the large mammal faunas (Klein 1980) which tend to be dominated either by grazers, in particular equids and alcelaphines, or by small browsing antelope, especially Raphicerus spp. (grysbok/steenbok). The correlation between grazers and cooler temperatures is further confirmed by the observation that the mean individual size of Canis mesomelas, the black-backed jackal, is much larger in the grazer-dominated carnivore accumulation at Swartklip than it is in Holocene and present-day samples (Klein 1983). Modern black-backed jackal samples show a clear correlation between cooler mean annual temperatures and larger body size.

The most complete sequence for the Last Interglacial in the Cape is that from the Klasies River Mouth (KRM) cave complex (Singer and Wymer 1982). These sites provide some of the best dating for the MSA artefact sequence and firmly establish its position relative to Late Pleistocene sea level changes. The base of the MSA sequence overlies a beach deposit 6-8 m above the present sea-level and therefore post-dates the warmest conditions of the Last Interglacial about 125 000 years ago (Butzer 1978). Aspartic acid dates confirm the age of the deposits with determinations of 115 000 and 90 000 BP for the MSA I at the base, 89 000 BP for the overlying Howiesons Poort levels and 70 000 BP for the uppermost MSA IV (Singer & Wymer 1982:192). An ionium date on flowstone within the MSA II deposit confirms the general antiquity (Vogel, unpublished), as does a similar deposit and a date from Herolds Bay cave to the west (Brink and Deacon 1982).

At KRM, warm conditions are indicated for the base of the sequence (MSA I and II) by the predominance of browsing antelope in the larger mammals hunted by the people occupying the site (Klein 1976) and by the high incidence of shellfish remains which show that the sea shore was never far from the site. The development of flowstone at the back of Cave 1 has been interpreted by Butzer (1978) as evidence for external decalcification and (?) moister conditions.

In the overlying Howiesons Poort and MSA III levels the larger mammal fauna indicates cooler conditions with a higher incidence of equids and alcelaphines. Analysis of the community structure of the micromammals also shows cool and wet conditions in the Howiesons Poort levels, with declining temperatures from MSA II to Howiesons Poort indicated by a gradual increase in the mean body size of the red musk shrew (Crocidura flavescens) (Avery 1982b). Cool temperatures in the Howiesons Poort are confirmed by the oxygen isotope composition of a small sample of marine shells (Shackleton 1982), although analysis of the species composition of the shell middens (Voigt 1973, 1982) suggests that the Howiesons Poort was relatively warm. Recent research in the site surrounds show, however, that changes in coastal ecology probably have been important in determining species composition of the marine shell component and this may go some way towards explaining the anomaly (HJ Deacon, pers. comm.).

On the basis of KRM and other cave sediments and palaeosols in the southern Cape, Butzer (1978, this volume) suggests that most of oxygen isotope stage 5 was moist relative to stage 4, and that the Howiesons Poort (5c/5d) was cooler than MSA I and II (late 5e) and MSA III and IV (5a/4). In the western Cape, however, substages 5e and 5b appear to have been drier than substages 5d and/or 5c, while the beginning of stage 4 at Die Kelders in the south-western Cape is shown by both cave sediments and micromammals to have been cool and wet (Tankard and Schweitzer 1974; Avery 1982b). By contrast, both larger and micromammal samples from Boomplaas Cave in the southern Cape, which is at present 80 km inland, indicate moderate conditions becoming cooler and drier at the onset of stage 4.

In both the western and southern Cape there is evidence for a widespread period of pedogenesis post-dating the Howiesons Poort phase of the MSA and pre-dating the earliest Later Stone Age deposits dating to before 20 000 BP. This indicates precipitation as high as that of the present at some stage within this and/or the next time zone. It is manifested in the southern Cape by the formation of ferruginous crusts overlying the MSA at Boomplaas, Nelson Bay, Kangkara and Buffelskloof (Butzer 1973; Opperman 1978; Deacon 1979, 1983), and in the south-western Cape by a BFe horizon of a truncated podzol capping the Middle Pleistocene fossil horizon at Elandsfontein (Deacon 1964).

TIME ZONE 3
Inter Pleniglacial. Oxygen isotope stage 3, about 40 - 25 000 BP

Both the biological and non-biological data for this time period show evidence for temperatures intermediate between those of the Last Glacial Maximum and the Present Interglacial. This is particularly clear where sequential changes have been measured as in the isotope analyses of speleothems (Heaton 1981; Vogel 1983).

During this period in the Kalahari most sites for which information is available appear to have experienced conditions which were much wetter than the present, probably with substantially increased rainfall and possibly with somewhat lower temperatures. Radiocarbon dates on lacustrine deposits and their associated mollusca from the Makgadikgadi Depression indicate high lake levels, possibly at the 920 m level, during the period 31 - 24 000 BP (Heine 1978). Although a tectonic influence on inflow to such palaeolakes cannot be discounted (Cooke 1980), complementary evidence for other sites in northern Botswana suggests that higher levels were maintained by substantially increased rainfall (possibly twice present amounts or 800 - 1 000 mm per annum).

At Kwihabe, cave sinters formed from 34 - 29 000 BP indicating a more humid and possibly cooler climate (Cooke 1975). Nearby, extensive freshwater lakes existed in the Dobe valley prior to 31 000 BP suggesting humid regional climates (Helgren and Brooks 1983). This interpretation is supported by the deposition of tufas at Xangwa as a result of the increased discharge by karstic groundwaters from the Aha Hills.

In the southernmost Kalahari, dunes were fixed by vegetation between 32 000 and 28 000 BP and organic soil horizons formed in interdune areas, whilst water levels were high in pans 24 - 23 000 BP (Heine 1982).

At Wonderwerk Cave, on the south-eastern margins of the Kalahari, cave sediments suggest a moist and cold climate from 35 000 to 24 000 BP with frost shattering in very cold conditions prior to 30 000 BP (Butzer, this volume). The record from the Gaap escarpment (Butzer, Stuckenrath et al. 1978) is similar, with greater cold from 35 000 to 30 000 BP suggested by the presence of angular frost shattered debris below tufa Vb. Although dating is uncertain, there may have tufa deposition in wetter climates from 30 000 to 25 000 BP.

In the Namib a number of sites record evidence for significantly wetter conditions. Cave sinters from near Rossing Mountain have dates ranging from before 41 500 BP to 26 000 BP suggesting to Heine and Geyh (this volume) increased rainfall and runoff in this now hyperarid area. Close to the Kuiseb valley, reed beds at Khommabes record higher groundwater levels in a semi-enclosed basin 30-27 000 BP. Carbonate pedotubules at Narabeb some 40 km to the south have dates of 30-28 000 BP implying plant growth and increased moisture availability at this time. At Sossus Vlei, flood silts with a radiocarbon date of 24 800 BP suggest increased flooding of the Tsauchab River and higher rainfall in its mountain catchment. At Koichab pan, near Luderitz, reed beds have dates around 23 000 BP indicating high groundwater tables prior to this date. The interpretation of the many radiocarbon dates on calcretes which fall in this time zone is controversial, but suggest perhaps some re-precipitation of calcium carbonate by increased rainfall during this period.

In the eastern interior, roof spall horizons interpreted as the result of frost action are present at Rose Cottage and Border Cave at the end of stage 3, but for the most part sediments assigned to stage 3 at these sites and at Bushman Rock Shelter and Heuningneskrans are interpreted as the result of mild temperatures and humid to subhumid conditions (Butzer, this volume). Speleothems in Wolkberg and other Transvaal caves indicate both that precipitation levels were high enough through stages 3 and 2 to promote speleothem growth and that temperatures between 30 000 and 20 000 BP were about 8-9 degrees C cooler than at present (Talma et al. 1974). Moister conditions are also indicated by the formation of spring tufas near Pietersburg in the Transvaal between 42 800 and 14 600 BP.

Pollen data from Wonderkrater in the Transvaal (Scott 1982) indicate relatively moist but only slightly cooler conditions associated with expanded forest during most of stage 3, with drier conditions developing before the onset of stage 2.

Micromammals at Border Cave indicate relatively mild conditions around 38 000 BP (Avery 1982a) and a number of larger mammal assemblages from the same general region indicate no significant difference between the range of mammals in the fossil deposits and those present in the region today (Klein 1980), confirming earlier observations that the larger mammals in the interior of southern Africa were not greatly affected by Pleistocene temperature changes except in the extension of the range of certain species.

In the Cape ecozone, on the other hand, samples of larger mammals from the time range of stage 3 were clearly different from the present day in that they were dominated by grazers rather than browsers (Klein 1978, 1980, 1983). Boomplaas Cave includes the most complete sequence for this time and here the micromammals confirm that temperatures were cooler and suggest generally moister conditions with drier intervals (Avery 1982b). Shrubland taxa dominate the charcoal samples from the 32 000 BP unit in contrast to the Olea woodland represented in the underlying OLP unit dated to greater than 42 000 BP. This shift suggests cooler and drier climatic conditions prevailed around 32 000 BP (Deacon et al. 1983; Deacon et al., this volume). The community structure of micromammals at Die Kelders in the south-western Cape also suggests cool and drier conditions there in deposits with a bone apatite date of 31 000 BP (Avery 1982b).

TIME ZONE 4
The Last Glacial Maximum. Oxygen isotope stage 2, 19 000 $\pm$ 3000 BP

All the palaeoclimatic indicators securely dated to this time zone point to temperatures significantly colder than at any other time during the last 125 000 years, and in most instances the evidence also points to drier conditions.

In the northern Kalahari this period is one of transition at the end of the preceding humid phase. In the Makgadikgadi, lake levels fell rapidly after 20 000 BP although water still was locally present around 19 000 BP as evidenced by the accumulation of organic deposits at this time. Lake levels then remained low until 12 000 BP. At Gi, valley-wide lakes with radiocarbon dates of 23 - 22 000 BP were succeeded by an environment drier than today in the period spanning the Last Glacial Maximum. This may correlate with the aeolian sand invasion of caves at Kwihabe between deposition of sinters SII and SIII.

In the south-western Kalahari stratigraphic evidence and radiocarbon dates indicate that fluvial activity ceased in the lower Molopo valley by 19 000 BP (Heine 1982). Conditions in the south-eastern Kalahari appear to have remained more humid than today. At Wonderwerk, major erosion of cave deposits occurred in conditions of heavy runoff, whilst along the Gaap escarpment, the lower Vaal basin and the western Orange Free State there is evidence for accelerated spring activity, tufa deposition, development of soil under moister conditions and greater spring discharge that continued throughout the period 22 000 to 14 000 BP (Butzer, this volume).

In the Namib this period saw a major aggradational episode in the Kuiseb River, with the deposition of the 20-30 m thick Homeb silts between 23 000 and 19 000 BP (Vogel 1982), probably as a floodplain deposit in conditions of low stream energy, aided perhaps by blocking of the valley by dunes. Coeval pond and reed bed deposits near Gobabeb suggest higher water tables, perhaps controlled by the aggradation of the Kuiseb. Farther south, at Apollo 11 Cave, stable carbon isotope analyses of zebra teeth show that winter rainfall did not penetrate to this latitude during this time zone (Vogel 1983).

In the eastern interior, frost spall horizons in Rose Cottage cave and Wonderwerk are interpreted by Butzer (this volume) as indicative of cooler conditions while the sediments at Border Cave between 29 000 and 13 000 BP are interpreted as the result of a cool and dry climate. Cooler temperatures are reflected also in oxygen isotope analyses of speleothems from Wolkberg cave in the Transvaal (Talma et al. 1974) where a reduction of 7,5 degrees C at 18 000 BP has been calculated. The effect of cooler temperatures on the vegetation surrounding Wonderkrater in the Transvaal (Scott 1982) was to favour more open heathland species in contrast to the present-day woodland.

In the Cape ecozone a firm estimate of the scale of temperature change has been calculated from the isotopic composition (O_2, N_2 and Ar) of groundwater in the Uitenhage aquifer that is of the order of 5,5 degrees C cooler than the present (Heaton 1981; Vogel 1983). This is confirmed by oxygen isotope analyses of a Cango Caves speleothem which indicate a temperature of 5 degrees C cooler than the present at 18 000 BP (Vogel 1983).

The low species diversity of micromammals, and community structure and mean individual size in particular species of both larger and smaller mammals at Boomplaas and Nelson Bay caves (Klein 1980; Avery 1982b), point to colder and drier conditions in the southern Cape at the Last Glacial Maximum. This is confirmed by the low diversity of species represented by charcoal and pollen samples at Boomplaas which are dominated by composites and include no trees (Deacon et al., this volume). Cooler temperatures, with no evidence for changes in rainfall, are shown by grazer-dominated larger mammal samples at Melkhoutboom in the eastern Cape and Elands Bay in the western Cape (Klein 1980).

At Boomplaas, the only site with a reasonably complete sequence covering this time period, the coldest interval dates between c. 21 220 $\pm$ 195 (Pta-1810) and some time between 18 600 and 14 200 $\pm$ 240 BP (UW-301). At Nelson Bay Cave the coldest interval dates between 18 660 $\pm$ 110 BP (GrN-5884) and 16 700 $\pm$ 240 BP (I-6516) with no older deposits of suitable age to give a lower limit to this interval. The Cango Caves speleothem places the period of maximum cold between about 20 000 and 16 000 BP, while the groundwater samples from the Uitenhage aquifer place it between about 22 000 and 16 000 BP.

Biological and sedimentological data have been used to infer precipitation in the vicinity of Nelson Bay Cave, now on the coast but about 80 km inland at the Last Glacial Maximum. Klein (in prep.) has shown that fluctuations in the mean individual size of small herbivores appears to vary with precipitation rather than with temperature and they tend to be smaller in regions of lower rainfall. Using such a scale, the Nelson Bay Cave small herbivores indicate very dry conditions in the units dating between 18 660 and 16 700 BP and this is confirmed by the community structure of micromammals in the same units (Avery 1982b). The cave sediments of the YSL horizon dated to 16 700 BP, on the other hand, have been interpreted by Butzer (1973) as indicative of a perennially moist cave microenvironment because roof spall material is impregnated with hydrated iron compounds.

In contrast to the evidence for cold and dry Glacial Maximum conditions in the southern Cape, palaeosols in the western and south-western Cape dating between c. 25 000 and 15 000 BP are interpreted as indicating conditions at least as moist as during the mid-Holocene when geomorphic processes were highly active in the littoral zone (Butzer, this volume). It was suggested during discussion that the western and south-western Cape, i.e. the winter rainfall area, may have been out of phase with the rest of the Cape ecozone as regards rainfall in the past and that while the southern and eastern Cape may have been drier, the western Cape may well have been wetter. The apparent anomaly between biological and sedimentological data as well as the difference between the winter and all-year rainfall regions needs further investigation.

TIME ZONE 5
The Late Glacial. Oxygen isotope stage 2/1, c. 16 000 - 10 000 BP

At all sites where sequences cover this time period there is evidence for a rapid change from cold temperatures to near modern ones by 10 000 BP or soon thereafter. In the Kalahari dry conditions and low lake levels prevailed in the Makgadikgadi Depression, but a rise in water levels, possibly to the 912 m level, may have begun between 12 000 and 11 000 BP. Elsewhere in the Kalahari there is a considerable body of information which suggests higher, or more effective, precipitation or groundwater discharge in the earlier part of the period, with dessication thereafter. At Kwihabe deposition of the massive sinters of unit SIII took place from 16 - 13 000 BP (Cooke 1980), implying substantially increased rainfall in the area, although this event does not appear to be recorded in the sediments at Gi. Farther south at Urwi pan, spheroidal stromatolites formed in permanent wave-agitated waters around 16 250 - 15 610 BP at the end of a period of lacustral conditions in the pans (Lancaster 1979). Nearby, at Gobabis, spring deposits record increased groundwater discharge in the period 17 200 - 11 800 BP whilst massive deposits at Breed Been Kolk pan in the northern Cape suggest a wet period which ended around 13 900 BP (Vogel in prep.).

In the south-east Kalahari deposits at Kathu Pan are interpreted as evidencing a dry climate and low water table at the end of this period. At Wonderwerk Cave the period 13 - 11 800 BP was one characterized by little runoff and in situ corrosion, but probably cool with signs of frost activity (Butzer, this volume). On the Gaap escarpment deposition of tufa Vb continued to 13 500 BP. Spring discharge declined thereafter and a period of minor erosional activity and deposition followed in dry climates until c. 11 500 BP. Angular debris suggests increased frost action and lower temperatures c. 14 000 BP (Butzer, this volume).

In the Namib there is evidence of increased rainfall in the region, although conditions were probably not as wet as those in the period prior to 20 000 BP. Pollen analyses from Sossus Vlei suggest that, although rainfall increased sufficiently in the highland zones to cause increased flooding of the Tsauchab River, it did not change the composition of the vegetation significantly (Van Zinderen Bakker, in prep.). Flood silts were also deposited in the Tsondab valley 5-10 km west of its present end point. Higher groundwater levels and wetter conditions are also implied by the growth of reeds at Meob and Conception bays prior to 12 000 BP. Pedogenic calcretes in interdune desert pavements south of Gobabeb as well as on the gravel plains to the north have radiocarbon ages of around 10 000 BP suggesting to Vogel that this was the last significantly humid episode in the region.

Palynological results from Aliwal North in the north-eastern Cape (Coetzee 1967) and Wonderkrater in the Transvaal (Scott 1982) suggest that in the eastern interior temperatures ameliorated during this time zone. Dramatic environmental shifts, at least during the early part of this interval, are indicated by the accumulation of more aeolian sediments in the western Orange Free State and north-western Cape Province (Butzer, this volume). The Aliwal North pollen profile suggests the development of relatively dry karroid conditions during which the degree of humidity fluctuated, while pollen data from Wonderkrater show the return of relatively dry low savanna vegetation.

In the Cape ecozone larger mammal samples from Elands Bay Cave, Byneskranskop, Boomplaas, Nelson Bay Cave and Melkhoutboom show a gradual shift between 12 000 and about 10 000 BP from predominantly grazers to predominantly browsers (Klein 1980). In the micromammal faunas, however, the shift from glacial to interglacial communities took place before 14 200 BP at Boomplaas and thereafter there were only relatively minor changes in community structure, species diversity and the mean size of individuals in particular species (Avery 1982b). The timing of the micromammalian shift is similar to that noted in the vegetation surrounding Boomplaas as reflected in the charcoals from the site. These show an initial Late Glacial period between 14 200 and c. 12 000 BP that was characterized by the re-establishment of woodland taxa, notably Olea, in the valley and the virtual absence of composites characteristic of the Last Glacial Maximum and drier habitats. This vegetation mosaic of the Late Glacial suggests higher effective precipitation. In the overlying deposit dated to c. 10 000 BP there is an increase in composites and thicket taxa like Euclea/Diospyros and Maytenus/Pterocelastrus indicating a change to a warmer and drier environment (Deacon et al., this volume).

Thus, where different lines of evidence can be traced over a well-dated and limited time span at the same site, micromammals and vegetation are sensitive indicators of local habitat conditions affected by changes in temperature and rainfall. This is not the case with larger mammals which show a time lag of the order of 2 - 3 000 years in response to the amelioration of climate after the coldest interval of the Last Glacial Maximum. This can be explained as the result of both the mobility of these larger animals and of selection by hunters.

Climatic changes on land were accompanied by a rapid rise in sea level dated to this period by radiocarbon determinations on littoral shells dredged from the sea bed off Cape St Francis (Dingle and Rogers 1972). By 11 000 BP substantial shell middens were building up at sites now on the coast showing that by this time the coastline was close enough to warrant their being occupied while the seafood resources were being exploited, and, by inference, that the sea was at that time very close to its present position.

TIME ZONE 6
The Post Glacial. Oxygen isotope stage 1, the last 10 000 years

All the data point to higher temperatures in the last 10 000 years than prevailed during the preceding 100 000 years. In the Kalahari data points become more scattered for the northern region, but are more numerous in the south-eastern Kalahari. The better resolution suggests fluctuating climates with wetter and drier episodes, but no period as wet as that before 20 000 BP.

Lake levels were generally low in the Makgadikgadi Depression, but locally high levels in pans occurred between 1700 and 1500 BP indicating increased rainfall and runoff in this region. Lake Ngami, at the south-eastern margin of the Okavango delta, rose to a level 12 m above 20th century levels at around 6445 BP implying substantial changes in the hydrology of the area at this time. At Kwihabe further formation of cave sinter occurred between 2500 and 2200 BP again suggesting increased rainfall in this area. Increased flow and higher regional rainfall is also indicated in the fluvial sediments of the Auob valley in the south-western Kalahari between 4500 and 3500 BP (Heine 1982).

Similar short-term fluctuations in groundwater discharge are recorded in the Holocene deposits at Kathu Pan where pollen analysis confirms warmer and more humid conditions around 7000 BP but drier environments from 3750 to 2700 BP and again from 2400 to 600 BP (Van Zinderen Bakker, in prep.). At Wonderwerk sediments, pollens and larger and micro-mammals indicate higher effective precipitation between 7400 and c. 5000 BP, and again around 2900 BP, but lower rainfall before and after this time (Van Zinderen Bakker 1983). On the Gaap escarpment an Early Holocene period of increased wetness is recorded by increased spring discharge and tufa deposition from 9800 to 7600 BP and periods of tufa deposition of lesser magnitude took place again from 3200 - 2400 BP and from 1300 - 250 BP (Butzer, Stuckenrath et al. 1978).

Some evidence of wetter conditions occurs in the early part of this period in the Namib with deposition of silts near Tsondab and Sossus vleis between 9600 and 8640 BP and deposition of the Gobabeb gravels prior to 9500 BP. This early Holocene period also saw occupation of many sites on the eastern margin of the Namib (Vogel and Visser 1981). The rock shelter at Mirabib in the central Namib plains was occupied also at this time, and again around 6000 BP and 400 AD (Sandelowsky 1977). Micromammals from this site suggest that conditions in the area have been generally more favourable than they are at present for most of the last 5000 years, with climates similar to the present occurring around 5200 BP and in the last 500 years (Brain and Brain 1977).

The body of evidence for climatic conditions in the eastern and central interior during the Holocene is relatively extensive and chronometric control by radiocarbon dating is more reliable than in earlier time zones. Sediment studies from Alexandersfontein, Voigtspost, Deelpan, Rose Cottage Cave and the lower Vaal and Riet river drainage basins, and associated pollen analyses from Voigtspost and Deelpan, added to those from Kathu Pan, the Gaap escarpment and Wonderwerk described above, indicate two relatively moist phases of differing nature (Butzer, this volume). The first, dating between c. 7700 and 6300 BP, did not support as complete a vegetation cover as that which developed during the second one (c. 3600 - 1300 BP) which resulted in extensive soil formation.

Pollen analyses from various radiocarbon dated peat deposits in the Transvaal (Scott 1982, 1984) suggest that after a warm dry phase in the Early Holocene, warmer moist conditions prevailed between 7000 and 6000 BP resulting in the expansion of broadleaved bushveld species at Rietvlei and Wonderkrater. More open and slightly cooler bushveld developed later in the Holocene between about 4000 and 2000 BP.

In the Cape ecozone all Holocene faunas are dominated by small browsing antelope and ground game, but those dating to between 10 000 and 5000 BP include relatively higher numbers of vaalribbok, mountain reedbuck and roan antelope, whereas they are rarer in samples post-dating 5000 BP when the blue duiker appears for the first time (Klein 1980). Klein suggests that this shift implies an increase in forest and woodland habitats in the second half of the Holocene. Micromammalian data from Boomplaas and Byneskranskop indicate relatively minor adjustments in community structure during the Holocene with a tendency towards cooler conditions within the last 2 000 years at Boomplaas and somewhat drier conditions at Byneskranskop between c. 6500 and 3500 BP. In the Late Holocene micromammalian samples from Elands Bay and Die Kelders it was wetter around 3000 BP at the former and after 2000 BP at the latter.

Pollen spectra from several southern Cape sites such as Groenvlei (Martin 1968) and Norga (Scholtz, in prep.) indicate vegetation communities similar to those of the present in the Late Holocene, with an increase in forest elements between about 7500 and 2600 BP. Drier and somewhat cooler conditions led to forest decrease after 2600 BP. Butzer (this volume) interprets the record of pedogenesis and erosion as indicating a relatively dry Early Holocene with modest pedogenesis between 8000 and 7000 BP, followed by moister conditions between 5000 and 1300 BP and a return to drier conditions thereafter.

POINTS FROM GENERAL DISCUSSION

The emphasis of the Workshop was to present the evidence for climatic change, but this cannot be done successfully without some theoretical framework, nor without critical assessment of the nature of the evidence to be evaluated. In his concluding address, HJ Deacon reviewed the classes of palaeoclimatic data presented at the meeting and noted the need for complementary lines of evidence in support of inferences from single sets of observations. He highlighted the necessity for critical evaluation of data and stressed points raised in general discussions such as the need for care in the interpretation of clastic deposits in caves and the inadvisability of using calcretes to date past climatic events. Several delegates had urged caution in using dates on calcretes because there is a lack of agreement about the origin of the carbon-14 content of calcretes which in turn complicates palaeoclimatic inferences. In designing future research projects he strongly recommended that modern analogues be used to model past climates and that biological data be used as checks against inferences drawn from other lines of evidence.

The geomorphic data, for which the dating controls are generally poor, gave information mostly on past rainfall changes, whereas the biological data from elsewhere were more useful for studying changes in temperature. Apparent discrepancies between these two sets of data need to be investigated further, but when all the evidence is seen in time perspective time zones 1, 2 and 3 show a gradual reduction in temperatures after the Last Interglacial with relatively warm intervals and generally relatively moist conditions becoming drier. A shift from browsing to grazing fauna occurred in the Cape ecozone indicating an increase in grassland, but elsewhere the larger mammal communities did not change drastically. The most sensitive indicators of habitat changes are micromammals and plant remains in the form of pollens and charcoals. During the period 32 000 to about 20 000 BP there is widespread evidence for higher effective precipitation in the Kalahari and adjacent regions than at any other time in the last 50 000 years, but during the Last Glacial Maximum, dated to between 20 000 and 16 000 BP, most of the evidence indicates colder and drier conditions than at any other time during the past 130 000 years. In the southern Cape and in the interior the dry conditions of the Last Glacial Maximum were followed by another period of higher rainfall which lasted to some time after 12 000. Climatic shifts during the Holocene were not as marked as those of the Late Pleistocene but the warmest temperatures occurred between about 7000 and 5000 BP when it was generally drier in the interior and moister along the Cape coast. There is some evidence for cooler temperatures in the Cape within the last 2 000 years, and for rainfall fluctuations of relatively low magnitude through the Holocene.

Delegates from other southern hemisphere countries offered their impressions of the evidence presented and urged caution in the use of the northern hemisphere oxygen isotope chronology and glacial sequence terminology for southern African terrestrial sequences. They felt it was preferable to accumulate empirical data on past temperatures and rainfall in order to build up a local chronology rather than to attempt to fit isolated data points into the northern hemisphere sequence and assume parity. C Heusser stressed the importance of quantifying pollen data and encouraged the search for long pollen sequences along the Drakensberg where the vegetation would have been sensitive to temperature changes in the past.

MJ Salinger described the analysis of modern weather patterns and their usefulness as analogues in developing models for past climates in New Zealand and Australia and urged that similar studies be applied to southern Africa. In a post-Workshop address, Tyson demonstrated such patterns in climate and weather noted from modern records which stress, for example, a tendency for rainfall fluctuations in the summer and winter rainfall regions to be out of phase with each other. Future palaeoclimatic modelling will surely be based on data of this kind.

REFERENCES

Avery DM 1982a. The micromammalian fauna from Border Cave, KwaZulu, South Africa. J. archaeol. Sci. 9:187-204.
Avery DM 1982b. Micromammals as palaeoenvironmental indicators and an interpretation of the late Quaternary in the southern Cape Province, South Africa. Ann. S. Afr. Mus. 85:183-374.
Beaumont PB, De Villiers H and Vogel JC 1978. Modern man in sub-Saharan Africa prior to 49 000 years B.P.: a review and evaluation with particular reference to Border Cave. S. Afr. J. Sci. 74:409-419.
Brain CK and Brain V 1977. Microfaunal remains from Mirabib: some evidence of palaeoecological changes in the Namib. Madoqua 10:285-293.
Brink J and Deacon HJ 1982. A study of a last interglacial shell midden and bone accumulation at Herolds Bay, Cape Province, South Africa. Palaeoecol. Afr. 15:31-9.
Butzer KW 1973. Geology of Nelson Bay Cave, Robberg, South Africa. S. Afr. archaeol. Bull. 28:87-110.
Butzer KW 1978. Sediment stratigraphy of the Middle Stone Age sequence at Klasies River Mouth, Tsitsikamma coast, South Africa. S. Afr. archaeol. Bull. 33:141-151.
Butzer KW 1984. Late Quaternary palaeoenvironments in South Africa. In: RG Klein (ed), Southern African palaeoenvironments and prehistory. Balkema, Rotterdam.
Butzer KW, Beaumont PB and Vogel JC 1978. Lithostratigraphy of Border Cave, KwaZulu, South Africa: a Middle Stone Age sequence beginning c. 195 000 B.P. J. archaeol. Sci. 5:317-341.
Butzer KW, Stuckenrath R, Bruzewicz A and Helgren DM 1978. Late Cenozoic paleoclimates of the Gaap Escarpment, Kalahari margin, South Africa. Quartern. Res. 10:310-339.
Coetzee JA 1967. Pollen analytical studies in East and Southern Africa. Palaeoecol. Afr. 3:1-146.
Cooke HJ 1975. The paleoclimatic significance of caves and adjacent landforms in western Ngamiland, Botswana. Geogr. J. 141:430-444.
Cooke HJ 1980. Landform evolution in the context of climatic change and neo-tectonism in the middle Kalahari of north central Botswana. Trans. Inst. British Geogr. (N.S.). 5:80-99.
Cruz-Uribe K 1983. The mammalian fauna from Redcliff cave, Zimbabwe. S. Afr. archaeol. Bull. 38:7-16.
Deacon HJ 1964. Report on the excavations at Elandsfontein, January 1964. Unpublished MS.
Deacon HJ 1979. Excavations at Boomplaas Cave - a sequence through the Upper Pleistocene and Holocene in South Africa. World Archaeol. 10:241-257.
Deacon 1983. Another look at the Pleistocene climates of South Africa. S. Afr. J. Sci. 79:325-328.

Deacon HJ, Scholtz A and Daitz LD 1983. Fossil charcoals as a source of palaeoecological information in the fynbos region. In: HJ Deacon, QB Hendey and JJN Lambrechts (eds), Fynbos palaeoecology: a preliminary synthesis. S. Afr. Nat. Scient. Progr. Rep. 75:174-182.

Dingle RV and Rogers J 1972. Pleistocene palaeogeography of the Agulhas Bank. Trans. R. Soc. S. Afr. 40:155-165.

Heaton THE 1981. Dissolved gases: some applications to groundwater research. Trans. geol. Soc. S. Afr. 84:91-87.

Heine K 1978. Radio carbon chronology of Late Quaternary lakes in the Kalahari, southern Africa. Catena 5:145-149.

Heine K 1982. The main stages of the Late Quaternary evolution of the Kalahari region, southern Africa. Palaeoecol. Afr. 15:53-76.

Heine K and Geyh MA 1984. The age of the Rossing cave sinters in the Namib Desert and their palaeoclimatic significance. Proceedings International Symposium on Late Cenozoic Palaeoclimates of the southern hemisphere, Swaziland 1983. Balkema, Rotterdam.

Helgren DM and Brooks AS 1983. Geoarchaeology at /Gi, a Middle Stone Age and Later Stone Age site in the northwestern Kalahari. J. archaeol. Sci. 10:181-197.

Klein RC 1976. The mammalian fauna of the Klasies River Mouth sites, southern Cape Province, South Africa. S. Afr. archaeol. Bull. 31:74-98.

Klein RG 1977. The mammalian fauna from the Middle and Later Stone Age (Upper Pleistocene) levels of Border Cave, Natal Province, South Africa. S. Afr. archaeol. Bull. 32:14-27.

Klein RG 1978. A preliminary report on the larger mammals from the Boomplaas Stone Age cave site, Cango Valley, Oudtshoorn district, South Africa. S. Afr. archaeol. Bull. 33:66-75.

Klein RG 1980. Environmental and ecological implications of large mammals from Upper Pleistocene and Holocene sites in southern Africa. Ann. S. Afr. Mus. 81:223-283.

Klein RG 1983. Palaeoenvironmental implications of Quaternary large mammals in the fynbos region. In: HJ Deacon, QB Hendey and JJN Lambrechts (eds), Fynbos palaeoecology: a preliminary synthesis. S. Afr. Nat. Scient. Progr. Rep. 75:116-138.

Korn H and Martin H 1957. The Pleistocene in South West Africa. In: JD Clark (ed), Proceedings of the Third Pan-African Congress on Prehistory, Livingstone 1955. Chatto & Windus, London. p. 14-22.

Lancaster IN 1979. Quaternary environments in the arid zone of southern Africa. Occas. Pap. Univ. Wits. Dept. Geogr. & Env. Stud. 22:1-77.

Martin ARH 1968. Pollen analysis of Groenvlei lake sediments, Knysna (South Africa). Rev. Palaeobot. & Palynol. 7:107-144.

Opperman H 1978. Excavations in the Buffelskloof rock shelter near Calitzdorp, southern Cape. S. Afr. archaeol. Bull. 33:18-28.

Sandelowsky BH 1977. Mirabib - an archaeological study in the Namib. Madoqua 10:221-283.

Scott L 1982. A Late Quaternary pollen record from the Transvaal Bushveld, South Africa. Quatern. Res. 17:339-370.

Scott L 1984. Palynological evidence for Quaternary palaeoenvironments in southern Africa. In: RG Klein (ed), Southern African palaeoenvironments and prehistory. Balkema, Rotterdam.

Shackleton NJ 1982. Stratigraphy and chronology of the Klasies River Mouth deposits: oxygen isotope evidence. In: R Singer and J Wymer, The Middle Stone Age at Klasies River Mouth in South Africa. Chicago University Press, Chicago. p.194-199.

Singer R and Wymer J 1982. The Middle Stone Age at Klasies River Mouth in South Africa. University of Chicago Press, Chicago.

Talma AS, Vogel JC and Partridge TC 1974. Isotopic contents of some Transvaal speleothems and their palaeoclimatic significance. S. Afr. J. Sci. 70:135-140.

Tankard AJ and Schweitzer FR 1974. The geology of Die Kelders Cave and environs - a palaeoenvironmental study. S. Afr. J. Sci. 70:365-369.

Van Zinderen Bakker EM 1967. Upper Pleistocene and Holocene stratigraphy and ecology on the basis of vegetation changes in sub-Saharan Africa. In: WW Bishop and JD Clark (eds), Background to evolution in Africa. University of Chicago Press, Chicago. p. 125-147.

Van Zinderen Bakker EM . Pollen analytical studies of the Wonderwerk Cave, South Africa. Pollen et Spores 24:235-250.

Van Zinderen Bakker in press. Aridity along the Namib coast. Palaeoecol. Afr.

Vogel JC 1982. Age of the silt terrace at Homeb. Palaeoecol. Afr. 15:201-209.

Vogel JC 1983. Isotopic evidence for past climates and vegetation of South Africa. Bothalia 14:391-394.

Vogel JC and Visser E 1981. Pretoria radiocarbon dates II. Radiocarbon 23:43-80.

Voigt EA 1973. Stone Age molluscan utilization at Klasies River Mouth caves. S. Afr. J. Sci. 69:306-309.

Voigt EA 1982. The molluscan fauna. In: R Singer and J Wymer, The Middle Stone Age at Klasies River Mouth in South Africa. University of Chicago Press, Chicago. p. 155-186.

Ward JD, Seely MK and Lancaster N 1983. On the antiquity of the Namib. S. Afr. J. Sci. 79: 175-183.

The southern deserts

Ancient ergs of the Southern Hemisphere

D.S.G.THOMAS & A.S.GOUDIE
University of Oxford, UK

ABSTRACT. Since the 1950's it has become apparent from work on the margins of the Sahara and elsewhere that during various periods of the Pleistocene there were major arid phases when dune extension took place into areas that are now relatively humid. Such studies have been extended in recent years as a result of the increasing availability of remote sensing imagery, and as a consequence of more intensive field work. This paper draws attention to the ancient ergs of South America, Central and Southern Africa, and Australia, and to some of the other aeolian features with which they are associated.

INTRODUCTION

Active dunes, which form as a result of the movement of sand grains by creep and in saltation, only occur where vegetation cover is limited and there is an adequate supply of sand. On coasts and in the immediate vicinity of sand bed rivers that are seasonally dry, dunes may form under quite humid conditions, but for large, active continental dunes to occur mean annual rainfall may need to be less than 250 mm (Goudie 1983), and possibly as little as 25 mm (Sarnthein and Diester-Haass 1977).

However, examination of air and satellite photography in many low-latitude areas revealed the presence of degraded ergs in areas of relatively high precipitation (see, for example, Grove 1958, 1969). On the ground such fossil dunes appear to be relatively heavily vegetated, cultivated, stabilized by the development of deep soil and weathering horizons, degraded by fluvial action, overlain by estuarine and lake deposits, and scattered with human artefacts of some antiquity (Allchin et al. 1978). Their morphology contrasts strikingly with modern, active dunes, which are steep-sided, unstable and unvegetated.

The purpose of this paper is to describe the distribution, character and significance of the ancient ergs, and associated aeolian landforms in the Southern Hemisphere.

LATIN AMERICA

In Latin America rainforest covers extensive areas, but in the 1970's pollen analysis revealed that in Late Pleistocene times it contracted in extent and the area covered by more xerophytic types, such as savanna, expanded. Furthermore, Damuth and Fairbridge (1970), from cores in the Atlantic off Brazil, found that the Late Pleistocene sediments contained much more feldspar (25-60 %) and relatively less quartz than those of the Holocene, when the feldspar level averaged 17-20 %. They suggested that this implied the existence of less intense chemical weathering activity in the Amazon Basin in Late Pleistocene times, possibly because of reduced precipitation levels. Likewise, the studies of Bonatti and Gartner (1973) on cores in the Caribbean basin indicated that quartz levels were higher during arid glacials. The Amazon basin appears to have become choked by coarse debris, and under the modern hydrological conditions only the major rivers are able to rework the coarser alluvium that was deposited during more arid phases (Baker 1978).

There is thus a body of sound evidence that large parts of Latin America may have suffered Pleistocene conditions drier than the present, and this is confirmed by the existence of fossil ergs in at least four main areas (Fig. 1): the Llanos (current mean annual rainfall up to

Figure 1. Location of the ergs of South America.

1400 mm), the Sao Francisco Valley of N E Brazil (1200-1500 mm), and the Pampas of Argentina (800-1100 mm). The first two of these ergs were described briefly by Tricart (1974), the third by Klammer (1982) and the last by Tricart (1969). Associated with the Mato Grosso and Pampas dune fields are very large numbers of deflation depressions (pans), comparable to those in southern Africa and Australia, together with calcrete crusts which, because of their presence in areas of high rainfall (>1000 mm p.a.), may also well have considerable palaeoclimatic significance (Putzer 1958, 1962). The precise dates for the ancient Latin American ergs remain uncertain.

SOUTHERN AFRICA

The most apparent evidence of Quaternary climatic change in southern Africa is the presence of extensive systems of fixed sand dunes in areas where contemporary climates are not conducive to aeolian sand movement. These ancient ergs extend from the Molopo River in South Africa to Zaire in the north and from Zimbabwe in the east to the Atlantic coast of Namibia.

With the exception of the Namib Desert the dunes of the southern African subcontinent are formed from the surface sediments of the Kalahari Beds, the Kalahari Sands, which cover an area of 2.5 million km^2 between the Orange and Congo Rivers (Cooke 1964). These sands are primarily aeolian in origin although there is localized evidence for fluvial deposition (Rogers 1934; Savory 1965; Verboom 1974; Baillieul 1975; Mallick et al. 1981), and although their original deposition has been variously dated as Tertiary (Maufe 1930, 1935), Miocene (Cahan and Lepersonne 1952), Pliocene (Mabbutt 1957), Plio-Pleistocene (King 1962) and Pleistocene (Bond 1948), the fixed dunes locally represent the most recent aeolian reworkings.

Both longitudinal and barchan dunes are present in the Namib Desert. Longitudinal dunes are up to 275 m high (Jaeger 1939), possibly making them the highest dunes in the world (Goudie 1970), and are 1-3 km apart (Lancaster 1979). The Namib Sand Sea Dunes are the only active desert dunes in southern Africa today, but Lancaster (1979) has suggested that the main linear forms are out of phase with today's resultant dune forming winds and are being reworked.

The inactive ergs of the continental interior are dominated by longitudinal forms (Grove 1969; Flint and Bond 1968; De Dapper 1979, 1981; Lancaster 1980, 1981; Williams, in press; Thomas, in press 1 and 2) which commonly form parallel to prevailing or resultant wind directions (Bagnold 1941; Warren 1970; Fryberger 1979). Goudie (1969, 1970) noted though that the ridges in the south-west of Botswana form complex dendritic patterns, comparable to river drainage networks, and in eastern central Botswana (Grove 1969) and south-eastern Zaire (De Dapper 1981) they are little more than faint furrow patterns. Transverse ridges, formed obliquely to low velocity dune building winds under conditions of considerable sand supply (Glennie 1970), occur locally to the south-east of the Okavango Delta in Botswana and in the Nossob/Molopo area where Mallick et al. (1981) observed them to be overlain by longitudinal forms. Grove (1969) also identified 'linked barchans' in the Ntwetwe Pan of the Makgadikgadi depression in eastern Botswana, although there is some doubt as to whether these are of aeolian or lacustrine origin (Grey and Cooke 1977).

All the dunes of central southern Africa are degraded and vegetated, although there is some movement of sand on the sparsely vegetated ridge crests in south-western Botswana, where ridges are up to 10 m high and 0.5 to 2.9 km apart (Lancaster 1981). The ridges to the north and west of the Okavango Delta are up to 25 m high, 1 to 2.5 km apart, up to 200 km long and covered in open savanna woodland, those to the east of the

Makgadikgadi depression in eastern Botswana and western Zimbabwe up to 25 m high, 1.3 to 2.3 km apart and covered in a dense woodland vegetation dominated by the Rhodesian Teak with grassland in the intervening straats (Thomas, in press 2). The spacing of these last two sets of dunes is comparable to that of the active Namib linear dunes and may indicate that their heights were similar before degradation.

Lancaster (1979, 1980, 1981) classified the fixed dunes of Botswana into the northern and southern dunes on the basis of orientation differences, with the former being divided into two subsystems. By extending the mapping of the dunes on the northern Kalahari Sands Thomas (in press 2) has shown that there are two arcs of dunes in the northern areas and proposes that the fixed dunes of southern Africa be divided into southern, northern and eastern systems (Fig. 2) each with differ-

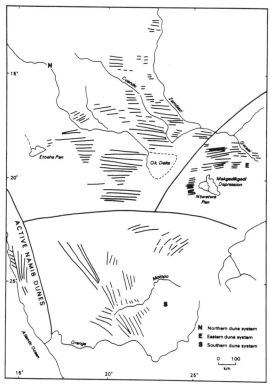

Figure 2. The dune systems of southern Africa.

ent orientation characteristics. These conform in general to the 'anti-clockwise wheel round of dunes' (Goudie 1970) but with the 'peak' of the northern dune arc along 21°E and the peak of the eastern arc along 27°E. Thus the northern system is centred to the north and west of the Okavango Delta, the eastern system in western Zimbabwe, and the southern system in the Nossob/Molopo area.

Heine (1982) suggested that the fixed dunes of southern Africa all formed at a time of greater anticyclonic circulation strengths coinciding with the Last Glacial maximum. Lancaster (1981) however proposed several periods of Late Pleistocene aeolian activity because of difference in pan-lunette and linear dune orientation in Botswana. From this evidence it may be inferred that the northern dune arc predated a humid

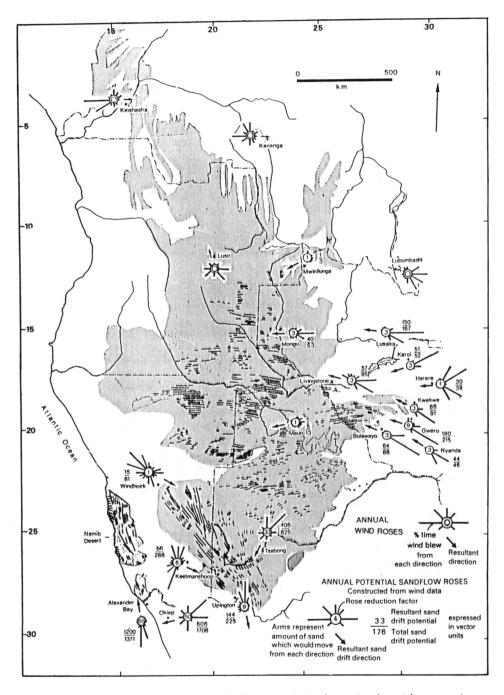

Figure 3. Annual potential sand flows and their relationships to the fixed dunes in central southern Africa.

phase from 30 000 to 20 000 BP, the eastern dune arc formed at the time of the Last Glacial maximum and the southern dunes from 12 000 BP to the Early Holocene. Thomas (in press 2) has also suggested on the grounds of greater dune ridge degradation and dune sand rubification that the northern dunes may predate the eastern dunes.

The comparison of palaeodune alignments and today's potential dune-building wind alignments has been used by Warren (1970) in the Sudan, Brookfield (1970) and Jennings (1975) in Australia and Lancaster (1980, 1981) in southern Africa to suggest that wind regimes have changed since the time of dune formation. The three differently orientated dune systems in southern Africa suggest that circulation conditions were different at each of these formations. Although active linear dunes exist today in areas with widely differing wind regimes, they are usually aligned parallel to the annual resultant drift direction of effective winds in the surrounding environment (Fryberger 1979). Using the method outlined by Fryberger, sand flow roses, resultant potential drift directions and total drift potentials were calculated from available data for 18 weather stations in proximity to the ergs of southern Africa (Fig. 3).

Resultant drift directions are broadly parallel with the nearest linear dunes at a number of localities but at Tsabong, Maun and Mwinilunga there are substantial divergences of present-day dune-building winds from palaeodune alignments. In Zimbabwe no weather stations are located on the Kalahari sand plateau, but there is a $30°$ difference between the resultant drift direction at Bulawayo and the nearest linear dunes.

The total drift potential (DP in Fig. 3) is a non-directional parameter of the total wind energy in the dune environment. According to Fryberger (1979) total drift potentials in active linear dune fields in the Sahara and Arabia are greater than 250. DP values in the southern and western Kalahari range from low (61 at Windhoek) to very high (825 at Tsabong), whilst in the northern Kalahari they are particularly low at Mwinilunga in Zambia and Maun in Botswana, supporting Heine's conclusion that circulation strengths in southern Africa were greater when the dunes were formed.

AUSTRALIA

About a third of the Australian continent is presently affected by seasonal aridity, but the presence of fixed sand dunes in semi-arid, temperate and tropical environments indicates that the arid zone had been considerably greater in extent during the Pleistocene. Even in the three ergs of today's arid core, the Simpson, Great Sandy and Great Victoria Deserts, aeolian sand movement is limited to dune crests, because of an almost continuous vegetation cover which led Madigan (1936) to describe the Simpson Desert as 'an ocean of Spinifex covered waves'. Dunes farthest from the arid core are vegetated by Eucalypt scrub in the south-east (Bowler 1976) and 'pindan' (Acacia) scrub in the tropical north-west (Jennings 1975).

As in southern Africa, longitudinal dunes are the most common form, being particularly regular in the Simpson and Great Sandy Deserts (Mabbutt 1968) where ridges are up to 35 m high, asymmetrical in cross-section, 300 to 500 m apart and can be over 100 km long, but are less regular in the Great Victoria Desert (Madigan 1936). Dune frequencies decrease with dune height (Twidale 1972; Breed and Breed 1980) in such a way that the total volume of sand per unit area is relatively consistent.

Mabbutt (1967) distinguished in Central Australia between sandy ergs and sandy plains, the latter having no dune form developments, possibly because of a higher silt and clay content in the sand, whilst Mabbutt (1968) classified the dunes of sandy ergs into several types. Apart from the longitudinal ridges of the Simpson and Great Sandy Deserts, he noted that the ridges of Northern Territory were further apart and had developed on a 'gibber' plain with ridges separated by reg rather than sand. Composite longitudinal forms also occur; Mabbutt (1968) suggested that bifurcating ridges could either represent an incomplete stage of ridge formation showing the effects of secondary winds or the modification of an older ridge pattern due to changes in prevailing wind direction. Other dune types are longitudinal ridges linked by transverse forms into trellises, polygonal dune networks and short, waved dunes where mobile crests occur upon fixed plinths and where morphology can vary from day to day depending on the direction of the prevailing wind (Twidale 1972). On the lee margins of flood plains in New South Wales (Bowler et al. 1976), Queensland and South Australia (Twidale 1972) short linear dunes extend upwind from the river channels. These 'source dunes' formed as a result of fluvially deposited sands being seasonally exposed to strong winds (Bowler et al. 1976) and may still be active today (Twidale 1972). In a similar fashion Twidale (1972) noted that many dune ridges in the Simpson Desert had their origins in playa lake fringing lunettes.

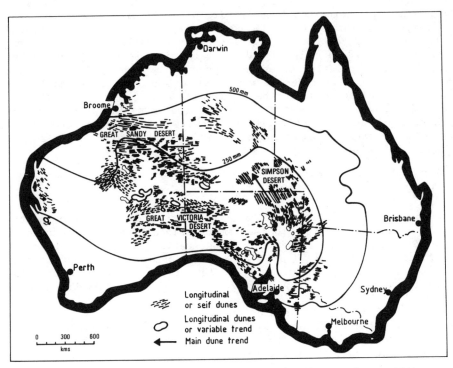

Figure 4. The desert dunes of Australia (after Jennings 1968).

In addition to the sandy ergs shown by Jennings (1968) (Fig. 4), Glasford and Killigrew (1976) proposed on the grounds of grain size distribution and stratigraphic relationships that the Perth Basin Sands of the south-west coast were of aeolian rather than marine origin.

There is remarkable agreement in the literature as to the age of the ergs of Australia, with dates being obtained from marine sediments overlying aeolian sands (Fairbridge 1964; Jessup 1968; Jennings 1975), from underlying alluvium and lacustrine sediments (Wopfner and Twidale 1967; Twidale 1972), from overlying calcareous palaeosols (Churchward 1963; Fill 1973) and underlying piedmont fans (Williams 1973). By these means the ergs of Australia have been attributed to a continent-wide arid period coinciding with the Last Glacial maximum (Galloway 1965) approximately 20 000 - 12 000 BP. Williams (1973) also noted that the arid core had seen aeolian activity on more than one occasion during the Holocene, indicated by the relatively fresher forms of the dunes, because arid core ergs would be less sensitive to climatic fluctuations. Twidale pointed out that they had been active for a longer period than peripheral ergs with sands being reworked from time to time, and perhaps still so today in certain geomorphological contexts.

Increased aridity and aeolian activity at the time of the Last Glacial would have been accompanied by eustatically lower sea levels. This is evidenced by the presence of calcite bearing palaeosols on the continental shelf off the north-west coast of Australia, 130 m below the present sea level and dated at 17 000 BP (Van Andel et al. 1967) and drowned relict dune stumps in Fitzroy Sound (Jennings 1975). It has already been noted that the ergs of Australia form an anti-clockwise pattern, and this coincides broadly with winds blowing around the winter anticyclone centred over the continent. Sprigg (1963) suggested that this cell was centred 5° further north when the dunes were formed, but Bowler (1976) pointed to its being in the same position but with a greater circulation strength, supported by Brookfield (1970) who found very little difference between dune orientation and today's resultant drift directions in the peripheral dunes of the north-west coast. This suggests that at the time of dune formation greater circulation strength and continentality due to lower sea levels reduced the penetration of the northern monsoonal winds leading to a domination of the easterly trades.

CONCLUSIONS

In the Southern Hemisphere fixed sand dunes are present in three continents in areas receiving up to 1400 mm of rainfall annually, extending across 42 degrees of latitude. In South America four ancient ergs are known to occur between 7° N and 37° S, in southern Africa sands which have been distributed predominantly by the wind stretch south from the Equator for over 3200 km covering 2.5 million km^2, with dune ridges known between latitudes 11° S and 20° S, and in Australia inactive dunes stretch from north to south coasts, 17° S to 37° S and cover the bulk of the continental interior. Whilst desert conditions prevail today along the west coasts of both South America and southern Africa, the only desert dunes which are currently subjected to aeolian processes occur in the Namib Sand Sea (Fig. 5).

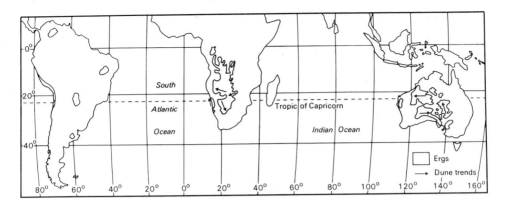

Figure 5. Ergs of the Southern Hemisphere.

No dates have yet been obtained for the formation of the ergs of South America, while there is a general consensus of opinion that the dunes of Australia were active continent-wide during the Last Glacial, 18 000 - 12 000 BP, with those of the arid core being subjected to aeolian processes at earlier and later dates and for longer periods. The extent of aridity at the Last Glacial in southern Africa is a matter of some debate: according to Heine (1982) all the dunes of the Kalahari were active at that time but Lancaster (1981) and others consider that only the eastern dune arc formed then, although arid conditions could have prevailed further south with dunes there being reworked later when the southern dunes were constructed with their present orientation characteristics. Certainly the differing orientations and degradations of the three dune systems suggest different ages (Lancaster 1980, 1981; Thomas, in press 2). The age of the northernmost Kalahari Sands extending beyond the dune systems which have been mapped is at present a matter of conjecture but could relate to the original deposition of the sands, which could be as early as the Tertiary (Maufe 1930, 1935).

It is clear that the formation of the ergs of the Southern Hemisphere relates to winds blowing around high pressure cells, and the differing orientations of the southern African ergs to the South African anti-cyclone having different positions at various times in the Quaternary (Lancaster 1980, 1981). Comparison of the wind environments of these ergs today and of current active-linear dune fields suggests that greater wind strengths prevailed at the time of their formation. Greater circulation strengths are proposed at the time of the Last Glaciation (e.g. Van Zinderen Bakker 1982) because of lower ocean temperatures (e.g. Hutson 1978) and therefore steeper thermal gradients (Heine 1982). Brookfield (1970) and Bowler (1976) propose similar increases in wind velocities when the Australian dunes were active. Reduced evaporation caused by lower temperatures and stronger winds, accompanied by greater continentality resulting from lower glacial sea levels would have created the necessary arid conditions conducive to dune formation, but other mechanisms need to be sought to explain continental dune formation at other times in the Quaternary.

REFERENCES

Allchin B, Goudie AS and Hegde K 1978. The prehistory and paleogeography of the Great Indian Desert. Academic Press, London. 370 pp.

Bagnold RA 1941. The physics of blown sand and desert dunes. Chapman and Hall, London. 265 pp.

Baillieul TA 1975. A reconnaissance survey of the cover sands of Botswana. J. sedim. Petrol. 45 (2):494-503.

Baker VR 1978. Adjustment of fluvial systems to climate and source terrain in tropical and subtropical environments. Can. Soc. Petrol. Geologists Mem. 5:211-230.

Bonatti E and Gartner S 1973. North Mediterranean climate during the last Würm Glaciation. Nature 209:984.

Bond GW 1948. The direction and origin of the Kalahari Sand of Southern Rhodesia. Geol. Mag. 85:305-313.

Bowler, JM 1976. Aridity in Australia: age, origin and expression in aeolian landforms and sediments. Earth sci. Rev. 12:279-310.

Bowler JM, Hope GS, Jennings JN, Singh G and Walker D 1976. Late Quaternary climates of Australia and New Guinea. Quat. Res. 6: 339-394.

Breed CS and Breed WJ 1979. Dunes and other windforms of central Australia (and a comparison with linear dunes on the Moenkopi Plateau, Arizona). In: F El Baz and DM Arner (eds.), Apollo-Soyuz Tedt Project Summary Science Report V2 earth observations and photography. U S NASA SP-112. 692 pp.

Brookfield M 1970. Dune trends and wind regime in central Australia. Z. Geomorph. Suppl. 10:121-158.

Cahan L and Lepersonne J 1952. Equivalence entre le système du Kalahari du Congo Belge et les Kalahari Beds d'Afrique australe. Mem. Soc. Belg. Palaeont. Hydrol. 8 (4):1-64.

Churchward HM 1963. Soil studies of Swan Hill, Victoria, Australia, IV Ground surface history and its expression in the array of soils. Aust. J. Soil Res. 1:242-255.

Cooke HBS 1964. The Pleistocene environment in southern Africa. In: DHS Davis (ed.), Ecological studies in southern Africa. Junk, The Hague p. 1-23.

Damuth JE and Fairbridge RW 1970. Arkosic sands of the last glacial stage in the tropical Atlantic off Brazil. Geol. Soc. Am. Bull. 81:189-206.

De Dapper M 1979. The microrelief of the sandcovered plateaux near Kolwezi (Shaba-Zaire). 1. The microrelief of the over-all dilungu. Geo-Eco-Trop. 3 (1):1-18.

De Dapper M 1981. The microrelief of the sandcovered plateaux near Kolwezi (Shaba-Zaire). 2. The microrelief of the crest dilungu. Geo-Eco-Trop. 5 (1):1-12.

Fairbridge RW 1964. African ice-age aridity. In: A E M Nairn (ed.), Problems in Paleoclimatology. Interscience, London. p. 356-363.

Flint RF and Bond G 1968. Pleistocene sand ridges and pans in western Rhodesia. Geol. Soc. Am. Bull. 79:299-314.

Fryberger SG 1979. Dune forms and wind regime. In: D McKee (ed.), A study of global sand seas, US Dept. Int. Geol. Surv. Prof. Pap. 1052:137-169.

Galloway RW 1965. Late Quaternary climates in Australia. J. Geol. 73:603-618.

Gill ED 1973. Second list of radiocarbon dates on samples from Victoria, Australia. Proc. R. Soc. Vic. 86:133-136.

Glassford DK and Killigrew LP 1976. Evidence for Quaternary westward extension of the Australian Desert into southwestern Australia. Search 7:394-395.

Glennie KW 1970. Desert sedimentary environments. Elsevier, Amsterdam. 222 pp.

Goudie AS 1969. Statistical laws and dune ridges in southern Africa. Geogr. J. 135 (3):404-406.

Goudie AS 1970. Notes on some major dune types in southern Africa. S. Afr. geogr. J. 52:93-101.

Goudie AS 1983. The arid earth. In: RAM Gardner and H Scoging (eds.), Megageomorphology. Oxford University Press, Oxford (in press). p. 152-171.

Grey DRC and Cooke HJ 1977. Some problems in the Quaternary evolution of the landforms of northern Botswana. Catena 4:123-133.

Grove AT 1958. The ancient erg of Hausaland, and similar formations on the south side of the Sahara. Geogr. J. 124:528-533.

Grove AT 1969. Landforms and climatic change in the Kalahari and Ngamiland. Geogr. J. 135:192-212.

Heine K 1982. The main stages of the late Quaternary evolution of the Kalahari region, southern Africa. Palaeocology of Africa 15:53-76.

Hutson WH 1978. Application of transfer functions to Indian Ocean planktonic foraminifera. Quat. Res. 9:87-112.

Jaeger F 1939. Die Trockenseen der Erde. Petermann's Mitt. Ergänzungsheft 236. 159 pp.

Jennings JN 1968. A revised map of the desert dunes of Australia. Aust. Geogr. 10:408-409.

Jennings JN 1975. Desert dunes and estuarine fill in the Fitzroy estuary (N W Australia). Catena 2:215-262.

Jessup RW 1968. Soil development in coastal South Australia in relation to glacioeustatic changes in sea level. Trans. 9th Int. Congr. Soil Sci., Adelaide, 1968. 4:641-649.

King LC 1962. The morphology of the earth. Oliver and Boyd, Edinburgh and London. 699 pp.

Klammer G 1982. Die Palaeowüste des Pantannal von Mato Grosso und die pleistozäne Klimageschichte der brasilianischen Randtropen. Z. Geomorph. 26:393-416.

Lancaster IN 1979. Quaternary environments in the arid zone of southern Africa. Dept. Geog. Environ. Studies, Univ. Witwatersrand. Occ. Pap. 22. 73 pp.

Lancaster N 1980. Dune systems and palaeoenvironments in southern Africa. Palaeont. Afr. 23:185-189.

Lancaster N 1981. Palaeoenvironmental implications of fixed dune systems in southern Africa. Palaeogr. Palaeoclim. Palaeoec. 33: 327-346.

Mabbutt JA 1957. Physiographic evidence for the age of the Kalahari Sands south west Kalahari. In: JD Clark and S Coles (eds.), Prehistory: Third Pan African Congress, Livingstone 1955 (London).

Mabbutt JA 1967. Denudation chronology in central Australia. In: JN Jennings and JA Mabbutt (eds.), Landform studies from Australia and New Guinea. A N U Press, Canberra. p. 144-181.

Mabbutt JA 1968. Aeolian landforms in central Australia. Aust. geogr. Stud. 6:139-150.

Madigan CT 1936. The Australian sand ridge deserts. Geogr. Rev. 26: 205-227.

Mallick DIJ, Habgood F and Skinner AC 1981. A geological interpretation of Landsat imagery and air photography of Botswana. Inst. of geol. Sciences, Overseas Geol. and Miner. Resour. 56. 35 pp.

Maufe HB 1930. Changes in climate in Southern Rhodesia during late geological times. S. Afr. geogr. J. 13:12-16.

Maufe HB 1935. Some factors in the geographical evolution of Southern Rhodesia and neighbouring countries. S. Afr. geogr. J. 18:3-21.

Putzer R 1958. Quartäre Krustenbildungen im tropischen Süd-Amerika. Geologisches Jb. 76:37-52.

Putzer H 1962. Geologie von Paraguay. Gebrüder Borntraeger, Berlin. 184 pp.

Rogers AW 1934. The build of the Kalahari. S. Afr. geogr. J. 17: 3-12.

Sarnthein M and Diester-Haas L 1977. Eolian sand turbidites. J. sedim. Petrol 47:868-890.

Savory BM 1966. Sand of Kalahari type in Sesheke district, Northern Rhodesia. In: GJ Snowball (ed.), Science and medicine in central Africa. Pergamon Press, Oxford. p. 189-200.

Sprigg RC 1963. Geology and petroleum prospects of the Simpson Desert. Trans. R. Soc. S. Aust. 86:35-60.

Thomas DSG, in press 1. Evidence of Quaternary palaeoclimates in western Zimbabwe - a preliminary assessment. Proc. southern Afr. Conf. Commonw. geogr. Bureau, Lusaka 1982.

Thomas DSG, in press 2. Ancient ergs of the former arid zones of Zimbabwe, Zambia and Angola. Trans. Inst. Br. Geogr.

Tricart J 1969. Actions éoliennes dans la Pampa deprimada. Revue Geomorph. dyn. 19 (4):178-189.

Tricart J 1974. Existence de periodes seches au Quaternaire en Amazonie et dans les regions voisines. Revue Geomorph. dyn. 23: 145-158.

Twidale CR 1972. Evolution of sand dunes in the Simpson Desert, central Australia. Trans. Inst. Br. Geogr. 56:77-106.

Van Andel TH, Heath GR, Moore TC and McGeary DFR 1967. Late Quaternary history, climate and oceanography of the Timor Sea, north western Australia. Am. J. Sci. 265:737-758.

Van Zinderen Bakker EM 1982. African Palaeoenvironments 18 000 BP. Palaeoecology of Africa 15:77-99.

Verboom WC 1974. The Barotse loose sands of the Western Province, Zambia. Zamb. Geogr. Ass. Mag. 27:13-17.

Warren A 1970. Dune trends and their implications on the central Sudan. Z. Geomorph. Suppl. 10:154-180.

Williams GE 1973. Late Quaternary piedmont sedimentation, soil formation and palaeoclimates in arid South Australia. Z. Geomorph. 17: 102-125.

Williams GJ, in press. A preliminary Landsat interpretation of the relict landforms of Western Zambia. Proc. southern Afr. Conf. Commonw. geogr. Bureau, Lusaka 1982.

Wopfner H and Twidale CR 1967. Geomorphological history of the Lake Eyre Basin. In: JN Jennings and JA Mabbutt (eds.), Landform studies from Australia and New Guinea. A N U Press, Canberra. p. 48-143.

Late Quaternary palaeoenvironments in the desert dunefields of Australia

R.J.WASSON
CSIRO, Canberra, Australia

ABSTRACT. Stratigraphic data from the extensive desert dunefields of Australia show that the last major phase of dune construction was between c. 25 000 and 13 000 BP with most radiocarbon dates lying between 20 000 and 16 000 BP. Late Holocene sand accumulation can also be identified on almost all dunes investigated, dating from between c. 3000 and 1000 BP. In the most arid region these Late Holocene sediments are still accumulating.

The Australian interior experienced lowered temperatures and precipitation during both periods of dune construction, the former more severely than the latter. An examination of the mobility threshold of dunes, judged from the amount of speed-up of wind required to mobilize lunettes under present conditions, showed that wind rather than vegetation is the limit on dune mobility. It is argued, therefore, that winds were stronger during at least the earlier dune building phase thereby increasing evapotranspiration and advection of heat at a time when seasonal contrasts may have been much greater.

There is an apparent conflict between inferences drawn from the pollen record about the degree of change of climate in the Mid- to Late Holocene and the existence of dunes of Late Holocene age. One solution lies in the perturbing effects of extreme events occurring during a period when recovery times of vegetation were increasing because of decreasing rainfall.

INTRODUCTION

The history of desert dune accumulation (aeolian sediment) and non-accumulation (palaeosols) during the last 30 000 years in Australia is reviewed in this paper. From an assessment of the factors responsible for dune construction an attempt is made to reconstruct some climatic variables for the major periods of dune accumulation : c. 25 000 to 13 000 and c. 3000 to 0 BP. One important conclusion to emerge is that the Late Holocene phase of dune accumulation differs radically from the earlier glacial phase and the two are only analogous in parts of the dunefields.

STRATIGRAPHIC SUMMARY

The following summary is organized according to area, with major locations shown on Fig. 1.

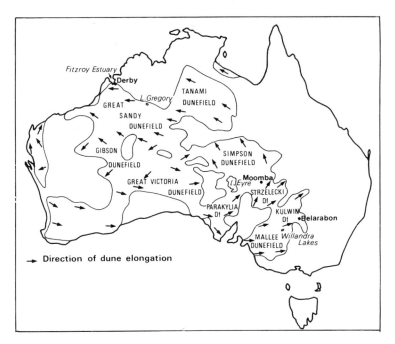

Figure 1. Boundaries of dunefields, average directions of dune elongation judged from y-junctions and crestal orientations, and dunefield names.

1. Southeastern Australia

Mallee Dunefield

Clayey sand units separated by calcareous palaeosols within longitudinal dunes of this area have been recognized by Hills (1939), described in detail by Churchward (1961) and dated by Bowler and Polach (1971). Radiocarbon evidence on almost certainly recrystallized soil carbonate shows that the last major phase of dune accumulation occurred before 15 500 BP at Nyah West. These longitudinal dunes are short, rounded in cross-profile and lack y-junctions. Clay occurs as sand-size aggregates (Bowler and Magee 1978) that strongly resemble clay pellets of the kind found in clay lunettes (Macumber 1970; Bowler 1973). All this evidence shows that these longitudinal dunes were never very mobile and, by analogy with coastal saline lagoons, derived their clay pellets and quartz grains by groundwater-controlled deflation. The identity of environmental conditions necessary for both clay lunette construction and the accumulation of these longitudinal dunes of the Mallee prompted Bowler and Magee (1978) to suggest that both dune types formed at the same time, that is, between c. 20 000 and 16 000 BP.

The well-dated sequence of lunette construction and lake-level fluctuations in the Willandra Lakes enabled Bowler (1976) to date the last phase of dune extension in that part of the Mallee by examining the relationships between shorelines, the large subparabolic quartzose dunes, and trim lines on the dunes. The extension and therefore accumulation of dune sand occurred after the high water phase that came to an end c. 25 000 BP (Bowler 1978). Similar dunes have not crossed the Willandra Creek overflow channel in the Willandra Lakes, and this channel has not functioned for 15 000 years. The subparabolic dunes of this area must have extended between 25 000 and 15 000 BP.

Subparabolic dunes near Lake Albacutya in the Mallee dunefield extended onto the western shore of the lake during the last period of lunette construction c. 16 000 BP (Bowler and Magee 1978). In Wyperfeld National Park near Outlet Creek, thermoluminescence dates on sand from the crest of a subparabolic dune indicate that the last phase of accumulation pre-dated 12 000 BP (A Ross and D Price, pers. comm. 1983). At Lake Wirrengren (Pine Plains) subparabolic dunes have not advanced since being trimmed by a high water stand c. 11 000 BP.

Source-bordering dunes along river channels of the Riverine Plain (Fig. 1) have been dated both directly and by stratigraphic association with other dated deposits (Bowler 1978). These dunes were accumulating while the channels were transporting large amounts of sand from before 30 000 BP to c. 15 000 BP. Dune accumulation ceased when the channels began to carry predominantly suspended load, a channel change possibly reflecting Post-glacial re-afforestation of the catchment of the streams. While the formation of these dunes was facilitated by the type of sediment carried in their source channels, the alluvium itself was determined by climatic conditions. Hence, the termination of dune accumulation is approximately synchronous with that determined for other dunes in the Mallee but the source-bordering dunes began to form much earlier than the regionally distributed longitudinal and subparabolic dunes; that is, before c. 25 000 BP.

Kulwin Dunefield

This area extends north of the Riverine Plain up the Darling River valley and onto the western side of the Cobar Plain (Fig. 1).

Radiocarbon dates of c. 15 000 BP on mussel shells within dune sand date source-bordering dunes near Menindee on the Darling River (J Hope, pers. comm.). Bowler et al. (1978) dated dune construction on the Darling flood plain near Tilpa by reference to radiocarbon dates from stratigraphically older or younger alluvium - the dunes accumulated between c. 20 000 and 16 000 BP.

An extensive area of longitudinal dunes on the Cobar Plain has been investigated at Belarabon Station (Wasson 1976). Prior streams last deposited alluvium between c. 6500 and 2500 BP, towards the end of which period dunes began to accumulate along the edges of the channels. The downwind ends of the dunes had all but ceased to extend by c. 600 BP, an age known from radiocarbon dating of charcoal buried in slopewash veneers on the flanks of dunes.

Southern South Australia

By radiocarbon dating sediments both older and younger than longitudinal dunes on the piedmont west of the Flinders Ranges, Williams (1973) placed dune building between c. 24 000 and 16 000 BP. The crests of these dunes consist of 1-2 m of moderately consolidated red sand containing Aboriginal artefacts of the 'small tool tradition', dating this dune sand to within the last 5000 years (PJ Hughes, pers. comm. 1982). The upper 2 m of dunes in the Parakylia Dunefield (Fig. 1) also contain microliths, overlying an undated calcareous palaeosol developed in dune sand.

2. Central Australia

Strzelecki Dunefield

Callen et al. (1983) have shown that radiocarbon dates determined for pedogenic carbonates segregated within units of dune sand east of Lake Frome are minimum estimates of the ages of dune stabilization. The last major phase of dune accumulation ceased well before 10 000 BP in this area. Bowler (1976) showed that the gypsum and clay-rich islands of Lake Frome are aeolian accumulations which formed between 20 000 and 16 000 BP under groundwater conditions such as those which prevailed at the same time at the Willandra Lakes. It seems likely that some of the longitudinal dunes east of the lake were constructed coevally with the islands.

In the flood basin of Cooper Creek in the northern Strzelecki Dunefield, two phases of dune construction have been dated in both longitudinal and transverse dunes. Near Strzelecki Creek, due east of Moomba camp, low and rounded longitudinal dunes stand in contrast to the much higher well-formed longitudinal dunes typical of this part of the dunefield. Charcoal bedded in the upper part of one of the low dunes yielded two dates of c. 2300 BP. Some of the downwind ends of the well-developed dunes consist of sands similar to those dated, suggesting dune extension during the Late Holocene.

The longitudinal dunes away from the creek have a cap of 1-2 m of well-bedded and periodically mobile sand. Near Moomba, organic remains at the base of the cap have 116 % modern C14 activity. Further south, near Moolawatana Bore, the equivalent young dune sands have yielded charcoal dated to between 1700 and 1000 BP. Substantial amounts of sand have accumulated on these dunes in the latest Holocene.

Beneath the Late Holocene cap lies a weakly developed calcareous palaeosol which is formed within a unit of dune sand that itself overlies another palaeosol. The youngest carbonate has been dated to 7600 BP near Moomba, giving an age comparable with the youngest palaeosol near Lake Frome. Charcoal within the upper part of this dune sand unit and below the 7600 BP palaeosol, at a site south of Moomba, was dated to 13 900 and 13 200 BP (Wasson 1983). The 7600 BP date is a minimum age for the dune sand, as was noted earlier. These longitudinal and transverse dunes ceased accumulation between 14 000 and 13 000 BP.

The pre-13 000 years old unit consists of quartz sand and up to 60 % sand-size clay pellets of the kind found in both clay lunettes and Mallee-type longitudinal dunes. The hydrologic conditions necessary for the production of these pellets in the northern Strzelecki were met by high water tables sustained by Cooper Creek flooding sandy mud onto the flats that separate the dunes.

The modern flood regine dominantly affects the lowest parts of the floodflats. The slightly higher parts consist of muddy fine sand up to 2 m thick overlying clean fine sandy alluvium. The muddy fine sand has yielded a basal C14 date of 22 300 BP while the top dates at 12 500 BP (Wasson 1983). The higher parts of the floodflats therefore were abandoned by frequent floods at c. 12 500 BP and the area of deposition contracted. This is confirmed by a date of c. 12 000 BP from the muddy fine sand that overlies channel sands in the centre of a floodflat north of Moomba. This flat continues to accumulate intermittently.

The basal sandy deposits are remarkably uniform over large areas suggesting that deposition was by a braided channel system. The change from this regime to deposition of muddy sand at c. 22 000 BP produced the conditions necessary for clay pellet formation. The finer alluvium was the overbank load of a series of meandering channels which had sandy bedloads. At c. 12 500 BP these channels became defunct, at about the same time as the dunes stopped accumulating. Alluvial deposition was then restricted to the lowest parts of the floodflats from which sediment, including clay pellets, was deflated in the Late Holocene.

The latest phase of accumulation is still going on and involves the reworking of older dune sands and their movement along the dunes. Like the Mallee dunes, the high water tables and salinization of swales are not part of the present landscape.

The dunes so far described from the Strzelecki Dunefield are all pale coloured and, as we have seen, required particular conditions for their formation. To the east of Strzelecki Creek lies a large area of red-brown quartzose dunes in which pellets are absent. These dunes have not yielded charcoal for dating, but dunes along Strzelecki Creek that are sedimentologically transitional between the pelletal and quartzose dunes provided a means of correlation between the two types of dune. The red-brown dunes have caps of periodically mobile sand that obviously correlate with the Late Holocene caps on the pale dunes. Beneath the red-brown caps lie up to 3 m of slightly consolidated almost carbonate free sand which appears to correlate with the dune unit dating 23 000 to 13 000 BP west of Strzelecki Creek. Below this unit are various palaeosols in dune sand, yielding apparent radiocarbon dates of c. 20 000 BP that are clearly minima.

Simpson Dunefield

This large area of predominantly longitudinal dunes can be subdivided into areas of pellet-rich pale dunes in the lowest areas, and pellet-free red-brown dunes on the slightly higher ground (Wasson 1983a, b). In the southern part of the pale dunes, near Lake Eyre, DLG Williams (pers. comm.) has obtained a date of c. 19 000 BP on emu egg shell from the upper part of a longitudinal dune. A date of 24 000 BP on bone from alluvium lying beneath dunes just east of Lake Eyre (Twidale 1972) is hard to accept as a maximum age for dune construction because of the problems attached to radiocarbon analyses of bone.

In the red-brown dunes of the central Simpson, a date of c. 7000 BP has been obtained for soil carbonate in the sand unit lying beneath the mobile cap. Quartzose dunes adjacent to Oolgawa Waterhole, on the western edge of the Dunefield, contain charcoal dated between c. 2200 and c. 2800 BP in the upper 8 m of carbonate-free sand lying beneath the well-bedded cap. At Allapallilla Waterhole, in the same area, a date of c. 750 BP supports the notion that in the southwestern Simpson the slightly consolidated sands lying beneath the mobile caps and overlying calcareous palaeosols on quartzose dunes are Late Holocene additions.

Western Australia

Well-organized longitudinal dunes of the Great Sandy Desert lie outside the former extensions of the former expanded Lake Gregory. Lower and more closely spaced dunes, within the area of this once extensive lake, were formed by the redistribution of the A horizon of pedologically altered lake sediments, The dunes post-date lacustrine molluscs with ages between c. 25 000 and 20 000 BP (JM Bowler, pers. comm.).

Jennings (1975) determined a minimum age of 8000 BP for longitudinal dunes that are partially buried by estuarine muds in the Fitzroy Estuary. The age is based upon both radiocarbon dates and the depth below present sea-level of dune sand relative to dated sea-level curves.

CONCLUSIONS FROM STRATIGRAPHIC SUMMARY

The evidence summarized above shows that the last major phase of dune construction was between c. 25 000 and 13 000 BP, with most dates lying between 20 000 and 16 000 BP. This chronology applies in the presently semi-arid Mallee and in the arid 'pelletal' Strzelecki Dunefield. In the area of clay pellet-rich dunes in the Strzelecki, the end of dune accumulation was controlled in part by a hydrologic change experienced by Cooper Creek. In the quartzose dunes in both the Strzelecki and Simpson, it seems that dune accumulation may have extended beyond 13 000 BP, perhaps intermittently till the present. Late Holocene sand accumulations can be identified on almost all dunes investigated. In most areas the Late Holocene was a time of reworking of older sediments but at Belarabon new dunes were built (Wasson 1976).

The present semi-arid zone records periods of both accumulation and stability in dunes, while the present arid zone records both periods of non-accumulation, when dunes were intimately associated with groundwater, and intermittent accumulation to the present, when groundwater has been unimportant.

The period of dune accumulation between 25 000 and 13 000 BP straddled the Last Glacial Maximum taken to be 18 000 BP (CLIMAP 1976), thereby showing broad synchroneity with other mid-latitude deserts (Sarnthein 1978). The Late Holocene dune sand accumulations date from c. 3000 to 1000 BP and in some cases they continue to form, post-dating the Climatic Optimum. We have sufficient knowledge of the palaeoenvironments of these time periods to state, at the broadest level, that the Last Glacial Maximum was, in mid-latitudes, a time of lowered precipitation, lowered annual temperatures and stressed vegetation, resulting in increased 'aridity'. Regarding the Late Holocene there is less unanimity among workers, but in eastern Australia this was a time when lake levels and possibly annual temperatures fell from their peak levels of the Mid-Holocene (Thom and Wasson 1982; Chappell and Grindrod 1983). It is clear that the Late Holocene was not as 'arid' as the Last Glacial Maximum. But these statements are very general and now have reasonably wide currency amongst Quaternarists. In what follows I shall attempt to define the palaeoenvironments of the Australian dunefields with greater precision, beginning with modern conditions and then examining palaeoenvironmental evidence from both Australia and elsewhere.

MODERN CONDITIONS IN THE DUNEFIELDS AND A MOBILITY THRESHOLD

Mobile sand is not common on Australian dunes, the scene in a dunefield being dominated by shrubby vegetation. There is a gradient of vegetation cover versus mobile sand from the semi-arid to the arid parts of the dunefields. There is no sharp break between active and inactive dunes, the slope of the gradient changing with the vagaries of rainfall and fire. A new approach to the analysis of wind/vegetation/topography interactions (Ash and Wasson 1983) has shown that the vegetation is not a major limit to modern sand movement, except in the semi-arid zone and in the southern winter rainfall zone of mallee eucalypts. It appears that an increase in windiness, expressed as the percentage of days experiencing winds above the threshold of sand movement, would mobilize most dunes.

A compilation by Bowler and Wasson (this volume) of vector resultants of modern sand-shifting winds (from Brookfield 1970; Jennings 1975; Fryberger 1979; Sprigg 1982; Hyde and Wasson 1983; and data analysed by the author) shows that the resultants diverge markedly only from the longitudinal dune trends in the Mallee, southern Strzelecki and the coastal part of the Great Sandy Dunefield. It must be concluded that the Last Glacial dune-forming resultants did not diverge significantly from modern winds.

The dunes form an anti-clockwise whorl (Jennings 1968) that superficially resembles the mean winter anticyclone. Brookfield's (1970) careful analysis demonstrated that this correlation is not sound and concluded that most dune-forming winds blow during the seasonal latitudinal migration of the anticyclones. During summer the mid-latitude anticyclones move eastwards with their mean axes near the southern coast. This produces southernly to southeasterly sand-moving winds over the Strzelecki and southern Simpson, and southwesterly winds over the Mallee. In the northern Simpson, Tanami and Great Sandy Dunefields the outflow from the anticyclones is from the south-east to east. The winds in the northern dunefields are often interrupted by northwesterlies as the ITCZ periodically sweeps southwards to bring rain in monsoonal disturbances and cyclones. As winter approaches the mid-latitude anticyclones move northwards, the steeper pressure gradients and strongest winds penetrating progressively from the southern coast. In the Mallee the sand-moving winds become westerly, with strong northwesterlies before cold fronts and southwesterlies after cold fronts. Over the central dunefields the dune-forming winds are more southwesterly and in the north the trade winds are again either southeasterly or easterly.

The modern pattern of sand-moving winds is a composite of the major climatological influences: the seasonal migration of the mid-latitude anticyclones with their intervening cold fronts, the seasonal movement of the ITCZ, and tropical cyclones. There is a tendency for most sand movement to occur in winter in the north and in the summer in the south, in anticorrelation with the wettest season.

While vegetation is an important element on Australian desert dunes, it is clear from the analysis of Ash and Wasson (1983) and the very recent radiocarbon dates for the mobile caps of dunes, that many dunes in the arid zone still accumulate sand periodically on both their crests and flanks. The periodicity of sand movement is controlled by both disturbances of the vegetation and periods of high winds. The effect on potential sand movement of rainfall variability is illustrated in a hypothetical example in Fig. 2. Two widely spaced wet/dry cycles occurred within the first 16 years, producing destruction and gradual regrowth of vegetation. The recovery of the vegetation affects potential sand discharge (q) according to a relationship derived by Ash and Wasson (1983, Fig. 6), and this has been used in constructing Fig. 2. Following the two long-lived cycles are a series of closely spaced lower amplitude wet-dry cycles signifying greater variability, the effect of which is to increase (q)some four times.

This highly idealized example shows that a more variable rainfall regime can have a dramatic effect on sand movement, especially if extreme wind gusts occur during dry times. The role of fire is less certain but can aid sand movement if followed by strong winds (Conacher 1971). The ideal concatenation of dry periods (and/or fire) and strong winds dramatically reduces the recurrence interval of major sand-moving

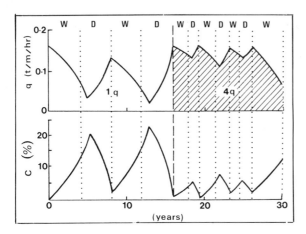

Figure 2. Schematic variation of the potential sand discharge (q) with vegetation cover (C) during wet (W) and dry (D) periods.

episodes but this reduction may be counteracted by increased variability.

Under the present conditions, there is no evidence of clay pellet formation in the dunefields. The closely adjusted groundwater conditions and evaporative concentration of salts in swale surfaces no longer occurs. Groundwater is tens of metres below the surface, salts are no longer concentrated (except as a result of modern irrigation practices) and vegetation grows contentedly between and on the dunes. In areas of both pellet-rich dunes and quartzose dunes of the arid zone, dunes are mobilized by strong winds, by winds during droughts or after fire, as indicated above. These periods of reactivation may only last a year (perhaps a season) or two and in the areas of pellet-rich dunes do not resemble the conditions of 25 000 - 13 000 BP (see Bowler and Wasson, this volume). In the quartzose dunes these modern periods of re-mobilization more closely resemble those of the Last Glacial Maximum.

The gradient of vegetation and percentage of mobile sand, on dunes samples in the semi-arid to arid zone, makes the establishment of a mobility threshold rather difficult. Where wind is accelerated up transverse dune (lunette) flanks, the fractional speed-up ratio for dunes with mobile sand on the crest can be used to estimate the mobility threshold necessary to remobilize all the longitudinal desert dunes (Ash and Wasson 1983). The original mobility index was expressed as a function of windiness and a measure of average annual vegetation growth potential E_a/E_p, where E_a is actual evapotranspiration and E_p is potential evapotranspiration.

A better and more readily evaluated relationship, and one that is more applicable for palaeoenvironmental reconstruction, has been derived:

$$M = 0.21 (0.13W + \ln E_p/P) (\geq 1)$$

where P is annual average precipitation, E_p is annual potential evapotranspiration (Ceplecha 1971), W is % of days experiencing sand-shifting winds (P in Ash and Wasson), and the mobility index M indicates mobile dunes if it exceeds unity. A graph of this relationship is

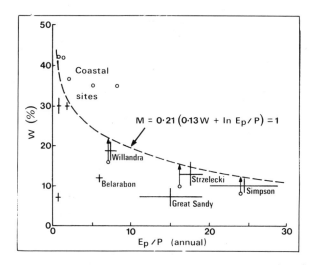

Figure 3. Dune mobility threshold expressed in terms of relative vegetation growth potential (Ep/P) and windiness (W). The arrowed lines indicate the minimum change of W necessary to mobilize dunes judged from speed-up calculations on lunettes.

given in Fig. 3, with data extending from coastal sites to the driest part of the continent in the Strzelecki and Simpson deserts.

On this diagram acceleration of wind up lunettes is shown as an open circle with an arrow and these values for acceleration, along with some from coastal sites, have been used to derive the curve given by the mobility threshold equation. Dunes and almost certainly swales are remobilized at $M > 1$ in areas where deflation is not controlled by groundwater.

It is possible that W and/or Ep/P varied in numerous ways to mobilize dunes. The combinations of changes in these variables that will result in $M = 1$ are shown in Fig. 4. With no change in W the maximum change occurs in Ep/P, and vice versa. Each of the curves in this diagram is derived from a point within the range of values for the variables given in Fig. 3 but as far from $M = 1$ as possible. In the case of the Simpson Dunefield the point of reference is $W = 8\%$ and $Ep/P = 20$.

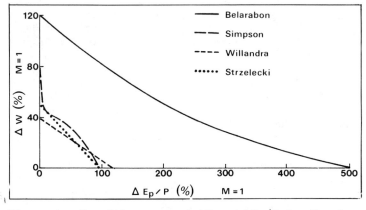

Figure 4. Possible range of ΔW and $\Delta Ep/P$ for $M = 1$.

Parts of both the Strzelecki and Simpson Dunefields have modern values of $M > 1$ and so most of the dunes in these areas are close to the stability threshold. The Mallee dunes of the Willandra area are also close to $M = 1$ but the Late Holocene dunes at Belarabon are distant from the mobility threshold. Values in W and Ep/P need to be little different from those of today to remobilize the Strzelecki and Simpson but significant changes are required at Belarabon. The degree of this change is not in line with evidence for only slight aridification during the Late Holocene relative to the present (Thom and Wasson 1982).

One origin of this discrepancy probably lies in the use of annual values of P and EP in the mobility index. As seen earlier, variations on time scales of years are critical in sand mobility, and seasonal variations may also be important. The conditions that prevailed at Belarabon between 2500 and 600 BP (Wasson 1976) may have been more variable than they are now. One prolonged drought in a time of strong winds and slightly lowered annual average rainfall may allow the growth of the initial instability of deflated surfaces. Feedback between sand mobility and plant seedling death may lead to exponential growth of the instability in the way described by Scheidegger (1983). But for this form of remobilization the average value of M must be greater than unity.

PALAEOENVIRONMENTS

From the discussion of the mobility threshold in the preceding section it is evident that increases in W and/or Ep/P (Fig. 3) will remobilize dunes. This statement applies to quartzose dunes in particular but is applicable also to the reworking of pellet-rich dunes. The supply of sediment from swales is not adequately addressed by a diagram such as that in Fig. 3 but it seems likely that sandy swales would be mobilized under conditions of $M \geqslant 1$. More poorly sorted swales would offer more resistance to deflation so that $M \gg 1$ would be necessary.

The supply of sediment to the clay pellet-rich dunes has been discussed by Bowler and Magee (1978) and Wasson (1983). On the analogy of pellet formation in coastal lagoons (Bowler 1973) the necessary conditions for pellet formation are regionally high water tables that bring salts into surface layers of muddy sand, high evaporation rates to concentrate the salts and thereby pelletize the sediment as the salts crystallize, and finally strong winds to break the pellets from their neighbours and carry them to nearby dunes. In clay lunettes the bedding of pellets is low-angle, presumably because the clays are hygroscopic and not highly mobile in locations where either dew or rain is moderately frequent (Bowler 1973). The low-angle flanks of the Mallee dunes suggest that the pellets were similarly immobilized in these longitudinal dunes. While the pellet-rich layers in dunes of the Strzelecki and Simpson Dunefields are low angle, many avalanche sets have up to 20 % pellets. The atmosphere in these regions during the Last Glacial dune building phase was presumably much drier than in the Mallee.

Global annual average temperatures were depressed during the Last Glacial Maximum and so from the curve relating saturation vapour pressure to temperature we can conclude that annual average evaporation potential was lower than modern values (Manabe and Hahn 1977). This conclusion contradicts the argument for high evaporation rates apparently needed to keep the lakes dry and form evaporatic salt minerals (Bowler and Wasson, this volume).

Two changes seem to have occurred to increase evaporation potential in the interior of Australia at the Last Glacial Maximum. Abundant evidence now exists for higher wind speeds, largely as a consequence of steeper meridional atmospheric temperature gradients (Newell et al. 1981; Petit et al. 1981; Sarnthein et al. 1981; Ash and Wasson 1983). Both mid-latitude westerlies and trades were increased but the strength of monsoon winds was probably reduced (Wasson et al. 1983). The degree of change in the westerlies and trades can be estimated very generally from an equation given by Newell et al. (1981). From estimates of sea surface temperature for 18 000 years ago and estimates of polar temperatures, Newell et al. calculated increases in wind speeds of 24 % in the northern winter, 124 % in the norther summer, and 17 % in both southern seasons. These authors argued that increased wind speeds can explain a large part of the sea surface temperature distribution of the Ice Age oceans, suggesting that change in wind climate was a pervasive force.

The second factor that may have increased evaporation at the Glacial Maximum is radiation. From Berger's (1978) calculations of solar radiation variations, following Milankovitch, Budd (in Chappel and Grindrod 1983) has shown that January radiation values were at least as high as those of today and may have been up to 1 % higher. The important role of summer radiation for evaporation is clear from monthly evaporation figures from any inland Australian station (Bowler and Wasson, this volume). The effect of cloud cover at the Glacial Maximum is hard to determine but it seems likely that average cloud cover in the middle latitudes was no greater than at present.

At 18 000 BP it seems that annual temperatures were depressed by as much as $6°C$ (Chappel and Grindrod 1983) but, as a result of high insolation, summer temperatures may have been as high as they are today (Bowler and Wasson, this volume). Wind speeds were as much as 30 % greater than at present (Ash and Wasson 1983).

From these estimates of important parameters it is possible to infer the effect of their change on potential evaporation, using the terms in the Penman (1963) equation. Quantitative estimates of E_p at various times during the last 30 000 BP are not possible because some of the Penman parameters cannot be specified. From the equation we can, however, deduce that holding all else constant, a $5°C$ annual average temperature fall will decrease E_p by about 4 %. Similarly, an increase of available radiant energy by 5 % will increase E_p by about 4 %, without adjusting for albedo changes.

Even if the errors in these estimates were 100 %, it still seems that the combination of the annual temperature decline and high January radiation at 18 000 BP would have had little net effect on E_p. This assumes that a dustier atmosphere had little effect on radiation at the ground. Changes in wind speed remain the only significant influence on E_p. Taking as an example the Willandra area, the present mean wind speed is 4 ms^{-1} and the maximum increase in W necessary to obtain M = 1 is 48 % (Fig. 4). This increase in W would produce an increase in E_p of 35 % annually if the radiation and temperature terms in the Penman equation cancelled each other in the way suggested. The other effect of increased wind speed would have intensified advection of heat, increasing E_p even further.

None of this discussion has treated change in precipitation adequately, and unfortunately there are no long pollen records from the interior to assist estimates of P. From more mesic sites, however, the results summarized in Chappel and Grindrod (1983) show a decline in P at the Glacial Maximum in eastern Australia. The modern relationship

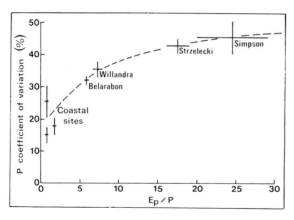

Figure 5. Relationship between Ep/P and coefficient of variation of P for modern conditions.

between Ep/P and the coefficient of variation of P is shown in Fig. 5. The increase in variability occurs with a decrease in P and an increase in Ep. We might suppose that while the present quantitative relationship between these variables did not apply at 18 000 BP, the sense of change was the same so that as the interior dried out from 30 000 BP to 18 000 BP so the variability of P increased. From Fig. 2 and the earlier discussion this should have had a considerable impact on dune mobility, both at the Glacial Maximum and during the Late Holocene when P declined also.

The changes in climatic variables sketched in the preceding discussion all enhanced dune accumulation and mobility. Increased wind speeds moved sand more readily and probably subdued seedling plants. The stronger winds also increased Ep and advection of heat at a time of reduced P, greater seasonal contrast in temperatures with hot summers and cold winters, and high values of radiant energy in summer. This must have been a vigorous climate for plants, animals, and man, but it was ideal for dune building.

REFERENCES

Ash JE and Wasson RJ 1983. Vegetation and sand mobility in the Australian desert dunefield. Z. Geomorph., N.F., Suppl. Bd. 45:7-25.
Berger AL 1978. Long-term variations of caloric insolation resulting from the Earth's orbital elements. Quat. Res. 9:139-167.
Bowler JM 1973. Clay dunes: their occurrence, formation and environmental significance. Earth-sci. Rev. 9:315-338.
Bowler JM 1976. Aridity in Australia: age, origins and expression in aeolian landforms and sediments. Earth-sci. Rev. 12:279-310.
Bowler JM 1978. Glacial age aeolian events at high and low latitudes: a Southern Hemisphere perspective. In: EM van Zinderen Bakker (ed.), Antarctic Glacial History and World Palaeoenvironments. Balkema, Rotterdam. p. 149-172.
Bowler JM and Magee JW 1978. Geomorphology of the Mallee region in semi-arid northern Victoria and western New South Wales. Proc. Roy. Soc. Vict. 90:5-26.

Bowler JM and Polach HA 1971. Radiocarbon analyses of soil carbonates: an evaluation from palaeosols in southeastern Australia. In: D Yaalon (ed.), Palaeopedology - Origin, Nature and Dating of Palaeosols. Int. Soc. Soil Sci. and Israel Univ. Press, Jerusalem. p. 97-108.

Bowler JM and Wasson RJ, this volume. Glacial Age Environments of Inland Australia.

Brookfield M 1970. Dune trends and wind regime in Central Australia. Z. Geomorph., Suppl. Bd. 10:121-153.

Callen RA, Wasson RJ and Gillespie R 1983. Reliability of radiocarbon dating of pedogenic carbonate in the Australian arid zone. Sed. Geol. 35:1-14.

Ceplecha VJ 1971. The distribution of the main components of the water balance in Australia. Aust. Geogr. 11:455-462.

Chappell JMA and Grindrod A 1983. CLIMANZ: a symposium of results and discussions concerned with Late Quaternary climatic history of Australia, New Zealand and surrounding seas. Dept. of Biogeography and Geomorphology, Aust. Nat. Univ. 2 vols.

Churchward HM 1961. Soil studies at Swan Hill, Victoria, Australia. 1. Soil layering. J. Soil Sci. 12:73-86.

CLIMAP project members 1976. The surface of the Ice-Age Earth. Science 191:1131-1137.

Conacher AJ 1971. The significance of vegetation, fire and man in the stabilisation of sand dunes near the Warburton Ranges, central Australia. Earth Sci. J. 5:92-94.

Fryberger SG 1979. Dune forms and wind regime. In: ED McKee (ed.), A Study of Global Sand Seas. U.S. Geol. Surv. Prof. Paper 1052: 137-170.

Hills ES 1939. The physiography of north-western Victoria. Proc. Roy. Soc. Vic. 51:293-320.

Hyde R and Wasson RJ 1983. Radiative and meteorological control on the movement of sand at Lake Mungo, N.S.W. Australia. In: ME Brookfield and TS Ahlbrandt (eds.), Eolian Sediment and Processes. Elsevier, Amsterdam. p. 311-323.

Jennings JN 1968. A revised map of the desert dunes of Australia, Aust. Geographer 10:408-409.

Jennings JN 1975. Desert dunes and estuarine fill in the Fitzroy estuary (north-western Australia). Catena 2:215-262.

Macumber PG 1970. Lunette initiation in the Kerang district. Min. Geol. J. Vic. 6:16-18.

Manabe S and Hahn DG 1977. Simulation of the tropical climate of an Ice Age. J. Geophys. Res. 82: 3889-3911.

Newell RE, Gould-Stewart S and Chung JC 1981. A possible interpretation of paleoclimate reconstructions for 18 000 BP for the region 60°N to 60°S, 60°W to 100°E. In: JA Coetzee and EM van Zinderen Bakker (eds.), Palaeoecol. Africa 13:1-20.

Penman HL 1963. Vegetation and Hydrology. Tech. communication No. 53 Commonwealth Bureau of Soils (Farnham Royal). 124pp.

Petit J-R, Briat M and Royer A 1981. Ice age aerosol content from East Antarctic ice core samples and past wind strength. Nature 293: 391-394.

Sarnthein M 1978. Sand deserts during glacial maximum and climatic optimum. Nature 272:43-46.

Sarthein M, Tetzlaff G, Koopmann B, Walter K and Pflaumann P 1981. Glacial and interglacial wind regimes over the eastern subtropical Atlantic and north-west Africa. Nature 293:193-196.

Scheidegger AE 1983. Instability principle in geomorphic equilibrium. Z. Geomorph.,N.F. 27:1-19.

Sprigg RC 1982. Alternating wind cycles of the Quaternary era and their influences on aeolian sedimentation in and around the dune deserts of southeastern Australia. In: RJ Wasson (ed.), Quaternary Dust Mantles of China, New Zealand and Australia. Dept. Biogeography and Geomorphology, Aust. Nat. Univ. p. 211-240.

Thom BG and Wasson RJ (eds.) 1982. Holocene Research in Australia. Dept. Geography, Univ. New South Wales, Duntroon.

Twidale CR 1972. Evolution of sand dunes in the Simpson Desert, central Australia. Trans. Inst. Br. Geogr. 56:77-109.

Wasson RJ 1976. Holocene aeolian landforms in the Belarabon area, S.W. of Cobar, N.S.W. J. and Proc. Roy. Soc. N.S.W. 109:91-101.

Wasson RJ 1983a. The Cainozoic history of the Strzelecki and Simpson dunefields (Australia), and the origin of the desert dunes. Z. Geomorph., N.F., Suppl. Bd. 45:85-115.

Wasson RJ 1983b. Dune sediment types, sand colour, sediment provenance, and hydrology in the Strzelecki-Simpson dunefield, Australia. In: ME Brookfield and TS Ahlbrandt (eds.), Eolian Sediments and Processes. Elsevier, Amsterdam. p. 165-195.

Wasson RJ, Rajaguru SN, Misra VN, Agrawal DP, Dhir RP, Singhvi AK and Kameswara Rao K 1983. Geomorphology, late Quaternary stratigraphy and palaeoclimatology of the Thar dunefield. Z. Geomorph., N.F., Suppl. Bd. 45:117-152.

Williams GE 1973. Late Quaternary piedmont sedimentation, soil formation and paleoclimate in arid South Australia. Z. Geomorph., N.F. 17: 102-125.

Aridity in southern Africa: Age, origins and expression in landforms and sediments

N.LANCASTER
University of Cape Town, South Africa

ABSTRACT. Today, arid and semi-arid climates have a wide distribution in southern Africa. Late Cenozoic landforms and sediments from the Namib suggest that although the area has experienced periods of increased humidity, these have been of limited extent and duration and that the region has experienced no climate more humid than semi-arid since the end-Miocene. Trends in the character of sediments and landforms suggest that there has been a progressive increase in aridity during the Late Cenozoic. The sedimentary record from the Kalahari is less clear, but similarly implies a long history of arid and semi-arid climates in this region. Thus, palaeolakes record periods of considerably increased humidity, whilst extensive systems of linear dunes, now fixed by vegetation, point to at least three periods when aridity was both more intense and of greater extent than at present.

INTRODUCTION

Today, the climates of southern Africa are dominated by seasonal or permanent aridity. Some 74 % of the area of the subcontinent is classified as arid or semi-arid (Paylore 1979) in a series of concentric zones of increasing aridity centered on the hyperarid Namib Desert along the west coast (Fig. 1). Two main sub-tropical arid regions can be distinguished on the basis of climate and physiography (Fig. 2): the hyperarid Namib Desert along the west coast and the arid to semi-arid Kalahari in the interior.

In recent years, there has been considerable interest in the age and Late Cenozoic history of aridity in southern Africa. Most of this has concentrated upon events in the Namib Desert and has been stimulated by considerations of the character and evolution of its distinctive fauna and flora (Koch 1961, 1962; Van Zinderen Bakker 1975; Seely 1978; Tankard and Rogers 1978; Endrody Younga 1978, 1982). Interest in the subject continues at a high level, as witnessed by the panel discussion at the 1981 SASQUA conference (Vogel, Rogers and Seely 1981) and recent papers by Ward et al. (1983) and Van Zinderen Bakker (in press). In contrast, discussion of palaeoenvironments in the Kalahari has concentrated on the Late Pleistocene and the chronology of periods of humid climates (Heine 1978, 1982; Lancaster 1979a, b).

The aim of this paper is to discuss the evidence contained in landforms and sediments for long-continued and persistent Late Cenozoic aridity in the Namib and the Kalahari.

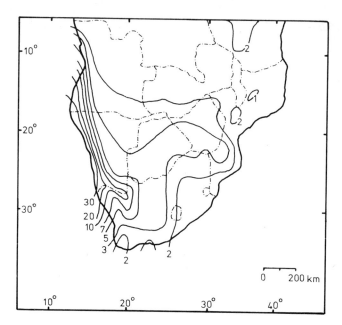

Figure 1. Distribution of arid climates in southern Africa, as shown by isolines of the Budyko-Lettau dryness ratio (the number of times the mean annual net radiation is able to evaporate the mean annual rainfall). Areas with values of 2 - 7 are considered semi-arid, 7 arid and 20 extremely arid.

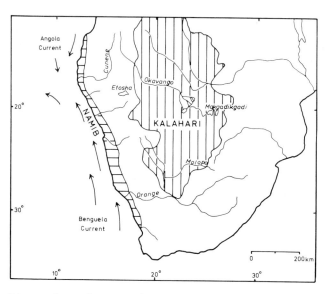

Figure 2. Major physiographic divisions of the subtropical and tropical arid zone in southern Africa.

TERTIARY ARIDITY IN THE NAMIB AND KALAHARI

The Namib

The evidence for the antiquity of aridity in the Namib has been fully discussed by Ward et al. (1983), who put forward the view that the character of most Tertiary sediments in the Namib suggests that they were deposited in arid or semi-arid conditions. The most striking evidence of Tertiary aridity in the region is contained in the thick red brown sandstones of the Tsondab Sandstone Formation and its probable equivalents, which are widely distributed in the central and southern Namib (Besler and Marker 1979; Ward et al. 1983). The sandstones contain both dune and sand sheet structures with large scale cross bedding dipping towards the northeast and northwest, implying deposition by winds from southerly directions (Besler and Marker 1979; Ward et al. 1983). They represent the accumulation of a major sand sea in the central and southern Namib over a period of some 20 to 30 m.y., prior to the Mid- to Late Miocene.

Although the Lower to Mid-Miocene fluvial gravels of the southern Namib appear to indicate a more mesic climate and savanna vegetation 12 to 18 m.y. ago (Corvinus and Hendey 1978), the sediments suggest deposition by seasonal or ephemeral streams (Ward et al. 1983). Further, in the central Namib at least, the upper members of the Tsondab Sandstone Formation were accumulating at the same time as gravels of apparently similar age to those which contain the Arrisdrift and Luderitz faunas were being laid down in the adjacent valleys (Yaalon and Ward 1982).

The Miocene fluvial gravels and the upper parts of the Tsondab Sandstone Formation and its equivalents were extensively calcreted during the end Miocene (Martin 1950). These thick pedogenic calcretes were interpreted by Yaalon and Ward (1982) as indicating some 500 000 years of landscape stability and a semi-arid climate with a summer rainfall of 350 to 450 mm. However, this evidence for more mesic climates in the Later Miocene represents an interval of no more than semi-aridity in the region and it appears that aeolian sand accumulation continued throughout the period, albeit possibly at a reduced level.

It appears that during the Pliocene, a climate of modern affinities was developing in the region. Seisser (1978, 1980) reports a variety of sedimentary, palaeontological and geochemical data from DSDP cores from the Walvis Ridge, which indicates that upwelling intensified significantly from the Late Miocene (7 to 10 m.y. BP) onwards and that the Benguela current developed progressively thereafter. The aridity of the Namib in the Pliocene appears to be confirmed by pollen from cores in the Walvis Ridge area, which indicate a hyperarid climate similar to that of today throughout the Pliocene (Van Zinderen Bakker, in press). Towards the end of the Pliocene, the Obib dunes of the southern Namib were formed and the main Namib Sand Sea began to accumulate (Ward et al. 1983), possibly aided by the Pliocene regression (Siesser and Dingle 1981).

The Kalahari

The Kalahari has acted as a major sedimentary basin since the Cretaceous (Tankard et al. 1982) and contains a considerable thickness (up to 500 m) of terrestrial sediments, mainly sands and marls, with playa deposits in some areas as described by Rogers (1934); Du Toit (1954); Boocock and Van Straten (1962); Money (1972); Smit (1977); Wright (1978); SACS (1980). Many of the sediments have been calcreted or silicified (Watts 1980; Summerfield 1983).

There are few exposures of these sediments, termed the Kalahari Group by SACS (1980), and even fewer detailed studies of their lithology and stratigraphy. Consequently, the age and depositional environments of the Kalahari Group remain uncertain. Cahen and Lepersonne (1952) and Mabbutt (1955) have attempted to assign ages to its members by correlating their occurrence with land surfaces of different antiquities. In the southwestern Kalahari, Mabbutt (1955) suggested that the pre-Kalahari Group surface was Cretaceous in age and that the Kalahari limestone and underlying calcareous sandstones were of Early Tertiary age, with the surface Kalahari sands predating the end-Tertiary incision of the River valleys. However, as Netterberg (1969) points out, the age given to the Kalahari limestone, and thus the age of the whole sequence, is based on slender palaeontological evidence. Martin (1950) has tentatively suggested that the Kalahari Group may be correlated with the Tsondab Sandstone Formation and its equivalents, with the Kalahari limestone possibly being equivalent to the end Miocene pedogenic calcretes of the eastern Namib. Most workers (e.g. Poldervaart 1957; Baillieul 1975) have suggested that the surface Kalahari sands are Late Tertiary to Recent in age, although Korn and Martin (1957) firmly assign them a Pliocene age.

Interpretation of depositional environments for the Kalahari Group is hampered by the lack of exposures, but there seems to be general agreement that its sediments accumulated in arid to semi-arid environments (Smit 1977; SACS 1980). This interpretation is supported by the widespread occurrence of calcretes, often with a high dolomite content (Watts 1980), which represents a series of episodes of pedogenesis in semi-arid climates since the Pliocene. Silcretes are also widely found (Summerfield 1983) and their distinct chemistry and mineralogy indicate that they were formed in alkaline pan margin environments. Late Tertiary stromatolites from Etosha Pan (Martin and Wilczewski 1972) indicate widespread saline lacustrine environments in this area.

Although the precise provenance of the surface Kalahari sands is uncertain, most workers have accepted that they have been redistributed by the wind a number of times during the Late Cenozoic (Poldervaart 1957; Grove 1969; Binda and Hildred 1973; Baillieul 1975). Thus, their widespread distribution in southern and south central Africa (Fig. 3) points to a number of periods of very extensive aridity in the Late Cenozoic.

THE QUATERNARY RECORD

The Namib

The Quaternary sedimentary geomorphic record for the Namib is one of continued aridity or extreme aridity, with considerable evidence to suggest that Quaternary climatic fluctuations in the region were of low magnitude and extent and were superimposed upon a trend of increasing aridity, as originally suggested by Korn and Martin (1957).

Accumulation of the Namib Sand Sea continued, with sand derived mostly from southern and western coastal source areas (Rogers 1977; Lancaster and Ollier 1983), fed ultimately by the sand fraction of the sediment load of the Orange River. This process was probably facilitated by eustatically lowered sea levels and higher wind speeds during Glacial Maxima. Further small sand seas and dune fields accumulated during the Late Pleistocene and Namaqualand (Tankard and Rogers 1978) and the Holocene in the northern Namib (Lancaster 1982). This process is still continuing. Stages in the development of the Namib Sand Sea are marked by wedges of dune arenites with steeply dipping large scale fore sets intercalated with fluvial conglomerates of probably Mid-Pleistocene age on the south bank of the Kuiseb river. These were interpreted by Ward (1982) as the deposits of large linear type dunes which abutted the river in a manner similar to those adjacent to the river today. Dune bedded sands are also present on the margins of a pan in the eastern Namib, the deposits of which contain remains of the Mid-Pleistocene Elephas reckii (Shackley 1980).

Continuing aridity in the Namib is also evidenced by the surface survival of calcrete palaeosols (Yaalon and Ward 1982) and calcareous lacustrine deposits (Selby et al. 1979). Quaternary pollens from DSDP cores on the Walvis Ridge are suggestive of a vegetation cover similar to that of today, with indications that the periods of increased humidity which occurred were not sufficient to change the composition of the vegetation (Van Zinderen Bakker, in press). The terriginous sediments from the same cores indicate a dominance of aeolian inputs and thus arid climates in the region over the last 200 000 years (Diester Haas, pers. comm.)

The valleys of the major Namib rivers from the Tsondab northwards contain sequences of deposits which record periods of aggradation of gravels and silts during the Pleistocene, with intervening periods of incision (Gevers 1936; Mabbutt 1952; Korn and Martin 1957; Ward 1982; Vogel 1982). Whether or not these sequences reflect climatic or eustatic changes in the Namib and its hinterland (Ward et al. 1983), it seems clear from the sedimentary character of the deposits that they were laid down in arid conditions, as suggested by the presence of large dunes on the southern bank of the Kuiseb during the deposition of the Oswater conglomerate (Ward 1982). Silt deposits accumulated in the Kuiseb valley 23 000 to 19 000 BP in conditions of reduced stream discharge and competence (Vogel 1982) which may have been accentuated by the blocking of the valley by dunes (Rust and Wienecke 1974; Marker and Muller 1978). Red sands contained in many marginal facies again suggest the close proximity of dunes to the Kuiseb river (Ward et al. 1983).

Deposits from the Uis river, in the central Namib, suggest a decline in stream competence from the Late Tertiary which is independent of tectonic and basin area changes (Korn and Martin 1957). A similar situation obtains in the Tsondab Valley, where an episodic decline in stream discharge and competence seems to have occurred throughout the Quaternary (Seely and Sandelowski 1974; Lancaster 1983). These changes in fluvial deposits suggest a progressive desiccation of the Namib region during the Quaternary.

Calcified reed bed and pan margin deposits with C14 ages clustering around 32 000 to 31 000, 29 000 to 27 000, 22 000 to 21 000 and 12 000 to 10 000 BP (Vogel and Visser 1981) are scattered through the central Namib. Although suggestive of moister conditions, their occurrence in proximity to modern groundwater seepages and river valleys suggests low amplitude climatic changes.

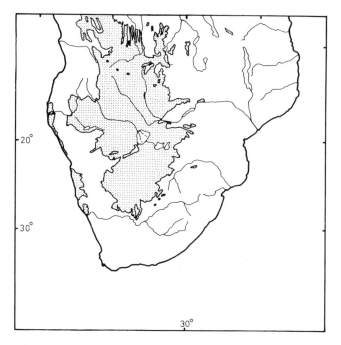

Figure 3. Distribution of sands of Kalahari type in southern Africa and major sand seas in the Namib.

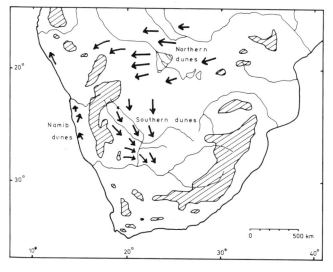

Figure 4. Generalized trends of major dune systems in southern Africa.

The Kalahari

The Quaternary sedimentary and geomorphic record for the Kalahari suggests that climates in the region fluctuated about a semi-arid mean. Periods of aridity of very considerable geographical extent alternated with widespread formation of pedogenic calcretes in intervals of semi-arid climates in the Mid-Pleistocene, Upper Pleistocene and Late Pleistocene - Recent (Netterberg 1969; Watts 1980). There were limited periods of sub-humid climates, as evidenced by shallow lakes in valleys and pan depressions (Lancaster 1979b; Helgren and Brooks 1983) and the deposition of sinter in caves (Cooke 1975). The palaeoclimatic significance of large palaeolakes in the Makgadikgadi Depression is uncertain, given the tectonic instability of the Okavango delta region and the exoreic nature of most of its drainage (Cooke 1980).

Microcosms of Late Quaternary climatic change in the southern Kalahari are contained in the deposits and landforms of the pans (Lancaster 1978), with periods of deflation in sub-arid to arid conditions alternated with shallow seasonal or permanent alkaline lakes formed in sub-humid climates. A rather similar record of alternating arid and sub-humid climates is provided by the alluvial, colluvial and lacustrine deposits of the Dobe and Xangwa valleys in Ngamiland (Helgren and Brooks 1983).

Extensive systems of parallel, mostly linear, dunes now fixed by savanna vegetation, occur throughout the Kalahari from the Orange river to southern Angola and southwestern Zambia (Lancaster 1981), forming an arc which corresponds approximately to the pattern of outblowing winds around the South African anticyclone (Fig. 4). The existence of dunes fixed by savanna vegetation and occurring in areas which today receive 450 - 650 mm of rainfall annually is a graphic illustration of the former extent of aridity in southern Africa. Currently active dunes are restricted to the Namib Desert where rainfall is less than 100 mm per year; whilst dune crests are active in some areas of the southwestern Kalahari where rainfall is between 100 and 150 mm. Comparisons of the present position of the 100 mm isohyet with the distribution of fixed dunes (Lancaster 1981) indicate that it lay up to 1000 - 1200 km northeast of its present position when the northern dunes were formed. Formation of the southern group of dunes involved a 200 to 300 km northeastward shift of the 100 mm isohyet.

On the basis of the alignment patterns of dunes of different morphologies, Lancaster (1981) identified three patterns of palaeosandflows and winds, whilst Mallick et al. (1981) identified two. It appears that periods of aridity and dune formation were associated with, and probably caused by, a larger and more intense anticyclonic circulation and higher windspeeds (Lancaster 1981).

Assigning ages to the periods of aridity is difficult. In the northwestern Kalahari Helgren and Brooks (1983) suggest that the large linear dunes were formed during the Early Pleistocene or even the Late Tertiary. Although it is difficult to credit the survival of recognizable dune forms from this length of time, these dunes are of some antiquity, as they are cut by one of the faults which forms the western boundary of the Okavango Delta (Mallick et al. 1981). The latest period of aridity recognized from the sedimentary and geomorphic record at Gi postdates a valley-wide lake in the Dobe valley and may span much of the Last Glacial Maximum period (Helgren and Brooks 1983). It may be tentatively correlated with the aeolian sand invasions of caves in the nearby Kwihabe Hills, which occurred prior to a period of extensive sinter deposition 17 000 to 13 000 (Cooke 1975).

In the Makgadikgadi Depression, Cooke (1980) and Helgren (in press) suggest intervals of Lower to Mid-Pleistocene aridity, with deflation of pan and lacustrine deposits. A major arid phase, with extensive dune formation, took place in the later Upper Pleistocene, after the drying of the 920 m lake. This period is possibly equivalent to that which formed the Group 2 dunes of Lancaster (1981) and may tentatively be correlated with a period of aridity and dune formation recognized from the Makgadikgadi area for the period coeval with the Last Glacial Maximum by Heine (1978). Widespread dry climates in the Northern part of the Kalahari sands region are recognized at this time in southern Zaire. Blow outs and deflation hollows formed on the Kwango Plateau (De Ploey 1963), whilst there is a hiatus in the archaeological record for the period 25 000 - 15 000 BP (Cahen 1978).

In the southwestern Kalahari, organic soil horizons below the dunes have C14 dates of 28 000 + 4900 - 3200 BP (Heine 1982), implying stabilized dunes at this time. Heine (1982) puts forward a variety of stratigraphic and sedimentary evidence to indicate a period of dune formation and much increased wind speeds in this area during the Late Glacial. There is further evidence from this area for increased aridity and dune extension during the Holocene, some 4500 to 3500 BP (Heine 1982), as dune sands are interstratified with fluvial and colluvial deposits in the Auob river near Gochas.

DISCUSSIONS AND CONCLUSIONS

The Tertiary sedimentary record for the Namib and Kalahari strongly suggests that sedimentation in the region during this period has taken place mostly under conditions of arid to extreme arid and semi-arid climates. Dominant depositional environments have been sand seas and dune yields, with proximal and distal alluvial fan, ephemeral flood plain and pan or playa facies. Periods of geomorphic stability were characterized by extensive pedogenic calcrete formation. Such conditions appear to have prevailed probably for much of the last 20 m.y. and possibly for the last 50 to 80 m.y. This view follows Ward et al. (1983), but is contrary to that expressed by most previous workers in the region (Tankard and Rogers 1978; Seely 1978; Lancaster 1979a; Deacon 1980), but supports the tentative suggestions for an Oligocene age for aridity in southern Africa put forward by Van Zinderen Bakker (1975). Unlike the Tertiary sedimentary and geomorphic record for Australia (Bowler 1976; Kemp 1978), there is no published evidence for humid climates during the Cenozoic in the region of the modern Kalahari and Namib deserts. It appears that the antiquity of arid climates in southern Africa may parallel that from the Sahara, where Sarnthein (1978) suggests that intervals of aridity occurred 38 to 34, 23 to 20 and 13 to 12 m.y. BP.

Comparison of the Quaternary sedimentary and geomorphic record for the Namib and Kalahari suggests an essentially similar pattern, with climatic oscillations being superimposed on a mean of generally arid to hyperarid climates in the Namib, and semi-arid climates in the Kalahari. There is some evidence to suggest that aridity may have intensified in the Namib during the period in question. The amplitude and extent of the climatic oscillations varied between the Namib and Kalahari. In the Namib, they were of low magnitude and extent, whilst in the Kalahari there is a much greater contrast between the massive expansions of aridity as evidenced by the distribution of fixed dunes and the widespread Late Glacial humid period in the southern Kalahari (Lancaster 1979b).

Whilst there is evidence for considerable expansions of the area of aridity at intervals, there is no reliable chronology of events, nor any agreement upon the nature of the changes in regional climatic patterns. For example, it is not known whether aridity and dune formation in the northern Kalahari resulted from a general expansion of the area of the arid zone as suggested by Lancaster (1981), or a latitudinal shift in the position of circulation patterns (Van Zinderen Bakker 1975; Tankard and Rogers 1978). A priority for future research in the palaeoclimatology of the arid zone in southern Africa should be to resolve these questions.

ACKNOWLEDGEMENTS

I wish to thank J Rogers, MK Seely and JD Ward for assistance and stimulating discussions of Late Cenozoic geology and palaeoenvironments in the Namib and Kalahari. I also wish to thank the CSIR (Research Grants Division and Cooperative Scientific Programmes) and Transvaal Museum for funding and the Directorate of Nature Conservation, SWA for facilities and permission to work in the Namib.

REFERENCES

Baillieu TA 1975. A reconnaissance survey of the cover sands in the Republic of Botswana. J. sediment. Petrol. 45:494-503.
Besler H and Marker ME 1979. Namib Sandstone, a distinct lithological unit. Trans. geol. Soc. S. Afr. 82:155-160.
Binda PL and Hildred PR 1973. Bimodal grain size distributions of some Kalahari type sands from Zambia. Sediment. Geol. 10:233-257.
Boocock C and Van Straten OJ 1962. Notes on the geology and hydrogeology of the central Kalahari, Bechuanaland Protectorate Trans. geol. Soc. S. Afr. 65:125-171.
Bowler JM 1976. Aridity in Australia, age, origins and expression in aeolian landforms and sediments. Earth-sci. Rev. 12:279-310.
Cahen D 1978. Vers une révision de la nomenclature des industries préhistoriques de l'Afrique centrale. L'Anthropologie 82:5-36.
Cahen L and Lepersonne J 1952. Equivalence entre le système du Kalahari du Congo Belge et des Kalahari Beds d'Afrique australe. Mém. Soc. Belg. géol. palaeont. hydrol. 8. 65 pp.
Cooke HJ 1975. The palaeoclimatic significance of caves and adjacent landforms in western Ngamiland, Botswana. Geogr. J. 141:430-444.
Cooke HJ 1980. Landform evolution in the context of climatic change and neo-tectonism in the middle Kalahari of north central Botswana. Trans. Inst. Br. Geogr. N S 5:80-99.
Corvinus G and Hendey QB 1978. A new Miocene locality at Arrisdrift in Namibia (South West Africa). Neues Jb. Palaeontol. Mh. 4:193-205.
De Ploey J 1963. Position géomorphologique, genèse et chronologie de certain dépôts superficiels au Congo occidental. Quaternaria 7:131-152.
Deacon HJ 1980. The comparative evolution of Mediterranean type ecosystems, a southern perspective. Pap. 3rd Int. Conf. on Mediterranean type ecosystems, Stellenbosch, September 1980. 72 pp.

Du Toit AL 1954. The Geology of South Africa, 3rd ed. Oliver and Boyd, London. 611 pp.

Endrody Younga S 1978. Coleoptera. In: MJA Werger (ed.), Biogeography and Ecology of Southern Africa. Junk, The Hague. p. 797-822.

Endrody Younga S 1982. The evidence of Coleoptera in dating the Namib Desert re-examined. In: Proc. 6th SASQUA Conf. Palaeoecol. Africa 15:217-223.

Gevers TW 1936. The morphology of western Damaraland and the adjoining Namib Desert. S. Afr. geogr. J. 19:61-79.

Grove AT 1969. Landforms and climatic change in the Kalahari and Ngamiland. Geogr. J. 135:191-212.

Heine K 1978. Radiocarbon chronology of late Quaternary lakes in the Kalahari, southern Africa. Catena 5:145-149.

Heine K 1982. The main stages in the Late Quaternary evolution of the Kalahari region, southern Africa. Palaeoecol. Africa 15:53-76.

Helgren DM and Brooks AS 1983. Geoarchaeology at Gi, Middle Stone Age and Later Stone Age site in the northwestern Kalahari. J. archaeol. Sci. 10:181-197.

Helgren DM, in press. Historical geomorphology and geoarchaeology in the southwestern Makgadikgadi basin, Botswana.

Kemp EM 1978. Tertiary climatic evolution and vegetation history in the south east Indian Ocean region. Palaeogeogr. Palaeoclimatol. Palaeoecol. 24:169-208.

Koch C 1961. Some aspects of abundant life in the vegetationless dunes of the Namib Desert. J. SWA. scient. Soc. 15:8-34. 77-92.

Koch C 1962. The Tenebrionidae of Southern Africa XXVI. Comprehensive notes on the Tenebrionid fauna of the Namib Desert. Ann. Transv. Mus. 24:61-106.

Korn H and Martin H 1957. The Pleistocene in South West Africa. In: JD Clark (ed.), Proc. 3rd Pan-African Congress, Prehist., Livingstone 1955. p. 14-22.

Lancaster IN 1978. Composition and formation of southern Kalahari pan margin dunes. Z. Geomorphol. 22:148-169.

Lancaster IN 1979a. Quaternary environments in the arid zone of southern Africa. Dept. Geogr. and Environmental Stud., Univ. Witwatersrand. Occ. Pap. 22. 77 pp.

Lancaster IN 1979b. Evidence for a widespread late Pleistocene humid period in the Kalahari. Nature 279:145-146.

Lancaster N 1981. Palaeoenvironmental implications of fixed dune systems in southern Africa. Palaeogeogr. Palaeoclimatol. Palaeoecol. 33:327-346.

Lancaster N 1982. Dunes on the Skeleton Coast, SWA/Namibia, geomorphology and grain size relationships. Earth Surf. Processes and Landforms 7:575-587.

Lancaster N, in press. Palaeoenvironments in the Tsondab Valley, central Namib Desert. Palaeoecol. Africa 16.

Lancaster N and Ollier CD 1983. Sources of sand for the Namib Sand Sea. Z. Geomorphol., Suppl. Bd. 45:71-83.

Mabbutt JA 1952. The evolution of the middle Ugab valley, Damaraland, SWA. Trans. R. Soc. S. Afr. 33:334-366.

Mabbutt JA 1955. Erosion surfaces in little Namaqualand and the ages of surface deposits in the southwestern Kalahari. Trans. geol. Soc. S. Afr. 58:13-29.

Mallick DIJ, Habgood F and Skinner AC 1981. A geological interpretation of LANDSAT and air photography of Botswana. Overseas Geol. and Min. Resour. 56. 35 pp.

Marker ME and Muller D 1978. Relict vlei silts of the middle Kuiseb river, South West Africa. Madoqua 11:151-162.
Martin H 1950. Südwestafrika. Geol. Rdsch. 38:6-14.
Martin H and Wilczewski N 1972. Algenstromatolithen aus der Etoscha-Pfanne Südwestafrikas. Neues Jb. Geol. Palaeontol. Mh. 12:151-162.
Money NJ 1972. An outline of the geology of western Zambia. Rec. geol. Surv. Zambia 12:103-123.
Netterberg F 1969. Ages of calcretes in southern Africa. S. Afr. arch. Bull. 24:88-92.
Paylore P 1979. Arid Lands Newsletter, April 1979.
Poldervaart A 1957. Kalahari Sands. In: JD Clark (ed.), Proc. 3rd Pan-African Congress, Prehist., Livingstone 1955. p. 106-114.
Rogers AW 1934. The build of the Kalahari. S. Afr. geogr. J. 17:3-12.
Rogers J 1977. Sedimentation on the continental margins off the Orange river and the Namib Desert. Joint Geol. Survey/UCT Marine Geoscience Group Bull. 7. 162 pp.
Rust U and Wienecke F 1974. Studies on gramadulla formation in the middle part of the Kuiseb river, SWA. Madoqua 3 :5-15.
SACS (S. Afr. Committee for Stratigraphy) 1980. Stratigraphy of S. Afr. Pt. 1 (comp. LE Kent). Lithostratigraphy of the Republic of S. Afr., SWA/Namibia and the Republics of Boputhatswana, Transkei and Venda. Handbk geol. Surv. S. Afr. 8.
Sarnthein M 1978. Neogene sand layers off NW Africa, composition and source environments. In: Y Lancelot and Seibold E (eds.), Initial Rep. of the Deep Sea Drill. Proj. 41:939-959.
Seely MK 1978. The Namib Dune Desert, an unusual ecosystem. J. Arid Environ. 1:117-123.
Seely MK and Sandelowsky BH 1974. Dating the regression of a river's end point. S. Afr. arch. Bull. Goodwin Ser. 2:61-64.
Selby MJ, Hendy CH and Seely MK 1979. A late Quaternary lake in the central Namib Desert, southern Africa and some implications. Palaeogeogr. Palaeoclimatol. Palaeoecol. 26:37-41.
Shackley ML 1980. An Acheulian industry with Elephas reckii fauna from Namib IV, South West Africa (Namibia). Nature 284:340-341.
Siesser WG 1978. Aridification of the Namib Desert, evidence from oceanic cores. In: EM van Zinderen Bakker (ed.), Antarctic Glacial History and World Palaeoenvironments. Balkema, Rotterdam. p. 105-113.
Siesser WF 1980. Late Miocene origin of the Benguela upwelling system off northern Namibia. Science 208:283-285.
Siesser WG and Dingle RV 1981. Tertiary sea level movements around southern Africa. J. Geol. 89:83-96.
Smit PJ 1977. Die geohidrologie in die opvanggebied van die Molopo-rivier in die Noordelike Kalahari. PhD thesis Univ. Orange Free State.
Summerfield MA 1983. Silcrete as a palaeoclimatic indicator, evidence for southern Africa. Palaeogeogr. Palaeoclimatol. Palaeoecol. 41:65-80.
Tankard AJ and Rogers J 1978. Late Cenozoic palaeoenvironments on the west coast of southern Africa. J. Biogeogr. 5:319-337.
Tankard AJ, Jackson MPA, Eriksson KA, Hobday DK, Hunter DR and Minter WEL 1982. Crustal Evolution of southern Africa. 3.8 Billion years of Earth History. Springer Verlag, New York. 523 pp.
Van Zinderen Bakker EM 1975. Origin and palaeoenvironments of the Namib Desert biome. J. Biogeogr. 2:65-73.
Van Zinderen Bakker EM, in press. Aridity along the Namibian coast. Palaeoecol. Africa 16.

Vogel JC 1982. Age of the silt terrace at Homeb. In: Proc. 6th SASQUA Conf. Palaeoecol. Africa 15:201-209.

Ward JD 1982. Aspects of a suite of Quaternary conglomeratic sediments in the Kuiseb valley, Namibia. In: Proc. 6th SASQUA Conf. Palaeoecol. Africa 15:211-216.

Ward JD, Seely MK and Lancaster N 1983. On the antiquity of the Namib. S. Afr. J. Sci. 79:175-183.

Watts NL 1980. Quaternary pedogenic calcretes from the Kalahari (southern Africa), mineralogy, genesis and diagenesis. Sedimentology 27:661-686.

Wright EP 1978. Geological studies in the northern Kalahari. Geogr. J. 144:235-249.

Yaalon DH and Ward JD 1982. Observations on calcrete and recent calcic horizons in relation to landforms in the central Namib Desert. In: Proc. 6th SASQUA Conf. Palaeoecol. Africa 15:183-186.

The development of the Namib dune field according to sedimentological and geomorphological evidence

HELGA BESLER
Universität Stuttgart, Germany

ABSTRACT. The Post-African planation surface of the Namib Plain also developed on the sandstone beneath the Namib dune field. On this ancient land surface five river systems have been active. After an erosional stage, decreasing run-off led to the accumulation of large alluvial fans. These were the main source of sand and the centres of erg development. Since the End-Pleistocene the wind regime shifted southward, only minor remodelling of the dune field has taken place.

INTRODUCTION

This paper is a brief summary of extensive research carried out mainly during 1976 and the following years and published in a monograph in German (Besler 1980). In view of the limited extent of the paper the presentation of techniques and methods of investigation is neglected in favour of results. However, as the Namib dune field covers an area of 34 000 km this cannot be more than a first attempt and much remains to be done.

PREREQUISITES FOR THE DEVELOPMENT OF THE DUNE FIELD

The Post-African Planation Surface

Ground control during the investigation of the Namib dune field (= Namib Erg) in 1976 (Besler 1976/77) revealed that the sand cover in the interdune valleys is not very thick and that the underlying Namib Sandstone (Besler and Marker 1979) can be found everywhere. Occasionally even the high dune ridges, where these embrace large dune pits, show sandstone at the base of the pits. It is particularly in the southern erg that the Namib Sandstone is exposed over large areas.
  Based on the only available preliminary contour maps (1:100 000 and 1:250 000) 38 cross-sections were drawn from west to east at a distance of 10 km. These profiles show the base of the dunes to be slowly rising towards the east. As the ground control had shown, the levels of the interdune valleys can be connected by straight lines to roughly represent the underlying sandstone surface. The inclination of the land surface was calculated from the cross-sections and transferred onto a map (Fig. 1.).

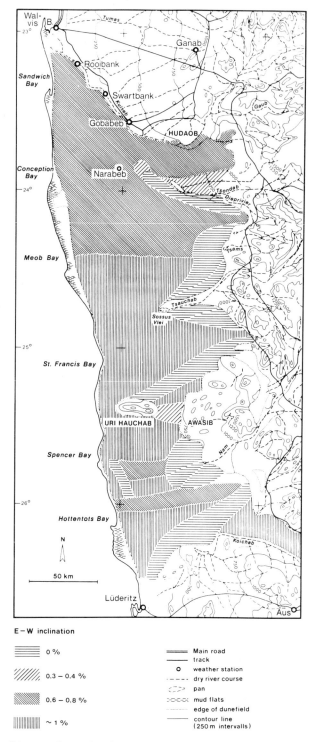

Figure 1. Gradients of the Namib Sandstone surface beneath the dune field calculated from 38 cross-sections (from Besler 1980).

Several points of interest should be noted:
1. The Namib Sandstone rises cliff-like up to 200 m above the crystalline basement near the coast (for example, near St Francis Bay and west of the Uri Hauchab inselberg) and up to 200 m above the foreland of the Great Escarpment in the east (for example, near the Tsauchab river and Nam Vlei). This sandstone pedestal gives the dune field its elevated appearance, exaggerating the heights of the dunes.
2. The Namib Sandstone thins out towards the north where the Kuiseb river approximately follows the contact between sandstone and crystalline rocks. It also thins out towards the south where the Koichab river marks the southern border.
3. With the exception of the northernmost part, the gradient of the sandstone surface is 1 % almost everywhere. This value is also given for the Namib Plain outside the dune field which according to King (1962) represents the Post-African Denudational Landsurface. Later King (1976) distinguished between Rolling = Post-African I Land surface (Mio-Pliocene) and Widespread = Post-African II Land surface (Plio-Pleistocene). Ollier (1978) suggests that this is the older Moorland Planation Surface (King 1976) because of the partly existing calcrete cover. But according to King (1962) the Moorland Surface is equivalent to the African Planation Surface which is characteristic for the South West African Highland. The important fact, however, is that the same planation surface is spread across the Precambrian of the Namib Plain and across the younger Namib Sandstone. The gradient of the crystalline basement beneath the sandstone is much lower and does not correspond to the Post-African Land surface.
4. Several marine benches are eroded into the coastal sandstone cliff. Unfortunately they cannot be traced everywhere and correlated because of the dune cover. Terraces were found at 25 m, 40 - 50 m, 80 - 100 m, 130 - 150 m, and even at 200 m. As King (1976) places the greatest uplift of the continent after formation of the Rolling Land surface in the Pliocene the Namib Sandstone surface beneath the dunes seems to represent the Post-African I erosion surface.
5. Lithologically the Namib Sandstone consists of only slightly cemented aeolian deposits that are reworked in some places. The dominating quartz grains are mostly subangular-pitted and poorly to moderately sorted, their mean grain size ranging from 0.08 mm to 0.38 mm with a frequency maximum around 0.2 mm. A high percentage of grains shows coatings of iron oxide.

Fluvial activity: the erosional stage

The gradient map shows horizontal parts of the sandstone surface near the eastern margin of the dune field and in the vicinity of still existing rivers. These surfaces either truncate the eastward rising sandstone or are incised into it. Whereas their westward inclination is zero, the gradient following the direction of probable former discharge is mostly 0.7 %. Especially near the Tsondab, the Tsauchab and the Nam rivers, the interpretation as ancient fluvial terraces is compelling. According to the gradient map five fluvial systems have thus been active on the Namib Sandstone surface in addition to the bordering rivers Kuiseb and Koichab. These are
1. the very complicated Tsondab-Dieprivier system
2. the Tsauchab-Tsams system with double terraces
3. the Bushman Hill-Uri Hauchab system without a visible river channel

4. the Nam system with double terraces

5. the Koichab northern branch reaching the bordering Koichab near Koichab Pan.

All fluvial terraces have been confirmed by aerial photographic interpretation. As the gradient in the direction of former discharge is 0.6 - 0.8 % in each case, it seems possible to explain the inclination of the northernmost part of the sandstone surface in terms of general fluvial truncation not confined to channels. The coincidence between the extent of the less inclined area and the relatively high undissected part of the Great Escarpment between the rivers Gaub and Tsauchab is remarkable. There also seems to be sedimentological evidence that the eroded sandstone of the east was redeposited fluvially on Namib Sandstone in the coastal area.

Further evidence is derived from pebble analysis on the fluvial terraces. The river systems of Tsondab-Dieprivier and Tsauchab-Tsams partly have their catchment areas in the Schwarzkalk Formation of the Escarpment, supplying conspicuous dark pebbles. These can be found everywhere on their terraces and even appear as dark patches on aerial photographs. In fact, some Schwarzkalk pebbles were also found on a terrace near Uri Hauchab, supporting the inference of a Bushman Hill-Uri Hauchab river system. The Kuiseb, on the other hand, has its catchment area in the Damara System and had no contact with the Schwarzkalk Formation. The majority of pebbles of the Kuiseb consists of quartz. Between Gobabeb-Narabeb and the highest marine terrace (200 m) quartz and Schwarzkalk pebbles are found together, indicating the existence of a formerly combined discharge, most probably in a very shallow and braided channel system.

The fluvial history was investigated in more detail in the more easily accessible northern erg. Incised into the described surface with an inclination of 0.6 to 0.8 % are the separate high terraces of the Kuiseb and Tsondab with the same gradient in the direction of discharge. They are covered by calcretes which supplied a series of C14 dates of $33\,350 \pm 2960$ BP (subsoil) to $24\,480 \pm 720$ BP (surface) for the Tsondab High Terrace east of Tsondab Vlei and of $26\,930 \pm 915$ BP (subsoil) for the Kuiseb High Terrace near Gobabeb. Here mixed pebbles of quartz and Schwarzkalk indicate that the Tsondab seems to have been a tributary to the Kuiseb at one stage. At this time there was enough local run-off on the Namib Sandstone to erode numerous small channels leading from the watershed between the Kuiseb and Tsondab (east of Tsondab Vlei) onto their respective high terraces. When local run-off ceased, the Kuiseb lost its southern tributary but could incise farther due to allochthonic water. This lower Kuiseb terrace branches off near Homeb where C14 dates ranges from $25\,040 \pm 475$ BP (subsoil) to $22\,770 \pm 635$ BP (surface). Farther west near Gobabeb the calcretes on the lower Kuiseb terrace contain rolled pieces of calcrete from the High Terrace and were dated $19\,860 \pm 590$ BP (subsoil). Unfortunately radiocarbon dates for calcretes are difficult to interpret, but the main erosion of the Kuiseb valley - especially east of Gobabeb - must have occurred after the calcrete formation.

Distinct channels of local run-off are also found on the high terraces of the Dieprivier, Tsams and Tsauchab where they lead into space and have no connection with the lower terraces. At this stage obviously only discharge from the Highland area provided the water. Yet the erosive capacity reached its maximum, cutting deep but narrow valleys into the sandstone and at the same time broadening the erosional moat below the Escarpment. As a consequence the broad high terraces of the river systems were cut off from their hinterland (with the exception of the Tsondab). Local run-off was only effective on crystalline bedrock outcrops which were separated from the surrounding sandstone by small erosional moats draining towards the tributaries of the greater river systems.

Fluvial activity: the aggradational stage

The application of sedimentological techniques to sand samples from the recent dune field and the interpretation of the results led to the conclusion that with decreasing discharge the rivers no longer reached the ocean but, during retreat, accumulated large alluvial fans on the formerly eroded Namib Sandstone surface. These fans can still be traced in the dune field because here the granulometric and morphoscopic qualities of sand grains are distinctly different, providing enhanced evidence for fluvial transport, although the sands were later incorporated into the dunes. Judging from these differences in the sand, the size of alluvial fans decreased from north to south. The combined Kuiseb and Tsondab accumulated the largest fan, reaching from west of Tsondab Vlei to the coast and from Walvis Bay to Conception Bay. The smaller Tsauchab alluvial fan still reached the coastline but the Nam river only deposited an interior fan. Practically no sands were accumulated in the Bushman Hill-Uri Hauchab system and in the Koichab northern branch. Both these systems seem no longer to have been active after the erosional stage. The Bushman Hill-Uri Hauchab system does not even possess lower terraces.

The history of the largest alluvial fan and the sand relations in the north were investigated in detail (Fig. 2).

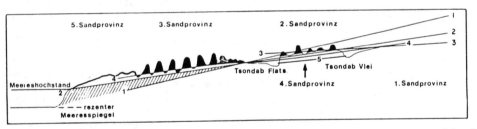

Figure 2 Sand provinces of the northern erg along the generalized cross-section no 8 (from Besler 1980)

Levels and gradients:
1. Post African Land surface (1.0%)
2. northern truncation surface (0.7%)
3. Tsondab Plain = High Terrace (0.3%)
4. ancient Tsondab (0.6%)
5. ancient Tsondab (0.6%)

Sand provinces:
1. red subangular sands from sandstone decomposition
2. brown rounded sands
3. brown mixed sands (1. and 2.)
4. grey calcareous sands from reworked calcretes
5. yellow mixed sands

1. The primary stage is represented by the Post-African Denudational Land surface with an inclination of 1 %.
2. The second surface with an inclination of 0.6 to 0.8 % consists of a truncated part in the east and an aggradational part near the coast.

3. Incised into the eastern part is the third surface with an inclination of 0.3 % still visible east of Tsondab Vlei where it is covered by calcretes. This is the High Terrace of the Tsondab river, the so-called Tsondab Plain. In this eastern part of the erg the modern red sands are simply weathered Namib Sandstone and therefore consist of subangular-pitted grains with a high percentage of iron oxide coating. These sands represent the first sand province. The river flowing on the High Terrace brought rounded sands from the Escarpment. These brown rounded sands were spread out west of Tsondab Vlei where they still form the second sand province.

4. The next stage of main valley erosion cut through the older deposits and deep into the Namib Sandstone. During river retreat the eroded sands were spread out west of the so-called Tsondab Flats in a large alluvial fan. They consist of subangular-pitted grains (from the Namib Sandstone) and rounded grains (from older deposits) and form the third sand province of brown mixed sands. These sands reach as far as the longitudinal dunes, that is, approximately to the highest marine terrace.

5. Another stage is represented by remnants of partly dune-covered fluvial terraces 10 - 20 m above the level of Tsondab Vlei. These terraces are covered by coarse grey calcareous sands representing the fourth sand province. The grey sands can be traced in a channel through the brown sands as far as north of Narabeb. They were derived from reworked calcretes of the High Terrace.

6. The sands in the coastal area covering the marine benches show the same qualities of grain shape and surface texture as the third sand province (brown mixed sands) but are better polished and lighter in colour. This could be due to the longer fluvial transport which destroyed the iron oxide coatings. Another reason could be the effects of the stronger aeolian mobility near the coast with grain abrasion. The yellow mixed sands represent the fifth sand province. There is no strong evidence for marine sands.

In fact, patina and the sorting of sand grains, increasing from the coast eastward, do not reflect aeolian transport from the coast but fluvial transport from the east, with increasing bleaching and mixing. During river retreat large alluvial fans were also accumulated in the moat between the Escarpment and the eastern sandstone cliffs, partly drowning the inselbergs. As a consequence, the lowest level (with small pans) is today found in basins close to the sandstone cliff sheltered from aggradation by inselbergs.

THE NAMIB ERG

Pleistocene formation

The alluvial fans of the retreating rivers were the source of blown sands and the centres of erg development. A rough mass-calculation comparing the dune volume west of the vleis and the volume of eroded sandstone in the Tsondab and Tsauchab channels showed that all the dune sands could have come from these river channels, without even considering erosion along the Escarpment. The separated sand provinces and distinguishable fan areas, in spite of the dunes, provide evidence for minimal aeolian sand movement over long distances. Otherwise the sands would have been mixed and show more typically aeolian qualities, such as pitted surfaces. This at first glance seems incompatible with the high dunes.

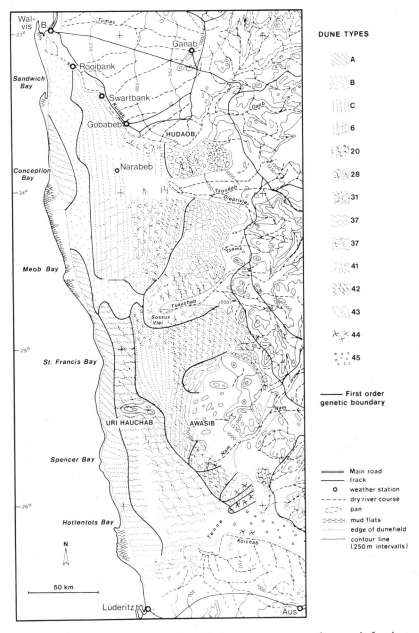

Figure 3. Dune types deduced from stereoscopic aerial photographs and ground control.
A.   transverse dunes
B.   transition forms
C.   longitudinal ridges un-
     differentiated
6.   network complex
20. branching longitudinal ridges
28. honeycomb structure with
     stellate dunes
31. lace dunes
37. zibar and zibar-silk system
41. high chaotic dunes
42. warty ridges with stellate
     dunes
43. giant honeycomb structure
44. pyramidal dunes
45. sand plain with craters

The fifteen varieties of high longitudinal dunes in the interior northern erg (Fig. 3) can all be reduced to one original form, the primary longitudinal dune ridge with one sharp undulating crestline and symmetrical flanks with an inclination of about 20 degrees. The heights range up to about 100 m with crest separation of 1.8 to 2.4 km. At present only one model exists which explains their formation without any contradicting evidence, that is, the Taylor-Görtler-movement within the Planetary Boundary Layer of the atmosphere. Under certain conditions parallel double vortices develop in the boundary layer (1 km dia.), eroding sand beneath their sinking branches and accumulating sand between their rising branches (Hanna 1969; Wippermann 1973) without sand transport over long distances. All necessary boundary conditions are fulfilled in the Namib, with the exception of frequent strong winds with velocities higher than 36 km/h. Today these are found only south of the Namib dune field (Lüderitz, Alexander Bay). But for the Pleistocene a meridional shift of wind systems and a higher acceleration can be assumed due to pressure-belt compression (van Zinderen Bakker 1975).

In the southern erg, with less alluvial sands, dunes have always been lower, displaying a zibar-sief (Warren 1972) or zibar-silk pattern (Mainguet and Callot 1978), zibars being large, weak undulations transverse to the wind.

Near the coast where marine benches cut the sandstone, two boundary conditions for vortex development were lacking: the undisturbed surface and the steady southern winds. Here strong additional sea winds moulded the sands into smaller transverse dunes.

The complex dunes around and east of the vleis cannot be explained by aeolian activity only. Interpretation of aerial photographs, ground control, and sand sample analysis provide evidence that these structures on the erosional part of the sandstone surface east of the alluvial fans are based on a fluvial sandstone topography. The sand is supplied by decaying sandstone (first sand province) which renders the distinction between sandstone base and dune cover even more difficult. The fluvial origin is most conspicuous in the north-eastern corner of the erg where the curved "dunes" are simply sandstone ridges between fluvial channels draining from a common watershed towards the Kuiseb and Tsondab High Terraces. On the High Terrace of Dieprivier the sand cover is poorer and the channels can be traced more easily. The conspicuous pyramidal dunes, for example near Nam Vlei and at the beginning of the Koichab northern branch, are all sitting on the rim of river terraces with one arm falling onto a lower platform. This descending arm is exaggerated in the case of the high dunes spilling into the Tsauchab channel east of Sossus Vlei. The dunes thus appear higher than they actually are since they are partly hiding a 200 m sandstone terrace. In fact, the eastern erg is not of aeolian but of fluvio-aeolian origin.

Holocene remodelling

After the End-Pleistocene southward shift of the wind regime the erg is modelled by a weaker, bimodal wind system consisting of trade winds (S-SW), sea breezes (SW), and easterly bergwinds. The southerly trade winds decrease towards the north but, together with bergwinds, seem still to be effective for the small sief or silk dunes in the southern erg. The transverse coastal dunes are still in equilibrium with the sea winds. But the high longitudinal ridges of the interior northern erg are remodelled according to their position within the erg and their exposure to the winds (see also Lancaster 1980). Barchan-like secondary dunes, transverse to the SW winds, cover the eastern flanks in the western erg. Towards the east they are gradually replaced by small secondary sief or silk dunes which need two obliquely blowing winds to form (SW and easterly bergwinds). Still further east these secondary dunes decrease in size with diminishing wind velocities and now cover the original ridges like small worms. In the NW corner near Swartbank strong NE winds constitute an additional element, forming numerous dune hollows (Besler 1976/77). The eastern erg, however, is practically fossilized. Mobility is only recorded where the sands, following the laws of gravity, fall onto lower platforms.

REFERENCES

Besler H 1976/77. Untersuchungen in der Dünen-Namib (Südwestafrika) - Vorläufige Ergebnisse des Forschungsaufenthaltes 1976. J. SWA Wiss. Ges. 31: 33 - 64.
Besler H 1980. Die Dünen-Namib: Entstehung und Dynamik eines Ergs. Stuttgarter geogr. Stud. 96. 241 pp.
Besler H and Marker M 1979. Namib Sandstone: a distinct lithological unit. Trans. geol. Soc. 82:155-160.
Hanna SR 1969. The formation of longitudinal sand dunes by large helical eddies in the atmosphere. J. Appl. Met. 8: 874-883.
King L 1962. The morphology of the earth. A study and synthesis of world scenery. Edingburgh. 699 pp.
King L 1976. Planation remnants upon high lands. Z. Geomorph., NF 20:133-148.
Lancaster N 1980. The formation of seif dunes from barchans - supporting evidence for Bagnold's model from the Namib Desert. Z. Geomorph., NF 24:160-167.
Mainguet M and Callot Y 1978. L'erg de Fachi-Bilma (Tchad-Niger). Contribution à la connaissance de la dynamique des ergs et des dunes des zones arides chaudes. Mém. et Docum., NS 18. 184 pp.
Ollier CD 1978. Inselbergs of the Namib Desert. Processes and history. Z. Geomorph., NF, Suppl. Bd. 31:161-176.
Van Zinderen Bakker EM 1975. The origin and palaeoenvironment of the Namib Desert biome. J. Biogeogr. 2:65-73.
Warren A 1972. Observations on dunes and bi-modal sands in the Ténéré-Desert. Sedimentology 19:37-44.
Wippermann F 1973. The orientation of vortices due to instability of the Ekman-Boundary-Layer. Annln. Met., NF 7:260-279.

A reappraisal of the Cenozoic stratigraphy in the Kuiseb valley of the central Namib desert

J.D.WARD
University of Natal, Pietermaritzburg, South Africa

ABSTRACT. Recent mapping of the surficial deposits in the Kuiseb Valley west of the Escarpment has facilitated a reappraisal of the Cenozoic succession of that region. The proposed formal nomenclature, as well as the probable sequence of formation, for the Cenozoic deposits is outlined briefly. The Cenozoic succession can be subdivided into: 1. deposits pre-dating the deep incision of the Kuiseb drainage system which include the Tsondab Sandstone Formation (proposed Early to Middle Tertiary age), the Karpfenkliff Conglomerate Formation (proposed Miocene age) and the Kamberg Calcrete Formation (End Miocene age); and 2. deposits post-dating the main incision event, including the Oswater Conglomerate Formation (proposed Early to Middle Pleistocene age), the Hudaob Tufa Formation (stratigraphical position and age uncertain), the Homeb Silt Formation (Late Pleistocene) and the Gobabeb Gravel Formation (proposed Terminal Pleistocene/Early Holocene age). The main Namib Sand Sea (Sossus Sand Formation) probably dated from the Pliocene and includes fossil pans that date to at least the Middle Pleistocene.

It is suggested that the Kuiseb River has been an effective northern boundary to the main Namib Sand Sea for at least the duration of the Quaternary except along the immediate coastal tract where the high energy, unidirectional south-southwesterly wind regime is dominant.

INTRODUCTION

The Kuiseb River, rising some 20 km southwest of Windhoek on the interior plateau of Namibia/South West Africa, traverses the Central Namib Desert in a rough southwest-west-northwest arc for some 300 km to reach the Atlantic Ocean in the vicinity of Walvis Bay. Within the relatively narrow Namib Desert strip, the Kuiseb River sharply delineates the northern boundary of the main Namib Sand Sea; a relationship first noted by Alexander (1838, reprinted 1967) and readily confirmed in recent years by the advent of satellite imagery.

Although considerable geomorphological and, to a lesser extent, geological attention has been accorded the Kuiseb Valley, no extensive field mapping of the surficial deposits had been carried out. The earliest reports date back to the last century, namely Stapff (1887) and Wilmer (1893), followed by the observations of Gevers (1936), Martin (1950, 1957, reprinted 1974), Logal (1960) and, more recently, the work of Goudie (1972), Barnard (1973, 1975), Rust and Wieneke (1974, 1980),

Sawyer (1976), Besler (1976/77, 1980), Selby (1976, 1977), Marker (1977, 1982), Ollier (1977, 1978), Marker and Müller (1978), McKee (1982), Vogel (1982) and Ward (1982). The general sequence in the accumulation of the surficial cover within the Kuiseb Valley was best appreciated by Martin (1950, 1957 and 1974) and subsequently by Ollier (1977) who outlined the geological and geomorphic history of the Central Namib Desert.

The recent mapping of the surficial deposits has added further detail to these earlier frameworks. This paper reports briefly on the sedimentary units identified in the Cenozoic succession within the Kuiseb Valley west of the Great Escarpment, their sequence of emplacement and introduces the proposed formal nomenclature for these surficial deposits.

Extensive field work undertaken in the Kuiseb-Gaub drainage system between the Great Escarpment and the Atlantic coast from December 1979 to December 1981 served to ground-check the initial photo-interpretation of Job 774/77 (scale c. 1:50 000) and Jobs 746/76 and 313/78 (scales c. 1:15 000). The field work was supplemented by reconnaissance observations and oblique aerial photography from a light aircraft. The surficial deposits were mapped and type localities selected, measured and described for the major units. Limited petrographic and X-ray diffraction analyses were undertaken to identify certain sediment types. Details of type localities, distribution etc., will appear elsewhere.

RESULTS AND DISCUSSION

The main sedimentary units and geomorphological events of the Cenozoic succession in the Kuiseb Valley west of the Great Escarpment are summarized in Table 1 relative to the framework outlined by Ollier (1977, 1978). The proposed formal nomenclature for the Cenozoic deposits, together with brief lithological descriptions, is presented in Table 2. An attempt has also been made, in Table 2, to place the major units into a geochronological framework. However, this attempt should be regarded with caution because of the apparent lack of suitable datable material, e.g. fossils, volcanics. The geochronological order is largely after the correlations proposed by Ward et al. (1983) for sedimentary deposits in the Namib Desert.

The Cenozoic deposits can be conveniently subdivided into two groups relative to the deep canyon incision of the Kuiseb River and its major tributaries:

1. deposits pre-dating the incision and resting unconformably on a planed bedrock surface developed across mainly Late Precambrian Damara Sequence rock types. These include the Kamberg Calcrete Formation, Karpfenkliff Conglomerate Formation, and Tsondab Sandstone Formation

2. deposits post-dating the incision and preserved mainly as terrace remnants. These include the Kuiseb River alluvium, Gobabeb Gravel Formation, Homeb Silt Formation, Hudaob Tufa Formation and Oswater Conglomerate Formation.

The development of the main Namib Sand Sea (Sossus Sand Formation is tentatively ascribed to this period.

The incision of the larger rivers along the west coast of Namibia/South West Africa has been attributed to epeirogenic uplift in the Late Tertiary (Korn and Martin 1957) which accords with the Plio-Pleistocene age for similar trends in South Africa (King 1951; Partridge 1982).

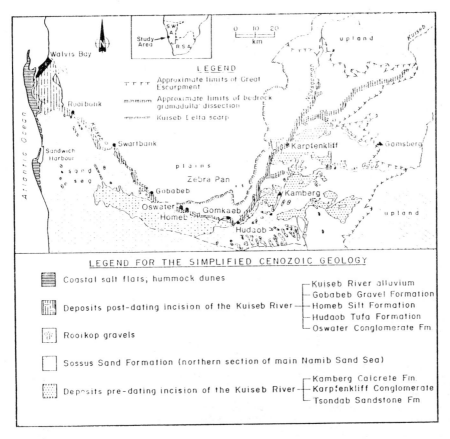

Figure 1. The distribution of Cenozoic deposits in the Kuiseb Valley, Central Namib Desert.

Therefore, the deposits pre-dating the incision are considered to be of a Tertiary age whereas those that post-date the incision belong possibly to the Late Pliocene and Quaternary.

The approximate distribution of these Cenozoic deposits in the Kuiseb Valley west of the Great Escarpment is shown in Fig. 1. The sequence of development of the Kuiseb deposits is briefly summarized below.

The fundamental datum of the surficial deposits is a marked unconformity formed by an extensively planed bedrock surface developed across mainly Late Precambrian Damara Sequence rock types. This bedrock pediplain extends back into embayments of a dissected Great Escarpment and is remarkably little weathered (Selby 1977; Ollier 1978). The planation has been interpreted as a consequence of the break-up of West Gondwana and the concomitant formation of the Great Escarpment by headward erosion in the Cretaceous (Martin 1975; Ward et al. 1983). The original Kuiseb Valley was a broad, up to 20 km wide, bedrock depression in which dominantly arenaceous sediments initially accumulated. These arenites (mainly calcite-dolomite cemented quartz sands) are interpreted here as a fluvially-deposited facies of the reddish dune and sand sheet arenites that characterize the Tsondab Sandstone Formation (SACS 1980).

Interbedded in both the "fluvial" and "aeolian" facies of the Tsondab

Table 1. Outline of the Late Mesozoic-Cenozoic geological and geomorphological history in the Kuiseb Valley west of the Great Escarpment.

OLLIER (1977, 1977, 1978)	PRESENT INVESTIGATION	
	Lithology	Name
12. Formation of minor terraces and present floodplain as well as north-south dunes south of the Kuiseb.	12. Present Kuiseb River; minor terraces and floodplain; coastal salt flats, pans, with low dune formation.	
11. Formation of the lower terraces.	11. Gravels, sometimes cemented; characteristic golden-yellow colour of quartz clasts in lower reaches.	Gobabeb Gravels
10. Re-excavation to bedrock, leaving remnants of Homeb Silts.	10.	
9. Deposition of Homeb Silts.	9. Micaceous silts with interbedded river- and dune-derived arenite lenses; silts with interbedded pan deposits in Delta region.	Homeb Silts
8. Erosion to bedrock at present river level.	8. Erosion to bedrock at present river level. position uncertain?	Hudaob tufa deposits
7. Deposition of gravels of Ossewater Conglomerate to form a terrace. Cementation.	7. Conglomerates; quartz arenites; interbedded palaeo-dune wedges in downstream outcrops.	Oswater Conglomerate
6. Cutting of canyon, almost to present level, tributaries form badlands on north bank.	6. Incision of Kuiseb drainage system and canyon formation.	? Onset of the accumulation of the main Namib Sand Sea (Sossus Sand Formation) with intermittent pan (playa) development. (Khommabes-type deposits).
5. Initiation of Kuiseb course on Tsondab Planation Surface or Namib Unconformity surface.	5.	
	Widespread pedogenic calcrete development on surficial deposits.	
4. Erosion of Tsondab Planation Surface; pediment with E-W drainage; calcrete conglomerates; possibly Middle to Late Tertiary.	4. Boulder conglomerates to lag gravels; scree and fan-like conglomerates.	Karpfenkliff Conglomerate
3. Deposition of the Tsondab Sandstone; dune origin.	3. Widespread arenites; palaeo-pans (-playas); palaeo-alluvial fan/colluvial.	Tsondab Sandstone Formation
2. Deposition of the Basal Conglomerate; locally, or breccia or may be absent.	2. Basal breccia	
1. Formation of the Namib Unconformity Surface; possibly pre-Upper Cretaceous; no evidence of deep chemical weathering.	1. Ditto; extensively planed bedrock surface with minimal chemical weathering; possibly End-Cretaceous surface.	

Sandstone are lenses of indurated carbonate (commonly dolomite-rich) that are interpreted as palaeo-pans (-playas). These carbonate units have been formally designated the Zebra Pan Carbonate Member. The accumulation of the Tsondab Sandstone Formation probably spanned the Early to Middle Tertiary and these arenaceous sediments suggest a predominance of desert to semi-desert conditions at that time (Martin 1950; Ward et al. 1983).

The Tsondab Formation arenites are overlain unconformably by rudaceous sediments, the Karpfenkliff Conglomerate Formation, in the broad proto-Kuiseb Valley. These calcified, dominantly conglomeratic, deposits wedge out from c. 60 m thick at the Chausib River to mere lag gravels west of Gobabeb. A decrease in maximum clast size also occurs from the Escarpment towards the coast. The Karpfenkliff Conglomerate records the first well-developed, integrated Kuiseb-Gaub drainage system. The clast assemblage is dominated by the resistant Damara metaquartzites and vein quartz as well as the Karoo-age Etjo Formation quartzites. The clasts are commonly rounded to well-rounded and may be scarred with arcuate percussion marks. The Karpfenkliff Conglomerate has been tentatively assigned a Middle Miocene age (Ward et al. 1983) and probably represents a large alluvial fan deposit.

Both the Tsondab Formation arenites on the interfluvial areas and the Karpfenkliff Conglomerate in the proto-Kuiseb Valley are commonly capped by a pedogenic calcrete, called here the Kamberg Calcrete Formation. This calcrete commonly exhibits a mature profile of laminar crust → hardpan → honeycomb nodular → host material and may be up to 5 m thick. Yaalon and Ward (1982) have suggested that this important Cenozoic stratigraphic marker represents landform stability spanning several hundreds of thousands of years and semi-arid conditions with a summer rainfall in the order of 350 - 450 mm. This calcrete formation is correlated with similar deposits in the Southern Namib (Ward et al. 1983) that are considered to be End Miocene in age (Stocken 1978; SACS 1980).

The Late Tertiary incision of the Kuiseb River and its major tributaries was followed by the aggradation of the Oswater Conglomerate Formation. These calcified rudaceous sediments thicken westwards away from the Escarpment and are largely confined to the canyon-like reaches of the Kuiseb River and its major tributaries (Ward 1982). Palaeodunes are interbedded in the conglomerates in the lowermost reaches of the Kuiseb canyon and imply the presence of dunes along the left (south) bank of the Kuiseb at the time of aggradation (Ward 1982). Significantly, the Oswater Conglomerate was calcified/cemented prior to the re-incision of the Kuiseb River and its major tributaries, although C14 dates of 28 000 - 33 000 BP (Vogel 1982) suggest a later remobilization of the cement (Vogel and Butzer, pers. comm. 1983). The Oswater Conglomerate has been cautiously assigned an Early to Middle Pleistocene age and would appear to reflect higher energy fluvial conditions than those currently experienced in the Kuiseb drainage system.

Following the re-excavation of the Kuiseb drainage system and an incision of some 50 m in the canyon reaches, a Late Pleistocene aggradation phase is recorded by the Homeb Silt Formation. These micaceous sediments, dated at 19 000 - 23 000 BP (Vogel 1982) probably represent floodplain sediments banked up into the tributaries (dominantly right, or north, bank) between Gomkaeb and about Gobabeb. The abundance of ripple-drift cross-lamination, interbedded arenites of Kuiseb River and dune origin and the presence of desiccation cracks in some layers tend to support a fluvial origin for the Homeb Silts. This interpretation contrasts with the dune dam (Rust and Wieneke 1974, 1980) and river end-point (Marker and Müller 1978; Vogel 1982) hypotheses proposed for the accumulation of the Homeb Silts.

The subsequent hydrological regime responsible for the re-excavation of the Kuiseb Valley resulted in the deposition of the Gobabeb Gravel Formation. These pebble-cobble size clasts of resistant quartzites and vein quartz are commonly uncemented and can be traced from about Homeb, some 20 km upstream from Gobabeb, as a lag gravel to the Kuiseb Delta. The vein quartz clasts in the Gobabeb Gravels are a characteristic golden-yellow colour. A C14 date on the vesicular carbonate formed below the stable surface of the lag gravels gave a value of 9600 BP (Vogel 1982).

The occurrence of tufa carapaces, up to c. 2 m thick and extending over some 30 - 50 m, has been interpreted as spring discharge events within the Kuiseb canyon. These tufas, known as the Hudaob Tufa Formation, appear to post-date the Oswater Conglomerate but pre-date the Gobabeb Gravels. However, their stratigraphic position is uncertain and these deposits may reflect at least several different phases of discharge relating to locally wetter conditions.

The occurrence of pan carbonates in the main Namib Sand Sea south of

Table 2. Principal rock types encountered in the study area - emphasis on the Cenozoic Succession.

ERA	PERIOD		PRINCIPAL GEOLOGICAL UNITS	LITHOLOGY
CENOZOIC	QUATERNARY	Recent	Kuiseb River alluvium. Sossus Sand Formation Gobabeb Gravel Formation*. including Khommabes Carbonate Member*.	· Quartz (>90%) sands · Pebble-cobble lag gravel; minor conglomerate; rolled MSA artefacts · Calcium carbonate; calcified reed stalks and leaves; fossils, including Elephas reckii, rare
		Pleistocene	Homeb Silt Formation*. Hudaob Tufa Formation*.	· Micaceous silts; minor quartz sand lenses and locally derived conglomerate · Calcium carbonate; traces of calcified vegetation
			Oswater Conglomerate Formation*.	· Boulder conglomerate (proximal to Great Escarpment); Cobble-boulder conglomerate and interbedded quartz arenite; Pebble conglomerate, arenite lenses and wedges of interbedded aeolianite (distal reach)
	TERTIARY		Unconformity	
		Neogene	Kamberg Calcrete Formation*.	· Calcium carbonate. In profile: laminar crust (surface)→hardpan calcrete+ honeycomb calcrete+nodular calcrete
			Rooikop gravels.	· Fluvio-marine littoral gravels; oysters (Striostrea margaritacea); gypsum
			Karpfenkliff Conglomerate Formation*.	· Boulder conglomerate (proximal to Great Escarpment); Cobble-boulder conglomerate and interbedded quartz arenite lenses; Pebble-cobble lag gravels (distal area)
			Unconformity	
		Palaeogene	Tsondab Sandstone Formation. Zebra Pan Carbonate Member *	Carbonate-cemented quartz arenite, red-brown; patterned ground and macro-fractures Calcite-dolomite cemented quartz arenite; rare pebble horizons Indurated, locally-developed carbonate; mainly dolomite Uncemented quartz arenite, red-brown; structureless to cross-stratified
			Gomkaeb Basal Breccia Member*	Calcite-dolomite cemented quartz-garnet breccia Basal 'quartz rich' breccia (locally absent)
MESOZOIC	CRETACEOUS		MAIN UNCONFORMITY	Namib Unconformity Surface
	JURASSIC		Dolerite dykes	Dolerite
			Intrusive	
			Etjo Formation (Karoo Sequence)	Silicified quartz sandstone (quartzites, original sand grains visible in hand speciment)
			MAIN UNCONFORMITY	
PRECAMBRIAN			Damara Sequence	Schist; metaquartzite (crystalline); marble; calcsilicate; amphibolite; intrusive granites (Cambrian)
			Gamsberg Suite	Granite
			Rehoboth Sequence	Metaquartzite; schist; metaconglomerate; quartz porphyry; intrusive metabasite dyke swarm, quartz diorite and granodiorite
			Abbabis (west); Mooirivier (east) Complexes	Paragneiss; metasedimentary rocks

the Kuiseb would appear to reflect a least temporarily standing water. These carbonates, identified as the Khommabes Carbonate Member of the Sossus Sand Formation may have formed at any time through the Late Tertiary to Quaternary. At one site, elephas reckii remains point to a Middle Pleistocene age (Shackley 1980) for at least one of these pan accumulations.

The results of a monitoring programme have shown that the large linear dunes advanced at rates of up to 2 m per yr in a general northerly direction into the Kuiseb River from May 1978 to December 1981. The crescentic coastal dunes moved northwards at rates of up to 6 m per yr into the southern channel of the Kuiseb Delta during that time.

CONCLUSIONS

The surficial deposits in the Kuiseb Valley west of the Great Escarpment form a Cenozoic succession that probably dates back to the Early Tertiary. A formal nomenclature, based on SACS (1980) guidelines, has been set up for the main sedimentary units of the Cenozoic succession and presented here.

Although the sedimentary history is relatively complex, the palaeo-environmental conditions appear to have been dominated by arid to semi--arid processes throughout the Cenozoic succession. The Cenozoic deposits fall into two broad categories, viz. those pre-dating the deep, canyon incision of the Kuiseb drainage system and those post-dating that incision. The deep incision of the Kuiseb drainage post-dates the Kamberg Calcrete Formation, which has been cautiously assigned an End Miocene age. The post-incision deposits appear to reflect an overall fluctuating, and possibly decreasing, from the Oswater Conglomerate → Homeb Silts → Gobabeb Gravels. This trend accords with the progressively aridifying trend noted in the Namib Desert following the full establishment of the Benguela Current and associated cold-water upwelling the the Late Miocene (Korn and Martin 1957; Siesser 1977, 1980; Tankard and Rogers 1978; Ward et al. 1983).

It appears that the Kuiseb River has played a significant role as the northern boundary of the main Namib Sand Sea since at least the aggradation of the Oswater Conglomerate, and thus probably for most of the Late Tertiary - Quaternary. The exception, however, can be found in the immediate coastal tract where the high-energy, unidirectional south-southwesterly wind regime has dominated, resulting in the formation of a narrow belt of littoral dunes from Walvis Bay northwards to the Swakop River.

ACKNOWLEDGEMENTS

This paper summarizes four years of work carried out under the auspices of the Kuiseb Environmental Project and I gratefully thank the many people and organizations who assisted me. The financial aid of the Co--operative Scientific Programmes, CSIR, and logistic support from Geological Survey, Windhoek, are acknowledged with gratitude. I particularly thank V von Brunn, CJ Ward, MC Ward, VA Gray, MK Seely, BJ Huntley, R McG Miller, K Hoffmann, D Scogings, H Martin, D Yaalon, D Bühmann, MJ Selby, T Partridge, C Stocken, N Lancaster, B Rimbault and L le Roux for their help. The Department of Nature Conservation, SWA, is thanked for permission to work in the Namib-Naukluft Park and for providing facilities at Gobabeb.

REFERENCES

Alexander JE 1838, repr. 1967. Expedition of discovery into the interior of Africa. 2 vol. Struik, Cape Town. 637 pp.

Barnard WS 1973. Duinformasies in die Sentrale Namib. Tegnikon: 2-13.

Barnard WS 1975. Geomorfologiese prosesse en die mens: die geval van die Kuiseb delta. SWA Acta geogr. 2:20-43.

Besler H 1967/77. Untersuchungen in der Dünen-Namib (Südwestafrika). J. S.W.A. scient. Soc. 31:33-64.

Besler H 1980. Die Dünen-Namib: Entstehung und Dynamik eines Ergs. Stuttgarter geogr. Stud. 96.

Geological Map of South West Africa/Namibia 1980. Comp. by Geol. Surv., Windhoek.

Gevers TW 1936. The morphology of Western Damaraland and adjoining Namib Desert of South West Africa. S.Afr. geogr. J. 19:61-79.

Goudie A 1972. Climate, weathering, crust formation, dunes and fluvial features of the central Namib Desert, near Gobabeb, South West Africa. Madoqua 1:15-31.

Harmse JT 1980. Die noordwaartse begrensing van die Sentrale Namib duinsee langs die Benede-Kuiseb. MA thesis, unpubl. Univ. Stellenbosch.

King LC 1951. South African Scenery, 2nd ed. Oliver and Boyd, Edinburgh.

Korn H and Martin H 1957. The Pleistocene in South West Africa. In: JC Clarke and S Cole (eds.), Proc. III Pan-Afr. Congr. on Prehistory. Chatto and Windus, London. p. 14-22.

Logan RF 1960. The Central Namib Desert, South West Africa. Publ. 758, Natn Acad. Sci. Natn Res. Coun., Washington, D.C.

Marker ME 1977. Aspects of the geomorphology of the Kuiseb River, South West Africa. Madoqua 10:199-206.

Marker ME 1982. Aspects of Namib geomorphology: a doline karst. In: JA Coetzee and EM van Zinderen Bakker (eds.), Palaeoecol. Africa 15:187-199.

Marker ME and Müller D 1978. Relict vlei silt of the middle Kuiseb Valley, South West Africa. Madoqua 11:151-162.

Martin H 1950. Südwestafrika. Geol. Rdsch. 38:6-14.

Martin H 1957, repr. 1974. The Sheltering Desert. John Meinert, Windhoek.

Martin H 1975. Structural and palaeogeographical evidence for an Upper Palaeozoic Sea between Southern Africa and South America. Proc. Pap. IUGS 3rd Gondwana Symp. 1973, Canberra. p. 37-51.

McKee ED 1982. Sedimentary structures in dunes of the Namib Desert, South West Africa, Geol. Soc. Am. spec. Pap. 188.

Ollier CD 1977. Outline geological and geomorphic history of the Central Namib Desert. Madoqua 10:207-212.

Ollier CD 1978. Inselbergs of the Namib Desert - processes and history. Z. Geomorph. N.F. 31:161-176.

Partridge TC 1982. A review of southern African geomorphology. Geol. Soc. S. Afr. Winter Fieldschool 7.

Rust U and Wieneke F 1974. Studies on gramadulla formation in the middle part of the Kuiseb River, South West Africa. Madoqua 3:69-73.

Rust U and Wieneke F 1980. A reinvestigation of some aspects of the evolution of the Kuiseb Valley upstream of Gobabeb, South West Africa. Madoqua 12:163-173.

South African Committee for Stratigraphy 1980. Stratigraphy of South Africa, Part 1. Comp. LE Kent. Handbk. geol. Surv. S. Afr. 8:578-609.

Sawyer EW 1976. Provisional legend of field sheets 2315CA Gobabeb and 2315CB Gorob. Geol. Surv., Windhoek.

Shackley M 1980. An Acheulean industry with Elephas reckii fauna from Namib IV, South West Africa. Nature 284:340.

Siesser WG 1978. Aridification of the Namib Desert: Evidence from oceanic cores. In: EM van Zinderen Bakker (ed.), Antarctic Glacial History and World Palaeoenvironments. Balkema, Rotterdam. p. 105-112.

Siesser WG 1980. Late Miocene origin of the Benguela upwelling system off northern Namibia. Science 208:283-285.

Stapff FM 1887. Karte des Untern !Khuisebthales. Petermann's geogr. Mitt. 33:201-214.

Stocken CG 1978. A review of the later Mesozoic and Cenozoic deposits of the Sperrgebiet. Unpubl. rep. Geol. Dept. CDM (Pty) Ltd.

Tankard AJ and Rogers J 1978. Late Cenozoic palaeoenvironments on the west coast of southern Africa. J. Biogeogr. 5:319-337.

Vogel JC 1982. The age of the Kuiseb river silt terrace of Homeb. In: JA Coetzee and EM van Zinderen Bakker (eds.), Palaeoecol. Africa 15:201-209.

Ward JD 1982. Aspects of a suite of Quaternary conglomeratic sediments in the Kuiseb Valley, Namibia. In: JA Coetzee and EM van Zinderen Bakker (eds.), Palaeoecol. Africa 15:211-216.

Ward JD 1983. Sand dynamics along the Kuiseb River. In: BJ Huntley (ed.), Kuiseb Environmental project: the development of a monitoring baseline. S. Afr. natn scient. Progrms, Rep. 68.

Ward JD, Seely MK and Lancaster N 1983. On the antiquity of the Namib Desert. S. Afr. J. Sci. 79:175-183.

Wilmer HC 1893. The relation of the sand dune formation on the south west coast of Africa to the local wind currents. Trans. S. Afr. phil. Soc. 5:326-329.

Yaalon DH and Ward JD 1982. Observations on calcrete and recent calcic horizons in relation to landforms, Central Namib Desert. In: JA Coetzee and EM van Zinderen Bakker (eds.),. Palaeoecol. Africa 15:183-186.

Radiocarbon dating of speleothems from the Rössing cave, Namib desert, and palaeoclimatic implications

KLAUS HEINE
Universität Regensburg, Germany

MEBUS A. GEYH
Niedersächsisches Landesamt für Bodenforschung, Hannover, Germany

ABSTRACT. About 2 km west of the Rössingberge in the Namib desert at about 340 m above sea level, a small cave system with stalactites, stalagmites, flowstone, popcorn and other sinter formations was investigated in 1978 and 1981. Speleothems have been formed since the Tertiary. As the cave is within a small catchment area, the speleothems reflect changing palaeohydrological and palaeoclimatological conditions in the Namib desert in the vicinity of the cave rather than those of the escarpment in the east.

According to the C14 dates, the climate after 41 500 BP can be divided into five phases. Until about 25 500 BP rather humid conditions prevailed as cave sinter was formed. Afterwards the climate remained dry in the study area. As from 19 000 BP, the interior of South West Africa also became arid.

INTRODUCTION

Some questions of the Late Quaternary climatic evolution in the Namib desert will be discussed briefly. During recent years many scientists have tried to find evidence for or against the penetration of hypothermal (Glacial Maximum) winter rainfall in the Namib. According to Rust and Schmidt (1981), for example, winter rainfall penetrated as far north as 20 °S during the last pleniglacial. Van Zinderen Bakker (1983a, b), on the other hand, argued that winter rainfall never extended so far north during the Late Quaternary. His pollen study indicates that the climatic conditions at Sossus vlei had not changed since at least 18 000 BP. Vogel (1982) concluded from dates obtained by him on silt, calcretes, and one wood sample taken at Homeb on the Kuiseb River, that the last humid period in the Namib desert ended about 28 000 BP.

About 2 km west of the Rössingberge (14° 48' E, 22° 31,5' S) at 338 m above sea level, a small cave with a variety of speleothems reflects the climatic evolution in the vicinity of the Rössingberge in the central Namib desert. Two Th230/U234 dates of speleothems exceeding 300 000 yr support the geomorphic evidence that the cave and the oldest sinter formation might be of Tertiary origin. In this paper attention is focused on the Late Quaternary only.

Speleothems from arid zones can be considered as "closed systems" with respect to carbon isotopes. Therefore, such samples are most suitable for C14 dating. The calibration uncertainty of the date is about ± 1000 years. In spite of this C14 ages seem to be more reliable than those of any other samples from the Namib desert, such as calcretes or fluviatile sediments which are often changed diagenetically.

THE CAVE AND PERIODS OF SPELEOTHEM FORMATION

The cave is situated in a narrow belt of calcareous rocks that emerges a few metres above the old Namib desert surface within a natural catchment area of a few km^2 (Fig. 1). Thus, the growth of speleothems was determined by the local precipitation in the vicinity of the cave on the western slopes of the Rössingberge rather than by that from the escarpment in the east.

The cave consists of three chambers situated parallel to the NNW/SSE direction of the folded rocks in the Namib. The chambers are about 10-20 m beneath the surface. Their widths range from a few metres to about 15 m, their lengths from about 10-25 m. The height of the chambers changes from several decimetres to c. 3 m.

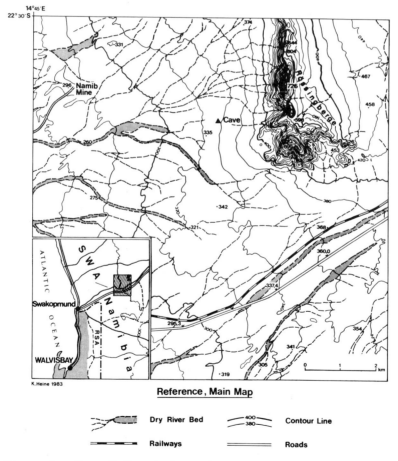

Figure 1. Map of the study area.

According to the geomorphologic situation the cave was formed by lime solution or erosion during pre-Pleistocene times. Only a few points in the cave display cave deposits as stalactites, stalagmites, flowstone, and popcorn. At present, the cave appears to be completely dry though drops of seeping water have been observed during extremely wet years.

In the northern part of the cave, a huge stalagmite (Fig. 2) stands beneath a big fissure which allowed fine aeolian sands from the Namib desert to enter the cave together with the dripping water. They are found in between the youngest flowstone layers (Fig. 2) surrounding the eastern and southern flanks of the stalagmite.

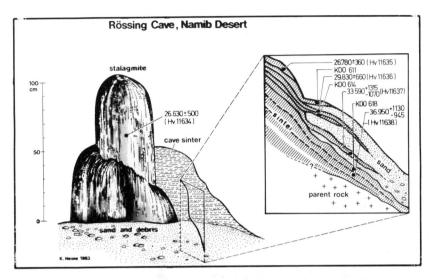

Figure 2. Speleothem morphology and sampling points for isotope analyses in the Rössing cave.

The conventional C14 ages of the uppermost flowstone layers (Fig. 2) range from 37 000 - 26 800 BP. The speleothems, formed until about 35 000 BP, consist of massive calcite deposits. Later on, sand horizons are intercalated. The conventional C14 dates of speleothem samples from other places in the cave (Table 1) also show that sinter formations terminated at 26 500 BP. This holds true also for the popcorn deposits consisting of irregularly alternating light and dark layers of aragonite and calcite. Their formation began before 41 000 BP. The outer part of a unique stalactite curtain yielded a date of 37 000 BP.

Only conventional C14 dates have been discussed up to now. It has been known, however, since the method for C14 dating of speleothems was introduced by Franke (1951) and proved to be applicable (Franke et al. 1958), that a "reservoir correction" of at least -1000 years is necessary to obtain actual ages. Labeyrie et al. (1967) even found a correction value of up to -3500 years for uncovered areas. However, as we do not know the geoecologic, pedologic and sedimentologic situation in the Namib desert around 30 000 BP, we accept Vogel's (1982) correction value of -1000 years for our geochronological evaluation. This does not exclude that the resulting time scale may have to be shifted further on by up to -2500 years. In this case, the end of speleothem formation might have coincided with the date of 23 000 BP obtained from wood found in silts at Homeb (Vogel 1982).

Table 1. Conventional C14 ages, C14 content (pmc) and $\delta C13$ values (‰) from speleothems of the Rössing cave in the Namib desert.

Hv	KOO	substance	RC age (years BP)	RC content (pmc)	$\delta C13$ (‰)
11634	610	stalagmite	26 630 ± 500	3.6 ± 0.2	-5.4
11635	613	flowstone	26 780 ± 360	3.6 ± 0.2	-7.9
11636	612	flowstone	29 830 ± 660	2.4 ± 0.2	-3.6
11637	616	flowstone	33 590 ± 1200	1.4 ± 0.2	-4.0
11638	619	flowstone	36 950 ± 1040	1.0 ± 0.1	-5.0
11639	621	stalactite curtain	37 000 ± 1700	1.0 ± 0.2	-6.8
11640	622	cave popcorn	41 500 ± 1280	0.6 ± 0.1	-5.3
9909	162	cave popcorn	29 700 ± 1360	2.5 ± 0.4	-4.5
9910	162a	cave popcorn	26 700 ± 540	3.6 ± 0.2	-5.7
9489	162(1)	cave popcorn	26 530 ± 920	3.7 ± 0.4	-2.0

PALAEOCLIMATIC IMPLICATIONS

Our results of radiometric datings together with sedimentologic observations at the sampling sites in the Rössing cave yield a rather differentiated picture of the Late Quaternary evolution of the climate in the Namib desert (Heine 1982). Various phases can be distinguished for the Middle Weichselian pluvial and other phases occurring up to the present. Once dates of the time before 41 000 BP become available, it will be possible to elongate the sequence by counting phases backwards:

PHASE 5: 40 000-34 000 BP

At the end of this phase, the Last Weichselian pluvial with compact sinter formation terminated within the Namib desert. Missing sand inclusions indicate that a closed plant cover might have existed. In view of the popcorn formation, the humidity must have been higher than later on. Various wall coatings may date back, however, to the Early Quaternary or even to the Tertiary.

PHASE 4: 34 000-27 000 BP

This phase is the beginning of the aridification in the central Namib. At least three noticeable climatic fluctuations occurred during which more humid conditions were replaced by more arid ones and vice versa. This is indicated by intercalating sand layers which account for a more windy climate and probably not completely closed plant cover. However, it was still humid enough for formation of compact sinter. This phase coincided with the calcrete formation on the 40 m terrace along the Kuiseb River (Vogel 1982) which lasted from 33 000 to 28 000 BP.

PHASE 3:   c. 27 000–25 500 BP

Humid conditions predominated once more and compact sinter without aeolian sand inclusions was formed. It is not known whether this was due to decreasing wind activity or once more to a completely closed plant cover in the Namib desert. The humidity dropped drastically at the end as popcorn formation ended. Phase 3 and 4 may be identical. During this phase the rainfall intensity in the interior increased and pebble deposits from the 40 m terrace of the Kuiseb River were removed (Vogel 1982).

PHASE 2:   25 000–19 000 BP

From this time onwards the Namib desert remained dry. No further sinter formation occurred in the Rössing cave. However, according to Heine (1982) the interior of SW Africa was more humid than it was after 19 000 BP. In the Kuiseb valley, dunes started to block the river bed and vlei silt was deposited. At the end of this phase the aridification of the Namib desert was complete (Van Zinderen Bakker 1983a, b).

PHASE 1:   after 19 000 BP

According to our own data from the Rössing cave as well as the results of other publications mentioned already, major climatic fluctuations no longer occurred. The Namib desert and the eastern escarpment remained under arid conditions contrary to the assumption by Rust and Schmidt (1981) that the area between 22 and 23 °S had been humid during the last Pleniglacial.

One of the most important findings of our study is that one has to be careful not to transpose chronostratigraphic sequences from one area to another as different palaeoclimatic events might have been responsible. Therefore, correlations between, for instance, different South African stratigraphies and their comparison with the marine oxygen isotope record can only make sense if at least the palaeogeographical situations are taken into account. Hence, our chronology of the climatic evolution within the Namib desert can be of regional importance only. Owing to the long history of the Rössing cave since the Tertiary, this cave with its speleothems seems to be one of the most suitable places for reconstructing the palaeoclimatic evolution of the Namib desert near 22° 30'S.

ACKNOWLEDGEMENTS

We thank Dr GJ Hennig, Institut für Kernchemie, Universität Köln, for the U/Th age determinations of two stalagmite samples and the Deutsche Forschungsgemeinschaft for financial support.

REFERENCES

Franke HW 1951. Altersbestimmungen an Kalzit-Konkretionen mit radioaktivem Kohlenstoff. Naturwissenschaften 38:527.

Franke HW, Münnich KO and Vogel JC 1958. Auflösung und Abscheidung von Kalk - C14-Datierungen von Kalkabscheidungen. Die Höhle 9:1.

Heine K 1982. The main stages of the Late Quaternary evolution of the Kalahari region, southern Africa. Palaeoecol. Africa 15:53-76.

Heine K 1983. Führt die Quartärforschung zu nicht-aktualistischen Modellvorstellungen in der Geomorphologie ? Colloquium Geographicum 16:93-121.

Labeyrie J, Duplessy JC, Delibrias G and Letolle R 1967. Etude des temperatures des climats anciens par la mesure de O18, de C13 et de C14 dans les concretions des cavernes. In: Radioactive Dating and Methods of Low-Level Counting. IAEA, Vienna. p. 153-160.

Rust U and Schmidt HH 1981. Der Fragenkreis jungquartärer Klimaschwankungen im südafrikanischen Sektor des heute ariden südlichen Afrika. Mitt. geogr. Ges. München 66:141-174.

Van Zinderen Bakker EM 1983a, in press. Aridity along the Namibian coast. Palaeoecol. Africa 16.

Van Zinderen Bakker EM 1983b, in press. A Late and Postglacial pollen record from the Namib desert. Palaeoecol. Africa 16.

Vogel JC 1982. The age of the Kuiseb river silt terrace at Homeb. Palaeoecol. Africa 15:201-209.

African faunal record

Horses, elephants and pigs as clues in the African later Cainozoic

H.B.S.COOKE
Dalhousie University, Halifax, Canada

ABSTRACT. In the later Cainozoic deposits of Africa, horses, elephants and pigs occur frequently and have proved particularly useful for correlation and relative dating. East Africa possesses several sequences of strata with good radiometric controls, providing a framework for calibrating evolutionary changes. Hipparionid horses invaded Africa in the mid-Miocene, close to 12 m.y.BP, with the last species ranging from a little over 2 m.y. BP to the end of the Middle Pleistocene. Equus first appears as an immigrant in East Africa close to 2 m.y. BP. The earliest true elephantine, Primelephas, occurs in the Late Miocene and gave rise in the early Pliocene to primitive species of three genera, Mammuthus, Elephas and Loxodonta. At about 3.5 m.y. BP a second species appears in each genus. Increasing hypsodonty and enamel folding occur in each line, especially after 2.5 m.y. BP, suggesting adaptation to a harsher diet. The suids comprise three major groupings. Nyanzachoerus and Notochoerus are derived from earlier African suids and are confined to the later Miocene and the Pliocene. The typical later Pliocene and Pleistocene suids are presumed to be descended from immigrant Sus-like forms. The one group, embracing Potamochoerus, Kolpochoerus and Hylochoerus have generally low-crowned cheek teeth. In the other group, the extinct Metridiochoerus tends to undergo premolar reduction and develops columnar third molars that resemble the very hypsodont teeth of Phacochoerus, which is derived from it. The rate of change in hypsodonty accelerated after 2.5 m.y. BP and it is likely that some significant environmental change took place at this time.

INTRODUCTION

In the later Cainozoic deposits of Africa, the most abundant fossils are usually those of the Bovidae, which have proved very useful in providing environmental information and have also contributed to the estimation of the probably relative ages of scattered deposits (see Gentry 1978; Gentry and Gentry 1978; Vrba 1976, 1982). However, the horses, elephants and pigs are also common as fossils and have been most useful for correlation and relative dating. The East African region has been particularly well studied and has furnished several well-calibrated stratigraphic sequences in which key horizons can be dated radiometrically. A suggested correlation for some of the major sites is incorporated in Fig. 1. The existence of this excellent

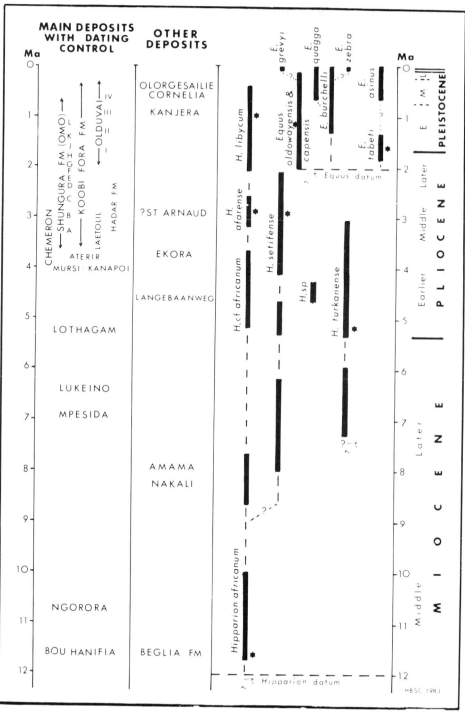

Figure 1. Age relations of the main African deposits with good faunas and time ranges of African fossil equids. The asterisks next to each range bar show the approximate date for the type specimen of that species. The open arrows denote immigration from Eurasia.

framework has made it possible to evaluate the evolution of the fossil groups in some detail and to use the East African succession as a "standard".

THE HORSES

The equids made their first appearance at Bou Hanifia in northern Algeria, where a tuff low down in the sequence has been dated at 12 m.y., indicating that Hipparion arrived in Africa at much the same time as it reached Europe. In East Africa, Hipparion first occurs in the Ngorora Formation, dated as close to 11 Ma. Known occurrences have been analysed by Churcher and Richardson (1978) and in a number of papers by Eisenmann (see especially 1977, 1980a, b, 1981, 1982).

The Bou Hanifia Hipparion was named H. africanum by Arambourg (1959) but Forstén (1968) suggested that it should be considered as a synonym of the European H. primigenium. Churcher and Richardson adopt this latter proposal but Eisenmann (1980) has demonstrated the existence of a number of consistent differences so that it seems better to retain Arambourg's specific name for the early African hipparionids. H. africanum was a medium to large hipparion with a moderately elongate face, broad nasal aperture and a pronounced preorbital fossa. The cheek teeth are of medium hypsodonty in the earlier forms but increase in height with the passage of geological time. In the lowers the ectostylid is only modestly developed in the earlier forms but it increases progressively. H. africanum appears to have given rise to a different species, H. afarense, at present only certainly known from good skulls at Hadar (c. 3 m.y. BP) in which the preorbital fossa is lacking. The cheek teeth are moderately small and in the lowers the ectostylids are well developed but do not normally reach the unworn crown. This species was considered as ancestral to the abundant H. libycum (which includes the northern ethiopicum and the southern variant steytleri) in which the ectostylids are strongly developed and hypsodonty is extreme. The skull of H. libycum lacks the preorbital fossa and in the lower jaw the lateral incisors are greatly reduced. This species is first recognized at Omo in Member F of the Shungura Formation and there are a few specimens in Member B to E that may possibly belong to H. afarense, but this is uncertain. H. libycum is thus most characteristically a Pleistocene species, although it is found also in the terminal Pliocene. It occurs finally in the Cornelia Beds and at Olorgesailie, thus having survived until almost the end of the Middle Pleistocene.

Other hipparionids are found in addition to the above lineage. H. setifense is a small form, known as early as Amama in Algeria (c. 8 m.y. BP) and as late as Member F of the Shungura Formation. It is possible that it is only a small variant of H. africanum, but the skull is not known. Another species, H. turkanense, is based on cranial material from Lothagam and seems to occur also in the Lukeino Formation and in the Chemeron Beds. H. turkanense lacks a preorbital fossa but, unlike H. libycum, it also lacks the ectostylids, and it is possible that it is a later immigrant rather than a descendant of H. africanum. An Hipparion from Langebaanweg (H. baardi) is of uncertain affinities.

Unfortunately the fossil record for all the hipparionids before about 4 m.y. BP is rather scanty. From an environmental viewpoint, the gradual increase in hypsodonty of H. africanum may be a response to a slow increase in grazing habits, while the rapid change in hypsodonty and the strong development of ectostylids that began at about 3 m.y. BP and was fully developed by 2 m.y. BP may indicate a more marked environmental change. With increasing reliance on the savanna habitat, the hipparionids could have suffered through competition with the rapidly diversifying bovids.

In East Africa, Equus appears for the first time in member G5 of the Sungura Formation of the Omo basin, with an age very close to 2 m.y. BP. In North Africa Equus numidicus is found at Aïn Boucherit, which is undated but may be late Pliocene in age and perhaps slightly earlier than in East Africa. In sub-Saharan Africa the first appearance of Equus may be inferred to indicate an age not greater than 2 m.y. BP. The North African E. numidicus is a moderate sized horse, about as big as a large zebra, and perhaps represents the immigrant stock from which arose the characteristic large forms of sub-Saharan Africa, E. oldowayensis in East Africa and E. capensis in southern Africa (probably synonyms). This large species ranges from 2 m.y. BP to the very end of the Pleistocene (Klein 1980). It possesses characters reminiscent of the living E. grevyi and is very possibly ancestral to it. Fossil material not separable from the living E. burchelli is widespread and commonly found along with E. oldowayensis/capensis at least as far back as the middle part of the Early Pleistocene. E. quagga has a very uncertain fossil record but is not known outside southern Africa and is not reported before the middle Pleistocene. E. asinus has a similar time range but is restricted to North Africa where a possible ancestor exists in the terminal Pliocene/Early Pleistocene E. tabeti (Eisenmann 1980b, 1981). E. zebra is extremely rare as a fossil, being known only in the south and essentially in Holocene settings. Equus is typically a grazer and its abundance suggests that grassland was plentiful throughout its time range. There is also a possibility that E. capensis/oldowayensis was, like the modern Grevy's Zebra, already adapted for survival in rather dry conditions.

THE ELEPHANTS

The proboscideans have a long history in Africa, dating back to the Eocene. The curious deinotheres are found in the early Miocene in East Africa (Prodeinotherium hobleyi) while a descendant form (Deinotherium bozasi) occurs in the Late Miocene and continues as an interesting element of the faunas right up to about 1.3 m.y. BP (Beden 1979). The mammutoid Zygolophodon occurs in the middle to Late Pliocene in North Africa but has not been reported south of the Sahara. Stegodon kaisensis is presumably a Late Pliocene immigrant from Asia and is a rare element in East Africa up to almost 3 m.y. BP in the lower part of the Shungura Formation. The gomphothere Anancus kenyensis ranges up to the very end of the Pliocene in East Africa and an allied species series, A. osiris persists into the Middle Pleistocene of North Africa.

The true elephants arose in Africa in the terminal Miocene and spread out in the Late Pliocene to become valuable chronometers in the Pleistocene of Eurasia and North America. The most primitive forms now placed in the Elephantidae are Stegotetrabelodon and Stegodibelodon, occurring in the terminal Miocene and earliest Pliocene of North and East Africa, but the earliest of the Elephantinae is Primelephas (Maglio 1973; Coppens et al. 1978). Primelephas is best known from Lothangam but occurs also in the lower Kaiso deposits and at Kanam East, as well as at Kolinga in Chad. By 4 m.y. BP, there occur representatives of each of the three well-known genera Mammuthus (M. subplanifrons),

Elephas (E. ekorensis) and Loxodonta (L. adaurora). The time ranges of these and descendant species are shown in Fig. 2. The teeth of the early species are rather similar and cannot always be distinguished in single isolated specimens, but cranial material shows clearly that the distinctive features of each genus were already developed. At about 3.5 m.y. BP further speciation occurred in each of the three genera, although the parent species apparently persisted for some time alongside the descendants.

In all the lineages there is a tendency for an increase in the number of plates in the molars, greater crown height and concurrent thinning of the enamel, accompanied by more folding. These trends are least marked in Loxodonta (see Maglio 1973 Figs. 40-42). Loxodonta atlantica succeeds L. adaurora and is first recorded at about 2.5 m.y. BP, ranging up to the Middle Pleistocene. It is rare in East Africa, where it seems to have been displaced by Elephas recki, but it occurs in North Africa and in South Africa, making its last appearance at Elandsfontein near the end of the Middle Pleistocene. The living species, L. africana, has a poor fossil record and is not certainly known before 1 m.y. BP but it seems more likely to have been derived from L. adaurora than from L. atlantica. It may well have been largely a forest-dweller and its spread across the continent seems to coincide with the decline of Elephas recki.

Mammuthus is not common as a fossil, although remains of M. subplanifrons occur in southern and eastern Africa. A skull from Hadar has teeth intermediate between those of typical M. subplanifrons and the characteristic Middle to Late Pliocene M. africanava of North Africa. Specimens attributed to the well-known Eurasiatic M. meridionalis have been reported from North Africa but it is not clear whether they are of African origin or were re-imported.

The most important fossil elephants in Africa belong to the genus Elephas. The primitive E. ekorensis, ranging from 4 m.y. to 3 m.y. BP, gave rise to E. recki before its own extinction. A thorough study by Beden (1979b) results in the recognition of five successive evolutionary stages, which have been given varietal names (see Fig. 2). Through the successive stages there is a slow increase in the number of plates (11-12 in r. brumpti to 15-18 in r. recki third molars) but the breadth remains fairly constant. Crown height increases slowly at first but changes more rapidly after 2 m.y. BP. Enamel thickness declines from 3.5 to 3.0 mm while the height is changing slowly but then remains fairly constant at about 2.8 mm during the phase of rapid increase in hypsodonty. There are also progressive changes in the degree of folding of the enamel so that a combination of these criteria can be used effectively for the identification of the different varieties, allowing time discrimination of the order of half a million years. The most advanced species of Elephas is E. iolensis, first recorded from North Africa but now known from Middle to Late Pleistocene sites in eastern and southern Africa.

Although one variety of Loxodonta africana is restricted to a forest environment, the species as a whole is adapted to a wide variety of habitats, ranging even into semi-desertic conditions and montane savanna. The living Elephas is also primarily a forest-dweller but its fossil representatives are common in savanna associations and it would be unwise to try to draw any close analogy between the fossil environments and the present restricted habitat. It seems likely that the development of the elephantine fore-and-aft method of chewing, as distinct from the rotary action of the gomphotheres, increased the shearing

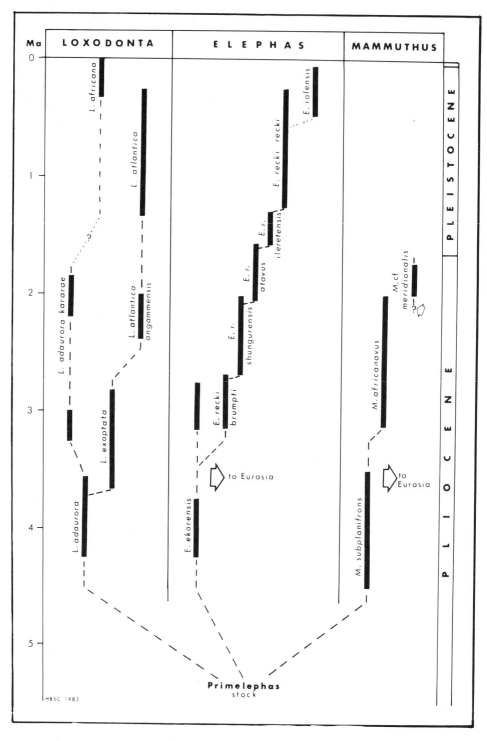

Figure 2. Time ranges of the species of Elephantinae in Africa.

action of mastication and opened the way for dealing with harsher diets. Hypsodonty and increased enamel folding also prolong tooth life with abrasive diets. The drying trend that is shown to have begun in the Omo region at 2.5 m.y. BP (Bonnefille 1980) may have played a role in the diversification that came about at the time, although the reasons are not clear.

THE PIGS

Remains of pigs are widespread and usually fairly abundant at most of the later Cainozoic sites and it is not uncommon for a single stratigraphic unit to yield remains of two, or even three different species, often representing different genera. Two slightly different taxonomic schemes have been proposed by Harris and White (1979) and by Cook and Wilkinson (1978) of which the latter forms the basis for Fig. 3, but the differences are in matters of detail and do not affect the phylogenetic picture as a whole.

The genus Nyanzachoerus appears to have evolved in Africa from a bunolistriodont ancestor such as the mid-Miocene Kubanochoerus massai, and it is confined to the later Miocene and the Pliocene. Nyanzachoerus is characterized by the possession of greatly enlarged third and fourth premolars and the species differ primarily in the relative development of the third molars. Ny. jaegeri probably gave rise to an important later Pliocene genus, Notochoerus, in which the premolars are slightly reduced and the third molars are enlarged and become progressively more and more hypsodont; those of advanced No. scotti resemble enlarged Phacochoerus teeth in proportions. Notochoerus does not persist into the Pleistocene.

The other two major groups of African fossil suids are thought to have been derived from Sus-like ancestors that entered Africa from Eurasia about 4 m.y. BP or a little later. The one group is fairly conservative in its dental patterns, with relatively low-crowned bunodont cheek teeth and a normal complement of premolars (three upper, four lower) except in the highly specialized living forest hog, Hylochoerus. This group includes the living bush pig, Potamochoerus porcus, but its fossil record is obscure, although Harris and White (1979) believe that it is already recognizable at 3 m.y. BP. The most abundant fossils belong to Kolpochoerus and the main series (afarensis - limnetes - olduvaiensis) exhibits progressive enlargement of the heel in the third molars and an increase in crown height that are useful for estimating probable geological age. The Middle Pleistocene K. paiceae of South Africa seems to be a separate species with some convergent resemblance in the third molars to those of the K. olduvaiensis.

The second major group of Late Pliocene to Pleistocene suids exhibits modification of the molars to a more columnar structure, accompanied by reduction in the premolars (except in the South African Potamochoeroides shawi) and progressive increases in hypsodonty in the Metridiochoerus andrewsi - jacksoni lineage. The climax of this group is reached in Stylochoerus compactus, which has teeth larger and proportionally higher crowned than in the warthog. Stylochoerus also has tusks superficially resembling those of an elephant and the upper tusks emerge at right angles to the axis of the skull and are directed diagonally upward and outward. The genus Phacochoerus appears to be derived from a fairly early member of the Metridiochoerus complex.

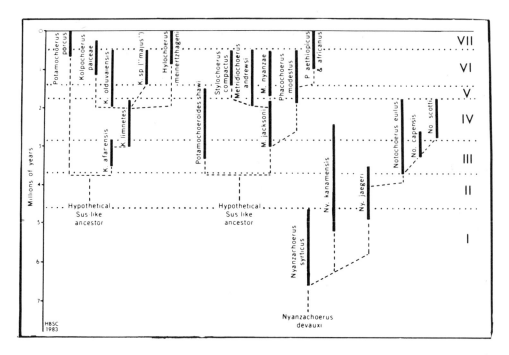

Figure 3. Time ranges of the fossil Suidae of Africa. The Roman numbers indicate species associations referred to in the text.

Because most sites carry more than one species of suid, it is possible to recognize certain groupings of taxa that can often be used almost as informal "zones" for dating purposes. Theseassociations are shown in Fig. 3. The oldest grouping (I) is dominated by Nyanzachoerus syrticus, which is replaced in Group II by an association of Ny. kanamensis (common) and Ny. jaegeri (rare), but not with Notochoerus euilus. Group III is characterized by No. euilus, often accompanied by Kolpochoerus afarensis but not by No. scotti. Group IV is typified by Kolpochoerus limnetes (sensu Cooke), Metridiochoerus jacksoni and Notochoerus scotti. A very substantial change, perhaps associated with some "event", seems to be indicated between 2 and 1.8 m.y. BP, when Group IV is replaced by Group V, now characterized by K. olduvaiensis, Metridiochoerus andrewsi (sensu Cooke), N. nyanzae (rare) and Phacochoerus modestus (very rare). In Group VI Stylochoerus compactus is prominent, accompanied by Kolpochoerus "Majus" (rare) along with Group V elements. Group VII is greatly impoverished and is characterized by the absence of most of the Group VI forms, which became extinct; Phacochoerus aethiopicus is the only major element in most of the known deposits. At lease in part, this sudden "extinction" is a consequence of lack of suitable deposits spanning the period. Each of the groupings covers a time span of about 1 million years, so it does not in itself provide a very refined chronology. However, in many of the species there are morphological changes through time, sometimes capable of being quantified, and these can be used for narrower time discrimination. Harris and White (1979) have used evolutionary data successfully to recognize 15 arbitrary divisions within which the various sites may be placed.

Ecological conditions clearly play a role in the occurrences of various taxa at different sites but it is not easy to attach environmental inferences to each species. However, Behrensmeyer (1976) was able to show that Kolpochoerus, with its low-crowned molars, was relatively more abundant in delta and flood-plain sedimentary environments at East Turkana, while the more hypsodont Notochoerus/Metridiochoerus were most abundant in channel deposits, almost to the level of mutual exclusion between the two groups.

By analogy with the living forms, it may be inferred that the species with brachyodont molars generally inhabited moist wooded environments where soft vegetation could be obtained by browsing, rooting, eating fruits and sometimes insects or small animals. In Nyanzachoerus, however, the relatively simple molars are accompanied by greatly enlarged premolars, the function of which is not clear. These premolars initially have a strong shearing action but later become abraded to the same plane as the other cheek teeth. Possibly the diet may resemble that of the hippopotamus, but this is very speculative.

The tendency towards higher crowned teeth with a multi-columnar structure and abundant cement is clearly an admirable adaptation to a diet of grass, with its harsh abrasive qualities. Stylochoerus compactus, Metridiochoerus andrewsi (sensu Cooke) and Phacochoerus aethiopicus/africanus are strongly convergent in third molar structure and are clearly grazers. While this may be a response to climatic changes leading to spread of grassland about 2 m.y. BP, Notochoerus scotti achieved a similar state and began its rapid evolution a million years earlier, thus suggesting that the climatic trend had already begun before 2.5 m.y. BP. This accords with the palynological evidence which Bonnefille (1980) considers to show that forest was present at 3.0 m.y. BP but between 2.5 and 1.0 m.y. BP there was a dominance of Graminae, fluctuating between desertic steppe and wooded savanna. It would seem from several lines of evidence that there was a period of environmental stress about 2.5 - 2.6 m.y. BP which resulted in considerable morphological changes in the faunas.

REFERENCES

Arambourg C 1959. Vertébrés continentaux du Miocène supérieur de l'Afrique du Nord. Mem. Publ. Serv. Carte Géol. Algerie, n.s.Paléont. 4:1-161.

Beden M 1979a. Données récentes sur l'évolution des Proboscidiens pendant le Plio-Pléistocene en Afrique Orientale. Soc. Géol. France, Bul.7e ser. 21(2):271-276.

Beden M 1979b. Les Éléphants (loxodonta et elephas) d'Afrique Orientale: systematique, phylogénie, intérêt biochronologique. Thesis Univ. Poitiers. v.1:1-223-39; v.2:226-367.

Behrensmeyer AK 1976. Fossil assemblages in relation to sedimentary environments in the East Rudolf succession. In: Y Coppens, FC Howell, GL1 Isaac and REF Leakey (eds.), Earliest Man and Environments in the Lake Rudolf Basin. Univ. Press, Chicago. p. 383-401.

Bonnefille R 1980. Vegetation history of savanna in East Africa during the Pleistocene. Proc. IV Int. Palyn. Conf. Lucknow (1976-77), 3:75-89.

Cooke HBS and Wilkinson AF 1978. Suidae and Tayassuidae. In: VJ Maglio and HBS Cooke (eds.), Evolution of African Mammals. Harvard Univ. Press, Cambridge. p. 435-482.

Churcher CS and Richardson ML 1978. Equidae. In: VJ Maglio and HBS Cooke (eds.), Evolution of African Mammals. Harvard Univ. Press, Cambridge. p. 379-422.

Coppens Y, Maglio VJ, Madden CT and Beden M 1978. Proboscidea. In: VJ Maglio and HBS Cooke (eds.), Evolution of African Mammals. Harvard Univ. Press, Cambridge. p. 336-367.

Eisenmann VE 1977. Les Hipparions africains: valeur et signification de quelques charactères des jugales inférieurs. Mus. Hist. Nat. Bull. 3e ser. No. 438:69-87.

Eisenmann VE 1980a. Charactères specifique et problèmes taxonomique relatif certains hipparions africaine. In: RE Leakey and BA Ogot (eds.), Proc. 8th Pan. Afr. Congress Prehist. and Quat. Studies, Nairobi 1977. Nairobi. p. 77-81.

Eisenmann VE 1980b. Les Chevaux (Equus sensu lato) fossiles et actuels: Crânes et dents jugales superieur. C.N.R.S. Cahiers de Paléontologie. p. 1-186.

Eisenmann VE 1981. Étude des dents jugales inférieurs des Equus (Mammalia, Perissodactyla) actuels et fossiles. Palaeovertebrata 10(3-4):127-226.

Eisenmann VE 1982. La phylogénie des Hipparion (Mammalia, Perissodactyla) d'Afrique d'après les charactères cranien. Non. Ned. Akad. Weten. Proc. B 85(2):219-227.

Forstén A 1978. Bovidae. In VJ Maglio and HBS Cooke (eds.), Evolution of African Mammals. Harvard Univ. Press, Cambridge. p. 540-572.

Gentry AW and Gentry A 1978. Fossil Bovidae (Mammalia) of Olduvai Gorge, Tanzania. Brit. Mus. (Nat. Hist.) Bull., Geol. Ser. 29(4):269-446 and 30(1):1-83.

Harris JM and White TD 1979. Evolution of the Plio-Pleistocene African Suidae. Trans. Am. phil. Soc. 69(2):1-128.

Klein R 1980. Environmental and ecological implications of large mammals from Upper Pleistocene and Holocene sites in southern Africa. Ann. S. Afr. Mus. 81(7):223-283.

Maglio VJ 1973. Origin and evolution of the Elephantidae. Trans. Am. phil. Soc. n.s. 63(3):1-149.

Vrba ES 1976. The fossil Bovidae of Sterkfontein, Swartkrans and Kromdraai. Transvaal Mus. Mem. 21:1-229.

Vrba ES 1982. Biostratigraphy and chronology, based particularly on Bovidae, of southern hominid-associated assemblages: Makapansgat, Sterkfontein, Taung, Kromdraai, Swartkrans; also Elandsfontein (Saldanha), Broken Hill (now Kabwe) and Cave of Hearths. Proc. 1st Int. Congress Human Palaeontol., Nice October 1982. C.N.R.S., Paris. Vol. 2:707-752.

Paleoclimatic framework for African hominid evolution

N.T.BOAZ
New York University, New York, USA

L.H.BURCKLE
Lamont-Doherty Geological Observatory, Columbia University, USA

ABSTRACT. World-wide ice volume/temperature fluctuations as recorded by the oxygen isotope curve for the past 5 m.y., relative changes in sea-level, and major tectonic and eustatic events are correlated with Pan-African records of paleoclimatic change by means of paleomagnetic stratigraphy, potassium-argon dates and bio-stratigraphy. First discovery datum (FDD) and last discovery datum (LDD) points for hominid taxa are designated. Although the data do not yet allow causal connections to be drawn between paleoclimatic change and events in hominid evolution, the synthesis is a step in this direction.

The continent of Africa has been the scene of origin and much subsequent evolution of the family Hominidae. Paleoanthropologists have long realized that African climatic history forged many events in that evolution, yet a pan-African view of paleoclimate and evolution has still to be formulated, despite important work in this direction (Bartholomew and Birdsell 1953; Howell and Bourliere 1963; Andrews and Van Couvering 1975; Behrensmeyer 1975; Cooke 1978; Axelrod and Raven 1978; Bernor 1983). This paper provides a framework for initiating a continent-wide Plio-Pleistocene reconstruction of African climatic, vegetational and biogeographic patterns within a context of global climatic, tectonic and eustatic events. An absolute time scale based on paleomagnetic reversal chronology provides a continuous baseline on which to place the deep-sea oxygen isotope record and discontinuous but temporo-geographically controlled paleoclimatic records from terrestrial sites in Africa. By organizing all available data, a start can be made towards understanding continent-wide patterns of environmental change as a background to hominid evolution.

That climatic history is important, and indeed crucial, to understanding hominid evolution was made clear by Raymond Dart (1925) in the first description of Australopithecus africanus from Taung. Dart pointed out that the proximity of the site to the Kalahari Desert, and the absence of forest-adapted fauna, indicated a habitat for Australopithecus substantially less wooded than that of the forest-adapted modern African great apes. Since 1925 paleoecological research on hominid sites in southern and eastern Africa has confirmed that all preserve a fauna, and in some cases, a flora, broadly characteristic of open, savanna to savanna-woodland, conditions (Vrba 1975; Boaz 1977; Behrensmeyer 1982). We must now look to smaller scale changes within this environmental mosaic, and to the whole of the African continent, to understand events in Plio-Pleistocene hominid evolution.

Before presenting the data on which our paleoclimatic chart is based, we discuss the relevance of paleoecological information to evolutionary problems at two extremes of the hominid career.

The origin of Hominidae involves essentially the same question as the initial evolutionary divergence of the African ape and hominid lineages. This divergence not only involves a change in morphology and behaviour, but a differentiation of habitat, probably from a forest adaptation to savanna or woodland. This change is exemplified by a comparison of the early Miocene forest of the Kenyan locality of Rusinga, containing hominoids probably ancestral to both great apes and hominids, with the Plio-Pleistocene peri-lacustrine savanna of the nearby Olduvai Gorge, northern Tanzania, where two sympatric hominid species, but no apes, are found in Beds I - II (Andrews and Van Couvering 1975; Hay 1976).

We presently have no knowledge of the forest biomes in Africa, past the Early Miocene, nor (probably not coincidentally) do we have any fossil remains of the ancestors of the African apes, which inhabit forest today. That the vicissitudes of the forest belts of Africa have played an important role in African hominoid speciation events, including probably the ape-human split, is clear from a consideration of modern mammalian biogeography. Kingdon (1971) and others have outlined three forest refugia in Africa, based on endemic mammalian species, which are likely the result of late Cenozoic contractions of forest, due to aridification. Subspecies of the common chimpanzee, Pan troglodytes verus, P. t. troglodytes, and P. t. schweinfurthi, inhabit the three refugia (Fig. 1), and subspecies of the gorilla inhabit the Cameroon-Gabon and Central refugia, i.e. those with mountain ranges. The timing of the contractions and distributional changes of African forest inferred from paleoclimatic data can provide important information on the timing of cladogenetic events in these hominoids.

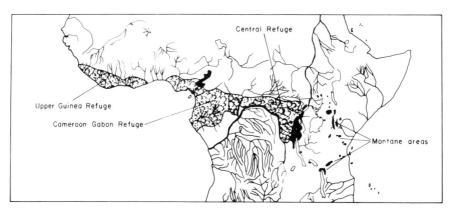

Figure 1. Biogeographic forest refugia of Africa. The Upper Guinea Refuge corresponds to the range of Pan troglodytes verus; Cameroon Gabon Refuge to P. t. troglodytes and Gorilla gorilla gorilla; and the Central Refuge to P. t. schweinfurthi; G. g. beringei, and G. g. graueri. These refugia likely represent islands of forests when aridity became widespread during isotopic shifts in the late Pliocene and Pleistocene. The pygmy chimpanzee, Pan paniscus, is the only African hominoid except man which does not inhabit a forest refuge - south and west of the Zaire River.

One African ape, the pygmy chimpanzee or bonobo (Pan paniscus), like early hominids, does not occur in a forest refuge. Its range is south and west of the Zaire River, in present-day forest underlain by Kalahari Sands (Lepersonne 1974). A hypothesis to be tested is whether P. paniscus evolved allopatrically in refuge regions in southern East Africa, perhaps in an analogous manner to the shrew, Petrodomus tetradactylus (Kingdon 1971: 66), a subspecies of which re-invaded the forest of the Zaire River basin from southeastern Africa. Pan paniscus may provide us with an example of the type of environmental change with attendant genetic isolation which accounted for the initial ape-human split.

We do not have either the required fossil evidence or the dates of evolutionary divergence of African apes from hominids to posit paleoclimatic scenarios for this evolutionary period. However, with the present evidence two mutually exclusive hypotheses can be entertained:

1. Kenyapithecus, a woodland-savanna hominoid from the Middle Miocene of Kenya, represents the common ancestor of African apes and hominids. Kenyapithecus wickeri from Fort Ternan lived in relatively open conditions (Gentry 1970; Shipman 1977). Accepting it as the ancestral species, which is not precluded on morphological grounds (Boaz 1983), requires that the living African apes moved back exclusively into forests.

2. An as yet undiscovered forest-living hominoid, perhaps a late-surviving dryopithecid, represents the common hominid-African ape ancestor. Hominids may have arisen in a manner analogous to the differentiation of Pan paniscus as a savanna-adapted species. Kenyapithecus is an evolutionary deadend. Current work in the Western Rift Valley may shed some light on the likelihood of this hypothesis.

A second area of hominid evolution where paleoclimatic data provide a useful framework for interpreting evolutionary events is the well-known Late Pleistocene Neanderthal record. Because there is abundant information on paleotemperature, pollen and fossil remains for the last glaciation, Boaz, Ninkovich and Rossignol-Strick (1982) undertook a paleoclimatic reconstruction of Weichselian-aged hominids in the circum-Mediterranean area. These authors proposed a mechanism for semi-isolation of Homo sapiens neanderthalensis, by glacial ice, related bodies of water and montane forests beginning around 100 000 BP, and subsequent replacement of this human subspecies in Europe at 34 000 - 40 000 BP by Homo sapiens sapiens. The spread of arid, desertic conditions with attendant extinction of large mammalian fauna in North Africa and Southwest Asia may have acted as a pump, pushing anatomically modern human populations into Europe. This mechanism may have been an important aspect of evolutionary change, effecting population dispersal roughly every 100 000 years during much of the Pleistocene, if the oxygen isotope curve can be taken to indicate similar paleoclimatic change. It remains to consider if hominid populations may have been similarly affected in sub-Saharan Africa.

The data to be considered here are for the period of hominid evolution between the ape-human divergence and the appearance of Homo sapiens. Figure 2 presents available data for worldwide paleoclimatic events for the last 5 m.y. The oxygen isotope record from foraminifera in deep-sea cores indicates relative changes in world ice volume and temperature through time. During Northern Hemisphere glaciations ocean water is enriched in the O18 isotope because the lighter O16 isotope is trapped preferentially in continental ice. Excursions of the curve in Fig. 2 to the left represent relative enrichment of ocean water in O18,

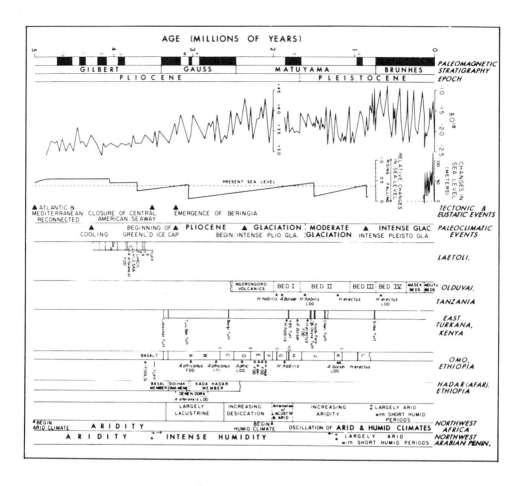

Figure 2. Chart summarizing major paleoclimatic events and hominid discovery datum points in Plio-Pleistocene time. Time scale is following Mankinen and Dalrymple (1979). Oxygen isotopic record is from Shackleton and Opdyke (1973, 1976) and Shackleton (pers. comm.). Sea level curve is from Vail et al. (1977). Hominid site stratigraphic records are from Hay (1976), Brown (pers. comm.) and de Heinzelin (1983). Paleoclimatic records for Hadar are from Gasse et al. (1980); for Northwest Africa from Diester-Haass (1977); and for Northwest Arabian peninsula from El-Sayari and Zotl (1978).

greater accumulation of continental ice and most likely decreased world temperature. Evidence from the Late Pleistocene suggests that these cold periods correlate with increased aridity and even desertification in Africa. Fluctuations in the oxygen isotope record become of progressively greater amplitude from the early Pliocene through the Pleistocene.

Four Plio-Pleistocene hominid sites are figured, each with the temporal extent of its sedimentary deposits. The first discovery datum (FDD) and last discovery datum (LDD) points for hominid taxa in Africa are shown. General climatic data for Hadar, Northwest Africa and the northeastern Arabian peninsula are summarized on the right of the chart. Figure 3 presents an overview of the paleoclimatic data available from the most intensively investigated of the early hominid sites, Omo, Ethiopia.

Terrestrial paleoclimatic indicators show a shift toward aridity, probably associated with greater seasonality, beginning at about 2.3 m.y. BP. The oxygen isotope curve records a major shift towards colder conditions worldwide, just after the Matuyama-Gauss boundary dated at 2.41 m.y. BP. This period of decreasing temperatures and increasing aridity is associated with the LDD of Australopithecus africanus (Member D of the Shungura Formation at Omo), and the FDD's of both Homo habilis and the robust australopithecine, Australopithecus boisei. Correlation does not necessarily indicate causation. One hypothesis is that changing ecological conditions may have effected a burst of evolutionary change in the A. africanus lineage, culminating in the appearance of H. habilis and A. boisei (Tobias 1980). Alternatively, H. habilis may have evolved from A. africanus under the stimulus of these environmental changes, with these same changes allowing the extension of the range of A. boisei into eastern Africa from elsewhere (Boaz 1977).

Increased amplitude of isotopic excursions is apparent between the Olduvai and Jaramillo Events. More common occurrence of bush fires at Omo, greater fluctuations in the bovid curve and supporting evidence from northern Africa and Arabia during this period, indicate a markedly fluctuating climate and vegetation. The biologically most important aspect of these fluctuations was probably periods of marked aridity. Homo erectus first appears during this time, at about 1.5 m.y. BP.

Just before the Brunhes-Matuyama boundary, at about 800 000 BP an isotopic shift towards even more widely fluctuating conditions occurs. This is near the LDD's of A. boisei and H. erectus. The A. boisei lineage became extinct after this point; H. erectus had evolved into H. sapiens probably by 500 000 BP.

Data at this point do not allow causal connections to be drawn between paleoclimatic indicators and events in hominid evolution. Several lines of information are needed to frame hypotheses. Dietary adaptations must be known for Plio-Pleistocene hominid species. A better knowledge of areal distributions of these species in Africa is required. A more complete and continuous sampling of paleoclimatic data through time from western, central and southern Africa is needed. Finally, a reliable paleoclimatic model to relate terrestrial, deep-sea, and extra-African data is necessary. We have presented an initial attempt to point out some of the patterns emerging from the data.

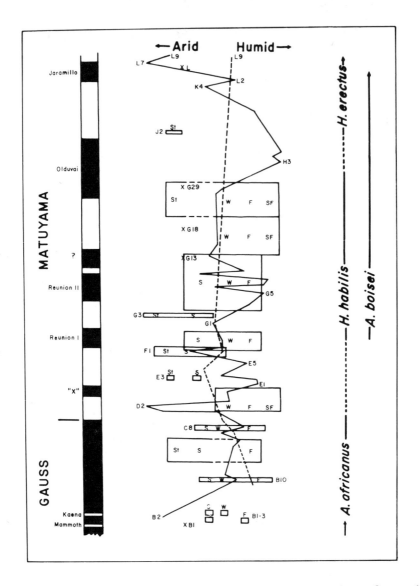

Figure 3. Summary diagram of paleoclimatic data from the Shungura Formation, lower Omo basin, Ethiopia, related to the paleomagnetic record (Brown et al. 1978). Dashed line indicates relative percentages of grass to arboreal pollen (Bonnefille, in press), and boxes are inferred environments through time based on fossil wood (Dechamps and Maes, in press) and micromammals (Wesselman 1982). Solid line represents percentages of alcelaphine and antilopine bovids to reduncine and tragelaphine bovids in stratigraphic units yielding at least 25 identifiable specimens (Boaz, in press: identifications by A. Gentry). Abbreviations are: SF - swamp forest, F - forest, W - woodland, S - savanna, St - steppe, X - fossil wood evidence of bush fire. Stratigraphic units are designated for major excursions of the environmental indicators. The hominid taxa at Omo are indicated on the right. Background information is provided by Coppens et al. (1976), Boaz (1977), Brown (1981) and de Heinzelin (1983).

REFERENCES

Andrews PJ and Van Couvering JH 1975. Palaeoenvironments in the East African Miocene. In: FS Szalay (ed.), Approaches to Primate Paleobiology. Contrib. Primatol. 5:62-103.

Axelrod DI and Raven PH 1978. Late Cretaceous and Tertiary vegetation history of Africa. In: MJA Werger (ed.), Biogeography and Ecology of Southern Africa. Monogr. Biol. 31:77-130.

Bartholomew GA and Birdsell JB 1953. Ecology and the protohominids. Amer. Anthrop. 55:481-498.

Behrensmeyer AK 1975. The taphonomy and paleoecology of Plio-Pleistocene vertebrate assemblages east of Lake Rudolf, Kenya. Bull. Mus. Comp. Zool. 146(10):473-578.

Behrensmeyer AK 1982. The geological context of human evolution. Ann. Rev. Earth Planet. Sci. 10:39-60.

Bernor RL 1983. Geochronology and zoogeographic relationships of Hominoidea. In: RL Ciochon and RS Corruccini (eds.), New Interpretations of Ape and Human Ancestry. Plenum, New York. pp. 21-64.

Boaz NT 1977. Paleoecology of early Hominidae in Africa. Kroeber Anthrop. Soc. Papers 50:37-62.

Boaz NT 1983. Morphological trends and phylogenetic relationships from middle Miocene hominoids to late Pliocene hominids. In: RL Ciochon and RS Corrucini (eds.), New Interpretations of Ape and Human Ancestry. Plenum, New York. pp. 705-720.

Boaz NT, Ninkovich D and Rossignol-Strick M 1982. Paleoclimatic setting for Homo sapiens neanderthalensis. Die Naturwissenschaften 69:29-33.

Brown FH 1981. Environments in the lower Omo basin from one to four million years ago. In: G Rapp and CF Vondra (eds.), Hominid Sites: their Geologic Settings. AAAS Selected Symp. 63. Westview, Boulder. pp. 149-163.

Brown FH, Shuey RT and Croes MK 1978. Magnetostratigraphy of the Shungura and Usno Formations, southwestern Ethiopia: New data and comprehensive reanalysis. Geophys. J.R. Astr. Soc. 54:519-538.

Cooke HBS 1978. Faunal evidence for the biotic setting of early African hominids. In: CJ Jolly (ed.), Early Hominids of Africa. St. Martin's Press, New York. pp. 267-281.

Coppens Y, Howell FC, Isaac GL and Leakey REF (eds.) 1976. Earliest Man and Environments in the Lake Rudolf Basin. Univ. Chicago Press. Chicago.

Dart RA 1925. Australopithecus africanus: the man-ape of South Africa. Nature 115:195-199.

Diester-Haass L 1977. Influence of carbonate dissolution, climate, sea-level changes and volcanism on Neogene sediments off Northwest Africa. In: Y Lancelot, E Seibold et al (eds.), Initial Reports of the Deep Sea Drilling Project, Vol. 41. US Govt. Printing Office, Washington. pp. 1033-1047.

El-Sayari SS and Zotl SG (eds.) 1978. Quaternary Period in Saudi Arabia. Springer Verlag, New York.

Gasse F, Rognon P and Street FA 1980. Quaternary history of Afar and Ethiopian rift lakes. In: MAJ Williams and H Faure (eds.), The Sahara and the Nile. Balkema, Rotterdam. pp. 361-400.

Gentry AW 1970. The Bovidae (Mammalia) of the Fort Ternan fossil fauna. In: LSB Leakey and RJG Savage (eds.), Fossil Vertebrates of Africa, Vol. 2. Academic Press, New York. pp. 243-323.

Hay RL 1976. Geology of the Olduvai Gorge. Univ. California Press, Berkeley.

De Heinzelin J (ed.) 1983. Sedimentary Formations of Pliocene and early Pleistocene age in the Omo basin, Ethiopia. Univ. Gent, Gent.

Howell FC and Bourliere F (eds.) 1963. African Ecology and Human Evolution. Aldine, Chicago.

Kingdon J 1971. East African Mammals: An Atlas of Evolution in Africa. Vol. 1. Academic Press, London.

Lepersonne J 1974. Carte geologique du Zaire. Mus. Roy. Afr. Centrale, Tervuren.

Mankinen E A and Dalrymple GB 1979. Revised geomagnetic polarity time scale for the interval 0 - 5 m.y. BP. Journ. geophys. Res. 84:615-626.

Shackleton NJ and Opdyke ND 1973. Oxygen isotope and paleomagnetic stratigraphy of equatorial Pacific core V28-238: Oxygen isotope temperatures and ice volumes on a 10^5 year and 10^6 year scale. Quat. Res. 3:39-55.

Shackleton NJ and Opdyke ND 1976. Oxygen isotope and paleomagnetic stratigraphy of Pacific core V28-239, late Pliocene to latest Pleistocene, Geol. Soc. Amer. Mem. 126:449-464.

Shipman PL 1977. Paleoecology, taphonomic history, and population dynamics of the vertebrate fossil assemblage from the middle Miocene deposits at Fort Ternan, Kenya. PhD Diss. New York Univ.

Tobias PV 1980. Australopithecus afarensis and A. africanus: a critique and an alternative hypothesis. Palaeontol. Afr. 23:1-17.

Vail PR, Mitchum RM and Thompson S 1977. Seismic stratigraphy and global changes of sea level. Part 4. Global cycles of relative changes of sea level. In: CE Payton (ed.), Seismic stratigraphy - Applications to Hydrocarbon Exploration. Amer. Assoc. Pet. Geol. Tulsa, Mem. 26:83-97.

Vrba ES 1975. Some evidence of chronology and palaeoecology of Sterkfontein, Swartkrans and Kromdraai from the fossil Bovidae. Nature 254:301-304.

Wesselman HB 1982. Pliocene micromammals from the lower Omo valley, Ethiopia: Systematics and paleoecology. PhD Diss., Univ. California, Berkeley.

The Terminal Miocene Event: A critical environmental and evolutionary episode?

C.K.BRAIN
Transvaal Museum, Pretoria, South Africa

ABSTRACT. A "conventional" reconstruction of palaeoclimate suggests that temperatures in high southern latitudes decline progressively through the Cainozoic Era to reach a low point at 6.5 - 5 m.y. BP when the West Antarctic ice sheet suddenly built up. This Terminal Miocene event coincided with a global sea-level drop and the salinity crisis in the Mediterranean. The event is thought to have affected the African environment dramatically and to have stimulated evolutionary change.

Another interpretation which has gained some recent support is that the event was not unique, but has been preceded by a number of others of similar effect at intervals since the Cretaceous.

INTRODUCTION

In a recent publication (Brain 1981), it was speculated that the Terminal Miocene Event, between 6.5 and 5 m.y. BP, was a highly significant episode in regard to African environmental change. In terms of human evolution, this time interval appears to have covered the split in the lineages leading to man and the African great apes, and also the time when hominid bipedalism was acquired. The purpose of the present contribution is to reconsider the significance of the Terminal Miocene Event in the light of important new evidence that has been published recently.

A "CONVENTIONAL" SCENARIO OF GLOBAL COOLING

The "conventional" reconstruction of global cooling has been reviewed in detail (Brain 1981), so that only an outline will be presented here. In essence it is suggested that, during the last 50 m.y., there has been a progressive cooling of the sea and atmosphere first in southern high latitudes, much later in the Northern Hemisphere and that this has also exerted an influence on the climate of lower latitudes through indirect effects. It is suggested furthermore that, in the course of this cooling process in the Southern Hemisphere, a critical threshold was crossed at the end of the Miocene, between 6.5 and 5 m.y. ago when, for the first time in Cainozoic history, African environments were seriously affected by rapidly changing climate. The changes were apparently caused by the Terminal Miocene Event, during which the West Antarctic ice sheet accumulated for the first time and the Messinian salinity crisis affected the Mediterranean.

It is assumed that at the beginning of the Cainozoic Era the world was free of ice sheets and that temperature gradients from equator to poles were very much less steep than they are today. A record of global temperature change has been constructed through oxygen isotope analyses of foraminifera preserved in deep sea sediments. Shackleton and Kennett (1975), for instance, worked on three cores from the floor of the southern ocean south of Australia, which appeared to provide a record of surface and deep water temperatures for most of the Cainozoic Era, or approximately the last 55 m.y. At the beginning of this era it appears that surface temperature of the high latitude southern ocean was 18-20 °C, but that this dropped to about 7 °C at the beginning of the Oligocene, 38 m.y. BP. Warmer conditions then prevailed again, followed by another temperature drop between 16.5 and 13 m.y. BP (Savin 1977). The main decline however seems to have been at the end of the Miocene, 6.5 - 5 m.y. BP, when the Terminal Miocene Event had its world-wide consequences.

The progressive decline in southern ocean temperatures that has been inferred for the course of the Miocene was apparently strongly influenced, if not mediated, by the isolation of Antarctica through continental drift. As Kennett et al. (1974) have documented, Australia began its northward drift from Antarctica in the early Eocene, about 55 m.y. ago. Throughout the ensuing time, Antarctica apparently remained relatively stable in its current polar position.

Kennett (1978) has pointed out that the present South Tasman Rise, which incorporates the island of Tasmania, is a continental crustal feature that would have obstructed the passage between Australia and Antarctica for a considerable time. It was not until about 38 m.y. BP that Australia's northward drift had proceeded far enough for the South Tasman Rise to clear Victoria Land in Antarctica. Kennett suggested that during the interval 38-22 m.y BP a deep seaway developed between Antarctica and the South Tasman Rise.

The only other obstruction in the way of a circum-Antarctic current would have been the Drake Passage, between the southern tip of South America and Antarctica. Kennett (1978) provided evidence to suggest that this opening occurred no later than 21 m.y. ago, while Barker and Burrell (1977, 1982) suggested that spreading between the Antarctic Peninsula and Tierra del Fuego started about 29 m.y. ago, with deep circulation beginning 23.5 ± 2.5 m.y. BP, thereafter increasing in volume.

In their discussion of the development of the circum-Antarctic current, Kennett et al. (1974) concluded that this current, which today transports a larger volume of water than any other, has flowed for the last 30 m.y. Its development led to a fundamental change in the world's oceanic circulation, isolating Antarctica from contact with warmer sea water that formerly had penetrated southwards from lower latitudes. As a result of this isolation progressive cooling of Antarctica appears to have set in.

As Mercer (1983) has explained, Antarctica would consist of an East Antarctic continent and a West Antarctic archipelago, if its ice cover were removed. Thus today the Antarctic ice sheet may be divided into two principal but unequal portions. In East Antarctica there is currently about 10.2 million km^2 of ice grounded largely above sea level, while the much smaller West Antarctic ice sheet, covering about 1.6 million km^2, is grounded mainly below sea level and possesses floating ice-shelf extensions into both the Ross and Weddell seas (Drewry 1978). It has been suggested by Drewry and others that by Mid-Late Miocene times substantial ice sheets were developing in East Antarctica, but that the West Antarctic ice sheet was not yet present. Mercer (1978) has provided and discussed evidence to show that, at the end of the Miocene, the West Antarctic ice sheet developed rather rapidly, linking the ice-covered islands of the archipelago with the main mass of East Antarctica to form the continent of Antarctica as we know it today.

Thus it appears that progressive cooling occurred in southern high latitudes, causing first the build-up of an East Antarctic ice sheet about 15 m.y. BP, followed by the establishment of the West Antarctic ice sheet between 6 and 5 m.y. BP. The cold episode that was responsible for this latter ice sheet expansion was apparently a widespread one, affecting large areas of the globe. For instance, Loutit and Kennett (1979) record a distinct C13/C12 isotopic shift to lower values in a shallow marine sedimentary sequence of Late Miocene age at Blind River, New Zealand. This may be correlated with a similar shift in Late Miocene Deep Sea Drilling Project sequences throughout the Indo-Pacific, dated at about 6.2 m.y. BP. This shift is interpreted as resulting from a Late Miocene cooling event, together with a marked drop in sea level at 6.1 m.y. BP caused by a sudden build-up of Antarctic ice. The shallowest water depths recorded at Blind River occurred about 5.6 m.y. BP when the sea level drop was between 40 and 70 m.

A comparable sea level decline has been documented in sediments of South-western Spain where foraminiferal assemblages have been used as indicators of sea water depth at the time of the accumulation (Van Couvering et al. 1978, Berggren and Haq 1976). In the upper part of the Late Miocene Andalusian stage stratotype section, water depths are thought to have declined rather abruptly about 5.5 m.y. BP from about 70-100 m to about 20-30 m. This eustastic sea level change can almost certainly be correlated with that described at Blind River in New Zealand.

But certainly the most dramatic result of the terminal Miocene sea level decline appears to have been the salinity crisis in the Mediterranean which has been documented in detail by Hsü et al. (1977) and Adams et al. (1977). As a result of drilling operations by the Glomar Challenger in 1970, the presence of vast deposits of salt beneath the Mediterranean was demonstrated. Furthermore it was established that these evaporite deposits were of Messinian (Late Miocene) age and that they had been laid down in a comparatively brief period somewhere between 6.5 and 5 m.y. BP.

Although the terminal Miocene eustatic sea level drop was almost certainly responsible for the final isolation of the Mediterranean, it could not have had this effect if the previously existing Atlantic-Mediterranean connection had not been a very shallow one. In the model proposed by Van Couvering et al. (1976), the Atlantic-Mediterranean connection is thus visualized as being over a sill that was slowly lifted tectonically. Gradually while this happened, the Mediterranean water was transformed from normal sea water to highly concentrated brine, formed by refluxive concentration. When the terminal Miocene sea level drop cut off the Mediterranean's connection with the open Atlantic, the complete evaporation of the Mediterranean brine is thought to have occurred between 5.5 and 5.0 m.y. BP.

The almost unbelievable magnitude of the Mediterranean salt deposits has been discussed by Ryan (1973) who pointed out that these deposits had an estimated volume of more than one million cubic kilometres, or more than six per cent of the dissolved salts in the entire world's oceans. He wrote: "As a consequence of being drained of more than six per cent of its dissolved salts, the World Ocean turned more alkaline and its deep waters became noticeably more undersaturated in calcium carbonate. It is believed that the concurrent freshening of its surface water layer may have enhanced the formation of high-latitude sea ice in Late Miocene times, an event which could have led to the great burst of glaciation recorded on the continent of Antarctica at that time".

The complete drying up of so large a body of water as the Mediterranean over a brief geological period must have had dramatic physical and biological consequences, some of which have been mentioned by Hsü et al. (1977). These include the fact that the lowering of the base level of erosion led to the deep incision of rivers that drained into the Mediterranean, while a dry-land connection between Africa and Europe permitted free exchanges of fauna and flora. Hsü et al. (1977) visualized a cool, more arid climate around the desiccated Mediterranean and, to my knowledge, these authors were the first to speculate as to whether this aridity, caused by the Messinian salinity crisis, had not perhaps been responsible for the spread of East African savannahs, which in turn promoted the evolution of man from his pre-human ancestors.

There is abundant evidence from various parts of the world to indicate that the terminal Miocene sea level drop was short-lived on the geological time scale and that a warmer period with rapid regression of Antarctic ice followed. In the Andalusian Stage stratotype of southwestern Spain, for instance, Van Couvering et al. (1976) reported that by the beginning of the Pliocene at 5 m.y. BP the sea had returned to the level it had had before the terminal Miocene regression. The result was that a connection was once again established between the Mediterranean basin and the Atlantic, via the straits of Gibraltar.

Thunell (1979) has used Planktonic foraminifera from cores to monitor the climatic evolution of the Mediterranean during the last 5 m.y. He reports that, following the re-establishment of the Atlantic link, a gradual warming of Mediterranean surface waters occurred throughout the early Pliocene. This was followed by a 3 - 4 °C drop in surface water temperatures during the early part of the late Pliocene, the onset of this cooling being dated to approximately 3.2 m.y. BP, and thought to be related to initiation of Northern Hemisphere glaciation.

On evidence from a core raised in the equatorial Pacific, Shackleton and Opdyke (1977) concluded that before 3.2 m.y. BP, no Northern Hemisphere ice sheet had formed and that the world was experiencing a period of stable "interglacial" or "preglacial" climate. Following the establishment of the northern ice cap, however, which occurred between 3.2 and 2.6 m.y. BP, the pattern of oscillation between glacial and interglacial interludes was established. Evidence for the sequence of such glacial-interglacial cycles during the last 2 m.y. is now well established. For instance, Van Donk (1976) obtained evidence from a tropical Atlantic core covering the entire Pleistocene period which reflected 21 isotopically determined interglacial stages for the past 2.1 m.y., with an equal number of glacial or near-glacial stages. It is also possible to correlate the temperature recorded from deep sea cores with that preserved in European continental deposits, as Kukla (1977, 1978) has shown, making use of a remarkable record of alternating loess and forest soil at Krems in Austria. He concluded that during the last 1.7 m.y. the deposition of loess containing characteristic cold-resistant gastropods was interrupted at least 17 times by the development of temperate interglacial forests. This means that the last 17 glacial cycles had a mean duration of about 100 000 years each.

SOME NEW EVIDENCE AND NEW DOUBTS

In recent years the whole concept of progressive global cooling from ice-free Cretaceous times has been questioned. Barron et al. (1981) have quantitatively investigated the mechanisms that could explain the warm, equable climate that is believed to have been typical of the Cretaceous. They conclude that palaeoclimatologists should carefully re-examine their data and interpretations with regard to reconstructions of Cretaceous tropical and polar surface temperatures.

Likewise, Matthews and Poore (1980) have challenged the traditional interpretation of the oxygen isotope record and have argued that the abrupt enrichment of O18 in tropical surface water near the Eocene-Oligocene boundary marks a major increase in global ice volume. They do not exclude the possibility that significant quantities of land ice existed previously, well back into the geological record. How could this be when the oxygen isotope evidence appears to suggest otherwise? In a recent comment, Kerr (1983) has pleaded for caution in the interpretation of isotope ratios in foraminiferal tests from the older part of the deep sea record as the calcium carbonate in these tests may have undergone recrystallization. It appears that recrystallization could be 80 % complete after 60 m.y., a fact that could account for much of the observed trend in apparent global cooling. Not all palaeoceanographers agree that recrystallization is an important factor and the matter clearly is not yet resolved.

The deep sea sedimentation record shows an interesting series of breaks, or hiatuses, which have recently been studied and interpreted by Keller and Barron (1983). They record the presence of eight such hiatuses, at intervals during the last 23 m.y., each associated with a cold interlude.

Thus it appears that the timetable for Antarctic ice build-up, with the East Antarctic ice sheet being established about 15 m.y. BP and the West Antarctic equivalent developing 5 - 6 m.y. ago, may need to be reconsidered. Contradictory evidence comes from volcanic rocks known as hyaloclastites which were apparently formed below ice sheets. Le Masurier and Rex (1982) have recorded the presence in West Antarctica of subglacially erupted hyaloclastite 21 ±1 m.y. old and have concluded that an ice sheet was present in west Antarctica no later than the Middle Oligocene. Furthermore, they suggest that the ice sheet formed on a flat, prevolcanic subaerial erosion surface that has since been fragmented by block faulting.

The history of Cainozoic cooling and of the Antarctic ice sheets is obviously under extensive review, as Mercer (1983) has pointed out and conclusions reached during the next few years should be of great interest.

HOW UNIQUE WAS THE TERMINAL MIOCENE EVENT ?

The answer to this question will depend very much on how much validity the conventional interpretation of global cooling still has. Supposing that there HAS been progressive cooling of the southern oceans since the Eocene, with a major temperature decline at the end of the Miocene during which the West Antarctic ice sheet built up, then the Terminal Miocene Event could well have been a critical threshold, crossed for the first time in Cainozoic history. Alternatively, it could be that the cold, terminal Miocene interval was one of many which had affected the world periodically since Cretaceous times or before. In this case the effects felt in lower latitudes and particularly in African habitats, would have been felt before, but this does not mean to say that such effects could not have been significant. In fact, two features of the Terminal Miocene episode may have been unique, whether or not the low-temperature aspect had occurred previously.

The first unique feature of the event was the remarkable situation then obtaining in the Mediterranean, resulting in the salinity crisis and evaporation of the water in the basin. This disappearance of the Mediterranean presumably exerted a considerable influence on the climate of the surrounding areas - something not experienced before or since. In addition the dry-land connection was important in the interchange of animals and plants between Europe and Africa.

A second unique aspect was that the Terminal Miocene Event followed a period of active mountain building in the circum-Mediterranean and Himalaya areas, as well as of uplift and rifting in East Africa (Leporte and Zihlman 1983), so that the cold interlude interacted with a topography which was very different to that of earlier times.

POSSIBLE LINKS TO EVOLUTIONARY EVENTS IN AFRICA

A good deal of evidence has been collected to indicate that, during the last glacial at least, minimum temperatures over wide areas of Africa dropped by $5 - 10\,^{\circ}C$ (for a review see Brain 1981). In southern Africa this greatly enlarged the areas affected by serious frost, thus altering the distribution of certain plant species and perhaps promoting the spread of grasslands. The lower temperatures appear to have promoted tropical aridity and the shrinking of equatorial forest into separate refuges (e.g. Hamilton 1976). The possibility that these changes, occurring in Africa for the first time at the end of the Miocene, may have served as a significant evolutionary stimulus has been discussed elsewhere (Brain 1981). Whether this in fact was the case cannot be decided with certainty until the currently conflicting evidence on the thermal history of the southern oceans and Antarctica has been resolved.

REFERENCES

Adams CG, Benson RH, Kidd RB, Ryan WBF and Wright RC 1977. The Messinian salinity crisis and evidence of Late Miocene eustatic changes in the world ocean. Nature 269: 383-386.

Barker PF and Burrell J 1977. The opening of the Drake Passage. Mar. Geol. 25: 15-34.

Barker PF and Burrell J 1982. The influence upon Southern Ocean circulation, sedimentation and climate of the opening of the Drake Passage. In: C Craddock (ed.), Antarctic Geoscience 43. Univ. Wis. Press., Madison. p. 377-385.

Barron EJ, Thompson SL and Schneider SH 1981. An ice-free Cretaceous? Results from climate model simulations. Science 212: 501-8.

Berggren WA and Haq BU 1976. The Andalusian Stage (Late Miocene) biostratigraphy, biochronology and paleoecology. Palaeogeogr. Palaeoclimatol. Palaeoecol. 20: 67-129.

Brain CK 1981. The evolution of man in Africa: was it a consequence of Cainozoic cooling? Alex Du Toit mem. Lect. 17, Annex. Trans.geol.Soc. S. Afr. 84. 19 pp.

Drewry DJ 1978. Aspects of the early evolution of West Antarctic ice. In: EM van Zinderen Bakker (eds.), Antarctic Glacial History and World Palaeoenvironments. Balkema, Rotterdam. p. 25-32.

Hamilton AC 1976. The significance of patterns of distribution shown by forest plants and animals in Tropical Africa for the reconstruction of Upper Pleistocene palaeoenvironments: a review. Palaeoecol. Africa 9: 63-97.

Hsü KJ, Montadert L, Bernoulli D, Cita MB, Erickson A, Garrison RE, Kidd RB, Mèlierés F, Müller C and Wright R 1977. History of the Mediterranean salinity crisis. Nature 267: 399-403.

Keller G and Barron JA 1983. Paleoceanographic implications of Miocene deep-sea hiatuses. Geol. Soc. Am. Bull. 94: 590-613.

Kennett JP 1978. Cainozoic evolution of circumantarctic palaeo oceanography. In: EM van Zinderen Bakker (ed.), Antarctic Glacial History and World Palaeoenvironments. Balkema, Rotterdam. p. 41-56.

Kennett JP, Houtz RE, Andrews PB, Edwards AR, Gostin VA, Hajos M, Hampton MA, Jenkins DG, Margolis SV, Ovenshine AT and Perch-Nielsen K 1974. Development of the circum-Antarctic current. Science 186: 144-147.

Kerr RA 1983. Early climate data questioned. Science 220: 807.

Kukla GJ 1977. Pleistocene land-sea correlations. I. Europe. Earth-sci. Rev. 13: 307-374.

Kukla GJ 1978. The classical European glacial stages: correlation with deep sea sediments. Trans. Nebraska Acad. Sci. 6: 57-93.

Leporte LF and Zihlman AL 1983. Plates, climate and hominoid evolution. S. Afr. J. Sci. 79: 96-110.

Le Masurier WE and Rex DC 1982. Volcanic record of Cenozoic glacial history in Marie Byrd Land and western Ellsworth Land: revised chronology and evaluation of tectonic factors. Antarctic Geoscience 89: 725-734.

Loutit TS and Kennett JP 1979. Application of carbon isotope stratigraphy to Late Miocene shallow marine sediments, New Zealand. Science 204: 1196-1199.

Matthews RK and Poore RZ 1980. Tertiary ^{18}O record and glacial-eustatic sea level fluctuations. Geology 8: 500-514.

Mercer JH 1978. Glacial development and temperature trends in the Antarctic and in South America. In: EM van Zinderen Bakker (ed.), Antarctic Glacial History and World Palaeoenvironments. Balkema, Rotterdam. p. 73-94.

Mercer JH 1983. Cenozoic glaciation in the southern Hemisphere. Ann. Rev. Earth Planet. Sci. 11: 99-132.

Ryan WBF 1973. Geodynamic implications of the Messinian crisis of salinity. In: CW Drooger (ed.), Messinian events in the Mediterranean. North Holland, Amsterdam. p. 26-38.

Savin SM 1977. The history of the Earth's surface temperature during the past 100 million years. Ann. Rev. Earth Planet. Sci. 5: 319-355.

Shackleton NJ and Kennett JP 1975. Paleotemperature history of the Cenozoic and the initiation of Antarctic glaciation: oxygen and carbon isotope analyses in D.S.D.P. sites 279, 277 and 281. Initial Reports Deep Sea Drilling Project 29: 743-755.

Shackleton NJ and Opdyke ND 1977. Oxygen isotope and palaeomagnetic evidence for early Northern Hemisphere glaciation. Nature 270: 216-219.

Thunnell RC 1979. Climatic evolution of the Mediterranean sea during the last 5 million years. Sedim. Geol. 23: 67-79.

Van Couvering JA, Berggren WA, Drake RE, Aquirre E and Curtis GH 1976. The Terminal Miocene event. Mar. Micropalaeont. 1: 263-286.

Van Donk, J 1976. O^{18} record of the Atlantic Ocean for the entire Pleistocene Epoch. Mem. geol. Soc. Am. 145: 147-163.

Biogeography of Miocene-Recent larger carnivores in Africa

A.TURNER
Transvaal Museum, Pretoria, South Africa

ABSTRACT. The pattern of first and last appearances of felids, canids and hyaenids in Africa over the last five million years is examined. Major events in the fossil records of these families appear to correlate with the same climatic shifts that have been argued to provide evolutionary stimuli in the case of hominids and bovids.

INTRODUCTION

Current theories about evolutionary rates and mechanisms (Vrba 1980) suggest that larger carnivores in particular should be less likely to speciate over a given time-span than other mammals of similar body size. A predator is simply less likely to encounter new conditions and to suffer range fragmentation. And yet, as the fossil record shows, species of large carnivores do originate and, in many cases, subsequently become extinct. What circumstances correlate with these events and what constitute suitable conditions for the speciation or extinction of a large carnivore?

In an effort to address this problem, presence and absence data were considered for felids, canids and hyaenids from Miocene through to Pleistocene times from sites throughout Africa and examined for the patterns which emerge when appearances are plotted against time and the large-scale climatic and environmental shifts which occurred. At this stage this is very much an exploratory exercise. A number of deposits, particularly those from South Africa, have date ranges, and central estimates were made for the purpose of analysis. Moreover, because much of the information has been taken from the literature the exercise is somewhat constrained by the problems of taxonomic compatibility which result when several sources are consulted. A number of the problems, could be resolved, as discussed briefly below, but in other instances the species and genera have simply been listed, particularly in the case of the hyaenas and notwithstanding the recent review of the family by Soulianos (1982) which would shorten, or at least alter, the list given here.

CLIMATIC BACKGROUND AND SOME PREVIOUSLY OBSERVED CORRELATIONS

Brain (1981a) gives a useful synthesis of the data on climatic variations since the Miocene period, with particular emphasis on Africa. He

Table 1. Felidae, Canidae and Hyaenidae. Presence at various Miocene-recent sites. Dates and species list references as follows:
1. Pickford 1975, 2. Pickford 1978, 3. Smart 1976, 4. Hendey 1981, 5. Howell 1982, 6. Behrensmeyer 1976, 7. Howell and Petter 1976, 8. MD Leakey and Hay 1979, 9. MD Leakey et al. 1976, 10. Boaz et al. 1982, 11. Randall 1981, 12. Collings et al. 1975, 13. Ewer 1956b, 14. Vrba 1982, 15. MG Leakey 1976, 16. Cerling and Brown 1982, 17. Brain 1981b, 18. Kalb et al. 1982, 19. Petter 1973, 20. Hay 1976, 21. Walker et al. 1982, 22. Vrba 1981, 23. Vrba, in press, 24. Hendey 1974.
Schematic temperature curve after Brain (1981a).

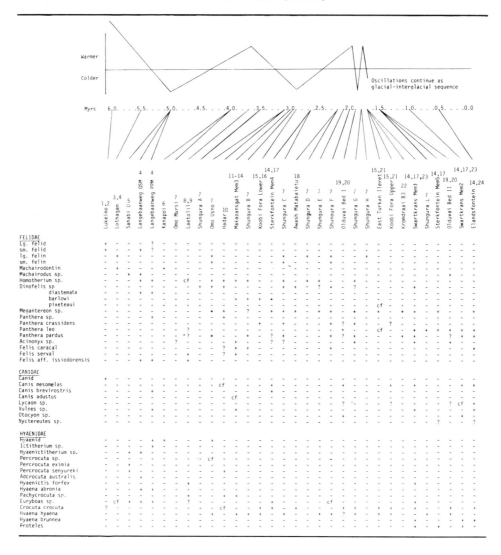

discusses the large-scale temperature drop at the end of the Miocene, at around 5.0 m.y., the return to higher temperatures at around 3.5 m.y., and the further subsequent drop in temperatures between 3.2 and 2.6 m.y. These variations are summarized schematically at the top of Table 1. Brain also stresses the correlation between major climatic changes and the apparent separation of hominid and pongid lineages at around 5.0 m.y. and the human and australopithecine lineages at around 3.2 - 2.6 m.y.

Vrba (in press) has pointed to the appearance of many new bovid tribes in Africa at around the end of the Miocene, to the extinctions of grazing species which coincide with the rise in temperatures at about 3.5 m.y., to the reversal of this pattern by around 2.6 m.y. and to the dramatic rise in the rates of bovid extinction and diversification rates in the period between 3.0 and 2.0 m.y. In the same paper she stresses the parallels to be seen in the correlations between evolutionary patterns and climatic shifts in the case of both the bovids and the hominids, and argues for similar mechanisms as an underlying cause.

DISTRIBUTION PATTERNS OF LARGER PREDATORS

The major features of larger predator appearances in time and space are shown in Table 1. In looking for patterns there seems little point in bemoaning the shortcomings of the data, but one point should be made. Many occurrences consist of relatively few specimens, and are often based on the identification of fragmentary material. If open-air sites preserve carnivore remains in something approaching their relative abundances in the palaeoecosystem, small sample size is to be expected. But the point remains that it often takes little variation in specimen numbers to turn presence at a given site into absence. Due caution must therefore be exercised in interpretations of first and last appearances. In particular: are the absences before and after such appearances respectively real or apparent? This problem is of course approached by looking at the larger patterning.

An obvious way to begin the examination of the pattern is to look for the first appearances of the still extant species. Thus the lion is first seen at Olduvai in Bed I at just after 2.0 m.y.; spotted hyaena is first clearly seen in the Lower Member of Koobi Fora just after 3.0 m.y.; leopard and striped hyaena occur in the Omo Usno levels just prior to that date; cheetah appears first at Makapansgat in Member 3 at around the same date; brown hyaena is not known until perhaps 1.6 m.y. in the Member 1 deposit at Swartkrans, although it should be emphasized that this deposit is less precisely dated and that the appearance could be closer to 2.0 m.y.

The first appearances of leopard and striped hyaena may correlate with lower temperatures around 3.0 m.y. The appearance of the lion, and perhaps also the brown hyaena, at just after 2.0 m.y., also seems to correlate with the generally colder and fluctuating climates of the Pleistocene, conditions which are likely to have produced open areas where numerous ungulates might be hunted. The spotted hyaena is recorded from around or perhaps very shortly after the time of first appearance of the leopard and striped hyaena, and the then continued presence of both hyaena types in most of the East African sites during this period may emphasize the mosaic nature of the vegetation. Open country is perhaps confirmed by the presence, from around 3.0 m.y., of the highly cursorial cheetah, poor though the fossil record is for that animal. Suggestions have recently been made, based on finds in North America which may date to 2.5 m.y. and perhaps earlier, that this species has a New World origin, in which case its appearance in Africa may reflect immigration rather than in situ speciations of the kind most likely, on present evidence, to have produced lion and leopard (Martin el al. 1977; Adams 1979; Kurtén and Anderson 1981). Similar early dates for the cheetah are reported from the Villafranchian of Europe (Azzarolli et al. 1982).

Dinofelis, Homotherium and Machairodus all appear to have been immigrants from Eurasia at or about the time of the terminal Miocene temperature drop (Hendey 1981; Howell 1982). At around the time of the appearance of the lion, both Dinofelis and Homotherium apparently became extinct. The Machairodus sp. canine fragment identified by Petter (1973: 91) from Bed II at Olduvai is most probably that of a Homotherium. Megantereon, a later immigrant, appears to have coexisted with the true felids for somewhat longer, although the time range is largely extended by the record from Elandsfontein and it is conceivable that this marks the presence there of a deposit earlier than the Middle Pleistocene one. The first appearance of Megantereon, at Omo around 3.0 m.y., may suggest a dispersal during initial stages of opening in the vegetation, perhaps like that of the cheetah. The number of sites in which Megantereon has been found, as indicated in Table 1, makes it difficult to see why Howell and Petter (1976: 323) implied that it was rare, albeit that it never occurs in great quantities in any instance. In fact all the machairodont and false-machairodont species appear to have coexisted with the leopard for some considerable time - about 1.0 m.y. - suggesting that they were capable of living in a variety of environments. Although a number of species are claimed by various authors for the sabre and false sabre-toothed cats, the remains are generally too fragmentary to permit adequate identification of speciation events. However, the distributions shown in Table 1 suggest that some resolution may be possible in the case of the Dinofelis species if just a few more good specimens should become available.

Whether the eventual extinctions of the machairodonts and Dinofelis relate to the appearance of the lion, or instead to the conditions which produced the lion, is however difficult to resolve. Both Hendey (1974) and Vrba (1981) have remarked on the incidence of bone pathology in specimens of machairodont post-cranial material, which perhaps imply some dietary deficiency. Such possible deficiencies could of course have resulted from feeding pattern interferences by lions which simply exacerbated an existing problem for the machairodonts. The involvement of lions might be further indicated by the extinction of Dinofelis, in which no such pathological conditions have been reported to my knowledge. This situation could be compared with that in North America during the Pleistocene, where lion and a number of machairodont cats seem to have coexisted almost into Post-Glacial times (Kurtén and Anderson 1980). It is possible that much of the reason for the differences in machairodont extinctions between the two areas may lie in the nature of prey availability following the rapid diversification of African bovids between 3.0 and 2.0 m.y. and the possible inability of those cats to hunt them efficiently.

A notable feature of the pattern in Table 1 is the poor fossil record of the canids. It was at first thought that this might be a taphonomic artefact, since the family is moderately well represented in the Sterkfontein Valley caves (neither the Kromdraai A nor the main Kromdraai B samples appear in Table 1 because of uncertainties over their stratigraphic position, but canids are reasonably well represented in both assemblages; for details see Brain (1981b)). However, the incidence of finds from Elandsfontein and the albeit patchy representation in a number of the East African open-air sites suggest that the explanation is unlikely to be quite that simple. The appearance in any numbers from after c. 2.5 m.y. suggests that we may simply be observing a relatively late immigration which perhaps resulted from the rather more open conditions then emerging. One interesting point concerning the canids, however, relates to a specimen from the Kromdraai A sample which Broom (1948) gave as Thos terblanchei and which Ewer (1956a) later renamed as Canis terblanchei. This specimen has a peculiar lobe on the angle of its mandible and may be an example of the Asian genus Nyctereutes, the raccoon dog (see Kurtén 1968), a possibility which will be discussed in detail elsewhere (Ficcarelli, Torre and Turner, in preparation). Because of this possible re-allocation other specimens given as examples of C. terblanchei have been placed under Nyctereutes in Table 1, although it should be stressed that the fragmentary nature of many of these specimens makes such an allocation tenuous whatever the correct identification of the Kromdraai material.

Two other items in Table 1 which may be discussed are Panthera crassidens, first named by Broom (1948) on the basis of material from Kromdraai A, and the maxillary fragment from Olduvai Bed II which Petter (1973) allocated to Metailurus sp. First-hand examination of Broom's Panthera crassidens material, a left maxillary fragment containing P^3 and P^4, and a right mandible with P_3, P_4 and M_1, shows the respective specimens to be a leopard and a cheetah (Turner, in press). East African specimens shown as Panthera crassidens in Table 1 have been so identified by other workers largely because they seem to fit Broom's original description of something not quite leopard and not quite cheetah. In a number of instances the specimens are isolated post-cranial remains which clearly cannot even be tied in with the material on which Broom reconstructed his new species, so that allocation to P. crassidens is unwarranted. I have not seen any specimens, but if they cannot be allocated to a known species then perhaps they should be left as Panthera sp. It is also possible that the "curious distribution" of leopard in East Africa to which Howell and Petter (1976: 325) refer, partly on the basis of the species' apparent absence from East Turkana deposits, may be due to misidentification of leopard remains from the Koobi Fora upper and lower members as Panthera crassidens.

The specimen from Olduvai Bed II, a right maxillary fragment with an incompletely erupted canine, identified as Metailurus sp., appears from Petter's description and figure to be, in all probability, that of a cheetah. It is listed as such in Table 1. It is important that such potential confusions be reduced when looking for patterns of speciation and extinction in the fossil record.

EVOLUTION AND DISPERSALS

Although the pattern is by no means clear-cut, at least some of the predator dispersals, extinctions and first appearances shown in Table 1 thus appear to correlate with the same set of major climatic shifts that have been argued to have acted as evolutionary stimuli in the case of hominids and bovids. These parallel correlations for the carnivores and hominids may be of further interest. Various arguments have appeared in the literature about the extent to which early hominids may or may not have eaten meat, and about the extent to which they may therefore be considered, ecologically, as another large carnivore. As Foley (1982) has recently stressed, much of the discussion ultimately focuses on the question of resource availability and packaging in given circumstances. Isaac (1982), while cautioning that the role of meat procurement in hominid behaviour should not be assumed, has also pointed out that ungulates form a very convenient source of food for a species able to exploit the possibilities. The potential ecological similarities between hominids and large predators may mean that broadly similar environmental shifts produced speciations which were not only coincident, as they appear to have been to some extent, but also occurred in the same area. If the appearances of new species of both carnivores and hominids did indeed take place in similar areas, then subsequent dispersals may also have taken place in similar directions. The possible value of such parallels in movement is suggested in a discussion of hominid dispersal into the northern temperate zones (Turner 1982). If the pattern of dispersals in the various carnivore species could be followed back to the point of speciation, then there might be increased opportunities to trace the genus Homo back to its place of origin, instead of having to rely solely on the vagaries of the hominid fossil record. At present, however, that possibility is remote. Certainly a number of the extant species appear to have their origins in East Africa, an area which would make a suitable point of origin for Homo, but as Table 1 shows, almost all the early sites occur in that area, and a first appearance there is therefore hardly surprising. A further complication is likely to be raised by the ability of the large predators to exist in a variety of environments and certainly within one in which a new species had just arisen. Rates of dispersal are therefore likely to have been rapid, and given the coarseness of even our best chronologies it may prove difficult or even impossible to resolve relative dates clearly.

REFERENCES

Adams DB 1979. The cheetah: native American. Science 205:155-8.
Azzaroli A, de Guili C, Ficcarelli G, Torre D 1982. Table of the stratigraphic distribution of terrestrial mammal faunas in Italy from the Pliocene to the Early Middle Pleistocene. Geogr. Fis. Dinam. Quat. 5:55-8.
Behrensmeyer AK 1976. Lothagam Hill, Kanapoi, and Ekora: a general summary of stratigraphy and faunas. In: Y Coppens, FC Howell, GL Isaac and REF Leaky (eds.), Earliest Man and Environments in the Lake Rudolf Basin. Univ. Chicago Press, Chicago. p. 163-70.
Boaz NT, Howell FC, McCrossin ML 1982. Faunal age of the Usno, Shungura B and Hadar Formations, Ethiopia. Nature 300:633-5.
Brain CK 1981a. The evolution of man in Africa: was it a result of Cainozoic cooling? Du Toit mem. Lect. Annex. Trans. geol. Soc. S. Afr. 84. 19pp.
Brain CK 1981b. The Hunters or the Hunted? Univ. Chicago Press, Chicago.

Broom R 1948. Some South African Pliocene and Pleistocene mammals. Ann. Transv. Mus. 21(1):1-38.

Cerling TC, Brown FH 1982. Tuffaceous marker horizons in the Koobi Fora region and the lower Omo Valley. Nature 299:216-21.

Collings GE, Cruikshank AR, Maguire JM, Randall RM 1975. Recent faunal studies at the Makapansgat Limeworks, Transvaal, South Africa. Ann. S. Afr. Mus. 71:153-65.

Ewer RF 1956a. The fossil carnivores of the Transvaal caves: Canidae. Proc. zool. Soc. London 126:97-119.

Ewer RF 1956b. Some fossil carnivores from the Makapansgat Valley. Paleontologia Africana 4:57-67.

Foley R 1982. A reconsideration of the role of predation on large mammals in tropical hunter-gatherer adaptation. Man 17:393-402.

Hay RL 1976. Geology of the Olduvai Gorge. Univ. California Press, Berkley.

Hendey QB 1974. The Late Cenozoic Carnivora of the south-western Cape province. Ann. S. Afr. Mus. 63:1-369.

Hendey QB 1981. Palaeoecology of the Late Tertiary fossil occurrences in 'E' Quarry, Langebaanweg, South Afica, and a reinterpretation of their geologic context. Ann. S. Afr. Mus. 84(1):1-104.

Howell FC 1982. Preliminary observations on carnivora from the Sahabi Formation (Libya). Garyounis scient. Bull. spec. Issue No. 4:49-61.

Howell FC and Petter G 1976. Carnivora from Omo Group formations, southern Ethiopia. In: Y Coppens, FC Howell, GL Isaac and REF Leakey (eds.), Earliest Man and Environments in the Lake Rudolf Basin. Univ. Chicago Press, Chicago. p. 314-31.

Isaac GL 1982. Some archaeological contributions towards understanding human evolution, Can. J. Anthropol. 3(2):233-43. Proc. Symp. Human Evolution. GH Sperber (ed.).

Kalb JE, Jolly CJ, Mebrate A, Tebedge S, Smart C, Oswald EB, Grainer D, Whitehead R, Wood CB, Conroy GC, Adefris T, Sperling L and Kana B 1982. Fossil mammals and artefacts from the Middle Awash Valley, Ethiopia. Nature 298:25-9.

Kurtén B 1968. Pleistocene Mammals of Europe. Weidenfeld and Nicolson, London.

Kurtén B and Anderson E 1980. Pleistocene Mammals of North America. Columbia Univ. Press, New York.

Leakey MD and Hay RL 1979. Pliocene footprints in the Laetolil Beds at Laetoli, northern Tanzania. Nature 278:317-23.

Leakey MD, Hay RL, Curtis GH, Drake RE, Jackes MK and White TD 1976. Fossil hominids from the Laetoli Beds. Nature 262:460-466.

Leakey MG 1976. Carnivora of the East Rudolf succession. In: Y Coppens, FC Howell, GL Isaac and REF Leakey (eds.), Earliest Man and Environments in the Lake Rudolf Basin. Univ. Chicago Press, Chicago. p.302-13.

Martin LD, Gilbert BM and Adams DB 1977. A cheetah-like cat in the North American Pleistocene. Science 195:981-2.

Petter G 1973. Carnivores Pleistocènes du Ravin d'Olduvai (Tanzanie). In: LSB Leakey, RJG Savage and SC Coryndon (eds.), Fossil Vertebrates of Africa, Vol. 3. Academic Press, London. p. 43-100.

Pickford M 1975. Late Miocene sediments and fossils from the northern Kenya rift valley. Nature 256:279-84.

Pickford M 1978. Stratigraphy and mammalian palaeontology of the Late-Miocene Lukeino Formation, Kenya. In: WW Bishop (ed.), Geological Background to Fossil Man. p. 263-78.

Randall RM 1981. Fossil hyaenidae from the Makapansgat Limeworks deposit, South Africa. Palaeontologia Africana 24:75-85.

Smart C 1976. The Lothagam 1 fauna: its phylogenetic, ecological and biogeographic significance. In: Y Coppens, FC Howell, GL Isaac and REF Leakey (eds.), Earliest Man and Environments in the Lake Rudolf Basin. Univ. Chicago Press, Chicago. p. 361-9.

Solounias N 1981. The Turolian fauna from the island of Samos, Greece. Contr. Vertebrate Evolution, Vol. 6. S Krager, Basel.

Turner A 1982. Hominids and fellow travellers. S. Afr. J. Sci. 78:231-7.

Turner A, in press. Panthera crassidens Broom, 1948. The cat that never was? S. Afr. J. Sci.

Vrba ES 1980. Evolution, species and fossils: how does life evolve? S. Afr. J. Sci. 76(2):61-84.

Vrba ES 1981. The Kromdraai australopithecine site revisited in 1980: recent investigations and results. Ann. Transv. Mus. 33(3):17-60.

Vrba ES 1982. Biostratigraphy and chronology, based particularly on Bovidae, of southern hominid-associated assemblages. Nice: Congrès Int. Paléontologie Humaine. p. 707-52.

Vrba ES, in press. Palaeoecology of early Hominidae, with special reference to Sterkfontein, Swartkrans and Kromdraai. In: Y Coppens (ed.), L'Environment des Hominidés.

Walker A, Zimmerman MR and Leakey REF 1982. A possible case of hypervitaminosis A in Homo erectus. Nature 296:248-50.

Preliminary radiometric ages for the Taung tufas

JOHN C. VOGEL
CSIR, Pretoria, South Africa

T.C. PARTRIDGE
University of Witwatersrand, Johannesburg, South Africa

ABSTRACT. Reassessment of the records and interpretations concerning the site at which the Taung hominid fossil was found, confirms the opinion that the locality was in a breccia-filled cavity within the Thabaseek tufa carapace at the Taung limestone quarry. The age of the fossil should therefore be younger than this carapace and probably older than the succeeding Norlim carapace. Ionium dating of the younger tufa deposits at the site as well as radiometric analysis of the Thaba Sikwa spring water and modern tufa, suggest that the $U234/U238$ ratio can be used to derive ages for the Norlim and Thabaseek tufas. These are 764 000 BP and 942 000 BP respectively. The subsequent introduction of secondary uranium could conceivably result in ages that are too young, but local circumstances do not make this likely.

INTRODUCTION

The tufa carapaces of Taung on the Ghaap Escarpment have been well known since the discovery there of a juvenile cranium in 1924 and the subsequent recognition of its crucial status by RA Dart (1925) in the face of almost universal criticism from his colleagues of the time. Although some useful first-hand information on the provenance of this find was forthcoming from the company then engaged in lime-quarrying operations at the site and from near-contemporary photographs, no detailed geological investigations were undertaken prior to the major study of FE Peabody in 1947 (Peabody 1954). By this time the site of the original discovery had been obliterated entirely by quarrying operations; indeed, when RB Young visited the site in February 1925, not long after the discovery, portion of the fossil deposit which had yielded the cranium had already been blasted away (Young 1926), and still less remained at the time of the visit by A Hrdlicka later that year (Hrdlicka 1925).

It is a pity that these early and generally well-trained observers did not deem it necessary to make detailed notes of the stratigraphic provenance of this unique specimen nor to record its precise location in relation to points of reference in the general area. Equally unfortunate is the absence of reliable records linking the extensive primate (baboon) fauna, recovered from various parts of the deposit since 1919, with the hominid specimen or with the fossil locality investigated by Hrdlicka in 1925 and by Peabody in 1947. It was left to Peabody and others to reconstruct the probable relationships from these limited, but nonetheless valuable, early records and from verbal accounts gathered more than twenty years later. Yet despite the uncertainties which surround the precise context of the discovery, a sufficient body of evidence exists to place the Taung hominid within the broad sequence of deposits recognized at the site by Peabody and confirmed by later workers.

In view of the general nature of the associated evidence it is scarcely surprising that early estimates for the age of this specimen varied widely. Indeed, it was not until a reasonably well established chronological framework was provided by radiometric and palaeomagnetic methods for the most important East African hominid and faunal sequences that any faunally based estimates could be ventured for the South African sites, including Taung. But because of the absence of direct evidence linking the Taung hominid with the baboon fauna recovered during the 1920's and the wider range of genera excavated at the Hrdlicka site by Peabody in 1947, estimates based on faunal calibrations (for a review of the faunal evidence see Tobias 1978:60-62) have been viewed with circumspection by most workers. The Taung child is the holotype of Australopithecus africanus, a species whose cranial morphology is now well known from the extensive sample of adult specimens recovered from Member 4 of the Sterkfontein Formation and Members 3 and 4 of the Makapansgat Formation (Partridge 1978, 1979). Evidence both from East Africa and the Transvaal suggests that A. africanus (or comparable gracile forms) are absent from the fossil record at all reasonably well dated sites after about 2 to 2.5 m.y. ago. Considerable controversy was, therefore, generated by the view put forward independently by TC Partridge (1973) and KW Butzer (1974), on the basis of geomorphic and regional stratigraphic evidence, that a case could be made for placing this specimen well within the Lower Pleistocene. In an attempt to resolve this issue, radiometric dating of the tufas has been undertaken.

THE CONTEXT OF THE HOMINID DISCOVERY

In order to plan the sampling programme and to interpret the results of the subsequent analyses, a careful reassessment of the early records and interpretations was essential. In summary, these sources reveal the following:

The specimen was blasted from the quarry face by a Mr de Bruyn late in 1924. It was entrapped in a hard, pinkish deposit which was situated a short distance below an almost circular body of loose, reddish-brown sand preserved within the quarry face; at the time of the discovery this had been quarried inwards (i.e. westwards) from the eastern edge of the limestone deposit for a distance of between 60 and 100 m and the fossiliferous matrix was located some 12-15 m below the natural surface of the deposit and a similar distance above the base of the quarry (Dart 1925; Hrdlicka 1925; Young 1926). As was observed by Young (1926), the quarry face consisted of calcareous tufa, comprising relatively pure calcium carbonate, deposited during localized streamflow over the edge of the Ghaap Escarpment. The mode of preservation of fossiliferous pockets within the limestone was well comprehended by Hrdlicka who commented: "The exposed parts of this limestone, or as Professor Young called it, calcareous tufa, became in the course of time pitted by many holes, crevices and caves, and some large caves as well as crevices and tunnels were also formed, probably by the action of water, in the interior of the mass. These internal cavities must have had external openings, for there is evidence that they were entered and perhaps inhabited by the various animals which eventually left in them their skeletal remains. Eventually, however, these cavities were all filled by washed-in and blown-in sand and earth, which in turn were permeated by water carrying lime in solution and these cemented the different materials, converting them from crumbly to very hard secondary rock. Thus the various cavities are now filled with a hard sand-limestone which, due to the fine reddish Kalahari Sand, is more or less pinkish in shade and can thereby be recognized from the surrounding whiter stone of the older formation" (Hrdlicka 1925:318).

That an extensive system of secondary, infilled caverns existed at the site is confirmed by GB Gordon, Geologist of the Northern Lime Company which was then conducting quarrying operations at the site: "The cave ...... where Australopithecus africanus was found is but one of a series originally reaching right to the east front of the deposit and being in some instances connected from nearly the present base level of the quarry to the upper surface of the deposit" (Gordon 1926:xxxviii). Gordon makes the further point that the great purity of the host tufa bodies, a characteristic which is clearly evident in the exposures preserved at the site today, contrasts dramatically with the impure nature of the cemented cavern fillings, a fact which argues strongly against the accumulation of these fillings concurrently with the deposition of the tufa.

An important link with the later studies of Peabody is provided by Hrdlicka, who examined fossiliferous deposits near the hominid site. These deposits of the so-called Hrdlicka site were systematically studied and excavated by Peabody in 1947, and their relationship to the hominid site is therefore of considerable importance. Hrdlicka records: "After examining what remains of the tunnel that gave the anthropoid skull and of the remnants of the large cave a short distance to the right of this, the writer was led to the northern wall of the quarry, where a part of the original limestone elevation abutting on the dry ravine to the north of it has been left standing. In this wall, a little over 30 feet from its base and about the same distance from the top, there could be seen the pinkish fillings of another tunnel, somewhat less than 3 feet in diameter" (Hrdlicka 1925:386). It was from this locality that Hrdlicka recovered fossil baboons and Peabody his more extensive faunal sample. From the accounts of eyewitnesses present at the time of the hominid discovery and Hrdlicka's visit, Peabody (1954) concluded that the Hrdlicka site was situated some 30 m to the north-east of the hominid site. Although definite contemporary observational evidence is lacking, Peabody argues that both deposits were probably part of a single, continuous series of cave fillings. Such a possibility is certainly supported by the comments of Gordon quoted previously.

The relative stratigraphic positions of the hominid and Hrdlicka sites were not considered in any detail by early observers, and the presently accepted subdivision of the main tufa deposits into four discrete units or carapaces was first proposed by Peabody (1954). These carapaces were named, from oldest to youngest, the Thabaseek, Norlim, Oxland and Blue Pool. Each has at its base an apron-like accumulation of tufa-cemented rubble consisting chiefly of talus, similar to that accumulating today below the cliffs of the Ghaap Escarpment, but also containing more rounded components and secondarily derived tufa clasts in the younger carapaces. In the case of the Norlim/Thabaseek and the Blue Pool/Oxland tufas a clear overlapping relationship confirms the chronological sequence, but the Oxland carapace is laterally separate from the Norlim and the succession is indicated by topographic rather than stratigraphic position. At some time after their accumulation, these carapaces were cut through by erosion, which produced spectacular gorges transecting both tufa and bedrock to depths of 20 m or more. Smaller scale pitting, fissuring and cavern formation bears testimony to the continued operation of solution and karst weathering on all of the carapaces. Peabody was able to show, on the basis of his own observations and those of earlier witnesses, that the filled caverns comprising the hominid and Hrdlicka sites were both located within the upper part of the oldest (Thabaseek) carapace.

Peabody's nomenclature formed the basis of the subsequent re-evaluation of stratigraphic relationships at the site by KW Butzer (1974), who confirmed Peabody's succession of tufa carapaces. He furthermore drew attention to Peabody's detailed description of the Hrdlicka site, in which specific reference was made to the fact that the cemented, fossiliferous cave deposit occupied a cavity which transgressed the steeply inclined foreset beds characteristic of the upper levels of each of the carapaces. The cave deposit itself consisted of a lower accumulation of pinkish, heavily cemented sand, followed by a unit comprising pinkish clayey silt with numerous sub-horizontal tufa laminae. Both units contained fossil faunas consisting predominantly of baboons, but the majority of these appear to have come from the lower sands. These deposits were followed by sterile, poorly cemented reddish sands and, finally, unconsolidated black earth containing MSA artefacts.

Both Peabody and Butzer were of the view that the caves at the Hrdlicka and hominid sites were part of a single system, as suggested by earlier accounts, and both considered that comparisons of the matrixes adhering to the various fossil specimens linked the hominid specimen with the pinkish clayey silt present at the Hrdlicka site rather than with the underlying, baboon-rich sands. Butzer (1974) suggested that the lower "baboon" sands accumulated during a dry phase separating moister periods during which the Thabaseek tufa and the succeeding Norlim tufa accumulated, while the overlying hominid deposit, with its clayey matrix and frequent tufa lenticles, belonged to an early phase in the accretion of the Norlim carapace. It must be pointed out however, that there is at present no direct evidence regarding the age of the cave filling. The high degree of cementation of the hominid matrix and the massive nature of the tufa intercalations within it are consistent with its accumulation during deposition of the Norlim tufa; later carapaces tend to be considerably softer and more vesicular.

There is no evidence that the hominid was obtained from a cave in the adjacent bedrock whose filling predated the formation of the Thabaseek tufa. Indeed, such a possibility is precluded by contemporary accounts that the discovery site was situated approximately half way up the 30 m high working face of the quarry which was then located entirely within the tufa of the Thabaseek carapace. Moreover, the pattern of quarrying at the site during the 60 years over which it was worked, beginning in 1916, is clearly evident both from the reports of the quarry managers and from the present outlines of the old excavations; workings were abandoned as soon as bedrock, chiefly shale of the Schmidtsdrift Formation, was encountered, and no attempt was made to work deposits preserved in caverns or irregularities within this bedrock surface. The quarrying programme was geared to the production of high-quality lime on a large scale from deposits of exceptional purity, and any impediment to this, such as non-calcareous bedrock or large zones of sandy contamination of the type associated with secondary cave infillings, resulted in a shift of operations to another area.

In the light of all the evidence the accumulation of the hominid deposit and the entrapment within it of the hominid specimen may be assumed, with confidence, to postdate the accumulation of the youngest (uppermost) parts of the Thabaseek tufa. It is also plausible that the hominid deposit may have been coeval with the accretion of the lower levels of the succeeding Norlim carapace, providing an upper age limit for the hominid specimen. Hence age determinations on the relevant tufas will provide a terminus post quem for the Taung hominid as well as a likely younger age limit.

THE PRESENT SAMPLING PROGRAMME

Samples of the Oxland tufa from above and below the Equus cave site were collected in 1979 by P Beaumont for radiometric analysis. In 1981 one of us (TCP) collected samples from defined positions within each of the other tufa carapaces. Care was taken to take only massive cryptocrystalline material well removed from voids, zones of obvious secondary recrystallization, areas of clastic contamination and from modern exposed surfaces. All sites were marked with paint. They are shown in their relation to the various tufa carapaces and the present outlines of the quarries in Fig. 1.

In order to provide baseline data a water sample was taken from the nearby Thaba Sikwa spring which is the source of the stream from which the tufas were deposited. Present-day tufa forming just below the spring was also collected for this purpose.

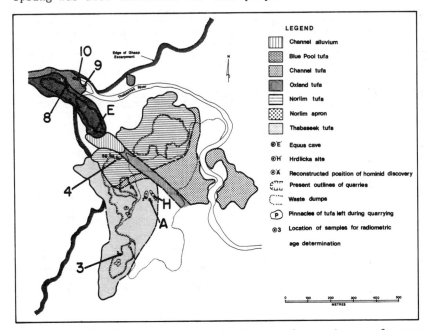

Figure 1. Sampling sites in relation to the various tufa carapaces and the present outlines of the quarries. Numbers refer to the sample numbers in the top row of Table 1.

URANIUM SERIES DISEQUILIBRIUM DATING

The circumstances of the calcium carbonate deposition from the dolomite spring water appear to offer excellent conditions for applying uranium series disequilibrium dating to the Taung tufas. The lime is more than 98 % pure and mostly hard and compact. Radiocarbon dates for Blue Pool tufa reported by Butzer et al. (1978) suggested that this carapace was formed during the Last Glacial period and would thus be well within the range of ionium (Th230) dating. The ages obtained for the Blue Pool and Oxland tufas are given in Table 1. The amount of Th232 in the samples is disturbingly high and this requires some adjustment to the dates to account for the presence of initial thorium, which introduces an additional uncertainty of several thousand years to the results. Nevertheless, the dates of 30 000 to 103 000 BP for the Blue Pool carapace and of 230 000 BP for a late stage of Oxland carapace are probably not far off the mark.

To obtain age estimates of the two older carapaces, use was made of the radioactivity disequilibrium between U234 and U238 (Vogel 1980). Experience has shown that in the semiarid environment of southern Africa extreme fractionation occurs between these two isotopes during the dissolution process. The disequilibrium in the U234/U238 ratio observed in groundwater in this country ranges from a factor of two to more than four (Vogel and Kronfeld unpubl.). After co-precipitation in limestone the α-particle activity ratio gradually reverts to unity with a half-life of 248 000 years, and ages of up to 1 or 1 m.y. can be deduced. The problem is, however, that the initial ratio is in

Table 1. Summary of radiometric analyses of samples from the Taung site. (Errors quoted are 1 sigma uncertainties).

Sample Anal No	Thaba water Pta-3244 U-258	Sikwa spring tufa Pta-3556 U-289	10 Blue pool Upper 1 Pta-3266 U-262	10 Upper 2 – U-285	9 Lower Pta-3265 U-260	E Oxland Outer Pta-2741 Ave[1]	8 Upper Pta-3264 U-274	4 Norlim base U-272	3 Thaba- seek top U-271
C14 age (ka)	(118.3%) ±.6	(95.2%) ±.5	23.6 ±.1	–	41.9 ±.8	>49.7	26.2 ±.3	–	–
[U][2] (ppm)	0.57 ±.01	0.666 ±.042	0.140 ±.007	0.127 ±.003	0.200 ±.006	0.084 ±.002	0.088 ±.003	0.062 ±.002	0.046 ±.001
$\frac{U234}{U238}$[3]	2.88 ±.04	2.52 ±.08	2.25 ±.11	2.49 ±.06	2.33 ±.07	1.82 ±.05	1.65 ±.06	1.20 ±.04	1.13 ±.03
$\frac{Th230}{Th232}$	–	0.63 ±.06	4.05 ±.42	3.79 ±.23	21.0 ±1.5	48	5.34 ±.27	3.00 ±.14	4.36 ±.30
$\frac{Th230}{U234}$	–	0.059 ±.007	0.306 ±.027	0.530 ±.025	0.670 ±.022	0.99 ±.02	1.12 ±.04	1.05 ±.04	1.05 ±.05
Th age[4] (ka)	–	0	30 ±3	61 ±5	103 ±6	230 ±17	>322	>325	>360
U age (ka)	–	–	114 ±32	47 ±15	86 ±21	256 ±21	343 ±30	764 ±65	942 ±90

[1] Average of four samples collected above, within and below the Equus cave by P Beaumont.

[2] Concentration in ppb

[3] The ratio of the α-particle activity of the respective isotopes.

[4] Th ages corrected for initial thorium assuming Th230 = Th232 initially.

general not known and this dating possibility is therefore of only limited use.

In the present instance the conditions of formation of the tufas from a dolomite spring over a long timespan, has provided the opportunity to test the constancy of the initial U234/U238 activity ratio for the younger tufas. This ratio for the Blue Pool and Oxland samples as well as that for the spring water and modern tufa is plotted against their ionium ages in Fig. 2. The data show that the initial U234/U238 ratio has remained between 2.4 and 3.0 over at least the past 300 000 years. It is then reasonable to assume that the situation did not change drastically in the preceding periods and the observation can be used to derive tentative ages for the two older carapaces.

The important fact that has emerged from the analysis of two samples from the Norlim and Thabaseek carapaces, is that the uranium isotopes are not yet in radioactive equilibrium (Table 1). Accepting an average initial ratio of 2.7 the measured values of 1.20±0.04 and 1.13±0.03, respectively, correspond to ages of 764 000 years and 942 000 years for basal Norlim tufa and younger Thabaseek tufa. The ages are not very precise, but even if the initial disequilibrium ratio was double, i.e. the activity ratio U234/U238 and 4.4 instead of 2.7, this would only increase the ages by one half-life of U234 or 248 000 years. The indications are, thus, that the Thabaseek carapace was still forming after at most 1.2 m.y. BP.

The implications of such a young age for the Thabaseek carapace warrant careful consideration of possible errors in the dating method. The only way in which a too young apparent age can be induced is if the uranium in the tufa was added at a later stage. The fact that the uranium content of the tufas decreases with age (Table 1) does, however, not make this seem likely.

IMPLICATIONS OF THE AGE ESTIMATES

The obvious implications of the dating of the hominid specimen as well as the fossils from Hrdlicka's cave to between 942 000 BP and 764 000 BP or, taking extremes, to 1.0±0.3 m.y. BP need not be stressed. The matter is of such import for our understanding specifically of the

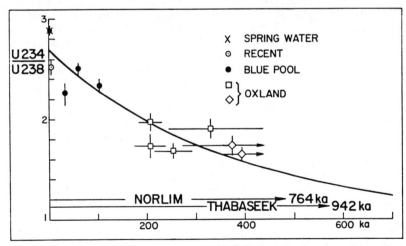

Figure 2. Activity ratios, U234/U238, of the samples plotted against their apparent Th ages. A single U234/U238 decay curve starting at 2.7±0.3 satisfies all the data.

hominid evolutionary process, that further analyses must be undertaken to either substantiate or disprove these preliminary findings.

The palaeoclimatic significance of the various tufa complexes occurring along the Gaap Escarpment have been considered by Butzer (1974) and Butzer et al. (1978). Their model suggests tufa accretion during major wet cycles accompanied by increased spring activity. As the authors themselves concede, this model does not explain both tufa accretion and profound gorge incision within the same or successive wet phases.

The dating evidence presented here indicates that the Blue Pool carapace formed during the Last Glacial, but ended before the Last Glacial Maximum. Current opinion suggests that this stage was indeed more humid than the present (Deacon et al., this volume). It may be possible to link earlier periods of tufa accretion with hemispheric climatic cycles once further dating provides more accurate estimates of the duration of their formation.

REFERENCES

Butzer KW 1974. Paleoecology of South African australopithecines: Taung revisited. Curr. Anthrop. 15:367-382.
Butzer KW, Fock GJ, Stuckenrath R and Zilch A 1973. Palaeo-hydrology of late Pleistocene lake, Alexandersfontein, Kimberley, South Africa. Nature 243:328-330.
Butzer KW, Stuckenrath R, Bruzewicz AJ and Helgren DM 1978. Late Cenozoic paleoclimates of the Gaap Escarpment, Kalahari margin, South Africa. Quat. Res. 10:310-339.
Dart RA 1925. Australopithecus africanus: the man-ape of South Africa. Nature 115:195-199.
Deacon J, Lancaster N and Scott L, this volume. Evidence for Late Quaternary climatic change in southern Africa. Sum. proc. SASQUA Workshop, Johannesburg, Sept. 1983.
Gordon GB 1926. Discussion of "The calcareous tufa deposits of the Campbell Rand, from Boetsap to Taungs Native Reserve" by RB Young. Proc. geol. Soc. S. Afr. 28 (1925): xxvii-xxxix
Hrdlicka A 1925. The Taungs ape. Am. J. phys. Anthrop. 8:379-392.
Partridge TC 1973. Geomorphological dating of cave opening at Makapansgat, Sterkfontein, Swartkrans and Taung. Nature 246:75-79.
Partridge TC 1978. Re-appraisal of lithostratigraphy of Sterkfontein hominid site. Nature 275:282-287.
Partridge TC 1979. Re-appraisal of lithostratigraphy of Makapansgat Limeworks hominid site. Nature 279:484-488.
Partridge TC 1982. The chronological positions of the fossil hominids of Southern Africa. Proc. 1st Int. Congress on Human Palaeont., Nice, October 1982. Paris, CNRS, vol 2. p. 617-675.
Peabody FE 1954. Travertines and cave deposits of the Kaap Escarpment of South Africa and the type locality of Australopithecus africanus Dart. Bull. geol. Soc. Am. 65:671-706.
Tobias PV 1978. The South African australopithecines in time and hominid phylogeny, with special reference to the dating and affinities of the Taung skull. In: C Jolly (ed.), Early Hominids of Africa. Duckworth, London. p. 45-84.
Vogel JC 1980. Dating possibilities for the South African hominid sites. Palaeont. Afr. 23:41-44.
Young RB 1926. The calcareous tufa deposits of the Campbell Rand, from Boetsap to Taungs Native Reserve. Trans. geol. Soc. S. Afr. 28 (1925): 55-67.

Climatic change and evolution – Evidence from the African faunal and hominid sites

Summary of the proceedings of the SASQUA Workshop held in Johannesburg, September 1983

PHILLIP V. TOBIAS
University of Witwatersrand, Johannesburg, South Africa

The object of the Workshop* was to explore whether a basis exists for a correlation between climatic change and biological evolution, with special reference to Africa of the Late Cainozoic. It was not the purpose of this Workshop to discuss the causes of climatic change. Nevertheless, the discussion touched on such aspects as

(a) the progressive northward shift of Africa, from a time when the equator was north of the horn of Africa to the recent disposition of the continent almost symmetrically about the equator with a maximum proportion of Africa lying in the tropics and sub-tropics;

(b) the marked continental uplift which has been of the order of 1200 m since the Miocene, of which some 700 m of uplift occurred towards the end-Pliocene, this uplift being especially clearly manifest along the eastern hinterland of the continent; for every 100 m of uplift the mean temperature drops about $1\,^{\circ}C$, so that appreciable uplift alone could have produced effects at least as great as a glaciation;

(c) the progressive separation of Africa from Antarctica, with changes in oceanic mean temperatures and their possible effects on African climates;

(d) the evidence of major cooling events at about 16.5 and 2.5 m.y. BP.

SOME PITFALLS IN THE DRAWING OF CLIMATIC INFERENCES FROM FAUNA

A number of features were considered here. First, the accurate dating of faunal deposits is fundamental. In this respect, the remarkable sequence of fossil-bearing members and volcanic tuffs in East Africa, especially at Omo in southern Ethiopia and East Turkana in Northern Kenya, is unique, not only in the African continent but in the world. It has furnished a well-controlled time-scale of biological phenomena in Africa. To this have recently been added the results of Frank Brown's comparisons of the geochemical composition of the various tuffs

---

*In this Workshop the panel of principal discussants comprised Prof. TC Partridge (Chairman/Moderator), Dr NT Boaz, Dr CK Brain, Emeritus Prof. HBS Cooke, Prof. PV Tobias and Dr Elizabeth Vrba. Other discussants included Dr RJ Clarke, Dr GH Denton, Mr TN Pocock, Dr A Turner, Mrs DM Avery, Prof. JA Coetzee, Dr ABA Brink and Miss M Levinson.

recognized in Ethiopia and Kenya, from which a series of correlations has emerged between specific tuffs of comparable age at widely separated localities.

The dolomitic cave deposits of southern Africa have long eluded precise dating; however, the recognition of a characteristic and definable part of the palaeomagnetic sequence within the Makapansgat Formation and its close correlation with the faunal stages represented in the members of that deposit, have brought an important breakthrough in the dating of southern African cave deposits. Some uncertainties remain; for example, the relative dating of the robust australopithecine-bearing member or members at Kromdraai and Swartkrans, and the dating of the tuffs at Buxton, Taung, and of the type specimen of Australopithecus africanus. New results obtained by TC Partridge and JC Vogel and first reported at the Swaziland Symposium on Late Cainozoic Palaeoclimates elicited much interest, but need confirmation before it may be accepted that the oldest tuff, from which the Taung hominid skull almost certainly emanated, is younger than a million years.

The Workshop considered some other problems bearing on climatic interpretations, such as the effects of faunal migrations, the vagaries of diverse taphonomic circumstances, the respective rôles of local phenomena and of more widespread regional or even global events, the world-wide association of hominids and carnivores which seem to have moved around the globe 'claw-in-hand', the effects of climate and of tectonic phenomena on biological evolution, the rôle of man in promoting extinction, especially after he became a confirmed and systematic hunter.

In elaboration of some of these problem-areas it was noted that a great faunal interchange between Asia and Africa took place at the Mio-Pliocene boundary with the desiccation of the Mediterranean basin in the Messinian 'crisis'. The importance of taphonomic data is illustrated by A Behrensmeyer's demonstration that the proportions of two genera of suids represented in fossil assemblages varied according to whether the respective deposits had been formed under deltaic, channel or floodplain conditions. Thus a difference in relative frequencies of two faunal elements might reflect differing local (e.g. tectonic or geomorphological) conditions, or climatic changes, or both. The interrelationship between hominids and carnivores may suggest some form of interdependence between them, for example, the hominids might have been scavenging on carnivore kills. It was agreed that it would be worthwhile to re-examine the question of whether scavenging might have occurred as a phase prior to systematic hunting by hominids. It was noted that hominids were a relatively static, immobile species, at least until about 2.3 m.y. BP with the emergence of the genus Homo. The suggestion was raised that material culture might have been a crucial factor, making for greater hominid mobility after 2.3 m.y. BP.

MAJOR EVENTS IN HOMINID EVOLUTION

The Workshop took note that remarkable changes had come about in the most widely agreed view of the pattern and timing of hominid emergence and evolution since the late 1970's. Throughout the 1960's and 1970's there was widespread agreement, though by no means unanimity, that hominids had appeared as early as 15 to 20 m.y. BP. This view was based almost entirely on the concept that Ramapithecus of the Indian sub-continent and its African counterpart were to be classified as hominid. This Early Divergence hypothesis tended to visualize one major branching between the lineage leading to man, all members of which were classified as Hominidae, and the lineage leading to the African and Asian great apes, lumped into a single family, the Pongidae.

Increasingly throughout that period, the accumulation of molecular evolutionary data revealed that modern man and the modern African great apes (chimpanzee and gorilla) were extremely closely related genetically; so closely, indeed, that a much more recent divergence of the hominid lineage from the African ape lineage seemed to be the only possible explanation. On the other hand the genetic distances among man, chimpanzee and gorilla were found to be far smaller than that between any one of these and the Asian great ape, orang-utan. For instance, chimpanzee is far closer to man than chimpanzee is to orang-utan, something not perceived in earlier studies. The careful application of molecular data has led to several recent syntheses in the early 1980's, from which it seems clear that two major branchings and not one must have occurred in the history of the modern higher primates, including man.

Between 10 and 16 m.y. BP, the orang line must have split off from the African hominoid lineage. Newer discoveries of more complete Ramapithecus specimens and re-evaluation of their morphology support the view the Ramapithecus may, after all, be on that orang lineage. Gigantopithecus, both the earlier bilaspurensis species from India and the later blacki species from China, could well have been associated with the orang lineage.

Late in the Miocene, probably between 7 and 5 m.y. BP, the evidence would indicate that the African hominoid lineage split into three lines, that of the hominids, and the lines leading to the chimpanzee and gorilla respectively. This was a critical nodal or punctuational event which generated, inter alia, the new family, the Hominidae, characterized initially by bipedal stance and gait.

A further major nodal event occurred on that hominid lineage late in the Pliocene. Between 2.5 and 2.0 m.y. BP the hominids seem to have split into several derivative lineages: one led to the South African robust australopithecines (A. robustus robustus and A. robustus crassidens or, alternatively, A. robustus and A. crassidens, according to how greatly one weights the morphological differences between the australopithecines of Kromdraai and of Swartkrans); another led, by stages not yet known and not yet identified in East Africa, to the hyper-robust australopithecine of East Africa, Australopithecus boisei, whilst a third engendered the genus Homo and its earliest known species, Homo habilis.

Although further speciation occurred along the lineage of Homo, leading from H. habilis to H. erectus and thence to H. sapiens, there is no evidence of major nodal events being apparent along that line of Homo between its origin, at about 2.3 m.y. BP and the first appearance of H. sapiens.

Thus, there are two striking nodal events in the genesis of man: one at 7-5 m.y. BP, marked by the emergence of the family of the hominids, and one at 2.5-2.0 m.y. BP, marked by the first appearance of the genus Homo. It is in these two crucial time-zones, the first terminal Miocene and the second terminal Pliocene, that one should seek to find possible climatic determinants. As a step in that direction, the Workshop sought evidence for other biological changes of significance at those two time-nodes.

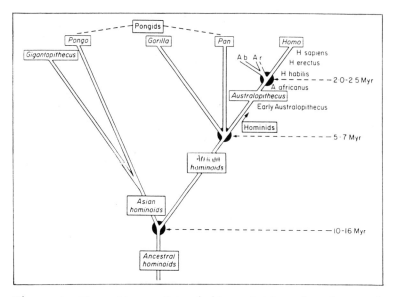

Figure 1. The pattern of hominid evolution, based on molecular and palaeontological evidence and reflecting a consensus achieved at several international discussions in the 1980's. Three major nodal points are highlighted in the reconstructed history of the hominoids. One, at 10 - 16 m.y. BP, represents the sundering of the lineage of the orang-utan, the Asian great ape, from that of the other hominoids. The second great node occurred between 5 and 7 m.y. BP and was marked by the splitting of an African hominoid lineage into the lines leading to the extant African great apes (chimpanzee and gorilla) and to man (the family of the hominids). The third dramatic episode occurred between 2.0 and 2.5 m.y. BP, when the lines of robust australopithecines and of the genus Homo appeared. The two points of cladogenesis or lineage-splitting, marked by explosive evolution at 5 - 7 and 2.0 - 2.5 m.y. BP, are characterized also by major changes of other faunal elements, of flora and of inferred climate: these are the most favourable occurrences at which to seek climatic correlations with biological evolutionary events in Africa. The former episode, however, is complicated by its coinciding with the Messinian 'crisis', when desiccation of the Mediterranean basin permitted a major faunal interchange to occur between Asia and Africa. To what extent this late Miocene immigration of Asian forms into Africa affected the emergence of, or even generated, the family of the hominids is not at present known. No such major faunal interchange occurred at the latest of the three nodes, 2.0 - 2.5 m.y. BP, and the major biological events of that end-Pliocene phase reflect what seems to have been largely autochthonous evolution within Africa. It was at this crucial point that the genus Homo appeared, in the form of its earliest known species, namely H. habilis.

Key to figure:
Hominoid : pertaining to the superfamily Hominoidea, which comprises the families of the apes and of man
Hominid : pertaining to the Hominidae or the zoological family of man
Pongid : pertaining to the Pongidae or the zoological family in which the African and Asian extant great apes and certain extinct apes are most commonly classified

BIOLOGICAL EVENTS TOWARDS THE END OF THE MIOCENE

The Workshop garnered evidence that, apart from the emergence of the family of the hominids, major biological events characterized this period. In the Cape Province the vegetation underwent a replacement of tropical and sub-tropical forest by 'fynbos' shrubland cover.

Changes in non-hominoid animals at this time include the last appearance, and subsequent disappearance, in southern Africa of the ursid or bear, Agriotherium; no fewer than 7 new tribes of bovids appear, of the 13 bovid tribes known, though the forms living then were quite different from those of the Quaternary; elephantids appeared in Africa; pigs diversified greatly; hyaenas and machairodonts (sabre-toothed cats) arrived on the African scene; the rodent Otomys occurred. It is likely that some of these changes, especially those of the bovids, elephantids, carnivores and rodents, flowed from the major faunal interchange accompanying the Messinian 'crisis'.

Hence, whether by autochthonous evolution or by immigration, the flora and fauna of Africa underwent drastic changes at the very time that the hominids are inferred from molecular data to have first appeared, namely between 7 and 5 m.y. BP.

BIOLOGICAL EVENTS TOWARDS THE END OF THE PLIOCENE

At the time when the hominid lineage split into the lines of robust australopithecines and of Homo, between about 2.5 and 2.0 m.y. BP, striking changes affected the flora and the non-hominoid fauna of Africa. The identified vegetation testifies to the supervention of more arid conditions. The era is marked by the opening of the woodland and the increase of savannah.

The approach of the end of the Pliocene was marked by the appearance of the modern genus of baboon, Papio; the disappearance of a number of elephantids; the vanishing of certain suids and the appearance of new, later suids; the disappearance of many archaic bovids and the appearance of many modern species of bovids; the disappearance of the hexaprotodont hippopotami (the hippos with six front teeth) and the appearance of the tetraprotodont hippos (with four front teeth); the vanishing of the machairodonts (or sabre-toothed cats); the disappearance of several kinds of rodents and the appearance of some new forms.

Clearly, at the time when the genus Homo appeared in Africa conspicuous other changes were occurring in the flora and non-hominid fauna of Africa as well. As there is less evidence of major Afro-Asian faunal interchange at this time than at 7-5 m.y. BP, it seems that vigorous biological evolution characterized the continent in the dying aeons of the Pliocene. Man, in the sense of Homo, was born to a world of drier conditions and of a rapidly changing biota.

Pongo	: the genus to which the orang-utan belong
Pan	: the genus to which the chimpanzee belongs
A.r.	: Australopithecus robustus, one of the 'robust' species of the genus Australopithecus, thus far known only from Kromdraai and Swartkrans in the Transvaal
A.b.	: Australopithecus boisei, the hyper-robust species of Australopithecus, so far known only from a number of East African sites.

(Modified after L Greenfield, 1983, and PV Tobias, 1983)

Thus, the strikingly revealed biological transformations in Africa at the two nodes, 7-5 m.y. and 2.5-2.0 m.y. BP, would seem to be especially favourable for the study of the interaction between physical environmental changes, particularly climatic changes, and biological evolutionary phenomena.